Machine Learning and AI for Precision Plant Epigenetics

Machine Learning and AI for Precision Plant Epigenetics

Edited by

Jen-Tsung Chen
National University of Kaohsiung
Kaohsiung
Taiwan

WILEY

This edition first published 2026

Registered Offices
John Wiley & Sons, Inc., 111 River Street, Hoboken, NJ 07030, USA
John Wiley & Sons Ltd, New Era House, 8 Oldlands Way, Bognor Regis, West Sussex, PO22 9NQ, UK

For details of our global editorial offices, customer services, and more information about Wiley products visit us at www.wiley.com.

The manufacturer's authorized representative according to the EU General Product Safety Regulation is Wiley-VCH GmbH, Boschstr. 12, 69469 Weinheim, Germany, e-mail: Product_Safety@wiley.com.

Wiley also publishes its books in a variety of electronic formats and by print-on-demand. Some content that appears in standard print versions of this book may not be available in other formats.

Library of Congress Cataloging-in-Publication Data
Names: Chen, Jen-Tsung editor
Title: Machine learning and AI for precision plant epigenetics / edited by Jen-Tsung Chen.
Description: First edition. | Hoboken, NJ : Wiley, 2026. | Includes index.
Identifiers: LCCN 2026003584 (print) | LCCN 2026003585 (ebook) | ISBN 9781394380282 cloth | ISBN 9781394380305 adobe pdf | ISBN 9781394380299 epub
Subjects: LCSH: Plant genetics–Technological innovations | Artificial intelligence–Biological applications
Classification: LCC QK981.M335 2026 (print) | LCC QK981 (ebook)
LC record available at https://lccn.loc.gov/2026003584
LC ebook record available at https://lccn.loc.gov/2026003585

Cover Design: Wiley
Cover Image: © Leonid Sorokin/Shutterstock

Set in 9.5/12.5pt STIXTwoText by Straive, Pondicherry, India

Contents

List of Contributors

Thiruvengadam Abarna
Department of Biochemistry
Shrimati Indira Gandhi College
Tiruchirappalli, Tamil Nadu, India

Hakima Achetoui
Department of Plant Protection
Phytopathology Unit
Ecole Nationale d'Agriculture de Meknès
Meknès, Morocco

Ioannis Adamopoulos
Department of Public Health Policy
Sector of Occupational & Environmental Health
School of Public Health
University of West Attica
Athens, Greece

Fatima-Zahra Akensous
Centre d'Agrobiotechnologie et Bioingénierie
Unité de Recherche labellisée CNRST (Centre AgroBiotech-URL-CNRST-05)
Université Cadi Ayyad
Marrakech, Morocco

Mohammad Ali
Department of Computer Science
Aligarh Muslim University
Aligarh, Uttar Pradesh, India

Mohd Anas
Institute of Laboratory Research on Geomaterials
Faculty of Natural Sciences
Comenius University in Bratislava
Bratislava, Slovakia

Abdelhamid Aouabe
Centre d'Agrobiotechnologie et Bioingénierie
Unité de Recherche labellisée CNRST (Centre AgroBiotech-URL-CNRST-05)
Université Cadi Ayyad
Marrakech, Morocco

Yamshi Arif
Department of Botany
Faculty of Life Sciences
Aligarh Muslim University
Aligarh, Uttar Pradesh, India

Burcu Arikan
Faculty of Science
Department of Molecular Biology and Genetics
Istanbul University
Istanbul, Türkiye

Vijaya Anand Arumugam
Department of Human Genetics and Molecular Biology
Bharathiar University
Coimbatore, Tamil Nadu, India

Maria Jerline Babu
Department of Biochemistry
Dr. R.A.N.M Arts and Science College
Erode, Tamil Nadu, India

Amarjeet Singh Bhogal
Department of Plant Breeding and Genetics
Assam Agricultural University
Jorhat, Assam, India

Shibani Ritusmita Borah
Department of Plant Breeding and Genetics
Assam Agricultural University
Jorhat, Assam, India

Abdelghani Bouchyoua
Research Unit of Plant Breeding and Plant Genetic Resources Conservation
Regional Agricultural Research Center of Meknes
National Institute of Agricultural Research
Rabat, Morocco

Laboratory of Natural Resources and Environment
Polydisciplinary Faculty of Taza
Sidi Mohamed Ben Abdallah University
Taza, Morocco

Neha Chaurasia
Department of Botany
Faculty of Life Sciences
Aligarh Muslim University
Aligarh, Uttar Pradesh, India

Manabendra Dutta Choudhury
Department of Botany
Rabindranath Tagore University
Hojai, Assam, India

Mohammad Danish
Department of Experimental Plant Biology
Institute of Botany
Plant Science and Biodiversity Centre
Slovak Academy of Sciences
Bratislava, Slovakia

Olympica Das
Department of Plant Breeding and Genetics
Assam Agricultural University
Jorhat, Assam, India

Sandip Debnath
Department of Genetics and Plant Breeding
Institute of Agriculture
Visva-Bharati University
Sriniketan, West Bengal, India

Ilham Dehbi
Department of Plant Protection
Phytopathology Unit
Ecole Nationale d'Agriculture de Meknès
Meknès, Morocco

Laboratory of Plant Biotechnology and Molecular Biology
Faculty of Sciences
Moulay Ismail University
Meknes, Morocco

Rinee Doley
Department of Plant Breeding and Genetics
Assam Agricultural University
Jorhat, Assam, India

Umer Fayaz
Accelerated Crop Improvement
International Crops Research Institute for the Semi-Arid Tropics
Patancheru, Telangana, India

Xianzhong Feng
Key Laboratory of Soybean Molecular Design Breeding
Northeast Institute of Geography and Agroecology
Chinese Academy of Sciences
Changchun, PR China

College of Advanced Agricultural Sciences
University of Chinese Academy of Sciences
Beijing, PR China

Arkan A. Ghaib
Information Technology Department
Management Technical College
Southern Technical University
Basrah, Iraq

Gnana Sowndariyan Gnanasekaran
Department of Human Genetics and Molecular Biology
Bharathiar University
Coimbatore, Tamil Nadu, India

Abhik Gogoi
Department of Agricultural Biotechnology
Assam Agricultural University
Jorhat, Assam, India

S. Gomathi
Department of Biochemistry
Shrimati Indira Gandhi College
Tiruchirappalli, Tamil Nadu, India

Akilandeswari Govindraj
Department of Biochemistry
CMS College of Science & Commerce
Coimbatore, Tamil Nadu, India

Anas Hamdani
Research Unit of Plant Breeding and Plant Genetic Resources Conservation
Regional Agricultural Research Center of Meknes
National Institute of Agricultural Research
Rabat, Morocco

Dikshita Hazarika
Department of Plant Breeding and Genetics
Assam Agricultural University
Jorhat, Assam, India

Nazneen Hussain
Department of Biosciences
Assam Don Bosco University
Tapesia, Assam, India

A. K. M. Aminul Islam
Department of Genetics and Plant Breeding
Gazipur Agricultural University
Gazipur, Bangladesh

Abdul Jalil
Department of Botany
Rabindranath Tagore University
Hojai, Assam, India

Antony Prabhu Jeyabal Philomenathan
Department of Human Genetics and Molecular Biology
Bharathiar University
Coimbatore, Tamil Nadu, India

Neslihan Turgut Kara
Faculty of Science
Department of Molecular Biology and Genetics
Istanbul University
Istanbul, Türkiye

Sanjivani Karki
Department of Agronomy
Uttar Banga Krishi Viswavidyalaya
Pundibari, West Bengal, India

Thirunethiran Karpagam
Department of Biochemistry
Shrimati Indira Gandhi College
Tiruchirappalli, Tamil Nadu, India

Fredrick Kayusi
Department of Environmental Studies, Geography and Planning
Maasai Mara University
Narok, Kenya

Moahmed Kouighat
Research Unit of Plant Breeding and Plant Genetic Resources Conservation
Regional Agricultural Research Center of Meknes
National Institute of Agricultural Research
Rabat, Morocco

Rachid Lahlali
Department of Plant Protection, Phytopathology Unit
Ecole Nationale d'Agriculture de Meknès
Meknès, Morocco

Nelofar Lone
Division of Genetics and Plant Breeding
Sher-e-Kashmir University of Agricultural Sciences & Technology of Kashmir
Jammu and Kashmir, India

Asma Majid
Division of Genetics and Plant Breeding
Sher-e-Kashmir University of Agricultural Sciences & Technology of Kashmir
Jammu and Kashmir, India

A.K. Mall
Division of Crop Improvement
ICAR-Indian Institute of Sugarcane Research
Lucknow, Uttar Pradesh, India

Manoj Mani
Department of Human Genetics and Molecular Biology
Bharathiar University
Coimbatore, Tamil Nadu, India

Ramasamy Manikandan
Department of Biochemistry
Shrimati Indira Gandhi College
Tiruchirappalli, Tamil Nadu, India

Poovarasan Manjeeswaran
Department of Human Genetics and Molecular Biology
Bharathiar University
Coimbatore, Tamil Nadu, India

Ather Manzoor
Division of Genetics and Plant Breeding
Sher-e-Kashmir University of Agricultural Sciences & Technology of Kashmir
Jammu and Kashmir, India

Abdelilah Meddich
Centre d'Agrobiotechnologie et Bioingénierie
Unité de Recherche labellisée CNRST (Centre AgroBiotech-URL-CNRST-05)
Université Cadi Ayyad
Marrakech, Morocco

African Sustainable Agriculture Research Institute (ASARI)
Mohammed VI Polytechnic University (UM6P)
Laâyoune, Morocco

Harshit Mishra
Department of Agricultural Economics
College of Agriculture
Acharya Narendra Deva University of Agriculture and Technology
Ayodhya, Uttar Pradesh, India

Rashmi Mishra
College of Economics and Business Administration
University of Technology and Applied Sciences—Al Musannah
Muscat, Oman

Varucha Misra
Division of Crop Improvement
ICAR-Indian Institute of Sugarcane Research
Lucknow, Uttar Pradesh, India

Nahidha Parveen Mohammed Koya
Department of Human Genetics and Molecular Biology
Bharathiar University
Coimbatore, Tamil Nadu, India

Yassine Mouniane
Natural Resources and Sustainable Development Laboratory
Faculty of Sciences
Ibn Tofail University
Kenitra, Morocco

Elshan Musazade
Key Laboratory of Soybean Molecular Design Breeding
Northeast Institute of Geography and Agroecology
Chinese Academy of Sciences
Changchun, PR China

Kathiresan Nagaraj
Department of Human Genetics and Molecular Biology
Bharathiar University
Coimbatore, Tamil Nadu, India

Vivek Patel
Department of Human Genetics and Molecular Biology
Bharathiar University
Coimbatore, Tamil Nadu, India

Shobana Devi Paulraj
Department of Biochemistry
Shrimati Indira Gandhi College
Tiruchirappalli, Tamil Nadu, India

Bhanu Priya Pegu
Department of Botany
Rabindranath Tagore University
Hojai, Assam, India

Mohammed Radi
Department of Plant Protection, Phytopathology Unit
Ecole Nationale d'Agriculture de Meknès
Meknès, Morocco

Environment and Valorization of Microbial and Plant Resources Unit
Faculty of Sciences
Moulay Ismail University
Meknes, Morocco

Shyamalin Rajmedhi
Department of Plant Breeding and Genetics
Assam Agricultural University
Jorhat, Assam, India

Prioty Bandhan Roop
Department of Genetics and Plant Breeding
Institute of Agriculture
Visva-Bharati University
Sriniketan, West Bengal, India

Niharika Saharia
Department of Plant Breeding and Genetics
Assam Agricultural University
Jorhat, Assam, India

Debojit Sarma
Department of Plant Breeding and Genetics
Assam Agricultural University
Jorhat, Assam, India

Sangeeta Sarma
Department of Botany
Jagiroad College
Jagiroad, Assam, India

Ashutosh Sawarkar
Department of Genetics and Plant Breeding
IRDM Faculty Centre
Ramakrishna Mission Vivekananda Educational and Research Institute (RKMVERI)
Narendrapur, West Bengal, India

Naira Sbbar
Centre d'Agrobiotechnologie et Bioingénierie
Unité de Recherche labellisée CNRST (Centre AgroBiotech-URL-CNRST-05)
Université Cadi Ayyad
Marrakech, Morocco

Angappan Shanmugapriya
Department of Biochemistry
Shrimati Indira Gandhi College
Tiruchirappalli, Tamil Nadu, India

Isha Sharma
Department of Microbiology
Dr. Harisingh Gour Vishwavidyalaya (A Central University)
Sagar, Madhya Pradesh, India

Rekha Sharma
Department of Plant Breeding and Genetics
Assam Agricultural University
Jorhat, Assam, India

Aditya Pratap Singh
Department of Genetics and Plant Breeding
Bidhan Chandra Krishi Viswavidyalaya
Nadia, West Bengal, India

Siddhartha Singh
College of Horticulture & Forestry
Pasighat
Arunachal Pradesh
Central Agricultural University
Imphal, Manipur, India

Dhanawantari L. Singha
Department of Botany
Rabindranath Tagore University
Hojai, Assam, India

Fatima-Ezzahra Soussani
Centre d'Agrobiotechnologie et Bioingénierie
Unité de Recherche labellisée CNRST (Centre AgroBiotech-URL-CNRST-05)
Université Cadi Ayyad
Marrakech, Morocco

Faculty of Sciences
Laboratory of Botany
Biotechnology and Plant Protection
University Ibn Toufail
Kenitra, Morocco

Satwik Subhankar
Department of Plant Breeding and Genetics
Assam Agricultural University
Jorhat, Assam, India

Arpita Talukdar
Department of Botany
Rabindranath Tagore University
Hojai, Assam, India

Vivek Kumar Varshney
Department of Human Genetics and Molecular Biology
Bharathiar University
Coimbatore, Tamil Nadu, India

Nashra Waqar
Department of Biotechnology
Faculty of Science
Era University
Lucknow, Uttar Pradesh, India

Showkat A. Waza
Division of Genetics and Plant Breeding
Sher-e-Kashmir University of Agricultural Sciences & Technology of Kashmir
Jammu and Kashmir, India

Accelerated Crop Improvement
International Crops Research Institute for the Semi-Arid Tropics
Patancheru, Telangana, India

Suxin Yang
Key Laboratory of Soybean Molecular Design Breeding
Northeast Institute of Geography and Agroecology
Chinese Academy of Sciences
Changchun, PR China

College of Advanced Agricultural Sciences
University of Chinese Academy of Sciences
Beijing, PR China

Hajar Zennouhi
Department of Plant Protection
Phytopathology Unit
Ecole Nationale d'Agriculture de Meknès
Meknès, Morocco

Laboratory of Functional Ecology and Environmental Engineering
Sidi Mohamed Ben Abdellah University
Fez, Morocco

About the Editor

Dr. Jen-Tsung Chen is a professor of cell biology at the National University of Kaohsiung in Taiwan. He teaches cell biology, genomics, proteomics, plant physiology, and plant biotechnology. Dr. Chen's research interests include bioactive compounds, chromatography techniques, plant molecular biology, plant biotechnology, bioinformatics, and systems pharmacology. He is an active editor of academic books and international journals that promote the exploration of multidisciplinary knowledge involving plant physiology, plant biotechnology, nanotechnology, materials science, ethnopharmacology, and systems biology. He serves as an associate editor, editorial board member, and guest editor in reputed journals. Dr. Chen has published books in collaboration with international publishers, and he is currently handling book projects on diverse topics such as AI technologies, drug discovery, drug development, herbal medicine, medicinal biotechnology, nanotechnology, bioengineering, plant functional genomics, plant speed breeding, epigenetics, functional RNAs, and CRISPR-based genome editing. Dr. Chen is a productive author in academic publications and was recognized as one of the World's Top 2% Scientists in 2023, 2024, and 2025 by Elsevier and Stanford University. In 2025, Dr. Chen received the Springer Nature Editor of Distinction Award.

Preface

RNA-based approaches have become a life science frontier over the last decade. They are highly expected to create extraordinary innovations that impact various biological and biomedical fields, such as "epigenetic modifications" by different types of non-coding RNAs (ncRNAs), "gene silencing" by RNAi, "precision genome engineering" by CRISPR/Cas systems involving single guide RNA (sgRNA), and "vaccine delivery" by mRNA technology.

In recent years, artificial intelligence (AI) technology, including machine learning, deep learning, generative AI, and large language models (LLMs), has advanced exponentially through the innovation of accelerated computing and is expected to revolutionize all aspects of human life and science. In plant biology, studies have been conducted through the introduction and integration of AI models with RNA technologies; however, based on the survey, a lack of literature summary in the international book market was found. Hence, there is a high demand to organize a new volume to guide the use of AI models to uncover the roles of plant functional RNAs in complex regulatory networks and cellular activities, improve the resolution and dimension of the RNA research data, integrate plant RNA-based tools with high-throughput systems, and synthesize artificial RNAs for multi-faceted applications in plant biology and biotechnology.

This book provides a series of critical summaries highlighting recent advances and applications of AI-driven RNA technologies in plants and offers opinions and prospects. It covers combinatorial strategies showing how to integrate AI models with RNA-based technologies for advancing epigenetic and epigenomic modifications by ncRNA, fine-tuning gene silencing via RNAi, regulating cell-to-cell communication via mobile mRNA, directing CRISPR/Cas9-mediated precision genome engineering using sgRNA, developing biotechnological applications using synthetic RNAs, and enabling nano-delivery of functional RNAs for various applications. These upgraded intelligent approaches hold promise for developing sustainable agriculture and supporting food and nutrition security under a changing climate through future crops with high quality, high yield, and stress resilience, which fits the goal of SDGs by the United Nations.

This book is a pioneer reference providing emerging knowledge of AI-enabled plant RNA technology. It's an ideal reference that was designed for readers, covering students, researchers, teachers, and professors in plant biology and crop breeding, to gain an overview of emerging knowledge of such a critical topic. It can potentially inspire readers to design innovative future research related to AI models integrated with RNA-mediated functions and tools.

The editor, Dr. Jen-Tsung Chen, appreciates all contributors' innovative chapters and the publisher's guidance and assistance.

1

Machine Learning for Precision Epigenetic Modification in Plants

Elshan Musazade[1], Suxin Yang[1,2], and Xianzhong Feng[1,2]

[1] *Key Laboratory of Soybean Molecular Design Breeding, Northeast Institute of Geography and Agroecology, Chinese Academy of Sciences, Changchun, PR China*
[2] *College of Advanced Agricultural Sciences, University of Chinese Academy of Sciences, Beijing, PR China*

1.1 Introduction

The global population is expected to reach 10 billion by 2050 and will place an unparalleled burden on global agricultural systems that must produce 2.4% greater yield each year to feed the world's increasing population. Conversely, the average yield growth has reached only 1.3%, and almost half of the grain-producing areas in the world are currently experiencing stagnant yields (Bahar et al., 2020; Rose & Chilvers, 2018). This widening productivity gap underscores the urgency of developing high-yielding, stress-tolerant crop varieties through sustainable intensification and technological advancement (Daszkiewicz, 2022; Mansoor et al., 2025).

The recent progress in genome science, integrative multiomics technologies, and speed-breeding technologies has revolutionized the modern agriculture, as now the cultivar development can be accelerated with shorter generation times and a higher traits-selection efficiency (Altaf et al., 2023). However, the gradual improvements achieved with the use of the traditional breeding are not enough to address the ever-growing world food consumption (Hinterberger et al., 2022; Mahmood et al., 2023). The use of big data analytics and artificial intelligence (AI) with molecular breeding represents a new paradigm for precision crop improvement (Zhang et al., 2025). Through the combination of genomics, phenomics, and environmental information, AI-driven applications make it possible to design breeding strategies that are principled to enhance the yield potential, nutrient-utilization efficiency, and environmental adaptability (Fan et al., 2025).

Under the molecular paradigm, genetic, epigenetic, and environmental factors are combined in a complex relationship in the control of agronomic traits. The key point of contact between the genome and environmental inputs is epigenetic mechanisms, including DNA methylation, histone modifications, chromatin remodeling, and RNA-mediated regulatory pathways. These reversible, heritable changes fine-tune gene expression without modifying nucleotide sequences, and as a result, coordinate developmental events and facilitate adaptations to biotic and abiotic stresses (Fan et al., 2025; Schmitz et al., 2022; Yocca & Edger, 2022).

The recent advancements in machine learning (ML) and deep learning (DL) have radically transformed the field of epigenomic research, providing it with the previously unavailable ability to decipher and predict multilayer regulatory networks. DeepCpG (Angermueller et al., 2017) and DeepHistone (Yin et al., 2019) models predict DNA methylation and histone modification patterns, which are key determinants of chromatin accessibility and topology of the three-dimensional (3D) genome. Other attempts to simulate chromatin architecture have used high-resolution reconstruction models, such as the convolutional neural networks (CNN)-based HiCPlus (Zhang et al., 2018c), the generative-learning-based hicGAN (Liu et al., 2019b), and the graph-neural-network-based CAESAR (Feng et al., 2022). However, these methods mainly use the training and validation based on high-throughput chromosome conformation capture sequencing (Hi-C) experimental data. On the contrary, modern integrative models combine the information about the DNA sequence with epigenetic signatures to predict higher-order chromatin contacts and overall genome architecture (Piecyk et al., 2022; Yang & Ma, 2022). Major methods are CoRNN for A/B compartmentalization (Zheng et al., 2024), TAD-Lactuca (Gan et al., 2019), and pTADS (Wang et al., 2021b) for identifying topologically associating domains (TADs), and ChINN (Cao et al., 2021a), CharID (Shen et al., 2022), and 3DepiLoop (Al Bkhetan & Plewczynski, 2018) for detecting chromatin loops and open chromatin interactions. EpiMCI is an extension of these ideas and represents the first multiway chromatin-interaction prediction framework (Hong et al., 2024).

The high-throughput data revolution in epigenomics has served as a driver behind the adoption of DL as an innovative method in the field of computational biology. Whereas traditional ML frameworks require defining features, which need to be identified in the data before analysis, DL models can extract hierarchical and nonlinear patterns automatically, thus allowing the discovery of molecular marks, chromatin states, and multifaceted regulatory interactions to be identified without any explicit genomic annotation. This has made it possible to make significant advances in detecting DNA methylation (Angermueller et al., 2017; Li et al., 2021; Lv et al., 2020), histone modifications (Hoffman et al., 2019; Xu et al., 2017), RNA methylation (Sun et al., 2019; Wang & Wang, 2020), and chromatin interactions (Yang et al., 2020; Zhang et al., 2018c). However, most of the above DL systems have been built in non-plant systems, which remains unclear on whether they will be applicable to plant genomes with different epigenetic structures and methylation environments.

Nevertheless, more recent developments in plant-centric epigenomic ML have been developed despite this limitation. Indicatively, the 5mC predictor in balsam poplar (*Populus balsamifera*) based on DL (Champigny et al., 2020), the 6mA site predictors in rice (Chen et al., 2019; Lv et al., 2019; Zhang et al., 2020), and the Smart Model for Epigenetics in Plants (SMEP) (Wang et al., 2021a) represent the increased use of AI to understand the plant-specific epigenetic architecture. Such models allow predicting patterns of methylation and histone modifications directly based on the DNA sequence data, which is a significant move toward making computational tools applicable to the task of guiding epigenetic engineering in crops.

This review provides a comprehensive synthesis of advances at the intersection of epigenetics and ML, focusing on the applications of ML and DL in precision epigenetic modification and smart plant breeding. The review begins by providing a description of the molecular processes behind epigenetic regulation, before explaining how the ML and DL models are reshaping the interpretation, modeling, and manipulation of regulatory layers. With the combination of genomic, epigenomic, and phenomic data, AI models can determine important epigenetics that determine gene expression, stress adaptation, and heritable memory. The review also discusses existing limitations of ML and DL models, such as model interpretability, cross-species transferability, and multiomics integration, and suggests approaches to improve the model performance and biological significance. Overall, the AI-epigenetics convergence will allow creating a new generation of data-driven, smart breeding designs that can create stress-resistant, high-yielding, and sustainable crops to respond to the needs of the fast-evolving world.

1.2 Epigenetic

Epigenetics represents a fast-evolving subtopic of genetics, which substantially expands our understanding of gene-trait interactions, which has been of interest to biologists for many years (Varotto et al., 2020). It does not rely on classical models of genetics but examines how genes can be regulated by molecular and biochemical processes that can respond to environmental changes without changing the underlying DNA sequence (Waddington, 2012). Modern epigenetic research focuses on DNA-related regulatory pathways, and the growing role of such results is manifested in the ability to use hereditary epigenetic characteristics to improve crop functions and resistance (Slotkin, 2016).

Importantly, compared to Mendelian inheritance, epigenetic modifications such as DNA methylation, histone modification, chromatin remodeling, and noncoding RNA (ncRNA) regulation by microRNAs (miRNAs), small interfering RNAs (siRNAs), and long ncRNAs (lncRNAs) are both reversible and heritable and play a role in phenotypic plasticity and environmental adaptation (Figure 1.1) (Dobránszki et al., 2025; Varotto et al., 2020). These multilayered regulatory systems allow for both transient and persistent stress responses and enable plants to respond to dynamic fluctuations in environmental conditions. The reversible and dynamic epigenetic regulation is facilitated by three major classes of enzymes, namely, writers (methyltransferases), erasers (demethylases), and readers (recognition proteins), which, respectively, create, remove, and decipher epigenetic marks to maintain transcriptional homeostasis (Liang et al., 2020; Mehta & Jeffrey, 2015).

Furthermore, among these regulatory mechanisms, DNA methylation is a major factor determining crop viability and influences seed dormancy, flowering time, and yield. Besides, the partial heritability of DNA methylation highlights its importance in evolution and domestication (Song et al., 2017; Zhang et al., 2013). A thorough understanding of methylation processes can help breeders determine the beneficial epigenetic profiles and reduce the number of unintended changes, thus promoting sustainable changes in crop production and adaptation (Han et al., 2018; Lin et al., 2019).

The epigenetic regulation provides the plants with developmental plasticity and environmental adaptability (He & Li, 2018; Hou et al., 2022). The recent advances have deepened the understanding of histone post-translational modifications (PTMs) and DNA methylation. New research has also provided insights into the regulatory functions of RNA chemical modifications, namely 5-methylcytosine (m^5C) and N6-methyladenosine (m^6A), representing another dimension of epigenetic regulation. These RNA modifications were

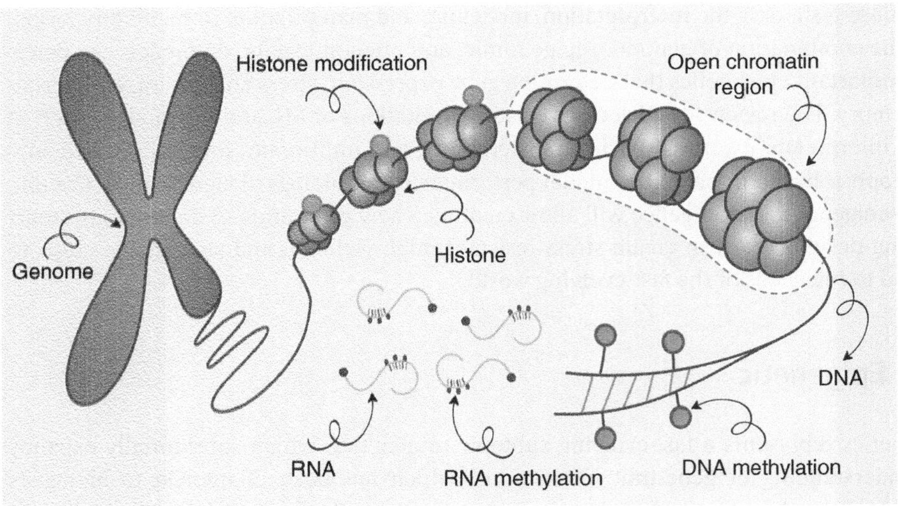

Figure 1.1 Components of the epigenome. Schematic representation of the major layers of epigenetic regulation in eukaryotic genomes. The DNA molecule is wrapped around histone proteins to form nucleosomes, which undergo various histone modifications (e.g., acetylation and methylation) that influence chromatin accessibility. DNA methylation involves the covalent addition of methyl groups to cytosine residues, affecting gene expression without altering nucleotide sequences. RNA methylation represents an additional layer of post-transcriptional regulation that modulates RNA stability and translation efficiency. Together, these interconnected mechanisms coordinate chromatin structure and gene activity within open chromatin regions of the genome.

first discovered in *Arabidopsis thaliana* and present new insights into transcriptional and post-transcriptional regulation (Mansoor et al., 2025). Similar changes have been noted in key crop species, including tomato (*Solanum lycopersicum*) and black poplar (*Populus trichocarpa*). Nevertheless, DNA methylation is the major controller of important agronomic characteristics, such as heterosis, fruit ripening, and stress resistance (An et al., 2017; Wang et al., 2016; Xing et al., 2015). Moreover, DNA methylation has been identified to have important functions during fertilization, gametogenesis, and inter-epigenetic communication between histone modifications. Therefore, it plays multifaceted roles in plant development, reproduction, and environmental adaptation (Johannes et al., 2009; Verhoeven et al., 2010). Taken together, these results highlight the enormous potential of epigenetic manipulation to increase crop resilience, accelerate breeding programs, and enable quick adjustment to changing environmental conditions.

1.3 Epigenetic Modifications

1.3.1 DNA Methylation Dynamics

DNA methylation is one of the key epigenetic processes involved in the maintenance of genome stability, silencing transposable elements, and transcriptional regulation in plants. It arises from the incorporation of a methyl group at the fifth carbon atom of cytosine

(5-methylcytosine, m^5C) in three different sequence contexts (CG, CHG, CHH) and influences developmental plasticity, environmental stress response, and transgenerational inheritance (Li et al., 2022a; Mishra et al., 2024). Environmentally induced changes in methylation patterns cause changes in chromatin structure to either promote or suppress gene expression. Reducing methylation levels in certain loci generally increases plant resistance against pathogens by activating salicylic acid (SA)-mediated signaling. However, such effects mainly depend on the type of tissue, developmental stage, and stress intensity (Dobránszki et al., 2025).

Cytosine methylation is established and maintained through an integrated enzyme network. Domains Rearranged Methylation Transferases (DRM1 and DRM2) in *A. thaliana* catalyze de novo methylation through the RNA directed DNA methylation (RdDM) pathway, while DNA methyltransferase 1 (MET1) and Chromomethylases (CMT2/CMT3) maintain methylation in CG and non-CG contexts (Law & Jacobsen, 2010; Matzke & Mosher, 2014; Zhang et al., 2018a). Active demethylation is mediated by DNA glycosylases like Demeter (DME) and Repressor of Silencing 1 (ROS1) that remove methylated cytosines, creating abasic sites that are repaired with unmethylated cytosines to reactivate stress and defense-related genes. Therefore, the constant balance of methylation and demethylation allows plants to adjust their transcriptional responses to environmental fluctuations and maintains stable epigenetic memories during repeated stress conditions (Farkas & Dobránszki, 2024; Halter et al., 2021; Schumann et al., 2019).

Besides m^5C, DNA N^6-methyladenine (m^6A) has emerged as another epigenetic mark that performs complementary functions. While 5mC is present in both plant and animal genomes, 6mA is less common and its biological role only recently became understood in eukaryotes (Zhang et al., 2018b). Advanced detection methods, such as liquid chromatography-tandem mass spectrometry (LC-MS/MS) and m^6A immunoprecipitation (m^6A-IP) followed by third-generation sequencing have enabled the accurate mapping of m^6A sites showing that mA exists in multiple plant species such as *A. thaliana*, rice (*Oryza sativa*) and soybean (*Glycine max*) (Liang et al., 2020). The ratio of m^6A to adenine (m^6A/A) was measured at 0.048% in *Arabidopsis*, 0.55% in rice, and 0.0264% in soybeans which is significantly less than those found in fungal and bacterial systems (Liang et al., 2020; Yuan et al., 2020; Zhang et al., 2018b; Zhou et al., 2018).

In rice, alpha-ketoglutarate-dependent dioxinase Alkb homolog (ALKBH) family members, particularly ALKBH1, act as DNA m^6A demethylases removing methyl groups without affecting the methylation state of RNA (Zhou et al., 2018). Although m^6A binding recognition proteins have not yet been identified in plants, functional analyses have shown that m^6A methylation may participate in plant development and stress-response pathways. For example, rice *ddm1a/1b* mutants showed low m^6A levels and exhibited a dwarf phenotype and reduced seed-setting rates (Zhang et al., 2018b), whereas ALKBH1-mediated demethylation delayed heading date in rice (Zhou et al., 2018). However, recent evidence suggests that m^6A interacts with 3D chromatin organization to mediate transcriptional reprogramming in response to heat stress (Liang et al., 2021).

Collectively, m^5C and m^6A represent two complementary and evolutionarily conserved layers of DNA methylation to secure genomic integrity and to regulate gene expression. The coordinated action of methyltransferases, demethylases, and RdDM components enables plants to reach a dynamic equilibrium between stability and flexibility, allowing heritable but adaptive regulation of gene expression required for growth, development, and environmental resilience.

1.3.2 RNA Modification Dynamics

Over 170 chemical modifications of eukaryotic RNA molecules are known, with m^6A and m^5C being the common and biologically important marks (Roundtree et al., 2017). These epitranscriptomic marks have been shown to regulate nearly all post-transcriptional processes, such as RNA splicing, processing, translation, and decay, and have an enormous impact on plant growth, development, and adaptation to stress (Shen et al., 2021).

The m^6A modification, the most abundant internal mRNA mark, is dynamically and reversibly regulated by a conserved set of methyltransferases, demethylases, and recognition proteins that decode methylation signals (Hou et al., 2021; Shao et al., 2021; Song et al., 2021). The human fat mass and obesity-associated (FTO) protein, the first identified m^6A demethylase (Jia et al., 2011), established the reversible nature of this modification. Although direct FTO homologs have not been identified in plants, related ALKBH family enzymes, such as AtALKBH9B and AtALKBH10B, in *Arabidopsis* serve as functional demethylases, though their evolutionary conservation and substrate specificity remain partially unresolved (Duan et al., 2017; Martínez-Pérez et al., 2017). The m^6A methyltransferase complex exhibits partial conservation between animals and plants, with METTL3, METTL14, WTAP corresponding to MTA, MTB, FIP37 in *Arabidopsis* and rice (Shen et al., 2016; Zhang et al., 2019a). In addition, YTH-domain-containing proteins (the YTHDF/ECT family) function as m^6A recognition proteins across kingdoms, recognizing m^6A-modified transcripts to regulate RNA metabolism and gene expression. In *Arabidopsis*, ECT2, ECT3, and ECT4 recognize m^6A-modified transcripts, regulating alternative polyadenylation and 3′-UTR processing in the nucleus while promoting mRNA decay in the cytoplasm (Arribas-Hernández et al., 2018; Wei et al., 2018).

Functionally, m^6A is indispensable for plant developmental and metabolic plasticity. It regulates embryogenesis, leaf morphogenesis, flowering transition, male meiosis, tapetum degradation, nitrate metabolism, and root development (Cheng et al., 2022; Hou et al., 2021; Xu et al., 2022; Zhang et al., 2019a). In *Arabidopsis*, the methyltransferase FIP37 stabilizes *WUSCHEL* and *SHOOT MERISTEMLESS* transcripts, controlling shoot apical meristem fate (Shen et al., 2016). In rice, OsFIP37 and its interacting partner OsFIP37-associated protein 1 regulate early microspore degeneration and male meiotic progression (Cheng et al., 2022; Zhang et al., 2019a). The METTL16 homolog FIONA1 (FIO1) also contributes to flowering regulation by adding m^6A marks to specific pre-mRNAs (Wang et al., 2022; Xu et al., 2022). Furthermore, the *Arabidopsis* Cleavage and Polyadenylation Specificity Factor 30 (CPSF30) protein, another m^6A recognition protein, mediates alternative polyadenylation and modulates nitrogen uptake and flowering (Hou et al., 2021; Song et al., 2021).

In addition to its role in regulating development, m^6A is an important component of stress responses. For example, the cooperative action of MTA, MTB, VIRILIZER (VIR), and HAKAI regulates salt tolerance in *Arabidopsis* (Hu et al., 2021). Similarly, m^6A was shown to be involved in modulating rice virus interactions as well (Zhang et al., 2021). Also, m^6A

remodeling is involved in tomato fruit ripening (Zhou et al., 2019). In the *colorless non-ripening* (*cnr*) mutant, which has elevated DNA methylation, it was found that the levels of m^6A were elevated in more than 1,100 transcripts, and expression of the RNA demethylase SlALKBH2 and stability of SlDML2 mRNA was reduced, which are necessary for the proper ripening of fruits. In tomato, the m^6A recognition protein SlYTH2 promotes ripening and the formation of aromas through the RNA-protein condensate and enhancing the translation of ripening related-genes (Bian et al., 2024). Similarly, transgenic expression of the human FTO demethylase in potato (*Solanum tuberosum*) resulted in an increase of 50% in yield and biomass compared to control plants, showing the potential of RNA methylation engineering to improve the production of crops (Yu et al., 2021).

Another important mRNA methylation mark is m^5C, which is catalyzed by homologs of the human NSUN2 methyltransferases, identified in rice and *Arabidopsis* (Cui et al., 2017; Tang et al., 2020). TET2 is the first identified m^5C demethylase in mammals. m^5C is functionally conserved and affects transcript stability and translation (Shen et al., 2016). In *Arabidopsis*, m^5C has been reported to stimulate protein synthesis and mRNA transport to target cells throughout root formation (Cui et al., 2017). Methyltransferase OsNSUN2 regulates oxidative stress response and increases thermotolerance in rice (Gao & Fang, 2021; Tang et al., 2020). Finally, m^5C-regulated transcripts are involved in ethylene biosynthesis and ethylene signaling in tomato, which is essential in fruit maturation (Guo et al., 2022a).

Collectively, m^6A and m^5C modifications form a dynamic post-transcriptional regulatory layer that combines developmental and environmental signals to orchestrate plant growth, reproduction, and stress tolerance. These modifications control gene expression programs that control development and resistance through the regulation of mRNA stability, transport, and translation, thus providing appealing molecular targets for the breeding of high-yielding and climate-resilient cultivars.

1.3.3 Histone Modifications and Chromatin Dynamics

The basic building block of chromosomes, chromatin, is made of DNA circular around histone octamers to create nucleosomes. The histones are extensively post-transcriptionally modified in their N-terminal tails, affecting the state of chromatin organization and the level of gene expression (Mierziak & Wojtasik, 2024; Pu & Sung, 2015). PTMs, including methylation, acetylation, phosphorylation, adenylation, ubiquitination, SUMOylation, and ADP-ribosylation, have context-dependent effects on gene expression, plant growth, and adaptation to environmental stresses (Cui et al., 2017).

Histone acetylation, which is mainly catalyzed by histone acetyltransferases (HATs), is the neutralization of the positive charge of lysine residues and consequently promotes chromatin relaxation and enhanced accessibility for transcription factors and RNA polymerase. Conversely, histone deacetylases (HDACs) remove acetyl groups, restoring the compaction of chromatin and silencing transcription (Ramirez-Prado et al., 2018). This type of regulatory circuit is reversible and is observed in rice, where the HDAC HDA704 works with the transcriptional factor OsWR2 to deacetylate H4K8 at the OsABI5 promoter during drought, thereby inhibiting OsABI5 expression and increasing drought tolerance (Guo et al., 2023).

One of the most studied PTMs, histone methylation, is catalyzed by histone methyltransferases (HMTs) and can take place on lysine and arginine residues in a mono-, di-, or trimethylated state (me1, me2, me3). Methylation has a functional impact on the target residue and extent of methylation. Methylation of K4 and K36 in the histone H3 (H3K4me3 and H3K36me3) is typically linked with active transcription, whereas K9 and K27 (H3K9me2 and H3K27me3) methylation is linked with transcriptional repression (Kolasinska-Zwierz et al., 2009; Liu et al., 2010; Pien et al., 2008; Zhang et al., 2007). These changes are dynamically controlled by the action of methyltransferases, demethylases, and recognition proteins and are spatiotemporally specific, which forms the basis of developmental and physiological plasticity (Pu & Sung, 2015).

Along with acetylation and methylation, histone ubiquitination, specifically monoubiquitination of histone H2B, also plays an important role in transcriptional regulation, chromatin regulation, and adaptation to stress (Chen et al., 2020; Ma et al., 2019). The interaction of these PTMs enables the optimization of growth, defense, and metabolic stability in plants in reaction to differing environmental conditions (Dobránszki et al., 2025). Interaction between histone-modifying enzymes, HDACs, and other chromatin modifiers promotes dynamic chromatin remodeling and subsequently affects agronomically relevant characteristics like fruit ripening, tuberization, and resistance to stress (Liu et al., 2010; Zhang et al., 2024).

Collectively, histone modifications form a hierarchical epigenetic signal, which interprets transcriptional outputs by controlling chromatin accessibility and positioning of nucleosomes. Dynamic control combines the developmental programs and external signals, thus allowing the plants to optimize gene expression to produce high adaptation and productivity.

1.3.4 Chromatin Remodeling and Nucleosome Dynamics

Histone modifications are chemical signals that control chromatin access, while adenosine triphosphate (ATP)-dependent chromatin remodelers physically reposition or remove nucleosomes to modulate DNA accessibility for transcriptional machinery. Chromatin remodeler enzymes utilize energy from ATP hydrolysis to alter chromatin structure and gene expression control (Huang & Jin, 2022). In plant systems, the SNF2 helicase-like members, including BRM (BRAHMA), SYD (SPLAYED), and DDM1 (decrease in DNA methylation 1), are key elements in the regulation of transcription, especially under stress and defense conditions (Bhadouriya et al., 2021). BRM and SYD directly promote the expression of stress-responsive genes, and DDM1 allows DNA methyltransferases to approach heterochromatic compartments and thereby remodel defense-related gene expression by epigenetic silencing.

These remodelers work together with DNA methylation and histone modifications to tightly regulate chromatin organization and rapidly reorganize the position of nucleosomes at specific sites of the genome associated with environmental stress signals, hormone regulation, and pathogen defense responses. Such reorganization allows for the rapid activation of silent genes in response to environmental stresses while also ensuring long-term epigenetic memory and maintaining a balance between transcriptional plasticity and genomic stability (Dobránszki et al., 2025).

1.4 ML in Plant Epigenetics

ML has become a paradigm of specific epigenetic manipulation in plants, allowing to develop predictive models and modify epigenetic states, such as DNA methylation, histone, and chromatin remodeling, and small RNA-based silencing (Figure 1.2 and Table 1.1). ML frameworks offer a predictive perspective of interactions between regulatory factors that mediate stress adaptations, developmental plasticity, and heritable gene-expression reactions, which returns to be a result of integration between multiomics data including methylomes, transcriptomes, and chromatin accessibility data (Dobránszki et al., 2025). This can be further boosted by DL models, which can use nonlinear associations of molecular features to perform accurate inferences and regulate epigenetic marks at a single locus. All these approaches combine to give the mathematical foundation of the information-based approaches to reprogramming epigenetic states, as a means of improving crop resistance, yield stability, and adaptive phenotype to changing environmental conditions.

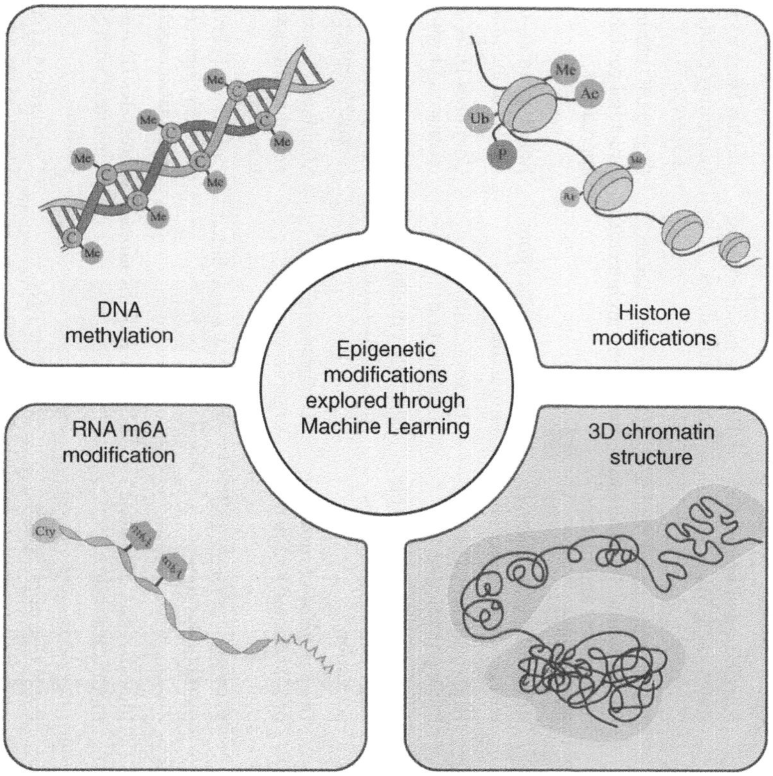

Figure 1.2 Epigenetic modifications deciphered using ML approaches. ML approaches have been employed to investigate multiple layers of epigenetic regulation, including DNA methylation, histone modifications, RNA m^6A modification, and 3D chromatin structure. By integrating diverse datasets and predictive models, ML algorithms can identify modification patterns, infer regulatory relationships, and enhance understanding of how epigenetic dynamics influence gene expression, development, and stress responses.

Table 1.1 AI-based tools for analyzing epigenetic modifications.

AI-based tool/model	Primary application/function	Type of epigenetic modification or data	Key features and achievements	Reference
ChIPr	Prediction of higher-order chromatin interactions	3D chromatin organization	Integrates DNA sequence and epigenetic features to model chromatin topology and interaction dynamics.	Abbas et al. (2024)
ChroModule	Annotation of epigenetic states using ML	Histone modifications; chromatin states	Hidden Markov model (HMM)-based framework for annotating promoters, enhancers, and regulatory elements.	Won et al. (2013)
ChromFormer	Transformer-based model for predicting 3D chromatin interactions and histone methylation	Histone methylation; chromatin architecture	Predicts 3D chromatin architecture from Hi-C data with structural confidence estimation, outperforming existing approaches.	Valeyre et al. (2022)
ChromHMM	Chromatin state discovery and genome segmentation	Histone modification; chromatin states	Integrates multi-ChIP-seq datasets using multivariate HMMs to automate functional genome annotation.	Ernst and Kellis (2017)
ChromaFold	High-resolution chromatin-folding prediction	3D chromatin conformation	Combines DNA sequence and epigenomic features to reconstruct high-resolution chromatin architecture.	Gao et al. (2024a)
CRISPRoff/ CRISPRon	Programmable, reversible epigenetic silencing and activation	CpG DNA methylation; transcriptional control	Fuses DNMT3A/3L with dCas9-KRAB for targeted methylation, while CRISPRon recruits demethylases for reversible epigenetic regulation.	Nuñez et al. (2021)
C.Origami	Deep generative modeling of 3D genome architecture	3D chromatin organization	Simulates chromatin folding and predicts long-range genomic interactions using sequence and epigenetic features.	Tan et al. (2023)
DeepCpG	Prediction of single-cell methylation states	m^5C DNA methylation	Deep neural network predicts missing CpG methylation states in single cells and reveals interpretable sequence-methylation relationships.	Angermueller et al. (2017)
DeepHistone	Identification of histone modification sites	Histone modifications	Integrates DNA motif and chromatin accessibility data, consistent with established transcription factor binding profiles.	Yin et al. (2019)
DeepM6A	Prediction of m^6A sites	m^6A DNA/RNA methylation	Detects 6mA at single-nucleotide resolution and identifies cis-regulatory motifs associated with 6mA deposition.	Tan et al. (2020)

DeepMod	Detection of base modifications from nanopore sequencing	DNA methylation and related marks	Employs bidirectional LSTM networks to generate strand-specific modification profiles at single-base resolution.	Liu et al. (2019a)
DeepSignal-Plant	Nanopore-based methylation detection in plants	m^5C (CG/CHG/CHH)	Corrects low-frequency biases, validated in *Arabidopsis thaliana*, *Oryza sativa*, and *Brassica nigra*, and identifies novel methylation loci.	Ni et al. (2021)
DeepCBA	Prediction of gene expression from 3D chromatin data	Chromatin interaction; DNA methylation	Integrates CNN, BiLSTM, and self-attention mechanisms (PCC up to 0.929) to identify key regulatory genes in maize (*Zea mays*).	Wang et al. (2024b)
EPCOT-COP	Multiway chromatin-interaction prediction	3D chromatin organization	Integrates DNA sequence and epigenetic features to improve chromatin contact prediction accuracy.	Zhang et al. (2023)
Epic2	Peak calling for histone-mark enrichment domains	Histone modification peaks (ChIP-seq)	Reengineered Spatial Clustering for Identification of ChIP-Enriched Regions (SICER) algorithm optimized for high speed, low memory consumption, and broad peak detection.	Stovner and Sætrom (2019)
EpiCas-DL	Prediction of sgRNA efficiency for CRISPR epi-editing	Epigenetic editing (DNA methylation, chromatin accessibility)	Integrates gene expression, DNA methylation, and TSS proximity (AUC = 0.87) to optimize sgRNA design for epigenetic editing.	Yang et al. (2023a)
EpiPredictor	Prediction of Polycomb-group (PcG) targets	Histone methylation (PcG-mediated repression)	Models transcription factor network interactions using nonlinear kernels to enhance Polycomb target gene prediction accuracy.	Zeng et al. (2012)
GENIE3	Network inference of lncRNA-gene interactions	ncRNA-mediated epigenetic regulation	Random forest (RF) algorithm infers directional lncRNA-gene regulatory networks, applied to *Solanum tuberosum* immunity.	Cao et al. (2021b)
iRNAm5C-PseDNC	Identification of RNA m^5C sites	m^5C RNA methylation	Utilizes pseudo dinucleotide composition with ten physicochemical features for high-accuracy m^5C site prediction.	Qiu et al. (2017)
LncADeep	Identification and annotation of lncRNAs	lncRNA regulation	Deep belief network integrates sequence and structural features with KEGG/Reactome enrichment for functional annotation.	Yang et al. (2018)

(*Continued*)

Table 1.1 (Continued)

AI-based tool/model	Primary application/function	Type of epigenetic modification or data	Key features and achievements	Reference
m^6Anet	Detection of m^6A RNA methylation from nanopore signals	m^6A RNA methylation	Multiple-instance learning framework that generalizes across species and quantifies read-level methylation stoichiometry.	Hendra et al. (2022)
Methyl-IT	Signal detection and methylome analysis	m^5C DNA methylation	Combines information thermodynamics and ML for sensitive discrimination of accurate methylation signals.	Sanchez and Mackenzie (2018)
MethylSeekR	Identification of active regulatory regions from bisulfite data	DNA methylation	Detects transcription factor binding footprints and identifies unmethylated regulatory regions across diverse tissues.	Burger et al. (2013)
MethyQA	Quality assessment of bisulfite sequencing	DNA methylation sequencing quality	Pipeline integrating open-source tools and custom scripts to assess bisulfite conversion efficiency and sequencing coverage.	Sun et al. (2013)
PHASIS	Characterization of phased siRNA (phasiRNA) loci	Small RNA epigenetic regulation	ML suite for phasiRNA locus discovery and trigger identification independent of genome or parallel analysis of RNA ends (PARE) data.	Kakrana et al. (2017)
PlantDeepMeth	Prediction and imputation of m^5C methylation from genomic sequence features	m^5C methylation (CG/CHG/CHH)	Combines CNN and RNN layers to capture local motifs and long-range dependencies ($AUC > 0.9$); validated across species for scalability.	Guo et al. (2025)
PLncPRO	Classification of coding vs. lncRNAs	lncRNAs	RF model leveraging sequence, homology, and 3-mer frequency features; validated across diverse plant species.	Singh et al. (2017)
Smart Model for Epigenetics in Plants (SMEP)	Prediction of methylation and histone modifications from DNA sequence	DNA methylation; histone modifications	AI-driven model decodes plant-specific epigenetic architectures, enabling computationally guided precision crop improvement.	Wang et al. (2021a)

1.4.1 ML in DNA Methylation Dynamics

DNA methylation is one of the most essential and active processes that regulate genome stability, gene expression, and plant development. By combining both multigenerational phenotypic and genome-wide DNA methylation data, AI-assisted predictive epigenomics can be performed to demonstrate the influence of methylation marks on plant stress memory and adaptive resilience (N'Diaye et al., 2020). Unlike the conventional statistical methods, ML and its DL extensions are effective to identify nonlinear and high-dimensional relationships, which would reveal subtle but biologically relevant methylation patterns with respect to environmental adaptation (Dobránszki et al., 2025).

Supervised ML algorithms such as extreme gradient boosting (XGBoost) (Chen & Guestrin, 2016), support vector machines (SVMs) (Scholkopf & Smola, 2001), random forests (RFs) (Breiman, 2001), and least absolute shrinkage and selection operator (LASSO) regression have proved to be highly effective in DNA methylation dynamics modeling (Zemlianskaia et al., 2022). The relationships between epigenetic features (i.e., the frequency of methylation or histone marks) and phenotypic traits are analyzed using linear and logistic regression approaches (Olive, 2017), while SVMs are used to classify epigenetic profiles or identify previously unknown subgroups based on methylation (Fernández & Miranda-Saavedra, 2012). RFs combine numerous decision trees to score feature significance and provide predictable locations of methylation that predict phenotypes (Chen & Ishwaran, 2012), and the LASSO regression helps to consider the most significant variables that cause developmental or stress-induced transitions (Brieuc et al., 2018).

ML algorithms have identified specific methylation signatures associated with drought resistance in rice (Smet et al., 2023) and have enhanced the classification of methylation patterns in a broad range of biological situations (Mavaie et al., 2021). Trained models based on similar methylome data showed the existence of conserved root-specific methylation states that persist upon stress recovery and that are associated with permanent activation of defense-related genes (Champigny et al., 2020; Zhang & Shafiq, 2024). In addition, ML methods are effective at identifying promoter and regulatory regions that experience stress-induced changes in the methylation state and thereby fine-tune gene expression during drought, heat, salinity, and pathogen stress (Kimotho & Maina, 2024; Rico-Chávez et al., 2022).

Other than sequence-level prediction, ML models have also associated variation in methylation with gene expression and adaptive phenotypes. N'Diaye et al. (2020) used a set of ML algorithms to study both cytosine methylation and transcriptional regulation in common wheat (*Triticum aestivum*). They were able to predict differentially expressed genes in leaf and root tissues using genome-wide profiles of methylation with an accuracy of 0.81. DNA methylation has been shown to be primarily localized in CG (60%), CHG (35%), and CHH (5%) regions, with weaker correlation in promoter regions compared to coding and exon regions. The integrated methylation pattern in promoter, exonic, and coding regions enables the capture of tissue-specific patterns of gene expression, notably those involved in photosynthesis and pigment synthesis. Thus, this study supports the idea of using multicontext methylation data to enhance the interpretative and predictive capacities of ML models on multifaceted polyploid genomes (N'Diaye et al., 2020).

DL models take advantage of these properties by discovering higher-order, nonlinear interactions in methylation landscapes. Recurrent neural networks (RNNs) are particularly effective in modeling sequential genomic data because they can utilize "memory" type processing to indicate relationships between loci of nucleotide sequences (Mienye et al., 2024). Wang et al. (2021a) developed the SMEP, which predicted the change in m^5C levels in rice, maize, and *A. thaliana* with an accuracy of 80–95% (Wang et al., 2021a).

DeepSignal-Plant is a pioneering framework designed to detect cytosine DNA methylation (m^5C) across all three types of contexts (CG, CHG, and CHH) from raw nanopore sequencing reads. DeepSignal-Plant incorporates denoising and sample-balancing methods to remove bias associated with the detection of low-frequency methylation events, specifically in the abundance of CHH contexts found in plant genomes. Validation studies were conducted on *A. thaliana*, *O. sativa*, and black mustard (*Brassica nigra*), and the model showed very high concordance with bisulfite sequencing while showing better resolution in repetitive and polyploid regions of the genome. Importantly, DeepSignal-Plant was also able to reveal previously unidentified methylated cytosines including clusters of 61 tRNA genes and 3 *Arabidopsis* NB-LRR genes (*At1g58807*, *At1g59124*, and *At1g59218*) demonstrating its ability to identify new methylation sites and enhance plant methylome characterization (Ni et al., 2021).

Complementary, PlantDeepMeth is a hybrid approach that combines both CNNs and RNNs to predict cytosine m^5C methylation in CG, CHG, and CHH contexts based on genomic sequence features. The combined use of CNNs and RNNs simultaneously captures both local sequence motifs and long-range dependencies, thus making it possible to impute missing methylation data in regions that are not sequenced completely. PlantDeepMeth was trained on *A. thaliana* and common mustard (*Brassica rapa*) and had a high predictive accuracy (AUC > 0.9) than their predecessor systems, including SMEP. Context-dependent sequence properties of hypo- and hypermethylation were identified by motif-enrichment studies and showed mechanistic understanding of methylation regulation and sequence conservation across species. Its strength and scalability were additionally validated by cross-species validation and made it a universal inference and interpretation tool used to involve large-scale methylome inferences and interpretations (Guo et al., 2025).

Parallel to cytosine methylation studies, DeepM6A can predict m^6A at a single-nucleotide scale using only primary DNA sequence data. *A. thaliana*, *Drosophila melanogaster*, and *Escherichia coli*, DeepM6A outperformed conventional k-mer-based methods, showing specific cis-regulatory motifs of m^6A deposition. Although only a few 6mA methyltransferases, such as roundworm (*Caenorhabditis elegans*) DAMT-1, are experimentally verified, the model revealed nonrandom sequence motifs that might be associated with transcriptional regulation and host–microbe epigenetic communication. DeepM6A is an attractive model to explain the regulatory and functional aspect of 6mA methylation in eukaryotic genomes with its incorporation of the SM-CAP motif discovery module and the ability to use a variety of tissue- or stage-specific datasets (Tan et al., 2020).

Besides the molecular-level modeling, DL has also been utilized in terms of relating the change of methylation to both phenotypic and transcriptional outcomes in plants. Champigny et al. (2020) showed that models that implement C-phosphate-G (CpG) methylation features can be used to predict complex quantitative traits in balsam poplar, including biomass, wood density, lignin, and mannose content, and to explain up to 57.5% of trait variance (Champigny et al., 2020). The derived features related to methylation also facilitated accurate classification of tissue type and population origin, reflecting both environmental and genetic influences. Notably, the model has identified a minimal predictive subset (<0.01% of total cytosines) and significantly outperformed traditional regression tools that also demand much more important features thus highlighting the effectiveness of DL in associating epigenomic signatures with quantitative traits (Champigny et al., 2020; Tahir et al., 2024).

Recent DL models have transformed the plant epigenomics by unraveling the molecular mechanisms behind the DNA methylation process with unprecedented precision. Based on the large-scale methylome data, DL models reveal latent sequence patterns and contextual dependencies, which allow high-resolution prediction of m^5C sites and regulatory outcomes (Chen et al., 2024; Wang et al., 2020). Models trained directly on plant genomic sequences perform strongly in identifying methylation-dependent gene regulation and stress-responsive epialleles (Guo et al., 2022b; Wang et al., 2021a). The combination of ML and DL into the area of methylation research creates a powerful computational framework to explain the process of epigenetic regulation, precision breeding, and the creation of stress-resilient and high-performing crop varieties.

Collectively, the application of XGBoost, RFs, CNNs, RNNs, and transformers has made AI one of the most important technologies to advance the research into plant methylomes. ML and DL algorithms combine sequence, signal, and context-based learning methods and transform the science of plant methylation from a descriptive profiling to a predictive, mechanistic, and integrative epigenomics. These convergences provide the data-driven basis of precision breeding where the generation of stress tolerant, high-performance varieties of crops can occur through the specific epigenomic editing of genes and regulatory pathways responsible for plant development, defense, and stress adaptation.

1.4.2 ML in Histone Modifications

DL, a branch of ML, has brought a paradigm shift in understanding patterns of histone modification and their biological purposes in the regulation of epigenetics. Its ability to encode high-dimensional data into multidimensional, nonlinear associations and latent regulatory signatures has placed it as a fundamental method to understand histone-mediated chromatin dynamics (Chen et al., 2024; Wang et al., 2020). Massive chromatin immunoprecipitation sequencing (ChIP-seq) data with multiomics data make it possible to identify, classify, and predict histone modification states that are relevant to either the activation or repression of genes or a bivalent configuration using DL models. Such models are essential to explain the functions of combinatorial histone marks in the regulation of transcription, developmental reprogramming, and stress adaptation in plants (Guo et al., 2022b; Wang et al., 2021a).

Deep belief networks (DBNs) are powerful probabilistic generative models of unsupervised learning paradigm, which can discover the latent relationships in complex biological systems (Hinton et al., 2006; Li et al., 2025). DBNs have been used to analyze epigenomic data to show nonlinear interactions between histone modifications and gene expression without prior training with prelabeled data. The evaluation of the performance is usually based on accuracy, sensitivity, specificity, and precision, and the best results are obtained on balanced datasets (Sokolova & Lapalme, 2009).

Self-organizing maps (SOMs) provide explanatory unsupervised visualization and clustering of multidimensional histone modification landscapes. SOMs are used to discover latent regulatory properties, clustering, and modeling of complex associations between histone modifications and transcriptional states (Javed et al., 2024; Nikkilä et al., 2002). Their functionality is not yet commonly used in plant epigenomics, but because of their capacity to model high-dimensional data on biologically relevant topologies, their future applications have the potential to be utilized (Dobránszki et al., 2025).

Complementing these probabilistic and clustering methods, recent advances in DL architecture have extended predictive modeling capabilities to sequence-based inference of histone modifications. Lv et al. (2025) developed a DL framework based on the Sei architecture to predict histone modifications and cis-regulatory features from DNA sequences in *A. thaliana*, *O. sativa*, and *Z. mays*. The models achieved high within-species performance (AUROC > 0.94 and AUPRC > 0.71) and strong concordance with ChIP-seq data but showed reduced generalization across distant lineages. Family-level models within Poaceae and Brassicaceae improved cross-species robustness and captured conserved promoter-associated H3K4me3 patterns. *In silico* mutagenesis verified the biological relevance of learned sequence motifs. While the framework enables genome-wide chromatin feature prediction and visualization, its reliance on ChIP-Hub datasets and absence of ATAC-seq, RNA-seq, and Hi-C integration limit broader applicability. The study highlights the need for transformer-based and multimodal architectures to enhance cross-species epigenomic prediction in plants (Lv et al., 2025).

Collectively, DBNs, SOMs, and sophisticated DL networks present a unified and powerful computation system of deciphering how histone modification dynamics change over time. Besides, these models have provided a global outlook of how histone codes are written, read, and erased in response to developmental or environmental signals, thus providing a base to predictive and precision-based applications in plant epigenetic engineering and chromatin state editing.

1.4.3 ML in Chromatin Remodeling

The high dimensionality and complexity of epigenetic data such as gene expression, chromatin accessibility, and three-dimensional organizational structures require the current state-of-the-art ML algorithms to identify biologically meaningful patterns that are otherwise not reachable by traditional approaches (Arslan et al., 2021). In contrast to supervised, which relies on labeled data to perform predictive tasks (e.g., control versus treatment), unsupervised learning is an autonomous method to identify latent structures in unlabeled datasets, which makes it invaluable in exploratory data analysis, clustering, and feature extraction (Bröker et al., 2024). The underlying biological subgrouping is readily determined through these methodologies, shedding some light on the regulatory hierarchies that drive chromatin remodeling and regulation of genes.

RNNs are one of the most appropriate DL architectures to process sequential genomic data, such as RNA and DNA sequences (Mienye et al., 2024). It allows the modeling of dependencies across several positions of nucleotides which makes them useful in predicting chromatin states and differential gene expression in different physiological or stress conditions. However, their performance can be hampered by vanishing gradients and a small memory bandwidth. Therefore, architectural improvements or hybrid methods are often necessary (Mienye & Swart, 2024).

Graph neural networks (GNNs) have become strong models of intercellular biological networks, such as gene regulatory, protein–protein interaction, and chromatin contact networks. GNNs have the capability to capture topological dependencies in chromatin architecture by modeling genes, transcription factors, enhancers, and epigenetic marks as graph nodes and their interrelations as graph edges. They have been especially successful in predicting 3D genome organization and remodeling events using Hi-C and other chromatin conformation measurements (Du et al., 2024; Gamarnik, 2023; Jiang et al., 2024). Although they have advantages, GNNs have scalability and over-smoothing computational issues with large and complex graphs, and where the node representations are indistinguishable after several layers of propagation.

Collectively, the application of ML and DL in chromatin remodeling represents a significant step toward resolving how epigenomic architecture dynamically interfaces with transcriptional regulation. These approaches are laying the groundwork for predictive modeling of chromatin dynamics, enabling precise dissection of regulatory mechanisms that drive plant adaptation, development, and stress resilience.

1.4.4 ML in Chromatin Modification Dynamics

Histone methylation and acetylation, alongside DNA methylation are an important epigenetic switch that activates plant stress memory, which is crucial to survival and adaptation (Gallusci et al., 2023; Sharma et al., 2022). The dynamically regulated gene accessibility through these chromatin marks is important to make sure that the stress-responsive genes are activated or repressed in time to face the subsequent environmental challenges. ML applications to ChIP-seq data have allowed predicting and interpreting histone modification patterns linked to stress responses in a significant way (Li et al., 2022b; Wan et al., 2024).

Activating histone marks have been shown to accumulate at the promoters of stress-induced genes to enhance transcriptional activation (Ueda & Seki, 2020), whereas suppressive marks are enriched in areas that need silencing so that a larger portion of energy is conserved and the organism can maintain homeostasis (Yung et al., 2021). Combined with transcriptomic and methylomic data, large-scale ChIP-seq datasets can provide a detailed picture of chromatin-mediated regulation during stress, allowing the identification of the principle regulatory loci to be targeted to improve crops. ML models trained on such datasets can predict histone modification signatures associated with plant resilience or vulnerability to certain stressors, including drought, salinity, or pathogen infection, through the analysis of genomic features, enrichment levels, and association with gene expression (Wu et al., 2019).

Beyond single-mark prediction, ML approaches increasingly model interactions among multiple epigenetic layers, integrating chromatin modifications with DNA methylation and ncRNA activity to achieve a systems-level understanding of chromatin regulation (Roychowdhury et al., 2023). Time-series analyses of ChIP-seq data from plants exposed to recurrent stress cycles further enable AI models to capture the temporal dynamics of chromatin remodeling, identifying histone marks and loci associated with long-term stress memory formation (Peng & Rajjou, 2024; Piecyk et al., 2022).

An example of such breakthroughs, which Wang et al. (2021a) proposed, is the SMEP. SMEP is a DL framework, which makes predictions of six main modifications of epigenetic markers, such as DNA m^5C, DNA m^6A, RNA m^6A, and the histone markers H3K4me3, H3K27me3, and H3K9ac, using genomic sequence alone. SMEP trained on *O. sativa* (Nipponbare) attained a success of achieved ~95% accuracy for m^6A and ~80% for other marks, compared to traditional ML techniques. Cross-species validation was performed in *A. thaliana*, maize, and indica rice, where strong generalization and weak overfitting were observed. The model correctly predicted more than 95% of heat-shock-dependent m^6A peaks and validated epigenetic locations of HsfA1, AREB1, and DREB2A, which supports the biological significance of the model. Therefore, SMEP is an easily accessible, scalable, and versatile system of multiepigenome prediction and specific genome engineering in plants (Wang et al., 2021a).

Complementary studies further highlight the expanding role of ML and DL in chromatin-level epigenomic prediction. Wu et al. (2019) integrated ChIP-seq and RNA-seq data within an ML framework to identify drought-responsive histone modification patterns in *A. thaliana*, later validated by ChIP-qPCR (Wu et al., 2019). Similarly, Peng and Rajjou (2024) applied DL to time-course ChIP-seq data in tomato, uncovering heat-induced H3K27me3 changes in heat-memory genes confirmed by RT-qPCR and phenotypic assays (Peng & Rajjou, 2024). Collectively, these findings demonstrate that ML approaches can detect subtle, condition-specific chromatin transitions that underpin plant stress adaptation.

Collectively, these studies demonstrate increasing utility of AI in understanding and explaining the mechanism of plant "remember" for precise use in chromatin-based memory systems by defining the specific histone marks that remain after a plant has been exposed to stressful conditions and therefore facilitate the rapid and effective response when the plant is once again subjected to those stressful conditions.

1.4.5 ML in Chromatin-Interaction Prediction

Accurate interpretation of noncoding genetic variation requires analysis within the framework of 3D genome organization. DeepC, a transfer learning-based neural network, predicts genome folding from megabase-scale DNA sequences with high precision. It identifies domain boundaries, learns sequence determinants of chromatin organization, and quantifies the effects of both large-scale structural and single-nucleotide variants, thereby improving the reconstruction of chromatin-interaction matrices from low-depth sequencing data (Schwessinger et al., 2020). Similarly, Orca employs a sequence-based DL framework to model multiscale 3D genome architecture directly from DNA sequence, encompassing kilobase to chromosome-level organization. Its cascading CNN with dilated convolutions effectively captures sequence determinants of chromatin compartments, TADs, and long-range interactions, enabling precise variant effect prediction and *in silico* exploration of genome folding (Zhou, 2022).

Inspired by sequence-driven modeling strategies, HiC-Reg (Zhang et al., 2019b) and Epiphany (Yang et al., 2023b) are central computational models within the reconstruction of chromatin-interaction matrices based on the prediction of epigenetic features as inputs.

The two approaches combine a breadth of epigenomic signals, such as histone modifications, transcription factor binding, and chromatin accessibility, to derive 3D genomic interactions, but they differ in methodology. The HiC-Reg uses the RF algorithms to predict the association of localized epigenetic states with the intensity of chromatin contacts using anchor loci as training instances, thus enabling the predictive capability of the model using limited data and maintaining high predictive accuracy over all genomic loci. By contrast, Epiphany combines CNNs, bidirectional long short-term memory (Bi-LSTM) networks, and generative adversarial networks (GANs) to run on fixed genomic windows. Dimensionality reduction via the use of Z-stripe allows the extraction of hierarchical dependencies inherent in chromatin structure. Despite the relatively small correlation improvement with the GAN element compared to experimental Hi-C data, it significantly improves the spatial resolution and structural fidelity of the reconstructed matrices, thus creating a more realistic genome architecture (Yang et al., 2023b; Zhang et al., 2019b).

Even more recent multimodal frameworks, such as EPCOT-COP (Zhang et al., 2023), ChIPr (Abbas et al., 2024), C.Origami (Tan et al., 2023), and ChromaFold (Gao et al., 2024a), further enhance chromatin-interaction prediction by using the combination of DNA sequence and epigenetics data. EPCOT-COP and ChIPr implement feature-integration approaches. EPCOT-COP is a concatenation of DNA sequence and chromatin accessibility measurements of 1 Mb genomic regions to single feature vectors, which are then inputted to CNNs and self-attention modules to learn local and distal dependencies. ChIPr uses heterogeneous features, which include RAD21, H3K27ac, H3K27me3, genomic distance, GC content, and CTCF motif orientation, with dense neural networks, RFs, and gradient-boosted models, effectively training interaction strength with a synergistic mixture of epigenetic and structural information (Abbas et al., 2024; Gao et al., 2024a; Tan et al., 2023; Zhang et al., 2023).

On the other hand, C.Origami and ChromaFold use preliminary feature-extraction methods. C.Origami extracts features related to CTCF narrow peak positions and chromatin accessibility and encodes DNA sequences in one-hot format. The parallel CNN modules then independently process genomic and sequence inputs, and then concatenate columns to enhance long-range interaction detection. Similarly, ChromaFold extracts features related to CTCF motif scores obtained from FIMO and scATAC-seq peaks. The independent CNN modules then obtain high-level representations from each modality and then concatenate their outputs to be subject to linear regression to predict chromatin-interaction matrices that are consistent with Hi-C data (Wang et al., 2024a).

In the context of plant genomics, Wang et al. (2024a) proposed the DeepCBA, a DL framework, where 3D chromatin architecture is utilized to predict gene expression in maize. Based on the data of the Hi-C interaction, the model combines CNNs, bidirectional LSTMs (BiLSTMs), and self-attention capabilities to focus on both local and distal regulatory relationships. DeepCBA outperformed sequence-only models, with a high predictive accuracy of promoter (Pearson correlation coefficient [PCC] = 0.818), promoter distal (PCC = 0.625), and combined datasets (PCC = 0.929). The framework has been able to find biologically relevant motifs in the available chromatin regions and has accurately predicted major regulatory genes such as ZmRap2.7, ZmTb1, ZmCLE7, and ZmVTE4. Transfer learning ensured strong cross-tissue and cross-genotype generalization (PCC = 0.85). DeepCBA creates a powerful framework to explain chromatin-mediated transcriptional regulation and predicts precision crop enhancement by modeling 3D interactions in the genome (Wang et al., 2024b).

Taken together, these DL models represent increasingly complex multimodal integration strategies of 3D genomics. The combination of sequence-derived data with epigenetic cues improves the accuracy, interpretability, and biological sensitivity of predictions about chromatin structure and thus leads to a unified, data-driven view of genomic organization and its complex regulation.

1.4.6 ML in ncRNA Prediction

ML and DL have become indispensable to understand the regulatory roles of ncRNAs in chromatin remodeling and gene expression. The retraining of ML models like CPAT, PLEK, and LncFinder on plant-specific datasets has been used to detect lncRNAs that play a role in chromatin and epigenetic regulation. Recently, a CPAT-plant + LncFinder-plant complex demonstrated the best overall performance among diverse species of plants (Tian et al., 2024).

Integrative ML systems also develop ncRNA research. The SVM and RF methods are used to predict regulatory ncRNAs that are involved in stress-responsive networks (Pradhan et al., 2023; Rico-Chávez et al., 2022). To describe the response of miRNAs to drought, salinity, heat, and cold stress in the model of stress responses in *A. thaliana*, Asefpour Vakilian (2020) used Gaussian SVM regression with a high coefficient of determination ($R^2 = 0.96$) and revealed the significant miRNAs in relation to stresses, such as miR169, miR159, miR393, and miR396 (Asefpour Vakilian, 2020). Cao et al. (2021b) identified 2,857 lncRNAs from RNA-seq data in potato, focusing on those differentially expressed in resistant varieties in response to potato blight (*Phytophthora infestans*) infection. They used the GENIE3 RF algorithm to infer directional lncRNA-target gene regulatory interaction. Network enrichment studies demonstrated that particular differentially expressed lncRNAs (DELs) play a role in augmented resistance, indicating that lncRNA-mediated defense control is a key factor in potato immunity (W. Cao et al., 2021b).

Recent DL models contributed significantly to the study of ncRNA by predicting the interactions between lncRNAs and chromatin remodeling complexes. Accordingly, the DL models have increased our knowledge of the role of ncRNAs in the regulation of gene expression by changing the chromatin dynamics (Tabe-Bordbar & Sinha, 2023). The combination of ncRNA datasets with transcriptomic and epigenomic profiles has brought systems-biology insights into ncRNA-regulated networks and showed that ncRNAs control complex stress-response pathways (Pradhan et al., 2023; Tabe-Bordbar & Sinha, 2023).

Computational advancements are building a base that enables researchers to explore crop ncRNA-mediated stress resilience through predictive, computationally driven experimentation. However, experimental validation remains essential to confirm computational predictions. The future direction of this research should focus on employing CRISPR/Cas9-based systems to modify gene expression of ncRNAs and assess their role in stress memory and transgenerational adaptation. Single-cell RNA-seq, high-throughput functional assays, and standardized annotation pipelines will help researchers verify ncRNA targets, ensure that results are reproducible, and enable cross-study comparisons within plant ncRNA research.

1.4.7 ML in Epitranscriptomics

In addition to DNA and histones, epitranscriptomics, which is the study of post-transcriptional RNA modifications, has become an important layer of regulation in plant stress and memory (Cai et al., 2025; Shoaib et al., 2022). RNA modifications such as m^6A enhance stress resilience by stabilizing stress-responsive transcripts, optimizing the efficiency of translation and preconditioning the subsequent stress exposure to molecular mechanisms (Dobránszki et al., 2025; Liu et al., 2024). Recent advancements in the amount of large-scale epitranscriptomic data (e.g., m^6A-seq and MeRIP-seq) has enabled AI-based analyses to discover some complicated patterns of modifications and regulation (Dobránszki et al., 2025).

The positive-sample-only learning in ML tools like the plant epitranscriptomic analysis (PEA) toolkit predicts Chemical RNA Modifications (CMRs) like m^6A with high fidelity on a global scale across plant transcriptomes, significantly outperforming previous alternatives in the case of *A. thaliana* (Zhai et al., 2018). Similarly, m^6Apred may operate at above 90% in prediction accuracy with sequence features and contextual information (Taguchi, 2023). ML and DL systems, such as RNNs and transformers, can be used to capture temporal dynamics of modification. An example is that RNNs trained on time-series m^6A data of the heat-stressed tomato properly detected modification hotspots in heat shock protein (HSP) transcripts, which are key for thermotolerance (Vakulenko & Grigoriev, 2021). Nanopore sequencing indicates can be read using the DL framework EpiNano to identify many different RNA modifications at once and identify m^6A by the deviations in signals of sequencing reads (Shen et al., 2023).

AI and ML offer a combined platform to assess epigenetic, epitranscriptomic, transcriptomic, and proteomic data and understand how RNA modifications, including methylation, regulate gene expression networks during stress and provide an idea of the role of RNA modifications in alleviating stress and memory responses to stress. Applications like 6mA boost, iRNA-methyl, and DL models that are trained on plant-specific data support the prediction and annotation of sites of modification and embed them into multiomics stress-response networks (Acera Mateos et al., 2023). Such developments will allow scientists to discover new stress-specific epitranscriptomic signatures and dynamic models of regulation that will further clarify plant resistance mechanisms.

New multiepiomics technologies, integrating epigenomics, epitranscriptomics, and epiproteomics, provide a potent model to resolve the stress-responsive control and phenotypic plasticity, with a broad implication on crop resilience and sustainable agriculture (Miglani & Kaur, 2025). However, there are still obstacles: existing RNA modification mapping techniques, like MeRIP-seq, are not single-nucleotide sensitive (Zhong et al., 2023), whereas nanopore sequencing still demands optimization of base-calling programs with ML assistance (Liu et al., 2021). Besides, the overall interpretation of plant stress memory requires integrative comparisons of RNA modifications with DNA methylation and histone modification datasets (Hu et al., 2022).

Overall, epitranscriptomics is a new field of research on plant stress biology that provides insight into the mechanisms of how plants sense, encode, and retain signals of environmental stresses. The combination of DNA methylation, histone modification, ncRNAs, chromatin remodeling, and RNA modification all contribute to an integrated epigenomic network that adds together transcriptional, post-transcriptional, and chromatin-level regulation. These are processes involved in short and long-term plasticity of stress responses, and in future studies, it will be important to not only enhance the existing technologies to study epitranscriptomics but also create CRISPR-based technologies to manipulate this network to improve crops.

1.4.8 ML in Epigenetic Genome Editing

A groundbreaking advance in precise biology, epigenetic genome editing (Epi-GED) allows the reversible, programmable, and hereditary regulation of gene expression without modifying the underlying DNA sequence. Epi-GED represents a superior version of genome editing by offering a more precise, nonmutagenic, and less toxic replacement to the existing paradigm of genome editing, which relies on double-strand breaks (DSBs) to modify sequence. This strategy represents the prospective of functional genomics and crop enhancement, allowing a specific regulation of genes and pathways controlling plant development, defense, and adaptation to stress (Dixit et al., 2024; Fadul et al., 2023).

CRISPR/Cas-based Epi-GED systems are currently the most widely used systems for sequence-specific epigenetic editing in plants (Dixit et al., 2024). The systems utilize nuclease deficient Cas (dCas) variants fused to epigenetic effector domains, i.e., methyltransferases, demethylases, and histone modifiers to write or erase epigenetic marks at desired loci without the formation of DSBs (Goell & Hilton, 2021). This reversible, nonmutagenic approach maintains genomic fidelity and offers a tight functional control, thus offering a powerful platform for mechanistic and molecular breeding applications.

In *A. thaliana*, targeted CRISPR/dCas9-mediated methylation and demethylation have enabled locus-specific editing of m^5C marks, producing heritable and reversible gene silencing or activation (Papikian et al., 2019). Similarly, zinc-finger-based fusion constructs, such as SU(VAR)3-9 HOMOLOG 9 (SUVH9) or Ten-Eleven Translocation1 (TET1) linked to regulatory domains, allow precise methylation or demethylation of the *FLOWERING WAGENINGEN* (*FWA*) locus, thereby modulating flowering time (Johnson et al., 2014; Papikian et al., 2019). While most current plant studies focus on m^5C modification, the extension of epi-GED tools to other marks, including DNA m^6A, RNA m^5C, and m^6A, as well as histone acetylation and methylation, represents a promising frontier for future research. A striking example of RNA-level epigenetic manipulation is the introduction of the mammalian RNA demethylase FTO into rice and potato, which promoted site-specific RNA m^6A demethylation. This modification enhanced tillering, root development, and biomass accumulation, resulting in yield increases of nearly 50% (Yu et al., 2021). Such findings underscore the transformative potential of epitranscriptomic regulation in optimizing agronomic performance.

ML and DL are powerful partners in the developing Epi-GED, which facilitates predictive modeling and optimization, as well as rational design of epigenetic modifications. These models predict efficiency and functional performance of editing, and design of guide RNA (sgRNA). However, unlike the previous methods of genome editing, Epi-GED operates at the regulatory architecture level rather than at the DNA sequence level (Machnicka & Wilczynski, 2020; Rauschert et al., 2020). A representative model is the DL-based

EpiCas-DL, combining gene-expression, DNA methylation, chromatin-accessibility, and proximity to transcription-start sites (TSS) to predict sgRNA activity in CRISPR-mediated epigenetic editing. EpiCas-DL demonstrated an increased effectiveness (AUC = 0.87) compared to traditional models for predicting the performance determining factors of sgRNA when activated or repressed (Yang et al., 2023a). These AI-based tools significantly improve the precision and predictability of epigenetic modulation and accelerate the development of systems with higher editing performance.

Among the most sophisticated CRISPR-based epi-GED frameworks developed in mammalian systems, CRISPRoff exemplifies a highly versatile and reversible platform. By fusing the DNA methyltransferase DNMT3A and its cofactor DNMT3L with a zinc-finger Krüppel Associated Box (KRAB) domain and catalytically dead Cas9 (dCas9), CRISPRoff induces targeted CpG methylation and stable transcriptional silencing that persists through cell division and differentiation. Its complementary system, CRISPRon, reverses this silencing by recruiting demethylases to restore gene expression, demonstrating the programmable reversibility of epigenetic regulation (Nuñez et al., 2021).

Although there has been considerable advancement in AI-based epi-GED, it is still a developing field relative to base or prime editing. There are many challenges associated with the highly dynamic, spatially and temporally specific, and context-dependent mechanisms of epigenetic regulation that include temporal variations in DNA methylation, histone modification, and chromatin accessibility (Dixit et al., 2024). Thus, moving forward, future studies will need to develop and apply interpretable ML and DL methods that can capture both spatiotemporal dependencies as well as chromatin topology and cross-layer epigenomic interactions.

Precision crop improvement is being revolutionized by an unprecedented integration of CRISPR-based epigenetic editing and AI-based modeling. The convergence of predictive computation with programmable molecular control to create crops that have improved yields and can endure environmental stresses is creating the infrastructure for "Smart Epigenetic Design" which will be based on data and will be used to create crop varieties that are both sustainable and able to withstand climate changes.

1.5 Challenges and Limitations

While AI and ML have been increasingly applied to precision epigenetic modification in plants, there are many technical and conceptual limitations that still restrict its development (Dobránszki et al., 2025). One major issue is data bias and imbalance. Most plant genomic data currently available are derived from only a few model plant species such as *Arabidopsis* and rice. By comparison, datasets from various other crop species and their wild relatives are very limited (Richards et al., 2017). This lack of diversity limits the ability to generalize results from ML models because each species has its own unique architecture of epigenetics, chromatin dynamics, and regulatory pathways (Moore et al., 2020). Additionally, differences in experimental methods, sequencing depth, and conditions introduce "batch-effects" which obscure the true biological signal. Thus, there is an urgent need to standardize the collection, processing, normalization, and sharing of data across institutions (Murmu et al., 2024).

An additional significant barrier is the time-dependent and location-dependent behavior of epigenetic regulation. Most AI models can utilize DNA methylation, histone modification, and chromatin accessibility as static or frozen points in time. Thereby, they fail to model how these characteristics can be dynamically regulated throughout stress and recovery, and even through transgenerational inheritance (Hemenway & Gehring, 2023). The failure to represent these characteristics in a dynamic manner limits the ability to model the regulatory plasticity and changes in chromatin states which occur because of stress memory. Creating temporally aware frameworks, which allow for the modeling of tissue-specific epigenetic states at different developmental stages and environmental conditions, will be crucial for making biologically relevant predictions (Dobránszki et al., 2025).

Another important issue is model interpretability, especially for DL models. DL models can obtain very good prediction accuracies, but they are "black boxes," which makes it difficult to understand how a certain modification of the epigenome correlates to a phenotype (Hemenway & Gehring, 2023; Qamar & Bawany, 2023). The emergence of Explainable AI (XAI) frameworks, which include attention-based neural networks and Shapley additive explanations (SHAP), offers potential for identifying important features such as sequence motifs or chromatin signatures. However, there has been little use of XAI frameworks for interpreting plant epigenomic data. Therefore, expanding the use of interpretable ML techniques will be key to linking predictive capability to mechanistic insights (Ibragić et al., 2025).

Another obstacle is integrating multiomics. There are several algorithms that do very well with single omics type datasets. However, there are relatively few algorithms that can successfully integrate multiple levels of epigenetic layers (i.e., the methylome, histone modification, chromatin accessibility, and the transcriptome) into a singular predictive framework. Plant datasets are usually low sample number with many dimensions, and this presents a higher risk of both overfitting and statistically noisy outputs, which can significantly reduce the robustness of the predictive models. A further complication arises when attempting cross-species prediction, due to significant genomic architectural, methylation landscape, and regulatory circuitry differences between species that make it difficult to accurately translate models across species (Lämke & Bäurle, 2017). Transfer learning has been proposed as a method to bridge the gap between species, but it requires careful calibration to consider the evolutionary divergence in the mechanisms underlying epigenetics (Singh et al., 2021).

In conclusion, while ML-based methods have dramatically improved our understanding of how plants are regulated by their epigenetics, there remain several challenges to overcome such as data limitations, time-based complexities, ability to interpret results, and transferring knowledge from one species to another. To begin to address these issues, we will need to generate several high-quality, standard large-scale, and temporally resolved datasets, develop AI-based methods that can provide interpretable results, and create multiomic models which integrate DNA methylation, ncRNA, and histone modification. These advances will be critical for developing precise breeding or genetic engineering strategies to identify adaptive epigenetic markers for the development of crops with enhanced stress tolerance for sustainable agriculture in the face of accelerating climate change.

1.6 Future Perspectives

Further studies need to be conducted to clarify the molecular mechanisms by which epigenetic memory is passed down from one generation to another through transgenerational inheritance, thereby developing a basis for epigenetic priming methods that would increase the ability of multiple generations to be resilient. The integration of this fundamental knowledge into a genetic breeding pipeline would allow for the development of new varieties with enhanced ability to maintain productivity in response to environmental stressors while reducing their ecological footprint (Ibragić et al., 2025; Sarkies, 2020).

The future integration of computational predictions with empirical validations will be essential to transform the AI generated discovery into a real-world application. The application of CRISPR-based functional assays with ML systems will enable researchers to directly test the methylation sites and confirm whether those sites are responsible for gene regulation and plant adaptation to environmental stresses. Additionally, the development of transparent AI architectures, such as attention-based network, and explainable AI architecture (SHAP) will increase the transparency of the AI system, enabling researchers to understand which molecular attributes contribute to predictive accuracy and enhancing the reliability of the AI-based epigenetic interpretations (Ibragić et al., 2025).

Another equally significant factor is creating a standardized, well-curated collection of multiomics datasets, which include genomic, epigenomic, transcriptomic, and phenotypic layers (Fan et al., 2025). The integration of these data types in a multiomics framework will allow models to be made generalizable across different species and different environments and provide an opportunity for ML systems to identify methylation patterns that are part of the larger regulatory network of plant stress memory. Additionally, it will be essential to apply ML methods to explore how stress-induced methylation marks are maintained, reset, and reprogrammed through multiple generations as they provide insights into mechanisms responsible for heritable plant stress resistance (Dobránszki et al., 2025).

The new architectures of DL, such as Transformer models and large language models (LLMs), encourage new directions in epigenetics. Their ability to handle large amounts of data and to maintain long-range dependencies makes them highly appropriate to multiomics integration and data-driven hypothesis generation. In contrast to the previous sequential models that sequentially processed the elements, transformer models analyze the input as a whole (Wang et al., 2023). Transformer-based models, including bidirectional encoder representations from transformers (BERT) and generative pretrained transformer (GPT), have further been applied to computer vision through vision transformers (ViTs) (Wang et al., 2025). Like how LLMs are developed using specific domains and their respective language corpora, omics data provides hierarchical, structured biological data that can be used as input through the development of models that seek to understand regulatory networks and cellular processes. This type of analysis will provide new insights into cellular diversity, developmental plasticity, and functional specialization by using ncRNA-seq data and analyzing the epigenetic regulation associated with it. While there are no examples of the direct use of LLMs in plant epigenetics, recent frameworks and conceptual models demonstrate how the next wave of AI will significantly advance our ability to understand complex plant regulatory landscapes (Gao et al., 2024b).

Through integration of Next-Generation Sequencing (NGS) technologies, interpretable ML frameworks, and validation via genome editing, the future will lead to AI-based epigenetic modeling transforming into an important tool for precision breeding. The advancement will increase the speed at which climate-resilient crops are developed with the goal of unifying productivity, adaptability, and sustainability and thereby provide greater global food security as environmental challenges escalate.

1.7 Conclusion

The modern agricultural sector is facing a critical crossroad with two major issues: climate change and an explosive world population. Therefore, food security requires a shift in the current crop improvement practices to maximize the food potential and to facilitate the crop to adapt to the changing environmental factors with minimum ecological harm caused by excessive resource use. The future of modern agricultural practices is only achievable by the synergistic formulation of precision agriculture technologies and new biological knowledge that will enable farmers to monitor the health of their crops in real time, optimize the use of inputs, and produce sustainably.

At the core of this transformation is the increased recognition of the shifting of epigenetics as a dynamic regulatory layer that mediates the interface between genotype and environment. All these DNA and RNA methylation, histone modifications and chromatin remodeling contributions play a role in the formation of the heritability of patterns of gene expression which dictate plant growth, memory of stress and environmental adaptation. These molecular signatures can be analyzed and manipulated to selectively adopt epialleles that determine agronomically significant traits, including yield stability, stress resistance, and nutrient efficiency. Integrations of various omics datasets such as epigenomic, transcriptomic, metabolomic, and phenomic datasets recorded in reaction to different environmental cues have significantly increased our knowledge base on the molecular pathways of plant adaptability and resilience.

The patterns of how AI, ML, and DL are changing the analysis, modeling, and exploitation of these multifaceted, multistage regulatory systems are being fundamentally reconstituted. Analytical pipelines powered by AI are used to discover and characterize complex interdependence between regulatory epigenetic marks and phenotypes by preprocessing, normalization and extracting features. Furthermore, the predictive model based on ML/DL demonstrates an unprecedented ability to unravel hierarchical epigenetic networks, discover new sites of modifications, and uncover essential molecular factors of adaptive responses. Combining computational inference and experimental validation will result in the emergence of data-based, mechanism-based breeding strategies, which can form a new paradigm called Smart Epigenetic Design.

Recently significant advancements in high-throughput sequencing, accuracy of detection, and epigenome mapping have greatly increased the number of identified modification sites which provides an additional array of targets to use when designing crops rationally. Simultaneously, the successful application of AI in both plant and molecular biology has also significantly advanced the development of predictive models of epigenetics. These predictive models are now able to simulate and quantify how certain modifications can affect a variety of desirable crop characteristics. These new advances will provide the opportunity to develop breeding programs guided by models. The use of these model-based breeding programs will allow breeders to design germplasm which will have multiple desirable traits, such as high yields, superior quality, enhanced ability to withstand environmental stresses, and improved ability to utilize resources efficiently.

Looking forward, one of the most important research areas is to understand the molecular mechanism of transgenerational epigenetic inheritance and apply this knowledge to generate priming methods that increase the genetic resilience through generations. By integrating the mechanistic insights into transgenerational epigenetic inheritance with interpretability from ML, advanced NGS techniques, and genome editing will allow AI-guided epigenetic modeling to become a fundamental part of precision breeding.

Finally, the combination of AI, multiomics strategies, and epigenetic engineering can bring a new paradigm in the field of plant science. An interdisciplinary capability to discover and control certain heritable regulatory marks will enable the creation of the next-generation crop phenotypes with greater climate adaptability and performance, which will consequently accelerate the technology-based, environmentally conscious agriculture and sustainable food-production infrastructure.

Author Contributions: E.M. and X.F. conceptualized the manuscript; E.M. prepared the initial draft; E.M., Y.S., and X.F. reviewed and revised the manuscript. All authors have read and agreed to the published version of the manuscript.

Funding: This research was supported by the Biological Breeding-National Science and Technology Major Project (2023ZD0403201) and the National Natural Science Foundation of China (U21A20215 and 32488102).

Institutional Review Board Statement: Not Applicable.

Informed Consent Statement: Not Applicable.

Data Availability Statement: Not Applicable.

Acknowledgments: We thank the editors and reviewers for their critical comments on the manuscript.

Conflicts of Interest: The authors declare no conflict of interest.

References

Abbas, A., Chandratre, K., Gao, Y., Yuan, J., Zhang, M. Q., & Mani, R. S. (2024). ChIPr: Accurate prediction of cohesin-mediated 3D genome organization from 2D chromatin features. *Genome Biology*, *25*, 15. https://doi.org/10.1186/s13059-023-03158-7

Acera Mateos, P., Zhou, Y., Zarnack, K., & Eyras, E. (2023). Concepts and methods for transcriptome-wide prediction of chemical messenger RNA modifications with machine learning. *Briefings in Bioinformatics*, *24*. https://doi.org/10.1093/bib/bbad163

Al Bkhetan, Z., & Plewczynski, D. (2018). Three-dimensional epigenome statistical model: Genome-wide chromatin looping prediction. *Scientific Reports, 8*, 5217. https://doi.org/10.1038/s41598-018-23276-8

Altaf, M. T., Liaqat, W., Nadeem, M. A., & Baloch, F. S. (2023). Recent trends and applications of omics-based knowledge to end global food hunger. In A. Prakash, C. S. Fiaz, S. Nadeem, M. A. Baloch, & F. S. Qayyum (Eds.), *Sustainable agriculture in the era of the OMICs revolution* (pp. 381–397). Springer International Publishing. https://doi.org/10.1007/978-3-031-15568-0_18

An, Y. C., Goettel, W., Han, Q., Bartels, A., Liu, Z., & Xiao, W. (2017). Dynamic changes of genome-wide DNA methylation during soybean seed development. *Scientific Reports, 7*, 12263. https://doi.org/10.1038/s41598-017-12510-4

Angermueller, C., Lee, H. J., Reik, W., & Stegle, O. (2017). DeepCpG: Accurate prediction of single-cell DNA methylation states using deep learning. *Genome Biology, 18*, 67. https://doi.org/10.1186/s13059-017-1189-z

Arribas-Hernández, L., Bressendorff, S., Hansen, M. H., Poulsen, C., Erdmann, S., & Brodersen, P. (2018). An m 6 A-YTH module controls developmental timing and morphogenesis in arabidopsis. *Plant Cell, 30*, 952–967. https://doi.org/10.1105/tpc.17.00833

Arslan, E., Schulz, J., & Rai, K. (2021). Machine learning in epigenomics: Insights into cancer biology and medicine. *Biochimica et Biophysica Acta (BBA)-Reviews on Cancer, 1876*, 188588. https://doi.org/10.1016/j.bbcan.2021.188588

Asefpour Vakilian, K. (2020). Machine learning improves our knowledge about miRNA functions towards plant abiotic stresses. *Scientific Reports, 10*, 3041. https://doi.org/10.1038/s41598-020-59981-6

Bahar, N. H. A., Lo, M., Sanjaya, M., Van Vianen, J., Alexander, P., Ickowitz, A., & Sunderland, T. (2020). Meeting the food security challenge for nine billion people in 2050: What impact on forests? *Global Environmental Change, 62*, 102056. https://doi.org/10.1016/j.gloenvcha.2020.102056

Bhadouriya, S. L., Mehrotra, S., Basantani, M. K., Loake, G. J., & Mehrotra, R. (2021). Role of chromatin architecture in plant stress responses: An update. *Frontiers in Plant Science, 11*. https://doi.org/10.3389/fpls.2020.603380

Bian, H., Song, P., Gao, Y., Deng, Z., Huang, C., Yu, L., Wang, H., Ye, B., Cai, Z., Pan, Y., Wang, F., Liu, J., Gao, X., Chen, K., Jia, G., Klee, H. J., & Zhang, B. (2024). The m 6 A reader SlYTH2 negatively regulates tomato fruit aroma by impeding the translation process. *Proceedings of the National Academy of Sciences, 121*, e2405100121. https://doi.org/10.1073/pnas.240510.121.

Breiman, L. (2001). Random forests. *Machine Learning, 45*, 5–32. https://doi.org/10.1023/A:1010933404324

Brieuc, M. S. O., Waters, C. D., Drinan, D. P., & Naish, K. A. (2018). A practical introduction to random forest for genetic association studies in ecology and evolution. *Molecular Ecology Resources, 18*, 755–766. https://doi.org/10.1111/1755-0998.12773

Bröker, F., Holt, L. L., Roads, B. D., Dayan, P., & Love, B. C. (2024). Demystifying unsupervised learning: How it helps and hurts. *Trends in Cognitive Sciences, 28*, 974–986. https://doi.org/10.1016/j.tics.2024.09.005

Burger, L., Gaidatzis, D., Schübeler, D., & Stadler, M. B. (2013). Identification of active regulatory regions from DNA methylation data. *Nucleic Acids Research, 41*, e155–e155. https://doi.org/10.1093/nar/gkt599

Cai, J., Shen, L., Kang, H., & Xu, T. (2025). RNA modifications in plant adaptation to abiotic stresses. *Plant Communications*, *6*, 101229. https://doi.org/10.1016/j.xplc.2024.101229

Cao, F., Zhang, Y., Cai, Y., Animesh, S., Ying, Z., Akincilar, S. C., Loh, Y. P., Li, X., Chng, W. J., Tergaonkar, V., & Kwoh, C. K. (2021a). Chromatin interaction neural network (ChINN): A machine learning-based method for predicting chromatin interactions from DNA sequences. *Genome Biology*, *22*, 226. https://doi.org/10.1186/s13059-021-02453-5

Cao, W., Gan, L., Wang, C., Zhao, X., Zhang, M., Du, J., Zhou, S., & Zhu, C. (2021b). Genome-wide identification and characterization of potato long non-coding RNAs associated with phytophthora infestans resistance. *Front Plant Science*, 12. https://doi.org/10.3389/fpls.2021.619062

Champigny, M. J., Unda, F., Skyba, O., Soolanayakanahally, R. Y., Mansfield, S. D., & Campbell, M. M. (2020). Learning from methylomes: Epigenomic correlates of populus balsamifera traits based on deep learning models of natural DNA methylation. *Plant Biotechnology Journal*, *18*, 1361–1375. https://doi.org/10.1111/pbi.13299

Chen, T., & Guestrin, C. (2016). XGBoost: A scalable tree boosting system. *Proc. 22nd ACM SIGKDD Int. Conf. Knowl. Discov. Data Min.*, New York, NY, USA: Association for Computing Machinery, pp. 785–794. https://doi.org/10.1145/2939672.2939785

Chen, X., & Ishwaran, H. (2012). Random forests for genomic data analysis. *Genomics*, *99*, 323–329. https://doi.org/10.1016/j.ygeno.2012.04.003

Chen, W., Lv, H., Nie, F., & Lin, H. (2019). i6mA-Pred: Identifying DNA N6-methyladenine sites in the rice genome. *Bioinformatics*, *35*, 2796–2800. https://doi.org/10.1093/bioinformatics/btz015

Chen, K., Tang, W.-S., Zhou, Y.-B., Xu, Z.-S., Chen, J., Ma, Y.-Z., Chen, M., & Li, H. Y. (2020). Overexpression of *GmUBC9* gene enhances plant drought resistance and affects flowering time via histone H2B monoubiquitination. *Frontiers in Plant Science*, *11*. https://doi.org/10.3389/fpls.2020.555794

Chen, L., Liu, G., & Zhang, T. (2024). Integrating machine learning and genome editing for crop improvement. *aBIOTECH*, *5*, 262–277. https://doi.org/10.1007/s42994-023-00133-5

Cheng, P., Bao, S., Li, C., Tong, J., Shen, L., & Yu, H. (2022). RNA N6-methyladenosine modification promotes auxin biosynthesis required for male meiosis in rice. *Developmental Cell*, *57*, 246–259.e4. https://doi.org/10.1016/j.devcel.2021.12.014

Cui, X., Liang, Z., Shen, L., Zhang, Q., Bao, S., Geng, Y., Zhang, B., Leo, V., Vardy, L. A., Lu, T., Gu, X., & Yu, H. (2017). 5-methylcytosine RNA methylation in arabidopsis thaliana. *Molecular Plant*, *10*, 1387–1399. https://doi.org/10.1016/j.molp.2017.09.013

Daszkiewicz, T. (2022). Food production in the context of global developmental challenges. *Agriculture*, *12*, 832. https://doi.org/10.3390/agriculture12060832

Dixit, S., Kumar, A., Srinivasan, K., Vincent, P. M. D. R., & Ramu Krishnan, N. (2024). Advancing genome editing with artificial intelligence: Opportunities, challenges, and future directions. *Frontiers in Bioengineering and Biotechnology*, *11*, 11. https://doi.org/10.3389/fbioe.2023.1335901.

Dobránszki, J., Vassileva, V., Agius, D. R., Moschou, P. N., Gallusci, P., Berger, M. M. J., Farkas, D., Basso, M. F., & Martinelli, F. (2025). Gaining insights into epigenetic memories through artificial intelligence and omics science in plants. *Journal of Integrative Plant Biology*, *67*, 2320–2349. https://doi.org/10.1111/jipb.13953

Du, H., Wang, J., Hui, J., Zhang, L., & Wang, H. (2024). DenseGNN: Universal and scalable deeper graph neural networks for high-performance property prediction in crystals and molecules. *npj Computational Materials*, *10*, 292. https://doi.org/10.1038/s41524-024-01444-x

Duan, H.-C., Wei, L.-H., Zhang, C., Wang, Y., Chen, L., Lu, Z., Chen, P. R., He, C., & Jia, G. (2017). ALKBH10B is an RNA N6-methyladenosine demethylase affecting arabidopsis floral transition. *Plant Cell, 29*, 2995–3011. https://doi.org/10.1105/tpc.16.00912

Ernst, J., & Kellis, M. (2017). Chromatin-state discovery and genome annotation with ChromHMM. *Nature Protocols, 12*, 2478–2492. https://doi.org/10.1038/nprot.2017.124

Fadul, S. M., Arshad, A., & Mehmood, R. (2023). CRISPR-based epigenome editing: Mechanisms and applications. *Epigenomics, 15*, 1137–1155. https://doi.org/10.2217/epi-2023-0281

Fan, B.-L., Chen, L.-H., Chen, L.-L., & Guo, H. (2025). Integrative multi-omics approaches for identifying and characterizing biological elements in crop traits: Current progress and future prospects. *International Journal of Molecular Sciences, 26*, 1466. https://doi.org/10.3390/ijms26041466

Farkas, D., & Dobránszki, J. (2024). Vegetal memory through the lens of transcriptomic changes—recent progress and future practical prospects for exploiting plant transcriptional memory. *Plant Signaling & Behavior, 19*, 19. https://doi.org/10.1080/15592324.2024.2383515.

Feng, F., Yao, Y., Wang, X. Q. D., Zhang, X., & Liu, J. (2022). Connecting high-resolution 3D chromatin organization with epigenomics. *Nature Communications, 13*, 2054. https://doi.org/10.1038/s41467-022-29695-6

Fernández, M., & Miranda-Saavedra, D. (2012). Genome-wide enhancer prediction from epigenetic signatures using genetic algorithm-optimized support vector machines. *Nucleic Acids Research, 40*, e77. https://doi.org/10.1093/nar/gks149

Gallusci, P., Agius, D. R., Moschou, P. N., Dobránszki, J., Kaiserli, E., & Martinelli, F. (2023). Deep inside the epigenetic memories of stressed plants. *Trends in Plant Science, 28*, 142–153. https://doi.org/10.1016/j.tplants.2022.09.004

Gamarnik, D. (2023). Barriers for the performance of graph neural networks (GNN) in discrete random structures. *Proceedings of the National Academy of Sciences, 120*, e2314092120. https://doi.org/10.1073/pnas.2314092120.

Gan, W., Luo, J., Li, Y. Z., Guo, J. L., Zhu, M., & Li, M. L. (2019). A computational method to predict topologically associating domain boundaries combining histone marks and sequence information. *BMC Genomics, 20*, 980. https://doi.org/10.1186/s12864-019-6303-z

Gao, Y., & Fang, J. (2021). RNA 5-methylcytosine modification and its emerging role as an epitranscriptomic mark. *RNA Biology, 18*, 117–127. https://doi.org/10.1080/15476286.2021.1950993

Gao, V. R., Yang, R., Das, A., Luo, R., Luo, H., McNally, D. R., Karagiannidis, I., Rivas, M. A., Wang, Z. M., Barisic, D., & Karbalayghareh, A. (2024a). ChromaFold predicts the 3D contact map from single-cell chromatin accessibility. *Nature Communications, 15*, 9432. https://doi.org/10.1038/s41467-024-53628-0

Gao, Z., Liu, Q., Zeng, W., Jiang, R., & Wong, W. H. (2024b). EpiGePT: A pretrained transformer-based language model for context-specific human epigenomics. *Genome Biology, 25*, 310. https://doi.org/10.1186/s13059-024-03449-7

Goell, J. H., & Hilton, I. B. (2021). CRISPR/Cas-based epigenome editing: Advances, applications, and clinical utility. *Trends in Biotechnology, 39*, 678–691. https://doi.org/10.1016/j.tibtech.2020.10.012

Guo, S., Zheng, Y., Meng, D., Zhao, X., Sang, Z., Tan, J., Deng, Z., Lang, Z., Zhang, B., Wang, Q., Bouzayen, M., & Zuo, J. (2022a). DNA and coding/non-coding RNA methylation analysis provide insights into tomato fruit ripening. *The Plant Journal, 112*, 399–413. https://doi.org/10.1111/tpj.15951

Guo, W., Liu, H., Wang, Y., Zhang, P., Li, D., Liu, T., Zhang, Q., Yang, L., Pu, L., Tian, J., & Gu, X. (2022b). SMOC: A smart model for open chromatin region prediction in rice genomes. *Journal of Genetics and Genomics, 49*, 514–517. https://doi.org/10.1016/j.jgg.2022.02.012

Guo, Y., Tan, Y., Qu, M., Hong, K., Zeng, L., Wang, L., Zhuang, C., Qian, Q., Hu, J., & Xiong, G. (2023). OsWR2 recruits HDA704 to regulate the deacetylation of H4K8ac in the promoter of OsABI5 in response to drought stress. *Journal of Integrative Plant Biology, 65*, 1651–1669. https://doi.org/10.1111/jipb.13481

Guo, Z., Fan, W., Cai, C., Zhang, K., Hou, X., Li, Y., & Cheng, F. (2025). Plantdeepmeth: A deep learning model for predicting DNA methylation states in plants. *Plants, 14*, 1724. https://doi.org/10.3390/plants14111724

Halter, T., Wang, J., Amesefe, D., Lastrucci, E., Charvin, M., Singla Rastogi, M., & Navarro, L. (2021). The arabidopsis active demethylase ROS1 cis-regulates defence genes by erasing DNA methylation at promoter-regulatory regions. *eLife, 10*, 10. https://doi.org/10.7554/eLife.62994.

Han, Z., Crisp, P. A., Stelpflug, S., Kaeppler, S. M., Li, Q., & Springer, N. M. (2018). Heritable epigenomic changes to the maize methylome resulting from tissue culture. *Genetics, 209*, 983–995. https://doi.org/10.1534/genetics.118.300987

He, Y., & Li, Z. (2018). Epigenetic environmental memories in plants: Establishment, maintenance, and reprogramming. *Trends in Genetics, 34*, 856–866. https://doi.org/10.1016/j.tig.2018.07.006

Hemenway, E. A., & Gehring, M. (2023). Epigenetic regulation during plant development and the capacity for epigenetic memory. *Annual Review of Plant Biology, 74*, 87–109. https://doi.org/10.1146/annurev-arplant-070122-025047

Hendra, C., Pratanwanich, P. N., Wan, Y. K., Goh, W. S. S., Thiery, A., & Göke, J. (2022). Detection of m6A from direct RNA sequencing using a multiple instance learning framework. *Nature Methods, 19*, 1590–1598. https://doi.org/10.1038/s41592-022-01666-1

Hinterberger, V., Douchkov, D., Lueck, S., Reif, J. C., & Schulthess, A. W. (2022). High-throughput imaging of powdery mildew resistance of the winter wheat collection hosted at the German federal ex situ genebank for agricultural and horticultural crops. *GigaScience, 12*, 12. https://doi.org/10.1093/gigascience/giad007.

Hinton, G. E., Osindero, S., & Teh, Y. W. (2006). A fast learning algorithm for deep belief nets. *Neural Computation, 18*, 1527–1554. https://doi.org/10.1162/neco.2006.18.7.1527

Hoffman, G. E., Bendl, J., Girdhar, K., Schadt, E. E., & Roussos, P. (2019). Functional interpretation of genetic variants using deep learning predicts impact on chromatin accessibility and histone modification. *Nucleic Acids Research, 47*, 10597–10611. https://doi.org/10.1093/nar/gkz808

Hong, Y., Liu, L., Feng, Y., Zhang, Z., Hou, R., Xu, Q., & Shi, J. (2024). mHapBrowser: A comprehensive database for visualization and analysis of DNA methylation haplotypes. *Nucleic Acids Research, 52*, D929–D937. https://doi.org/10.1093/nar/gkad881

Hou, Y., Sun, J., Wu, B., Gao, Y., Nie, H., Nie, Z., Quan, S., Wang, Y., Cao, X., & Li, S. (2021). CPSF30-L-mediated recognition of mRNA m6A modification controls alternative polyadenylation of nitrate signaling-related gene transcripts in arabidopsis. *Molecular Plant, 14*, 688–699. https://doi.org/10.1016/j.molp.2021.01.013

Hou, Y., Yan, Y., & Cao, X. (2022). Epigenetic regulation of thermomorphogenesis in arabidopsis thaliana. *aBIOTECH, 3*, 12–24. https://doi.org/10.1007/s42994-022-00070-9

Hu, J., Cai, J., Park, S. J., Lee, K., Li, Y., Chen, Y., Yun, J. Y., Xu, T., & Kang, H. (2021). N6-methyladenosine mRNA methylation is important for salt stress tolerance in arabidopsis. *Plant Journal, 106*, 1759–1775. https://doi.org/10.1111/tpj.15270

Hu, J., Cai, J., Xu, T., & Kang, H. (2022). Epitranscriptomic mRNA modifications governing plant stress responses: Underlying mechanism and potential application. *Plant Biotechnology Journal, 20*, 2245–2257. https://doi.org/10.1111/pbi.13913

Huang, C.-Y., & Jin, H. (2022). Coordinated epigenetic regulation in plants: A potent managerial tool to conquer biotic stress. *Frontiers in Plant Science, 12*. https://doi.org/10.3389/fpls.2021.795274

Ibragić, S., Dahija, S., & Karalija, E. (2025). The good, the bad, and the epigenetic: Stress-induced metabolite regulation and transgenerational effects. *Epigenomes, 9*, 10. https://doi.org/10.3390/epigenomes9020010

Javed, A., Rizzo, D. M., Lee, B. S., & Gramling, R. (2024). Somtimes: Self organizing maps for time series clustering and its application to serious illness conversations. *Data Mining and Knowledge Discovery, 38*, 813–839. https://doi.org/10.1007/s10.18-023-00979-9

Jia, G., Fu, Y., Zhao, X., Dai, Q., Zheng, G., Yang, Y., Yi, C., Lindahl, T., Pan, T., Yang, Y. G., & He, C. (2011). N6-Methyladenosine in nuclear RNA is a major substrate of the obesity-associated FTO. *Nature Chemical Biology, 7*, 885–887. https://doi.org/10.1038/nchembio.687

Jiang, W., Han, H., Zhang, Y., Wang, J., He, M., Gu, W., Wang, J.', Mu, J., & Cheng, X. (2024). Graph neural networks for routing optimization: Challenges and opportunities. *Sustainability, 16*, 9239. https://doi.org/10.3390/su16219239

Johannes, F., Porcher, E., Teixeira, F. K., Saliba-Colombani, V., Simon, M., Agier, N., Bulski, A., Albuisson, J., Heredia, F., Audigier, P., Bouchez, D., Dillmann, C., Guerche, P., Hospital, F., & Colot, V. (2009). Assessing the impact of transgenerational epigenetic variation on complex traits. *PLoS Genetics, 5*, e1000530. https://doi.org/10.1371/journal.pgen.1000530

Johnson, L. M., Du, J., Hale, C. J., Bischof, S., Feng, S., Chodavarapu, R. K., Zhong, X., Marson, G., Pellegrini, M., Segal, D. J., & Patel, D. J. (2014). SRA- and SET-domain-containing proteins link RNA polymerase V occupancy to DNA methylation. *Nature, 507*, 124–128. https://doi.org/10.1038/nature12931

Kakrana, A., Li, P., Patel, P., Hammond, R., Anand, D., Mathioni, S. M., & Meyers, B. C. (2017). Phasis : A computational suite for de novo discovery and characterization of phased, siRNA-generating loci and their miRNA triggers. https://doi.org/10.1101/158832

Kimotho, R. N., & Maina, S. (2024). Unraveling plant–microbe interactions: Can integrated omics approaches offer concrete answers? *Journal of Experimental Botany, 75*, 1289–1313. https://doi.org/10.1093/jxb/erad448

Kolasinska-Zwierz, P., Down, T., Latorre, I., Liu, T., Liu, X. S., & Ahringer, J. (2009). Differential chromatin marking of introns and expressed exons by H3K36me3. *Nature Genetics, 41*, 376–381. https://doi.org/10.1038/ng.322

Lämke, J., & Bäurle, I. (2017). Epigenetic and chromatin-based mechanisms in environmental stress adaptation and stress memory in plants. *Genome Biology, 18*, 124. https://doi.org/10.1186/s13059-017-1263-6

Law, J. A., & Jacobsen, S. E. (2010). Establishing, maintaining and modifying DNA methylation patterns in plants and animals. *Nature Reviews Genetics, 11*, 204–220. https://doi.org/10.1038/nrg2719

Li, Z., Jiang, H., Kong, L., Chen, Y., Lang, K., Fan, X., Zhang, L., & Pian, C. (2021). Deep6mA: A deep learning framework for exploring similar patterns in DNA N6-methyladenine sites across different species. *PLoS Computational Biology, 17*, e1008767. https://doi.org/10.1371/journal.pcbi.1008767

Li, C., Xu, M., Cai, X., Han, Z., Si, J., & Chen, D. (2022a). Jasmonate signaling pathway modulates plant defense, growth, and their trade-offs. *International Journal of Molecular Sciences, 23*, 3945. https://doi.org/10.3390/ijms23073945

Li, Y., Quan, L., Zhou, Y., Jiang, Y., Li, K., Wu, T., & Lyu, Q. (2022b). Identifying modifications on DNA-bound histones with joint deep learning of multiple binding sites in DNA sequence. *Bioinformatics, 38*, 4070–4077. https://doi.org/10.1093/bioinformatics/btac489

Li, Z.-P., Du, Z., Huang, D.-S., & Teschendorff, A. E. (2025). Interpretable deep learning of single-cell and epigenetic data reveals novel molecular insights in aging. *Scientific Reports, 15*, 5048. https://doi.org/10.1038/s41598-025-89646-1

Liang, Z., Riaz, A., Chachar, S., Ding, Y., Du, H., & Gu, X. (2020). Epigenetic modifications of mRNA and DNA in plants. *Molecular Plant, 13*, 14–30. https://doi.org/10.1016/j.molp.2019.12.007

Liang, Z., Zhang, Q., Ji, C., Hu, G., Zhang, P., Wang, Y., Yang, L., & Gu, X. (2021). Reorganization of the 3D chromatin architecture of rice genomes during heat stress. *BMC Biology, 19*, 53. https://doi.org/10.1186/s12915-021-00996-4

Lin, W., Xiao, X., Zhang, H., Li, Y., Liu, S., Sun, W., Zhang, X., & Wu, Q. (2019). Whole-genome bisulfite sequencing reveals a role for DNA methylation in variants from callus culture of pineapple (Ananas comosus L.). *Genes (Basel), 10*, 877. https://doi.org/10.3390/genes10110877

Liu, C., Lu, F., Cui, X., & Cao, X. (2010). Histone methylation in higher plants. *Annual Review of Plant Biology, 61*, 395–420. https://doi.org/10.1146/annurev.arplant.043008.091939

Liu, H., Begik, O., & Novoa, E. M. (2021). EpiNano: Detection of m6A RNA modifications using Oxford nanopore direct RNA sequencing, p. 31–52. https://doi.org/10.1007/978-1-0716-1374-0_3

Liu, C., Mentzelopoulou, A., Hatzianestis, I. H., Tzagkarakis, E., Skaltsogiannis, V., Ma, X., Michalopoulou, V. A., Romero-Campero, F. J., Romero-Losada, A. B., Sarris, P. F., Marhavy, P., Bölter, B., Kanterakis, A., Gutierrez-Beltran, E., & Moschou, P. N. (2024). A proxitome-RNA-capture approach reveals that processing bodies repress coregulated hub genes. *Plant Cell, 36*, 559–584. https://doi.org/10.1093/plcell/koad288

Lv, H., Dao, F.-Y., Guan, Z.-X., Zhang, D., Tan, J.-X., Zhang, Y., Chen, W., & Lin, H. (2019). iDNA6mA-rice: A computational tool for detecting N6-methyladenine sites in rice. *Frontiers in Genetics, 10*, 793. https://doi.org/10.3389/fgene.2019.00793.

Lv, H., Dao, F.-Y., Zhang, D., Guan, Z.-X., Yang, H., Su, W., Liu, M. L., Ding, H., Chen, W., & Lin, H. (2020). iDNA-MS: An integrated computational tool for detecting DNA modification sites in multiple genomes. *Iscience, 23*, 100991. https://doi.org/10.1016/j.isci.2020.100991

Lv, T., Han, Q., Li, Y., Liang, C., Ruan, Z., Chao, H., Chen, M., & Chen, D. (2025). Cross-species prediction of histone modifications in plants via deep learning. *Genome Biology, 27*, 20. https://doi.org/10.1101/2025.05.19.655006

Ma, S., Tang, N., Li, X., Xie, Y., Xiang, D., Fu, J., Shen, J., Yang, J., Tu, H., Li, X., Hu, H., & Xiong, L. (2019). Reversible histone H2B monoubiquitination fine-tunes abscisic acid signaling and drought response in rice. *Molecular Plant, 12*, 263–277. https://doi.org/10.1016/j.molp.2018.12.005

Machnicka, M. A., & Wilczynski, B. (2020). Machine learning and deep learning for the advancement of epigenomics. In D. Kabelitz & J. Bhat (Eds.), *Epigenetics immune system* (pp. 217–237). Elsevier. https://doi.org/10.1016/B978-0-12-817964-2.00009-5

Mahmood, M. A., Naqvi, R. Z., Rahman, S. U., Amin, I., & Mansoor, S. (2023). Plant virus-derived vectors for plant genome engineering. *Viruses, 15*, 531. https://doi.org/10.3390/v15020531

Mansoor, S., Karunathilake, E. M. B. M., Tuan, T. T., & Chung, Y. S. (2025). Genomics, phenomics, and machine learning in transforming plant research: Advancements and challenges. *Horticultural Plant Journal, 11*, 486–503. https://doi.org/10.1016/j.hpj.2023.09.005

Martínez-Pérez, M., Aparicio, F., López-Gresa, M. P., Bellés, J. M., Sánchez-Navarro, J. A., & Pallás, V. (2017). *Arabidopsis* m 6 A demethylase activity modulates viral infection of a plant virus and the m 6 A abundance in its genomic RNAs. *Proceedings of the National Academy of Sciences of the United States of America, 114*, 10755–10760. https://doi.org/10.1073/pnas.1703139114

Matzke, M. A., & Mosher, R. A. (2014). RNA-directed DNA methylation: An epigenetic pathway of increasing complexity. *Nature Reviews Genetics, 15*, 394–408. https://doi.org/10.1038/nrg3683

Mavaie, P., Holder, L., Beck, D., & Skinner, M. K. (2021). Predicting environmentally responsive transgenerational differential DNA methylated regions (epimutations) in the genome using a hybrid deep-machine learning approach. *BMC Bioinformatics, 22*, 575. https://doi.org/10.1186/s12859-021-04491-z

Mehta, S., & Jeffrey, K. L. (2015). Beyond receptors and signaling: Epigenetic factors in the regulation of innate immunity. *Immunology and Cell Biology, 93*, 233–244. https://doi.org/10.1038/icb.2014.101

Mienye, I. D., & Swart, T. G. (2024). A comprehensive review of deep learning: Architectures, recent advances, and applications. *Information, 15*, 755. https://doi.org/10.3390/info15120755

Mienye, I. D., Swart, T. G., & Obaido, G. (2024). Recurrent neural networks: A comprehensive review of architectures, variants, and applications. *Information, 15*, 517. https://doi.org/10.3390/info15090517

Mierziak, J., & Wojtasik, W. (2024). Epigenetic weapons of plants against fungal pathogens. *BMC Plant Biology, 24*, 175. https://doi.org/10.1186/s12870-024-04829-8

Miglani, G. S., & Kaur, M. (2025). Multi-omics and crop improvement for sustainable agriculture. In K. Sharma (Ed.), *Omics and genome editing: Revolution in crop improvement for sustainable agriculture* (pp. 15–35). Springer Nature. https://doi.org/10.1007/978-3-031-81639-0_2

Mishra, S., Roychowdhury, R., Ray, S., Hada, A., Kumar, A., Sarker, U., Aftab, T., & das, R. (2024). Salicylic acid (SA)-mediated plant immunity against biotic stresses: An insight on molecular components and signaling mechanism. *Plant Stress, 11*, 100427. https://doi.org/10.1016/j.stress.2024.100427

Moore, B. M., Wang, P., Fan, P., Lee, A., Leong, B., Lou, Y.-R., Schenck, C. A., Sugimoto, K., Last, R., Lehti-Shiu, M. D., Barry, C. S., & Shiu, S. H. (2020). Within- and cross-species predictions of plant specialized metabolism genes using transfer learning. *In Silico Plants, 2*. https://doi.org/10.1093/insilicoplants/diaa005

Murmu, S., Sinha, D., Chaurasia, H., Sharma, S., Das, R., Jha, G. K., & Archak, S. (2024). A review of artificial intelligence-assisted omics techniques in plant defense: Current trends and future directions. *Frontiers in Plant Science, 15*, 1292054. https://doi.org/10.3389/fpls.2024.1292054

N'Diaye, A., Byrns, B., Cory, A. T., Nilsen, K. T., Walkowiak, S., Sharpe, A., Robinson, S. J., & Pozniak, C. J. (2020). Machine learning analyses of methylation profiles uncovers tissue-specific gene expression patterns in wheat. *Plant Genome, 13*. https://doi.org/10.1002/tpg2.20027

Ni, P., Huang, N., Nie, F., Zhang, J., Zhang, Z., Wu, B., Bai, L., Liu, W., Xiao, C. L., Luo, F., & Wang, J. (2021). Genome-wide detection of cytosine methylations in plant from Nanopore data using deep learning. *Nature Communications, 12*, 5976. https://doi.org/10.1038/s41467-021-26278-9

Nikkilä, J., Törönen, P., Kaski, S., Venna, J., Castrén, E., & Wong, G. (2002). Analysis and visualization of gene expression data using self-organizing maps. *Neural Networks, 15*, 953–966. https://doi.org/10.1016/S0893-6080(02)00070-9

Nuñez, J. K., Chen, J., Pommier, G. C., Cogan, J. Z., Replogle, J. M., Adriaens, C., Ramadoss, G. N., Shi, Q., Hung, K. L., Samelson, A. J., Pogson, A. N., Kim, J. Y. S., Chung, A., Leonetti, M. D., Chang, H. Y., Kampmann, M., Bernstein, B. E., Hovestadt, V., Gilbert, L. A., & Weissman, J. S. (2021). Genome-wide programmable transcriptional memory by CRISPR-based epigenome editing. *Cell, 184*, 2503–2519.e17. https://doi.org/10.1016/j.cell.2021.03.025

Olive, D. J. (2017). *Linear regression*. Springer International Publishing. https://doi.org/10.1007/978-3-319-55252-1

Papikian, A., Liu, W., Gallego-Bartolomé, J., & Jacobsen, S. E. (2019). Site-specific manipulation of arabidopsis loci using CRISPR-Cas9 suntag systems. *Nature Communications, 10*, 729. https://doi.org/10.1038/s41467-019-08736-7

Peng, S., & Rajjou, L. (2024). Advancing plant biology through deep learning-powered natural language processing. *Plant Cell Reports, 43*, 208. https://doi.org/10.1007/s00299-024-03294-9

Piecyk, R. S., Schlegel, L., & Johannes, F. (2022). Predicting 3D chromatin interactions from DNA sequence using deep learning. *Computational and Structural Biotechnology Journal, 20*, 3439–3448. https://doi.org/10.1016/j.csbj.2022.06.047

Pien, S., Fleury, D., Mylne, J. S., Crevillen, P., Inzé, D., Avramova, Z., Dean, C., & Grossniklaus, U. (2008). ARABIDOPSIS TRITHORAX1 dynamically regulates FLOWERING LOCUS C activation via histone 3 lysine 4 trimethylation. *Plant Cell, 20*, 580–588. https://doi.org/10.1105/tpc.108.058172

Pradhan, U. K., Meher, P. K., Naha, S., Rao, A. R., & Gupta, A. (2023). ASLncR: A novel computational tool for prediction of abiotic stress-responsive long non-coding RNAs in plants. *Functional & Integrative Genomics, 23*, 113. https://doi.org/10.1007/s10142-023-01040-0

Pu, L., & Sung, Z. R. (2015). PcG and trxG in plants—friends or foes. *Trends in Genetics, 31*, 252–262. https://doi.org/10.1016/j.tig.2015.03.004

Qamar, T., & Bawany, N. Z. (2023). Understanding the black-box: Towards interpretable and reliable deep learning models. *PeerJ Computer Science, 9*, e1629. https://doi.org/10.7717/peerj-cs.1629

Liu, Q., Fang, L., Yu, G., Wang, D., Xiao, C.-L., & Wang, K. (2019a). Detection of DNA base modifications by deep recurrent neural network on Oxford Nanopore sequencing data. *Nature Communications, 10*, 2449. https://doi.org/10.1038/s41467-019-10168-2

Liu, Q., Lv, H., & Jiang, R. (2019b). hicGAN infers super resolution Hi-C data with generative adversarial networks. *Bioinformatics, 35*, i99–i107. https://doi.org/10.1093/bioinformatics/btz317

Qiu, W.-R., Jiang, S.-Y., Xu, Z.-C., Xiao, X., & Chou, K.-C. (2017). *Irnam5c*-psednc: Identifying RNA 5-methylcytosine sites by incorporating physical-chemical properties into pseudo dinucleotide composition. *Oncotarget, 8*, 41178–41188. https://doi.org/10.18632/oncotarget.17104

Ramirez-Prado, J. S., Abulfaraj, A. A., Rayapuram, N., Benhamed, M., & Hirt, H. (2018). Plant immunity: From signaling to epigenetic control of defense. *Trends in Plant Science, 23*, 833–844. https://doi.org/10.1016/j.tplants.2018.06.004

Rauschert, S., Raubenheimer, K., Melton, P. E., & Huang, R. C. (2020). Machine learning and clinical epigenetics: A review of challenges for diagnosis and classification. *Clinical Epigenetics, 12*, 51. https://doi.org/10.1186/s13148-020-00842-4

Richards, C. L., Alonso, C., Becker, C., Bossdorf, O., Bucher, E., Colomé-Tatché, M., & Grosse, I. (2017). Ecological plant epigenetics: Evidence from model and non-model species, and the way forward. *Ecology Letters, 20*, 1576–1590. https://doi.org/10.1111/ele.12858

Rico-Chávez, A. K., Franco, J. A., Fernandez-Jaramillo, A. A., Contreras-Medina, L. M., Guevara-González, R. G., & Hernandez-Escobedo, Q. (2022). Machine learning for plant stress modeling: A perspective towards hormesis management. *Plants, 11*, 970. https://doi.org/10.3390/plants11070970

Rose, D. C., & Chilvers, J. (2018). Agriculture 4.0: Broadening responsible innovation in an era of smart farming. *Frontiers in Sustainable Food Systems, 2*. https://doi.org/10.3389/fsufs.2018.00087

Roundtree, I. A., Evans, M. E., Pan, T., & He, C. (2017). Dynamic RNA modifications in gene expression regulation. *Cell, 169*, 1187–1200. https://doi.org/10.1016/j.cell.2017.05.045

Roychowdhury, R., Das, S. P., Gupta, A., Parihar, P., Chandrasekhar, K., Sarker, U., Kumar, A., Ramrao, D. P., & Sudhakar, C. (2023). Multi-omics pipeline and omics-integration approach to decipher plant's abiotic stress tolerance responses. *Genes (Basel), 14*, 1281. https://doi.org/10.3390/genes14061281

Sanchez, R., & Mackenzie, S. (2018). Example of methylome analysis with MethylIT using cancer datasets. https://doi.org/10.1101/261982

Sarkies, P. (2020). Molecular mechanisms of epigenetic inheritance: Possible evolutionary implications. *Seminars in Cell & Developmental Biology, 97*, 106–115. https://doi.org/10.1016/j.semcdb.2019.06.005

Schmitz, R. J., Grotewold, E., & Stam, M. (2022). Cis-regulatory sequences in plants: Their importance, discovery, and future challenges. *Plant Cell, 34*, 718–741. https://doi.org/10.1093/plcell/koab281

Scholkopf, B., & Smola, A. J. (2001). *Learning with kernels: Support vector machines, regularization, optimization, and beyond*. The MIT Press.

Schumann, U., Lee, J. M., Smith, N. A., Zhong, C., Zhu, J.-K., Dennis, E. S., Millar, A. A., & Wang, M. B. (2019). DEMETER plays a role in DNA demethylation and disease response in somatic tissues of arabidopsis. *Epigenetics, 14*, 1074–1087. https://doi.org/10.1080/15592294.2019.1631113

Schwessinger, R., Gosden, M., Downes, D., Brown, R. C., Oudelaar, A. M., Telenius, J., Teh, Y. W., Lunter, G., & Hughes, J. R. (2020). DeepC: Predicting 3D genome folding using megabase-scale transfer learning. *Nature Methods, 17*, 1118–1124. https://doi.org/10.1038/s41592-020-0960-3

Shao, Y., Wong, C. E., Shen, L., & Yu, H. (2021). N6-methyladenosine modification underlies messenger RNA metabolism and plant development. *Current Opinion in Plant Biology, 63*, 102047. https://doi.org/10.1016/j.pbi.2021.102047

Sharma, M., Kumar, P., Verma, V., Sharma, R., Bhargava, B., & Irfan, M. (2022). Understanding plant stress memory response for abiotic stress resilience: Molecular insights and prospects. *Plant Physiology and Biochemistry, 179*, 10–24. https://doi.org/10.1016/j.plaphy.2022.03.004

Shen, L., Liang, Z., Gu, X., Chen, Y., Teo, Z. W. N., Hou, X., Cai, W. M., Dedon, P. C., Liu, L., & Yu, H. (2016). N6-methyladenosine RNA modification regulates shoot stem cell fate in arabidopsis. *Developmental Cell, 38*, 186–200. https://doi.org/10.1016/j.devcel.2016.06.008

Shen, H., Ontiveros, R. J., Owens, M. C., Liu, M. Y., Ghanty, U., Kohli, R. M., & Liu, K. F. (2021). TET-mediated 5-methylcytosine oxidation in tRNA promotes translation. *The Journal of Biological Chemistry, 296*, 100087. https://doi.org/10.1074/jbc.RA120.014226

Shen, Y., Zhong, Q., Liu, T., Wen, Z., Shen, W., & Li, L. (2022). CharID: A two-step model for universal prediction of interactions between chromatin accessible regions. *Briefings in Bioinformatics, 23*. https://doi.org/10.1093/bib/bbab602

Shen, L., Ma, J., Li, P., Wu, Y., & Yu, H. (2023). Recent advances in the plant epitranscriptome. *Genome Biology, 24*, 43. https://doi.org/10.1186/s13059-023-02872-6

Shoaib, Y., Usman, B., Kang, H., & Jung, K.-H. (2022). Epitranscriptomics: An additional regulatory layer in plants' development and stress response. *Plants, 11*, 1033. https://doi.org/10.3390/plants110810.3

Singh, U., Khemka, N., Rajkumar, M. S., Garg, R., & Jain, M. (2017). PLncPRO for prediction of long non-coding RNAs (lncRNAs) in plants and its application for discovery of abiotic stress-responsive lncRNAs in rice and chickpea. *Nucleic Acids Research, 45*, e183–e183. https://doi.org/10.1093/nar/gkx866

Singh, A., Jones, S., Ganapathysubramanian, B., Sarkar, S., Mueller, D., Sandhu, K., & Nagasubramanian, K. (2021). Challenges and opportunities in machine-augmented plant stress phenotyping. *Trends in Plant Science, 26*, 53–69. https://doi.org/10.1016/j.tplants.2020.07.010

Slotkin, R. K. (2016). Plant epigenetics: From genotype to phenotype and back again. *Genome Biology, 17*, 57. https://doi.org/10.1186/s13059-016-0920-5

Smet, D., Opdebeeck, H., & Vandepoele, K. (2023). Predicting transcriptional responses to heat and drought stress from genomic features using a machine learning approach in rice. *Frontiers in Plant Science, 14*. https://doi.org/10.3389/fpls.2023.1212073

Sokolova, M., & Lapalme, G. (2009). A systematic analysis of performance measures for classification tasks. *Information Processing & Management, 45*, 427–437. https://doi.org/10.1016/j.ipm.2009.03.002

Song, Q., Zhang, T., Stelly, D. M., & Chen, Z. J. (2017). Epigenomic and functional analyses reveal roles of epialleles in the loss of photoperiod sensitivity during domestication of allotetraploid cottons. *Genome Biology, 18*, 99. https://doi.org/10.1186/s13059-017-1229-8

Song, P., Yang, J., Wang, C., Lu, Q., Shi, L., Tayier, S., & Jia, G. (2021). Arabidopsis N6-methyladenosine reader CPSF30-L recognizes FUE signals to control polyadenylation site choice in liquid-like nuclear bodies. *Molecular Plant, 14*, 571–587. https://doi.org/10.1016/j.molp.2021.01.014

Stovner, E. B., & Sætrom, P. (2019). epic2 efficiently finds diffuse domains in ChIP-seq data. *Bioinformatics, 35*, 4392–4393. https://doi.org/10.1093/bioinformatics/btz232

Sun, S., Noviski, A., & Yu, X. (2013). MethyQA: A pipeline for bisulfite-treated methylation sequencing quality assessment. *BMC Bioinformatics, 14*, 259. https://doi.org/10.1186/1471-2105-14-259

Sun, P., Chen, Y., Liu, B., Gao, Y., Han, Y., He, F., & Ji, J. (2019). DeepMRMP: A new predictor for multiple types of RNA modification sites using deep learning. *Mathematical Biosciences and Engineering, 16*, 6231–6241.

Tabe-Bordbar, S., & Sinha, S. (2023). Integrative modeling of lncRNA-chromatin interaction maps reveals diverse mechanisms of nuclear retention. *BMC Genomics, 24*, 395. https://doi.org/10.1186/s12864-023-09498-9

Taguchi, Y.-H. (2023). Bioinformatic tools for epitranscriptomics. *American Journal of Physiology-Physiology, 324*, C447–C457. https://doi.org/10.1152/ajpcell.00437.2022

Tahir, M., Norouzi, M., Khan, S. S., Davie, J. R., Yamanaka, S., & Ashraf, A. (2024). Artificial intelligence and deep learning algorithms for epigenetic sequence analysis: A review for epigeneticists and AI experts. *Computers in Biology and Medicine, 183*, 109302. https://doi.org/10.1016/j.compbiomed.2024.109302

Tan, F., Tian, T., Hou, X., Yu, X., Gu, L., Mafra, F., Gregory, B. D., Wei, Z., & Hakonarson, H. (2020). Elucidation of DNA methylation on N6-adenine with deep learning. *Nature Machine Intelligence, 2*, 466–475. https://doi.org/10.1038/s42256-020-0211-4

Tan, J., Shenker-Tauris, N., Rodriguez-Hernaez, J., Wang, E., Sakellaropoulos, T., Boccalatte, F., Thandapani, P., Skok, J., Aifantis, I., Fenyö, D., Xia, B., & Tsirigos, A. (2023). Cell-type-specific prediction of 3D chromatin organization enables high-throughput in silico genetic screening. *Nature Biotechnology, 41*, 1140–1150. https://doi.org/10.1038/s41587-022-01612-8

Tang, Y., Gao, C.-C., Gao, Y., Yang, Y., Shi, B., Yu, J.-L., Lyu, C., Sun, B. F., Wang, H. L., Xu, Y., Yang, Y. G., & Chong, K. (2020). OsNSUN2-mediated 5-methylcytosine mRNA modification enhances rice adaptation to high temperature. *Developmental Cell, 53*, 272–286.e7. https://doi.org/10.1016/j.devcel.2020.03.009

Tian, X.-C., Chen, Z.-Y., Nie, S., Shi, T.-L., Yan, X.-M., Bao, Y.-T., Li, Z. C., Ma, H. Y., Jia, K. H., Zhao, W., & Mao, J. F. (2024). Plant-LncPipe: A computational pipeline providing significant improvement in plant lncRNA identification. *Horticulture Research, 11*, 11. https://doi.org/10.1093/hr/uhae041.

Ueda, M., & Seki, M. (2020). Histone modifications form epigenetic regulatory networks to regulate abiotic stress response. *Plant Physiology, 182*, 15–26. https://doi.org/10.1104/pp.19.00988

Vakulenko, S., & Grigoriev, D. (2021). Deep gene networks and response to stress. *Mathematics, 9*, 3028. https://doi.org/10.3390/math9233028

Valeyre, H., Pati, P., Gossi, F., Somnath, V. R., Martinelli, A., & Rapsomaniki, M. A. (2022). Chromformer: A transformer-based model for 3D genome structure prediction. https://doi.org/10.11.1/2022.11.15.516571

Varotto, S., Tani, E., Abraham, E., Krugman, T., Kapazoglou, A., Melzer, R., Radanović, A., & Miladinović, D. (2020). Epigenetics: Possible applications in climate-smart crop breeding. *Journal of Experimental Botany, 71*, 5223–5236. https://doi.org/10.1093/jxb/eraa188

Verhoeven, K. J. F., Jansen, J. J., van Dijk, P. J., & Biere, A. (2010). Stress-induced DNA methylation changes and their heritability in asexual dandelions. *New Phytologist, 185*, 1108–1118. https://doi.org/10.1111/j.1469-8137.2009.03121.x

Waddington, C. H. (2012). The epigenotype. *International Journal of Epidemiology, 41*, 10–13. https://doi.org/10.1093/ije/dyr184

Wan, H., Cao, L., Wang, P., Hu, H., Guo, R., Chen, J., Zhao, H., Zeng, C., & Liu, X. (2024). Genome-wide mapping of main histone modifications and coordination regulation of metabolic genes under salt stress in pea (Pisum sativum L). *Horticulture Research, 11*, 11. https://doi.org/10.1093/hr/uhae259.

Wang, J., & Wang, L. (2020). Deep analysis of RNA N6-adenosine methylation (m6A) patterns in human cells. *NAR Genomics and Bioinformatics, 2*. https://doi.org/10.1093/nargab/lqaa007

Wang, X., Xin, C., Cai, J., Zhou, Q., Dai, T., Cao, W., & Jiang, D. (2016). Heat priming induces trans-generational tolerance to high temperature stress in wheat. *Frontiers in Plant Science, 7*, 501. https://doi.org/10.3389/fpls.2016.00501

Wang, H., Cimen, E., Singh, N., & Buckler, E. (2020). Deep learning for plant genomics and crop improvement. *Current Opinion in Plant Biology, 54*, 34–41. https://doi.org/10.1016/j.pbi.2019.12.010

Wang, Y., Zhang, P., Guo, W., Liu, H., Li, X., Zhang, Q., Du, Z., Hu, G., Han, X., Pu, L., & Tian, J. (2021a). A deep learning approach to automate whole-genome prediction of diverse epigenomic modifications in plants. *New Phytologist, 232*, 880–897. https://doi.org/10.1111/nph.17630

Wang, Y., Liu, Y., Xu, Q., Xu, Y., Cao, K., Deng, N., Wang, R., Zhang, X., Zheng, R., Li, G., & Fang, Y. (2021b). TAD boundary and strength prediction by integrating sequence and epigenetic profile information. *Briefings in Bioinformatics, 22*, 22. https://doi.org/10.1093/bib/bbab139.

Wang, C., Yang, J., Song, P., Zhang, W., Lu, Q., Yu, Q., & Jia, G. (2022). FIONA1 is an RNA N6-methyladenosine methyltransferase affecting arabidopsis photomorphogenesis and flowering. *Genome Biology, 23*, 40. https://doi.org/10.1186/s13059-022-02612-2

Wang, D., Wang, Z., Chen, L., Xiao, H., & Yang, B. (2023). Cross-parallel transformer: Parallel ViT for medical image segmentation. *Sensors (Basel), 23*, 9488. https://doi.org/10.3390/s23239488

Wang, Y., Kong, S., Zhou, C., Yanfang, W., Zhang, Y., Fang, Y., & Li, G. (2024a). A review of deep learning models for the prediction of chromatin interactions with DNA and epigenomic profiles. *Briefings in Bioinformatics, 26*, 26. https://doi.org/10.1093/bib/bbae651.

Wang, Z., Peng, Y., Li, J., Li, J., Yuan, H., Yang, S., Ding, X., Xie, A., Zhang, J., Wang, S., Li, K., Shi, J., Xing, G., Shi, W., Yan, J., & Liu, J. (2024b). Deepcba: A deep learning framework for gene expression prediction in maize based on DNA sequences and chromatin interactions. *Plant Communications, 5*, 100985. https://doi.org/10.1016/j.xplc.2024.100985

Wang, Y., Deng, Y., Zheng, Y., Chattopadhyay, P., & Wang, L. (2025). Vision transformers for image classification: A comparative survey. *Technologies, 13*, 32. https://doi.org/10.3390/technologies13010.32

Wei, L.-H., Song, P., Wang, Y., Lu, Z., Tang, Q., Yu, Q., Xiao, Y., Zhang, X., Duan, H. C., & Jia, G. (2018). The m 6 a reader ECT2 controls trichome morphology by affecting mRNA stability in arabidopsis. *Plant Cell*, *30*, 968–985. https://doi.org/10.1105/tpc.17.00934

Won, K.-J., Zhang, X., Wang, T., Ding, B., Raha, D., Snyder, M., Ren, B., & Wang, W. (2013). Comparative annotation of functional regions in the human genome using epigenomic data. *Nucleic Acids Research*, *41*, 4423–4432. https://doi.org/10.1093/nar/gkt143

Wu, Z., Tang, J., Zhuo, J., Tian, Y., Zhao, F., Li, Z., Yan, Y., & Yang, R. (2019). Chromatin signature and transcription factor binding provide a predictive basis for understanding plant gene expression. *Plant and Cell Physiology*, *60*, 1471–1486. https://doi.org/10.1093/pcp/pcz051

Xing, M.-Q., Zhang, Y.-J., Zhou, S.-R., Hu, W.-Y., Wu, X.-T., Ye, Y.-J., Wu, X. X., Xiao, Y. P., Li, X., & Xue, H. W. (2015). Global analysis reveals the crucial roles of DNA methylation during rice seed development. *Plant Physiology*, *168*, 1417–1432. https://doi.org/10.1104/pp.15.00414

Xu, Y., Wang, Y., Luo, J., Zhao, W., & Zhou, X. (2017). Deep learning of the splicing (epi) genetic code reveals a novel candidate mechanism linking histone modifications to ESC fate decision. *Nucleic Acids Research*, *45*, 12100–12112. https://doi.org/10.1093/nar/gkx870

Xu, T., Wu, X., Wong, C. E., Fan, S., Zhang, Y., Zhang, S., Liang, Z., Yu, H., & Shen, L. (2022). FIONA1—mediated m 6 a modification regulates the floral transition in arabidopsis. *Advanced Science*, *9*. https://doi.org/10.1002/advs.202103628

Yang, M., & Ma, J. (2022). Machine learning methods for exploring sequence determinants of 3D genome organization. *Journal of Molecular Biology*, *434*, 167666. https://doi.org/10.1016/j.jmb.2022.167666

Yang, C., Yang, L., Zhou, M., Xie, H., Zhang, C., Wang, M. D., & Zhu, H. (2018). LncADeep: An ab initio lncRNA identification and functional annotation tool based on deep learning. *Bioinformatics*, *34*, 3825–3834. https://doi.org/10.1093/bioinformatics/bty428

Yang, H., Wei, Q., Li, D., & Wang, Z. (2020). Cancer classification based on chromatin accessibility profiles with deep adversarial learning model. *PLoS Computational Biology*, *16*, e1008405. https://doi.org/10.1371/journal.pcbi.1008405

Yang, Q., Wu, L., Meng, J., Ma, L., Zuo, E., & Sun, Y. (2023a). EpiCas-DL: Predicting sgRNA activity for CRISPR-mediated epigenome editing by deep learning. *Computational and Structural Biotechnology Journal*, *21*, 202–211. https://doi.org/10.1016/j.csbj.2022.11.034

Yang, R., Das, A., Gao, V. R., Karbalayghareh, A., Noble, W. S., Bilmes, J. A., & Leslie, C. S. (2023b). Epiphany: Predicting Hi-C contact maps from 1D epigenomic signals. *Genome Biology*, *24*, 134. https://doi.org/10.1186/s13059-023-02934-9

Yin, Q., Wu, M., Liu, Q., Lv, H., & Jiang, R. (2019). Deephistone: A deep learning approach to predicting histone modifications. *BMC Genomics*, *20*, 193. https://doi.org/10.1186/s12864-019-5489-4

Yocca, A. E., & Edger, P. P. (2022). Current status and future perspectives on the evolution of *cis*-regulatory elements in plants. *Current Opinion in Plant Biology*, *65*, 102139. https://doi.org/10.1016/j.pbi.2021.10.139

Yu, Q., Liu, S., Yu, L., Xiao, Y., Zhang, S., Wang, X., Xu, Y., Yu, H., Li, Y., Yang, J., Tang, J., Duan, H. C., Wei, L. H., Zhang, H., Wei, J., Tang, Q., Wang, C., Zhang, W., Wang, Y., ... Jia, G. (2021). RNA demethylation increases the yield and biomass of rice and potato plants in field trials. *Nature Biotechnology*, *39*, 1581–1588. https://doi.org/10.1038/s41587-021-00982-9

Yuan, D.-H., Xing, J.-F., Luan, M.-W., Ji, K.-K., Guo, J., Xie, S.-Q., & Zhang, Y. M. (2020). DNA N6-methyladenine modification in wild and cultivated soybeans reveals different patterns in nucleus and cytoplasm. *Frontiers in Genetics*, *11*, 736. https://doi.org/10.3389/fgene.2020.00736.

Yung, W.-S., Li, M.-W., Sze, C.-C., Wang, Q., & Lam, H.-M. (2021). Histone modifications and chromatin remodelling in plants in response to salt stress. *Physiologia Plantarum*, *173*, 1495–1513. https://doi.org/10.1111/ppl.13467

Zemlianskaia, N., Gauderman, W. J., & Lewinger, J. P. (2022). A scalable hierarchical lasso for gene–environment interactions. *Journal of Computational and Graphical Statistics*, *31*, 1091–1103. https://doi.org/10.1080/10.18600.2022.2039161

Zeng, J., Kirk, B. D., Gou, Y., Wang, Q., & Ma, J. (2012). Genome-wide polycomb target gene prediction in Drosophila melanogaster. *Nucleic Acids Research*, *40*, 5848–5863. https://doi.org/10.1093/nar/gks209

Zhai, J., Song, J., Cheng, Q., Tang, Y., & Ma, C. (2018). PEA: An integrated R toolkit for plant epitranscriptome analysis. *Bioinformatics*, *34*, 3747–3749. https://doi.org/10.1093/bioinformatics/bty421

Zhang, H., & Shafiq, M. O. (2024). Survey of transformers and towards ensemble learning using transformers for natural language processing. *Journal of Big Data*, *11*, 25. https://doi.org/10.1186/s40537-023-00842-0

Zhang, X., Clarenz, O., Cokus, S., Bernatavichute, Y. V., Pellegrini, M., Goodrich, J., & Jacobsen, S. E. (2007). Whole-genome analysis of histone H3 lysine 27 trimethylation in arabidopsis. *PLoS Biology*, *5*, e129. https://doi.org/10.1371/journal.pbio.0050129

Zhang, Y., Fischer, M., Colot, V., & Bossdorf, O. (2013). Epigenetic variation creates potential for evolution of plant phenotypic plasticity. *The New Phytologist*, *197*, 314–322. https://doi.org/10.1111/nph.12010

Zhang, H., Lang, Z., & Zhu, J.-K. (2018a). Dynamics and function of DNA methylation in plants. *Nature Reviews Molecular Cell Biology*, *19*, 489–506. https://doi.org/10.1038/s41580-018-0016-z

Zhang, Q., Liang, Z., Cui, X., Ji, C., Li, Y., Zhang, P., Liu, J., Riaz, A., Yao, P., Liu, M., Wang, Y., Lu, T., Yu, H., Yang, D., Zheng, H., & Gu, X. (2018b). N6-methyladenine DNA methylation in Japonica and Indica rice genomes and its association with gene expression, plant development, and stress responses. *Molecular Plant*, *11*, 1492–1508. https://doi.org/10.1016/j.molp.2018.11.005

Zhang, Y., An, L., Xu, J., Zhang, B., Zheng, W. J., Hu, M., Tang, J., & Yue, F. (2018c). Enhancing Hi-C data resolution with deep convolutional neural network HiCPlus. *Nature Communications*, *9*, 750. https://doi.org/10.1038/s41467-018-03113-2

Zhang, F., Zhang, Y.-C., Liao, J.-Y., Yu, Y., Zhou, Y.-F., Feng, Y.-Z., Yang, Y. W., Lei, M. Q., Bai, M., Wu, H., & Chen, Y. Q. (2019a). The subunit of RNA N6-methyladenosine methyltransferase OsFIP regulates early degeneration of microspores in rice. *PLoS Genetics*, *15*, e1008120. https://doi.org/10.1371/journal.pgen.1008120

Zhang, S., Chasman, D., Knaack, S., & Roy, S. (2019b). In silico prediction of high-resolution Hi-C interaction matrices. *Nature Communications*, *10*, 5449. https://doi.org/10.1038/s41467-019-13423-8

Zhang, P., Wang, Y., Chachar, S., Tian, J., & Gu, X. (2020). eRice: A refined epigenomic platform for japonica and indica rice. *Plant Biotechnology Journal*, *18*, 1642–1644. https://doi.org/10.1111/pbi.13329

Zhang, K., Zhuang, X., Dong, Z., Xu, K., Chen, X., Liu, F., & He, Z. (2021). The dynamics of N6-methyladenine RNA modification in interactions between rice and plant viruses. *Genome Biology*, *22*, 189. https://doi.org/10.1186/s13059-021-02410-2

Zhang, Z., Feng, F., Qiu, Y., & Liu, J. (2023). A generalizable framework to comprehensively predict epigenome, chromatin organization, and transcriptome. *Nucleic Acids Research*, *51*, 5931–5947. https://doi.org/10.1093/nar/gkad436

Zhang, Y., Zhong, J., Munawar, A., Cai, Y., He, W., Zhang, Y., Guo, H., Gao, Y., Zhu, Z., & Zhou, W. (2024). Knocking down a DNA demethylase gene affects potato plant defense against a specialist insect herbivore. *Journal of Experimental Botany*, *75*, 483–499. https://doi.org/10.1093/jxb/erad387

Zhang, Y., Huang, G., Zhao, Y., Lu, X., Wang, Y., Wang, C., Guo, X., & Zhao, C. (2025). Revolutionizing crop breeding: Next-generation artificial intelligence and big data-driven intelligent design. *Engineering (London)*, *44*, 245–255. https://doi.org/10.1016/j.eng.2024.11.034

Zheng, S., Thakkar, N., Harris, H. L., Liu, S., Zhang, M., Gerstein, M., Aiden, E. L., Rowley, M. J., Noble, W. S., Gürsoy, G., & Singh, R. (2024). Predicting A/B compartments from histone modifications using deep learning. *Iscience*, *27*, 109570. https://doi.org/10.1016/j.isci.2024.109570

Zhong, Z.-D., Xie, Y.-Y., Chen, H.-X., Lan, Y.-L., Liu, X.-H., Ji, J.-Y., Wu, F., Jin, L., Chen, J., Mak, D. W., Zhang, Z., & Luo, G. Z. (2023). Systematic comparison of tools used for m6A mapping from nanopore direct RNA sequencing. *Nature Communications*, *14*, 1906. https://doi.org/10.1038/s41467-023-37596-5

Zhou, J. (2022). Sequence-based modeling of three-dimensional genome architecture from kilobase to chromosome scale. *Nature Genetics*, *54*, 725–734. https://doi.org/10.1038/s41588-022-01065-4

Zhou, C., Wang, C., Liu, H., Zhou, Q., Liu, Q., Guo, Y., Peng, T., Song, J., Zhang, J., Chen, L., Zhao, Y., Zeng, Z., & Zhou, D. X. (2018). Identification and analysis of adenine N6-methylation sites in the rice genome. *Nature Plants*, *4*, 554–563. https://doi.org/10.1038/s41477-018-0214-x

Zhou, L., Tian, S., & Qin, G. (2019). RNA methylomes reveal the m6A-mediated regulation of DNA demethylase gene SlDML2 in tomato fruit ripening. *Genome Biology*, *20*, 156. https://doi.org/10.1186/s13059-019-1771-7

2

AI-Driven Precision Plant Epigenetic Regulation Under Changing Climate

Mohd Anas[1], Neha Chaurasia[2], Yamshi Arif[2], Mohammad Danish[3], Nashra Waqar[4], and Mohammad Ali[5]

[1] *Institute of Laboratory Research on Geomaterials, Faculty of Natural Sciences, Comenius University in Bratislava, Bratislava, Slovakia*
[2] *Department of Botany, Faculty of Life Sciences, Aligarh Muslim University, Aligarh, Uttar Pradesh, India*
[3] *Department of Experimental Plant Biology, Institute of Botany, Plant Science and Biodiversity Centre, Slovak Academy of Sciences, Bratislava, Slovakia*
[4] *Department of Biotechnology, Faculty of Science, Era University, Lucknow, Uttar Pradesh, India*
[5] *Department of Computer Science, Aligarh Muslim University, Aligarh, Uttar Pradesh, India*

2.1 Introduction

The European Parliament defines precision agriculture (PA) as an approach to farming management that is based on monitoring, quantifying, and reacting to crop variability within and between fields, as well as to characteristics of animal rearing (Zarco-Tejada et al., 2014). Growing yields, which boost productivity and profitability, are the main benefit of this activity. Because there is less contamination, we can explore other opportunities, such as improved working conditions, greater animal welfare, and better environmental outcomes. When employing PA, sensors streamline farming operations and procedures such as soil cultivation, planting, fertilization, harvesting, and even animal husbandry. PA's positioning provides an advantage. Large agricultural regions benefit most from these practices, especially during grain harvesting. PA results in increased yields. The two main pillars of the technology are self-driving systems and guided traffic management, or CTF. Variable dose application (VRA) techniques produce the best results (Gyarmati & Mizik, 2020). High yields of valuable items are already helping to enhance quality and boost revenues in the cases of fruits, vegetables, and vineyards. Additionally, it promotes the health of the animals, efficient irrigation, and water conservation. It is employed to keep an eye on the physical surroundings of animals as well as their behavior, welfare, and productivity (Zarco-Tejada et al., 2014). Farmers may better fulfill their role as suppliers of public goods, support specific initiatives, and adhere to good farming practices by measuring the environmental impact of their operations using PA. The environment in which plants live is ever-changing and frequently unfavorable or even hostile. Plants cannot actively avoid repeated violent confrontations since they are sessile organisms. Rather, they exhibited great phenotypic plasticity, which involves quick reactions to harsh environmental stimuli and environmental adaptations. This phenotypic flexibility resulted from changes in gene expression. Since epigenetic marks regulate gene expression, epigenetic diversity may play a significant role in how plants respond to stressors and adapt to their surroundings (Ashapkin et al., 2020). Sessile organisms, plants, are subject to a variety of environmental factors. As a result, plants evolved complex gene-regulation systems to guarantee their survival in the face of environmental changes. To adapt to environmental difficulties, plants sense their surroundings and convey those signals through a cascade of signal transduction. This process results in the buildup of transcription factors that stimulate gene expression (Mirouze & Paszkowski, 2011). Epigenetic regulation, which involves covalent changes of DNA and histones that alter chromatin transcriptional activity without altering DNA sequence, is another significant method of gene regulation in response to stress (Iwasaki & Paszkowski, 2014). DNA can be packaged in the nucleus, thanks to the nucleosomes that make up chromatin structure, which are created when histone proteins bind with DNA (Alberts et al., 2002). Since access to DNA is necessary for gene expression, the degree of chromatin condensation plays a significant role in this control. While heterochromatin is typically a transcriptionally repressed region with hypermethylation of DNA and certain histone modifications, euchromatin can be linked to transcriptionally active regions (Vaillant & Paszkowski, 2007). Three epigenetic markers have been identified by studies: short RNAs, histone changes, and DNA methylation. Crucially, short RNAs frequently cause chromatin remodeling and DNA methylation (Meyer, 2015).

2.2 Foundations of Plant Epigenetics

2.2.1 DNA Methylation and Histone Modifications

DNA methylation is the epigenetic phenomenon in plants that has been investigated the most (Feng & Jacobsen, 2011; Vanyushin & Ashapkin, 2011). Like in animals, a significant portion of methylated cytosine residues (m^5C) in plants are found in the symmetric CG sites. In contrast to animals, plants exhibit notable methylation in both the symmetric and asymmetric CHG sites (where H can be any nucleotide other than G). Protein-coding gene sequences are primarily methylated at CG sites, but repeat and transposable element (TE) sequences contain all three methylation contexts. DNA methyltransferase MET1 uses three variant in methylation (VIM) family members, VIM1–VIM3, to help carry out the maintenance methylation of CG sites (Woo et al., 2008). This methylation complex identifies and preferentially methylates hemi-methylated CG sites in the daughter strands during DNA replication. The maintenance methylation of symmetric CHG sites is carried out by a plant-specific DNA methyltransferase called chromomethylase 3 (CMT3). It most likely plays a role in the methylation of asymmetric CHH sites as well. Due to mutually promoting substrate-level interactions with H3K9-specific histone methyltransferases (HMTs), CMT3 is unable to recognize hemi-methylated sites, in contrast to MET1, and hence retains CHG-specific methylation (Du et al., 2012). The primary enzyme responsible for DNA methylation de novo at CG, CHG, and CHH sites is domains rearranged methyltransferase 2 (DRM2) (Cao & Jacobsen, 2002). Because CHH sites are asymmetric, repeated methylation de novo is required to maintain their methylation. It was recently discovered that CHH methylation is mediated by a different mechanism that is reliant on CMT2 (Stroud et al., 2014; Zemach et al., 2013). Like CMT3, CMT2 uses histone H3K9 methylation markers to target methylated regions. The RNA-directed DNA methylation (RdDM) mechanism, on the other hand, targets DRM2 to its methylation sites through the complementary interaction of 24-nt siRNAs (24-nucleotide small interfering RNAs) with sequences that need to be methylated from scratch (Matzke et al., 2015; Wendte & Pikaard, 2017). Through a base excision repair route, the particular m^5C-specific glycosylase enzymes repressor of silencing 1 (ROS1), demeter (DME), demeter-like 2 (DML2), and demeter-like 3 (DML3) can actively demethylate DNA in *Arabidopsis* (Gehring et al., 2009). It has been demonstrated that the net effect of repeated methylation by RdDM and demethylation by ROS1 is the DNA methylation status in several genomic locations (Tang et al., 2016). In addition to DNA methylation, plants and other eukaryotic organisms have a different set of epigenetic marks, which are covalent changes of different histone protein amino acid residues. The most extensively researched histone modifications in both plants and mammals were a few forms of methylation at lysine residues (Berr et al., 2011; Feng & Jacobsen, 2011). In *Arabidopsis*, the gene bodies (H3K4me1) and promoters (H3K4me2 and H3K4me3) of actively transcribed genes have three different forms of H3K4 methylation marks (mono/di/tri-methylation–H3K4me1, H3K4me2, and H3K4me3, respectively). While H3K4me1 and m^5C may coexist along the gene body, H3K4me2/3 and m^5C are mutually exclusive marks at the same promoter (Zhang et al., 2009). At particular locations, H3K27me3 exhibits a strong association with the suppression of gene transcription. During plant development, this epigenetic mark is known to affect a number of genes, mostly independent of other epigenetic mechanisms (Zhang et al., 2007a). Polycomb repressive complex 2 (PRC2), which contains an H3K27-specific HMT, mediates the regulation of gene transcription by H3K27me3 marks. When it binds to genomic loci that contain H3K27me3, the plant chromodomain protein like heterochromatin protein 1 (LHP1) most likely plays a role in mediating its regulatory effects (Zhang et al., 2007b). Another strong repressive marker found in plants is H3K9me2. This epigenetic mark is specific to heterochromatin, in contrast to H3K37me3. In genomic regions rich in repeats and transposons, H3K9me2 marks predominantly colocalize with methylated CHG and CHH sites, playing an essential role in non-CG DNA methylation by CMT3 and CMT2 (Bernatavichute et al., 2008).

2.2.2 Small RNAs and Noncoding RNA Regulation

Plant sRNAs (20–24 nt) are RNA species that have been identified as ubiquitous repressors of gene expression in both plants and animals, despite not coding for proteins. Given their diversity and abundance in plants, it is simple to conclude that one or more miRNAs regulate the majority of biological processes in plants. Because of their crucial roles in gene regulation, plant scientists have focused a lot of attention on these negative gene regulators, which are categorized according to their biogenesis: (1) miRNAs and (2) short interfering RNAs (siRNAs) (Kumar et al., 2018). Through the activity of DNA polymerase II, single-stranded primary miRNA (pri-miRNA) transcripts are generated from miRNA (MIR) genes and used to biosynthesize miRNAs. The precursor RNAs (pre-miRNA) that are formed from these pri-miRNA transcripts are either poorly paired hairpins or stem-loops (Chinnusamy et al., 2007). Through the actions of dicer-like 1 (DCL1) protein, hyponastic leaves 1 (HYL1, a dsRNA binding protein), and serrate (SE, a C2H2-zinc finger protein), the pre-miRNA hairpin precursor is subsequently transformed into a 20–22 nt miRNA–miRNA duplex (stem region of hairpin), where miRNA functions as the guide strand and miRNA as the degraded strand. The pre/mature miRNAs are subsequently exported from the nucleus to the cytoplasm by HATSY exportin proteins after the duplex is methylated at the 3′ end by HUA enhancer1 (HEN1). An unknown helicase unwinds the duplex into single-stranded mature-miRNA, after which miRNA joins the RNA-induced silencing complex (RISC) and uses sequence complementarities to direct RISC binding to appropriate target transcripts (Khraiwesh et al., 2010, 2012; Voinnet, 2009). The siRNAs are identical, 20–25 bp long double-stranded RNA molecules that are produced from long dsRNAs of both endogenous and external origins. Natural cis-antisense gene pair-encoded mRNAs, dsRNAs produced from heterochromatin and DNA repeats, and miRNA-directed cleavage products of ssRNAs that are subsequently transformed into dsRNAs by RNA-dependent RNA polymerases (RDR) are the main endogenous sources of dsRNAs. Certain RDR–DCL protein combinations act to produce distinct kinds of siRNAs. While natural antisense gene-generated siRNAs (nat-siRNAs) are obtained from dsRNAs synthesized from the mRNAs encoded by natural cis-antisense gene pairs, trans-acting (ta) siRNAs are siRNAs produced by miRNA-directed cleavage of ssRNAs/transgene mRNAs. RDRs and proteins that resemble DCLs process the dsRNAs that are so produced (Chinnusamy et al., 2007). Recent reputable studies have demonstrated that plants utilize sRNAs as sequence-specific epigenetic markers of transcriptional or post-transcriptional gene silencing in response to environmental stimuli (Fu et al., 2017). Furthermore, the expression, conservation, and divergence of miRNAs have been validated using next-generation sequencing (NGS) techniques (Budak et al., 2014). Tools and databases for predicting, identifying, and validating sRNAs, as well as accrediting their target mRNAs, are being developed at an increasing rate. As a result, sRNAs are considered a reliable platform for engineering plants' resistance to abiotic stress (Kumar et al., 2018).

2.2.3 Epigenetic Memory and Transgenerational Effects

Since plants are sessile creatures, they rarely flee from stress and have developed alternative adaptation strategies to ensure their survival. In contrast to genetic mutations, epigenetic modifications of genomes are reversible and can thus be removed in the absence of stress, making them essential for such adaptations (Shah, 2021). DNA methylation, histone modifications, chromatin alterations, and small RNA-based processes are examples of epigenetic modifications (Thiebaut et al., 2019). These alterations affect the phenotype because they are important modulators of gene expression. The majority of eukaryotes frequently experience epigenetic modifications. Even after several meiosis and mitosis, cells may be able to hold onto some of these alterations, which can then be passed on to the offspring. These genome alterations act as a coding language that might affect the offspring's phenotype; this is referred to as epigenetic memory (Mathieu et al., 2007). Transgenerational effects (TGE) can alter the phenotypes of subsequent generations, potentially having a significant impact on the ecology and evolution of numerous plant species. The majority of research on these effects has focused on sexually reproducing organisms and their generations. However, a significant number of plant species reproduce asexually, such as by clonal development. Although there is currently a lack of clear proof, TGE is believed to be made possible by heritable epigenetic alteration of DNA (Rendina et al., 2018).

2.3 Impacts of Climate Change on Plant Epigenomes

2.3.1 Abiotic Stress and Epigenomic Plasticity

Plant productivity is impacted by a variety of abiotic stresses that are frequently brought on by unpredictable climate change. These challenges mostly include extreme temperatures, dehydration, excessive salinity, inadequate nutrition, UV radiation, and heavy metal toxicity. Gene regulation in response to environmental stressors is significantly influenced by the dynamics of epigenetic coding (Zhang et al., 2018). For instance, by altering the chromatin state of stress-responsive genes, histone modification and DNA methylation have been demonstrated to change their expression at the transcriptional and post-transcriptional levels (Luo & He, 2020). For example, RNA-directed DNA methylation (RdDM) and transcription factors are important for controlling gene expression in plants under abiotic stress. DNA methylation mediated by the RdDM pathway regulates a few known heat stress genes (Popova et al., 2013). For instance, under normal circumstances, RdDM silences the *Arabidopsis* plant's MYB74 transcription factor; during salt stress, the transcription factor becomes desilenced (Xu et al., 2015). Furthermore, certain epigenetic alterations, like histone modifications, are important for stress memory and can be inherited by children (Friedrich et al., 2019). It is highly probable that intragenic epigenetic modifications control the expression of stress-responsive genes, for instance, by altering the alternative splicing of the transcripts since transposable elements (TEs) frequently insert into the gene body of stress-related genes in plant genomes. Following the removal of the stressor, epigenetic modifications typically revert to their prestress condition. However, it has been shown that in certain instances, the epigenetic modifications (or a portion of them) serve as stress memory for plants and persist even after the stress has been removed. Long-term adaptation and even evolution are aided by this stress memory. Genetically identical plants have variations in DNA methylation in response to different stressors. Apomictic *Taraxacum officinale* plants, for instance, were subjected to a variety of abiotic stressors, and DNA methylation varied significantly across all plants under all stressors (Verhoeven et al., 2010). The findings demonstrated that epigenetic inheritance in plant adaptation can occur independently of genetic diversity among individuals and showed that stress-induced variation in DNA methylation is widespread. Numerous studies have also revealed that the regulation of a large number of stress-responsive genes is impacted by stress-induced methylation patterns, which vary depending on the kind of stress, genotype, tissue, and organism (Bonasio et al., 2010; Mastan et al., 2012; Wang et al., 2011a).

2.3.2 Biotic Interactions in a Changing Climate

Much work has been achieved in the domain of range shifts and their implications on population dynamics at the individual species (population) level. However, it will be difficult to scale from populations to communities, much alone ecosystems (Berg et al., 2010; Fenton & Spencer, 2010; Kareiva et al., 1993; Schmitz et al., 2003; Tylianakis et al., 2008). According to Walther (2010), many species' population responses to climate change are unlikely to be merely additive, and their combinational dynamics are unlikely to be linear. Every species is a component of intricate networks of relationships. Despite this complexity, a number of studies have demonstrated that species interactions in various habitat

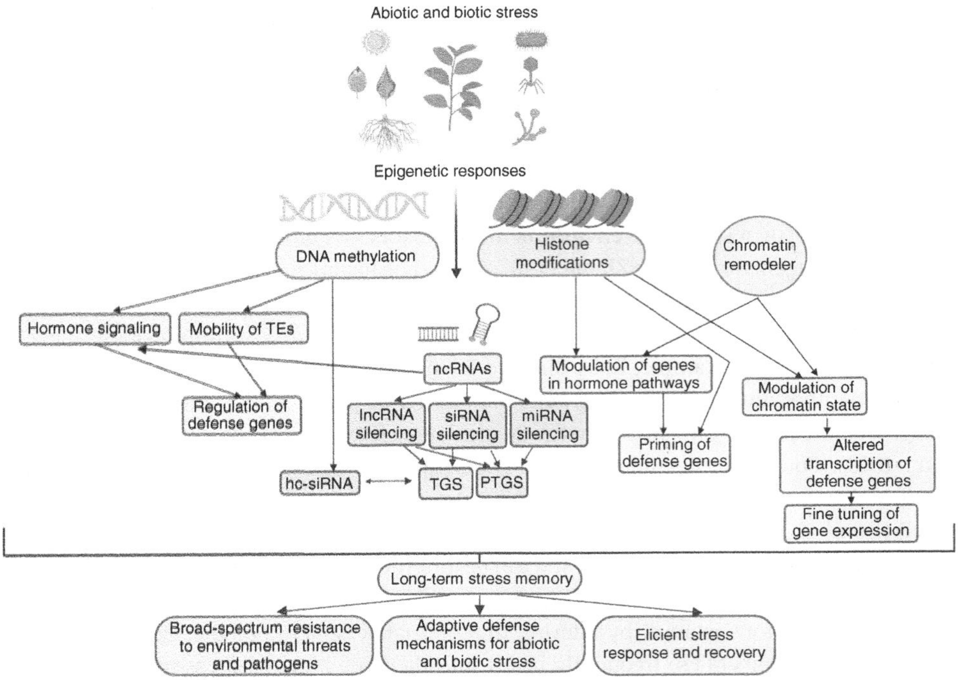

Figure 2.1 Epigenetic regulation of plant responses to abiotic and biotic stresses.

types follow universal patterns (Bascompte, 2009; Montoya et al., 2006; Pimm, 1991). For instance, most connections, such as those between predators and their prey, are often weak, and interaction networks are highly diverse—some species have a far higher number of interactions than would be predicted by chance (Montoya et al., 2006; Wootton & Emmerson, 2005). The stability of populations to recover from disturbances and the probable effects of local species extinctions on the remaining species in the interaction network are determined by these patterns. It is not well understood how climate change may affect those network features and the ecosystem services associated with them (Berg et al., 2010; Walther, 2010). An interesting illustration of how connectivity, one of the basic characteristics of the food web, may alter as temperatures rise is given by Petchey et al. (2010). Connectance is the ratio of actual trophic interactions to the total number of possible interactions if all species were connected. Using a theoretical model derived from metabolic theory and foraging biology, they demonstrate how rising temperatures would significantly impact connectance, which would have additional repercussions for the stability of species populations and community dynamics (Montoya & Raffaelli, 2010) (Figure 2.1).

2.3.3 Case Studies

2.3.3.1 Epigenomic Responses to Salinity

Both hyper- and hypo-DNA methylation are caused by environmental stressors. Studies indicate that under unfavorable circumstances, epigenetic mechanisms significantly influence gene modification. Methylation of promoters and gene bodies aids in genotypically and organ-specific regulation of gene expression during salt-induced stress. It has been shown that salt stress causes changes in the expression of several transcription factors in soybeans. Song et al. hypothesized that histone changes and DNA methylation would work together to influence genes that are triggered by stress (Song et al., 2012). Additionally, salt stress may change the expression of DNA demethylases, resulting in hypomethylation, according to Ferreira et al. (Ferreira et al., 2015). Similarly, when exposed to stress caused by salinity, salt-resistant wheat cultivar SR3 and its progenitor HBP1 had opposing alterations in cytosine methylation patterns. It was discovered that a particular demethylation gene works well during the epigenetic processes in rice roots under salinity stress (Wang et al., 2011b). Wheat's high-affinity potassium transporters, whose expression is regulated by genetic and epigenetic mechanisms, may exhibit altered expression levels in response to salt stress (Kumar et al., 2017). For many crop plants, chromatin structural changes are essential for granting salt tolerance. Kaldis et al. claim that *Arabidopsis thaliana*'s hypersensitivity to salt stress is caused by transcriptional adaptor ADA2b. As a result, salt stress affects both histone modifications and genome-wide DNA methylation, and these two processes are interconnected to allow for a coordinated salt stress response. It has been demonstrated that the shoots of wheat genotypes that are salt-sensitive and salt-tolerant have higher levels of DNA methylation, ranging from 30% to 40%, than the roots (Wada et al., 2004; Wang et al., 2015). When exposed to salt stress, sensitive genotypes of rape seeds exhibited higher levels of DNA methylation than intolerant genotypes (Eichten & Springer, 2015).

2.3.3.2 Epigenomic Responses to Drought

Clarifying regulatory mechanisms at the intricate molecular levels in plants subjected to a range of environmental restrictions has been the primary focus of research on plant stress biology. Histone changes and nucleosome density in the OR genes' promoters and ORFs are the main factors influencing drought-induced gene expression (To & Kim, 2014). When comparing the Rd20 and Rd29A areas under drought stress to ideal conditions, histone alterations such as H3K4me3 and H3K9ac were discovered (Kim et al., 2008). A detailed analysis of the post-drought stress recovery phase revealed that RNA polymerase II eventually cleared the H3K9ac sites after fast deacetylation at those locations. The genes responsible for this significant chromatin shifting were Rd29A, Rd20, and *Arabidopsis* galactinol synthase (AtGOLS2). At higher rehydration dosages, however, H3K4me3 was removed more slowly than H3K9ac (Banerjee & Roychoudhury 2017). The transcription of stress-sensitive genes in *A. thaliana* during drought was found to be dependent on increased H3K4 trimethylation and H3K9 acetylation on the gene promoter, as well as H3K23 and H3K27 acetylation on the coding regions (Kim et al., 2012). The gene encoding the ABA biosynthesis enzyme, 9-cis-epoxy carotenoid dioxygenase, was overexpressed by the *Arabidopsis* trithorax-like 1 HMT via increasing H3K4me3 modification, especially under drought stress (Ding et al., 2011). Under drought stress, DNA methylation shows tissue selectivity. Across various tissues, genotypes, and life stages, *Oryza sativa* underwent 12.1% methylation alterations as a result of drought. At the same developmental stage, the overall DNA methylation rate in roots was lower than in leaves, suggesting that roots are crucial in water shortage (Suji & Joel, 2010). Rice cultivars IR20, a drought-vulnerable variation, show hypomethylation during drought circumstances, but the resistant varieties show hypermethylation, demonstrating the link between DNA methylation and drought stress tolerance (Gayacharan & Joel, 2013).

2.3.3.3 Epigenomic Responses to Heat

With pronounced detrimental effects on plant development, physiology, and metabolism, heat stress is the primary abiotic stressor in plants (Shanker et al., 2020). Similar to other stressors, heat alters a plant's epigenetic makeup. The plants can withstand heat stress, thanks to these modifications. Numerous studies have proposed that these modifications help plants fend off heat stress. In *Arabidopsis* plants, heat stress results in increased methylation and a higher frequency of homologous recombination (Jewkes et al., 2011). DRM2, nuclear RNA polymerase D1, and NRPE1 overexpression may lead to increased genome methylation in *Arabidopsis* in response to heat stress (Naydenov et al., 2015). Global methylation rises in cork oak (*Quercus suber* L.) grown at 55°C. Following heat stress, the heat-sensitive genotype of Brassica napus exhibits higher levels of DNA methylation compared to the heat-tolerant genotype (Gao et al., 2014). OsFIE1 misregulation may be exacerbated in developing rice seeds when mild heat stress at 34°C for 48 hours reduces the DNA methylation level of fertilization-independent endosperm1 (OsFIE1), a component of the PRC2, and suppresses the transcript abundance of OsCMT3 (Folsom et al., 2014). Heat-stressed second-generation plants have higher levels of ribosomal RNA FtsJ-like methyltransferase and lysine-specific histone demethylase 1 (LSD1), which contribute to the demethylation of histone H3 lysine 4 (H3K4) me1/2. This illustrates how certain epigenetic markers cause genes in primed plants' progeny to be stimulated in order to produce tolerance (Wang et al., 2016).

2.4 Artificial Intelligence in Epigenetic Research

Artificial intelligence (AI) has transformed epigenetics from descriptive lists of marks to predictive, mechanism-based models (Guo et al., 2022). AI, through the acquisition of structure in high-dimensional, multiomics, and environmental data, enhances sensitivity to subtle signals, diminishes batch artifacts, and informs causal hypotheses that can be experimentally tested.

2.4.1 AI Tools and Algorithms for Omics Data Analysis

AI approaches translate naturally to the statistical and topological features of epigenomic data.

Sequence-to-Signal Modeling: Transformers and convolutional neural networks (CNNs) (Maurício et al., 2023) acquire regulatory grammars describing chromatin accessibility, histone modifications, and methylation states from sequence raw input, supplemented with k-mer or one-hot encoding. Medium-range enhancer syntax is captured with dilated CNNs and distal enhancer–promoter dependencies modeled with attention layers (Horlacher et al., 2023).

Graph-Aware Learning: GNNs model loci as nodes linked by Hi-C/HiChIP contacts or enhancer–target interactions. Message passing on these graphs predicts domain-level regulation (e.g., TADs and hubs) and aids imputation/denoising of sparse contact maps (Guan et al., 2025).

Latent-Variable and Probabilistic Models: Variational autoencoders (Kipf & Welling, 2016), topic models, and non-negative matrix factorization extract low-dimensional "epigenetic programs" across cell types and perturbations, allowing state discovery, bulk deconvolution, and trajectory inference.

Single-Cell–Specific Pipelines: Manifold learning and deep imputation (e.g., neural embeddings, followed by UMAP/PHATE) target dropout in scATAC-seq and single-cell methylomes. Cross-modal alignment (CITE-seq, scRNA–scATAC integration) transfers labels and reconstructs gene regulatory trajectories.

Harmonization and Robustness: Adversarial domain adaptation, contrastive self-supervision, and batch-effect critics enhance cross-study comparability. Uncertainty quantification (Bayesian heads, ensembles) conveys confidence in locus-level calls.

Common outputs are higher-fidelity peak calling, cross-assay imputation, enhancer–promoter association with confidence scores, motif discovery with TF attribution, and interoperable embeddings for downstream statistical testing.

2.4.2 Machine Learning Models for Predicting Epigenetic Changes

AI repositions epigenetic dynamics as supervised, semi-supervised, or causal prediction problems. One of the most important models is exposure-to-epigenome prediction (Pikaard & Scheid, 2014), which consists of gradient boosting, random forests, and deep networks that project environmental covariates (e.g., heat, drought, hypoxia, diet, and pollutants) and physiological context onto locus-specific methylation or histone modifications, allowing early-warning indices and risk stratification. The next method is the prioritization of noncoding and variant effects used for sequence-context models to make predictions for the impact of SNPs/indels on TF binding and chromatin states, filtering candidates for functional assessment and explaining GWAS signals with coherent regulatory explanations. Prediction of temporal and perturbation helps predict the recurrent networks, neural ODEs, and causal discovery (e.g., NOTEARS and invariant risk minimization) that forecast dynamics in response to stress or treatment, distinguishing between transient and long-lasting epigenetic remodeling and establishing windows for intervention. Finally, transfer and multitask learning is used for the transfer of knowledge from highly annotated species or tissues that enhance generalization to data-scarce systems (e.g., crops and forestry species), lowering the sample sizes for use in robust models. Assessment focuses on AUROC/AUPRC for classification, correlation and calibration for regression, external validation between cohorts/environments, and interpretability through saliency, motif attribution, TF importance, and Shapley values.

2.4.3 Multiomics and Environmental Data Integrative Platforms

Integrated platforms link genotype, epigenome, transcriptome, proteome, metabolome, phenotype, and exposome.

Unified Data Models and Knowledge Graphs: Multimodal autoencoders and knowledge graphs represent relationships between loci, marks, pathways, and traits time-stamping exposures from sensors or field logs, allowing end-to-end queries from environment → epigenome → phenotype.

Cross-Modal Alignment and Mediation: Contrastive learning, canonical correlation, and joint VAEs align expression and chromatin landscapes, whereas causal mediation estimates whether epigenetic marks convey environmental influence to traits.

MLOps and Governance: Containerized pipelines, model registries, and provenance tracking facilitate reproducibility; federated learning and differential privacy facilitate institution-to-institution collaboration while safeguarding sensitive human or proprietary agri data.

2.5 AI-Driven Precision Epigenetic Regulation

AI now directs interventions that reprogram epigenetic states with locus specificity, completing the loop from prediction to control. Through coupling causal inference with programmable editors, scientists are able to set priorities for targets, plan edits, and control safety.

2.5.1 Epigenetic Biomarkers for Stress Tolerance

Strong biomarkers are selection indices and intervention anchors. Some of them are discussed below:

First, feature extraction and simplification useful for stability selection, sparse regularization, and Shapley attribution recover sparse methylation/histone panels that distinguish tolerant from susceptible lines or persons. Second, extrapolation to novel environments determines domain generalization and hierarchical modeling maintain prediction accuracy across sites, seasons, and management regimes; uncertainty quantification identifies context-dependent markers. Third, mechanistic anchoring used for integration models connects biomarkers to upstream TF networks, 3D contacts, and downstream gene programs, favoring loci with convergent evidence over correlative signals alone. Finally, translation to breeding/diagnostics: Classifiers drive marker-assisted selection (MAS); targeted bisulfite sequencing, CUT&Tag/CUT&RUN panels, and multiplex qPCR put biomarker assays into practice at scale.

2.5.2 Targeted Epigenome Editing Strategy Design

AI speeds up the design–test–learn cycle for editing chromatin states without changing DNA sequence. Target nomination through causality is basically used in causality-based counterfactual thinking and in silico knock-downs order enhancers, promoters, and insulators whose state causally regulates stress-relevant activities. We can use effector selection and spatial tiling which works as target-effectors are matched by recommender systems (e.g., DNMT3A/TET1 for methylation; p300/HDACs for acetylation; PRC2/UTX modules for methyl-lysine states) and suggest gRNA tiling across regulatory elements for optimal effect size while minimizing spread. The other one is off-target and epigenetic-spread regulation used for hybrid sequence-chromatin frameworks, including accessibility, nucleosome occupation, and 3D proximity to reduce off-target edits and collateral heterochromatinization. Optimal experimental design encourages Bayesian optimization and chooses minimal perturbation sets (cell types, stages, or conditions) that are maximally informative, saving iterations and cost.

2.5.3 AI-Enabled CRISPR/dCas Systems for Epigenetic Regulation

Catalytically dead Cas (dCas9/dCas12) fused with epigenetic modifiers allows for programmable, noncutting control; AI facilitates design, delivery, and regulation.

- **Guide RNA (gRNA) Design**: Deep learning models forecast on-target activity based on local chromatin state, strand and PAM context, nucleosome location, and sequence content; multi-gRNA ensembles are optimized for cooperative effects and redundancy (Lee et al., 2020).
- **Dose and Schedule Optimization**: Reinforcement learning optimizes induction timing, effector dosage, and multiplex schedules to produce desired magnitude and persistence while constraining toxicity and off-target activity.
- **Logic-controlled Circuits**: AI generates orthogonal CRISPRa/i programs (AND/OR/NOT) that combine several environmental or developmental inputs to attain spatiotemporal specificity of epigenetic modulation (Martella et al., 2019).
- **Safety-by-design**: Predictive toxicology and immunogenicity screening, kill-switches or drug-inducible off-switches, and thorough off-target auditing are integrated into design pipelines. Model cards and data sheets record intended use, limitations, and uncertainty.

2.5.4 Reporting, Validation, and Ethics

2.5.4.1 Data and Code Transparency

Experiments should implement the FAIR principles by making the datasets and analytical artifacts findable, accessible, interoperable, and reusable. At a bare minimum, authors should publish full metadata summarizing sample provenance, extraction and assay protocols, and the environmental conditions under which measurements were made. Reproducibility by computation demands version-locked pipelines, direct documentation of software and hardware dependencies, and fixed random seeds for stochastic steps. To the extent that intellectual property or privacy constraints allow, trained model weights and lightweight inference scripts should be placed in reliable repositories to enable independent verification and reuse.

2.5.4.2 Performance and Interpretation

Beyond headline accuracy, reporting has to be more than just that. Discrimination metrics (e.g., AUROC and AUPRC) ought to be followed by calibration analyses for assessing probabilistic reliability and external validation in independent cohorts or environments to show generalizability. Uncertainty should be measured—through Bayesian methods, deep ensembles, or bootstrapping—and reported together with predictions. Critically, explanations for model outputs must be biologically rational: motif or saliency attribution, contact-domain context, and transcription factor significance enable reviewers to determine whether predictions match established regulatory principles or spurious correlations.

2.5.4.3 Validation Ladder

Strong statements regarding epigenetic regulation require a stepwise validation process beginning with in silico prediction, followed by directed perturbation experiments. Initial

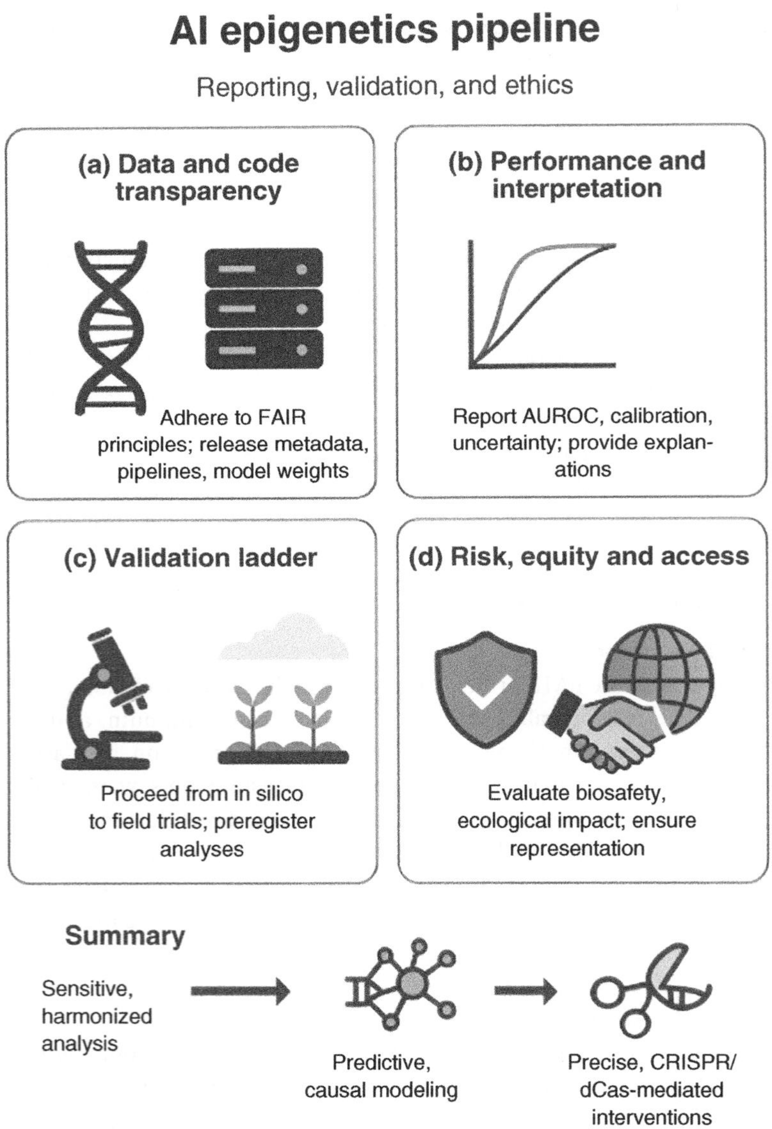

Figure 2.2 AI epigenetic pipeline summarizing data/code transparency, performance and interpretation, validation ladder, and risk–equity–access, culminating in precise CRISPR/dCas interventions.

observations must be validated by CRISPR/dCas-based in vitro modulation and then controlled environment trials that assess dose, timing, and persistence of epigenetic modification. Field or clinical studies should only be conducted after these steps provide reproducibility. Preregistered analysis plans, open-stopping rules, and independent replication across laboratories are critical measures against exaggerated effect sizes and selective reporting.

2.5.4.4 Risk, Equity, and Access

Precision epigenetic treatments introduce ecological, biosafety, and equity concerns that need to be considered early in the research pipeline. Authors should conduct formal risk assessments that consider possible off-target effects, epigenetic dissemination, and unforeseen ecological impacts. Model construction and comparison must consider understudied environments, populations, and species to minimize bias and ensure that instrumentation is universally applicable. For research on human data or research with local communities, consent processes, sharing of data agreements, and governance structures must be stated clearly and sensitive to local norms and rights. To orient the reader to our methodological rigor, Figure 2.2 synthesizes the reporting, validation, and ethics framework that governs our AI-enabled epigenetic workflow.

Collectively, the elements in Figure 2.2 operationalize reproducibility, interpretability, and safety, enabling responsible translation of AI-driven epigenetic insights from computation to laboratory, field, and clinical practice. AI now enables an entire epigenetic workflow from sensitive, harmonized analysis of multiomics and environmental data to predictive and causal modeling of epigenetic change and design and control of targeted CRISPR/dCas-mediated interventions. When technical innovations are complemented with strict reporting criteria, cross-environment validation, and safety-by-design principles, they facilitate reproducible, interpretable, and ethically informed translational advances in health, agriculture, and environmental resilience.

2.6 Applications in Crop Improvement

2.6.1 Enhancing Resilience in Strategic Crops

The stability of global food systems is intrinsically linked to the performance of a handful of strategic crops such as wheat, rice, maize, and soybean, which collectively provide a substantial proportion of the world's caloric intake. The escalating pressures of climate change, including increased frequency of droughts, floods, heatwaves, and soil salinization, pose a direct threat to the productivity of these vital plants. Consequently, enhancing their resilience is not merely an agronomic objective but a critical imperative for international food security. While conventional breeding techniques have historically driven yield improvements, the pace of these gains is now insufficient to match the rapid onset of environmental stresses. The emergence of sophisticated biotechnological tools offers a paradigm shift, enabling the development of robust, stress-tolerant crop varieties with unprecedented speed and precision (Varshney et al., 2021).

2.6.1.1 Precision Breeding Through Genomic Tools

A cornerstone of this new approach is precision breeding, exemplified by technologies like CRISPR-Cas9 genome editing. This powerful tool functions as a molecular scalpel, allowing scientists to make targeted modifications to specific genes responsible for agronomic traits. Unlike earlier genetic engineering methods that often involved introducing DNA from other species, CRISPR-Cas9 can be used to edit a plant's own genes, for instance, by knocking out a gene that increases susceptibility to drought or by fine-tuning the expression of a gene involved in nutrient uptake. This precision not only increases the efficiency of breeding but also alleviates many regulatory and public concerns associated with transgenic organisms, potentially smoothing the path from laboratory to field (Tester & Langridge, 2010). Complementing genome editing, MAS remains a vital component of the modern breeder's toolkit. MAS utilizes molecular markers—specific DNA sequences linked to desirable traits—to screen plants at the seedling stage. This allows breeders to identify and select individuals with the genetic potential for traits like disease resistance or abiotic stress tolerance long before these characteristics are physically expressed, dramatically shortening breeding cycles. The power of MAS is further amplified by genomic prediction, a statistical approach that uses genome-wide marker data to forecast the performance of potential breeding lines. The successful deployment of drought-tolerant maize in sub-Saharan Africa and submergence-tolerant rice in South Asia stands as a testament to the efficacy of these genomic-assisted breeding strategies, which have delivered tangible yield benefits to farmers operating in challenging environments (Bohra et al., 2022).

2.6.1.2 Systems Biology and Holistic Crop Management

The complexity of plant responses to stress necessitates a holistic understanding that transcends single-gene analysis. This is being addressed through multiomics approaches, which integrate data from genomics (DNA sequence), transcriptomics (gene expression), proteomics (protein abundance), and metabolomics (metabolite profiles). By examining the interplay between these different biological layers, researchers can construct a comprehensive systems-level model of how a plant responds to, for example, water deficit or high salinity. This integrated view reveals key regulatory networks and metabolic pathways, pinpointing the most critical molecular targets for genetic improvement. For instance, multiomics studies have been instrumental in unraveling the complex mechanisms that confer heat tolerance in wheat and salt tolerance in barley, providing a roadmap for developing the next generation of climate-resilient varieties (Backer et al., 2018). Beyond direct genetic manipulation, sustainable crop resilience also involves harnessing biological synergies. The use of beneficial microorganisms, such as plant growth-promoting rhizobacteria (PGPR) and arbuscular mycorrhizal fungi, represents a promising biological strategy. These microbes form symbiotic relationships with plant roots, enhancing the plant's ability to absorb water and essential nutrients like phosphorus and nitrogen. Furthermore, they can induce systemic resistance in the plant, priming its immune system against pathogens, and contribute to soil health by improving its structure and fertility. The application of these microbial inoculants offers a pathway to reduce reliance on synthetic fertilizers and pesticides, aligning crop production with the principles of ecological intensification (Araus et al., 2018).

2.6.1.3 Advanced Phenotyping for Precision Agriculture

The ability to accurately measure plant traits—a field known as phenotyping—is crucial for evaluating the success of resilience-enhancing interventions. Traditional phenotyping is slow and labor-intensive, but technological advancements are revolutionizing this domain. High-throughput phenotyping platforms utilizing drones equipped with multispectral and hyperspectral cameras can rapidly scan large fields, capturing detailed information on plant health, water status, and biomass. These are complemented by ground-based IoT sensors that continuously monitor soil moisture and microclimatic conditions. When this torrent of phenotypic data is analyzed using machine learning algorithms, it becomes possible to detect subtle signs of stress early, monitor crop performance in real time, and make data-driven decisions on irrigation and nutrient management, ultimately closing the loop between genotype and phenotype. The translation of these technological innovations from research labs to farmers' fields requires a concerted effort. Significant public investment in research and development is essential, alongside regulatory frameworks that are both scientifically rigorous and pragmatic. Equally important are capacity-building initiatives designed to equip extension workers and farmers, especially in vulnerable developing regions, with the knowledge and skills to adopt and benefit from these new tools. The challenge of enhancing crop resilience is as much about social and institutional innovation as it is about scientific discovery.

2.6.2 Epigenetic Breeding and AI-Guided Selection

The conventional paradigm of crop improvement has largely focused on altering the DNA sequence itself. However, a more nuanced layer of genetic regulation, known as epigenetics, is now emerging as a powerful new frontier. Epigenetic breeding involves modulating gene expression without changing the underlying DNA code, utilizing mechanisms such as DNA methylation, histone modification, and small RNA molecules (Zhang et al., 2018). These epigenetic marks act as a molecular switchboard, influencing how genes are read and expressed in response to environmental cues. Crucially, some of these modifications can be stable and heritable, meaning that plants can "remember" past stress events and pass on this adaptive advantage to their offspring. This provides a rapid mechanism for acclimatization and offers a nontransgenic approach to enhancing crop resilience, which is particularly advantageous in jurisdictions with restrictive GMO policies.

2.6.2.1 Mechanisms and Applications of Epigenetic Modulation

Research has begun to uncover specific epigenetic patterns associated with valuable agronomic traits. In wheat, for example, distinct DNA methylation profiles have been correlated with enhanced drought tolerance. Under water-limited conditions, certain genes involved in water conservation and root development may be epigenetically activated, leading to improved water-use efficiency (Li et al., 2020). Similarly, in rice, histone modifications have been linked to increased salinity tolerance, suggesting a potential strategy for breeding crops suitable for cultivation in increasingly saline soils. These examples illustrate how epigenetic breeding can work in concert with traditional and molecular breeding, offering an additional tool to address the multifaceted challenge of environmental stress.

2.6.2.2 The Confluence of Epigenetics and Artificial Intelligence

The true potential of epigenetic breeding is unlocked when combined with the analytical power of AI. The datasets generated by epigenetic studies are vast and complex, encompassing information from thousands of genomic locations across different tissues and environmental conditions. Machine learning algorithms, such as random forests and deep neural networks, are uniquely suited to decipher these complex patterns (Montesinos-López et al., 2021). AI models can be trained to identify specific epigenetic signatures that predict desirable outcomes, such as high yield under drought conditions. This enables AI-guided selection, where breeders can screen thousands of plants based on their epigenetic profiles and predict the optimal crosses to make, drastically accelerating the breeding process and reducing the need for extensive field trials. Industry-led AI platforms are already being adapted for agricultural applications. IBM's Watson Decision Platform for Agriculture can integrate epigenetic data with weather and soil information to provide farmers with personalized recommendations. Meanwhile, tools like Google's DeepVariant, originally developed for analyzing genetic variation, are being refined to detect epigenetic modifications with base-pair accuracy, allowing for an unprecedented level of detail in epigenetic analysis (Singh et al., 2022). These technologies are vital for decoding the intricate dialogue between the epigenome and the environment, which is key to developing truly climate-resilient crops.

2.6.2.3 Precision Epigenome Engineering

The convergence of epigenetics with genome editing has given rise to a new field: epigenome engineering. Using a modified CRISPR-Cas9 system where the DNA-cutting enzyme is deactivated (dCas9) and fused to epigenetic modifiers, scientists can now target specific genes to add or remove epigenetic marks (Gallego-Bartolomé, 2020). This allows for precise activation or silencing of genes without altering the DNA sequence. For instance, researchers have used this technology to silence genes responsible for susceptibility to a fungal pathogen in maize, thereby enhancing disease resistance. A significant advantage of this approach is its potential reversibility, offering a level of control that is not possible with permanent genetic deletions. The synergy of epigenetic breeding and AI presents a compelling alternative to conventional methods. It shortens breeding cycles, provides deeper insights into the regulatory networks controlling complex traits, and navigates the regulatory hurdles associated with GMOs. As the climate continues to change, this integrated approach will be increasingly valuable for creating crops that are not only high-yielding but also capable of withstanding the unpredictable stresses of the future. Realizing this full potential will depend on continued innovation in high-throughput epigenetic profiling, the development of more dynamic AI models, and the establishment of clear regulatory pathways for epigenetically enhanced crops.

2.6.3 Data-Driven Decision-Making for Climate-Smart Agriculture

The concept of climate-smart agriculture (CSA) has emerged as a holistic framework for transforming agricultural systems to sustainably increase productivity, enhance resilience, and mitigate climate change. At the heart of this transformation is the shift toward data-driven decision-making. The advent of big data analytics, the Internet of Things (IoT), and advanced remote sensing technologies is moving farming away from a reliance on tradition

Table 2.1 Predictive analytics framework in agriculture: integrating machine learning for proactive farm management, targeted interventions, and sustainable farming practices.

Focus area	Predictive analytics in agriculture	Use of data and machine learning to anticipate future events	Enables proactive farm management
Analytical approach	Machine learning models	Analyze historical data, weather patterns, and current crop conditions	Forecasts pest invasions and disease outbreaks
Operational shift	Reactive → proactive management	Farmers take preventive actions based on predictions	Reduces losses and improves timing of interventions
Application	Targeted control measures	Implemented only when and where needed	Minimizes pesticide use and operational costs
Environmental impact	Sustainable farming practices	Reduced pesticide and chemical usage	Lowers ecological footprint of farming operations

and intuition toward a practice grounded in empirical, real-time information (Lobell et al., 2020). This paradigm enables the optimization of inputs like water, fertilizers, and pesticides, thereby boosting efficiency, reducing environmental impact, and improving farm profitability (Table 2.1).

2.6.3.1 The Infrastructure of Precision Agriculture

Modern data-driven farming relies on a sophisticated network of sensing technologies. Satellite imagery provides a macroscale view of crop health and land use, while drones offer higher-resolution data for individual fields. These are supplemented by a growing array of in-situ IoT sensors that monitor critical variables such as soil moisture, temperature, humidity, and nutrient levels at a granular level (Kamilaris et al., 2017). Initiatives like the Food and Agriculture Organization's (FAO) WaPOR portal democratize access to such data by using satellite information to monitor water productivity across Africa and the Near East, providing invaluable insights for irrigation management in water-scarce regions (FAO, 2021). This integrated sensor network creates a digital twin of the farm, allowing for precise, variable-rate application of resources—applying the right input, in the right amount, at the right place and time.

2.6.3.2 Predictive Analytics for Proactive Management

Beyond real-time monitoring, data analytics is increasingly predictive. Machine learning models are being deployed to forecast pest invasions and disease outbreaks by analyzing historical data, weather patterns, and current crop conditions (Sharma et al., 2020). This predictive capability allows farmers to move from a reactive to a proactive stance, implementing control measures only when and where they are likely to be needed. This targeted approach minimizes pesticide use, reduces costs, and lessens the ecological footprint of farming operations. Furthermore, these predictive models can be integrated into farm management software, providing users with actionable alerts and recommendations.

2.6.3.3 Transparency and Efficiency Through Blockchain

The application of data-driven technologies extends beyond the farm gate to the entire agricultural supply chain. Blockchain technology offers a secure, decentralized, and transparent ledger for recording every step of a product's journey—from seed planting and harvest conditions to processing, transportation, and retail (Tripoli & Schmidhuber, 2018). This immutability reduces opportunities for fraud, improves traceability in the event of a food safety incident, and provides consumers with verifiable information about the origin and production practices of their food. By streamlining logistics and reducing information asymmetries, blockchain can also help minimize food waste and ensure that a greater share of the final product value returns to the primary producer. For these technologies to achieve widespread impact, significant investment in rural digital infrastructure and farmer education is paramount, especially for smallholder farmers who are often the most vulnerable to climate shocks. Governments and private sector actors must collaborate to develop affordable, user-friendly solutions and create an enabling policy environment. As climate variability intensifies, the ability to make informed decisions based on robust data will become a cornerstone of resilient and sustainable global food systems.

2.7 Challenges, Ethics, and Future Directions

2.7.1 Data Limitations and Model Interpretability

The integration of AI and big data into agriculture, while promising, is fraught with significant challenges related to the data itself and the complexity of the models it fuels. A primary obstacle is the "data divide" in global agriculture. High-quality, annotated, and standardized datasets are abundant in well-funded research programs and large-scale commercial farms in developed nations but are severely lacking in developing regions where agricultural diversity is the greatest (Kussul et al., 2017). This disparity creates an inherent bias in AI models. When machine learning algorithms are trained predominantly on data from temperate, large-scale monoculture systems, their predictive accuracy plummets when applied to the complex, diverse, and often smallholder farming systems common in tropical regions (Chlingaryan et al., 2018). Issues of inconsistent data collection protocols, missing metadata, and sample bias further compromise the reliability and generalizability of agricultural AI solutions (Hernández Hernández et al., 2025). A second, equally critical challenge is the "black box" problem associated with sophisticated AI models, particularly deep neural networks. These models can deliver highly accurate predictions, but the internal logic by which they arrive at a specific recommendation—for example, to apply a certain fertilizer—is often opaque to the end user (Samek et al., 2019). For a farmer or an agronomist, trusting and acting upon a recommendation requires understanding the reasoning behind it. A lack of interpretability fosters skepticism and hinders adoption. The emerging field of Explainable AI (XAI) aims to address this by developing techniques such as Shapley Additive Explanations (SHAP) and Local Interpretable Model-agnostic Explanations (LIME), which help to illuminate which input factors were most influential in a model's decision (Arrieta et al., 2020). However, the application of these XAI methods in agricultural contexts is still in its infancy. Finally, the ambition of creating truly holistic decision-support tools requires the integration of multimodal data—genomic, phenotypic, environmental, and socioeconomic—into a unified analytical framework. This integration poses immense computational and methodological challenges. Developing models that can effectively learn from these diverse data types and provide coherent, actionable insights remains a key frontier for research (van Klompenburg et al., 2020). Overcoming these hurdles will necessitate a concerted focus on creating open-access data repositories, advancing federated learning techniques that can train models on decentralized data without compromising privacy, and designing user-centric AI interfaces that translate complex model outputs into clear, practical advice for farmers.

2.7.2 Regulatory and Ethical Considerations

The breakneck speed of innovation in agricultural biotechnology and digital agriculture has outstripped the development of corresponding regulatory and ethical frameworks, creating a landscape of uncertainty and potential risk. A prominent issue is the global regulatory patchwork governing new breeding techniques like CRISPR-Cas9. There is no international consensus on how to classify and regulate gene-edited crops. The European Union, for instance, has ruled that they fall under the stringent regulatory purview of its GMO legislation, while countries like the United States, Argentina, and Brazil have adopted product-based approaches that do not automatically classify such crops as GMOs (Sprink et al., 2022). This disharmony disrupts international trade, creates market uncertainty, and can stifle research and development, particularly in the public sector (Eckerstorfer et al., 2019). The rise of digital agriculture brings its own set of ethical dilemmas, primarily centered on data rights and equity. As agribusiness corporations deploy sensors and platforms that collect vast amounts of farm-level data, critical questions arise: Who owns this data? How is it being used and by whom? There is a palpable risk of exploitation, where large companies monetize insights derived from farmers' data without providing adequate compensation or benefit, potentially exacerbating the power imbalance between multinational corporations and individual growers (Wiseman et al., 2019). Moreover, if the algorithms that power PA are trained on biased datasets that underrepresent certain farming systems or regions, they can perpetuate or even amplify existing socioeconomic disparities (Carbonell, 2016). Policymakers must therefore prioritize the development of frameworks that ensure data sovereignty for farmers, promote transparency in data use, and mandate algorithmic fairness. Environmental ethics also remain a central concern, particularly for genetically engineered organisms. The potential for gene flow from engineered crops to wild relatives, the impact on nontarget organisms, and the long-term ecological consequences demand rigorous, precautionary risk assessment and post-release monitoring (Hilbeck et al., 2015). Balancing the promise of innovation with the precautionary principle requires ongoing, transparent dialogue and interdisciplinary collaboration among scientists, ethicists, policymakers, and the public to ensure that technological advancements genuinely serve the collective good without compromising ecological integrity or social justice.

2.7.3 Prospects for Global Food Security

The technological advancements discussed throughout this review hold immense promise for achieving global food security in the face of a growing population and a changing climate. The United Nations projects the global population to reach 9.7 billion by 2050, demanding an estimated 70% increase in food production, all while minimizing the environmental footprint of agriculture (FAO, 2018). Precision agriculture, driven by AI, IoT, and data analytics, is pivotal to this "sustainable intensification", offering pathways to optimize resource use, drastically reduce water waste and fertilizer runoff, and lower greenhouse gas emissions (Pretty et al., 2018). The demonstrated ability of AI-driven irrigation systems to achieve water savings of up to 30% in water-stressed regions is a powerful example of this potential (Smith et al., 2020). However, technology alone is an insufficient guarantor of food security. The benefits of innovation must be equitably distributed. Smallholder farmers, who are responsible for over 80% of the food produced in many developing countries, often face insurmountable barriers to adopting these technologies, including cost, lack of infrastructure, and limited technical knowledge (Lowder et al., 2021). Bridging this gap requires proactive measures such as public-private partnerships, targeted government subsidies, and the development of low-cost, appropriate technologies, such as mobile phone-based advisory services, tailored to the needs and constraints of small-scale producers (Carbonara & Pellegrino, 2020). The future food system will need to be both high-tech and ecologically robust. The development of climate-resilient crop varieties—such as heat-tolerant wheat and salinity-resistant rice—is essential for adapting to increasingly volatile weather patterns (Zhao et al., 2017). Simultaneously, the adoption of regenerative agricultural practices, including agroforestry, cover cropping, and reduced tillage, can enhance soil health, increase biodiversity, and sequester atmospheric carbon, thereby contributing to climate change mitigation (Giller et al., 2021). Ultimately, achieving the Sustainable Development Goal of Zero Hunger (SDG 2) by 2030 will depend on a holistic strategy that seamlessly integrates cutting-edge science with supportive policies, equitable economic models, and grassroots community engagement. The journey toward a food-secure future is a complex one, but through the responsible and inclusive application of innovation, it is a goal within our reach.

2.8 Conclusion

The intersection of AI and plant epigenetics is a pivotal moment in modern agricultural science at a time when climate change is accelerating. As discussed throughout this chapter, the instability of the climate imposes a massive level of stress on plant systems, resulting in modifications to their physiological and genetic responses due to complex epigenetic changes. The understanding of these dynamic molecular activities (e.g., represented by DNA methylation, histone modifications, and noncoding RNA pathways) represents a significant advance in improving the adaptability and resilience of plants. At the same time, this complexity of multiomics data and the dynamic nature of epigenomic response require a computational framework that can distribute insights into areas that no human would normally consider. This is where AI has an opportunity to transform. Tools powered by AI, as well as machine learning models, are valuable for analyzing large datasets generated from next-generation sequencing, methylome mapping, and transcriptomic profiling. AI's role in multiomics integrative analysis expedites the discovery of epigenetic biomarkers related to stress tolerance, while creating predictive models of plant responses in different climate situations. Therefore, in conjunction with experimental biology, AI algorithms enable predictive models describing how plants regulate gene expression plasticity to cope with environmental extremes. Additionally, AI-enabled CRISPR/dCas systems are new and important for precision epigenome editing, in which stress-tolerant traits can be very accurately regulated with minimal off-target compared to other means of modifying plant genomes. The uses for AI-assisted epigenetic regulation range from molecular knowledge to applied crop improvement strategies. Precision breeding, coupled with an integrative systems biology approach and accompanied by advanced phenotyping, allows for real-time monitoring and management of crop health and performance and adaptation. Additionally, the use of AI-based predictive analytics interfaced with blockchain-supported transparency and data-driven decision-making will transfer agriculture to the realm of climate-smart, sustainable, equity-driven production of food. Epigenetic breeding, driven by AI-guided selection, can create cultivars that are genetically robust and epigenetically adaptable—offering resilience that can be passed on to future generations. Nevertheless, these advanced technological applications will not be straightforward. These measures require addressing key issues, particularly the interpretability of sophisticated AI models, data heterogeneity, higher-quality datasets that are not always easily accessible, and the ethical implications of genetic and epigenetic modifications. It is also critical that we establish regulatory frameworks designed for universal acceptance and equitable access to AI-led innovations in biotechnology to avoid marginalizing agricultural systems with low resource capacity. Along with these issues, maintaining their focus on data privacy, algorithmic transparency, and environmental biosafety in this emerging paradigm should remain of the utmost importance. As we look ahead, the fusion of AI, epigenetics, and PA presents an unparalleled opportunity to reshape the future of plant science and food security. This interdisciplinary merger positions us to transition from reactive to proactive crop management—predicting stress before it occurs and designing resilience at molecular and ecosystem scales. AI-enabled epigenomic insights will give us a holistic understanding of plant-environment interactions that will allow us to design crops that can withstand the multifaceted pressures of climate change. In summary, AI-enabled precision plant epigenetic regulation is not only a step change in agricultural technology but a fundamental shift toward sustainable and adaptive agriculture. By coupling data-informed intelligence with biological complexity, we are setting the stage for an agricultural revolution—predictive, resilient, and tilted to fit the ever-changing rhythms of our planet. The synergy between AI, epigenetics, and agriculture embeds our ability to sustainably monitor, maintain, and attain nutritional stability within the next generation of global food systems in this era of climate uncertainty.

References

Alberts, B., Johnson, A., & Lewis, J. (2002). *Molecular biology of the cell* (4th ed.). Garland Science.

Araus, J. L., Kefauver, S. C., Zaman-Allah, M., Olsen, M. S., & Cairns, J. E. (2018). Translating high-throughput phenotyping into genetic gain. *Trends in Plant Science*, *23*(5), 451–466. https://doi.org/10.1016/j.tplants.2018.02.001

Arrieta, A. B., Díaz-Rodríguez, N., Del Ser, J., Bennetot, A., Tabik, S., Barbado, A., García, S., Gil-López, S., Molina, D., Benjamins, R., Chatila, R., & Herrera, F. (2020). Explainable artificial intelligence (XAI): Concepts, taxonomies, opportunities, and challenges toward responsible AI. *Information Fusion*, *58*, 82–115. https://doi.org/10.1016/j.inffus.2019.12.012

Ashapkin, V. V., Kutueva, L. I., Aleksandrushkina, N. I., & Vanyushin, B. F. (2020). Epigenetic mechanisms of plant adaptation to biotic and abiotic stresses. *International Journal of Molecular Sciences, 21*(20), 7457. https://doi.org/10.3390/ijms21207457

Backer, R., Rokem, J. S., Ilangumaran, G., Lamont, J., Praslickova, D., Ricci, E., & Smith, D. L. (2018). Plant growth-promoting rhizobacteria: Context, mechanisms of action, and roadmap to commercialization. *Frontiers in Plant Science, 9*, 1473. https://doi.org/10.3389/fpls.2018.01473

Banerjee, A., & Roychoudhury, A. (2017). Epigenetic regulation during salinity and drought stress in plants: Histone modifications and DNA methylation. *Plant Genetics, 11*, 199–204. https://doi.org/10.1016/j.plgene.2017.05.011

Bascompte, J. (2009). Disentangling the web of life. *Science, 325*, 416–419. https://doi.org/10.1126/science.1170749

Berg, M. P., Kiers, E. T., Driessen, G., Van der Heijden, M., Kooi, B. W., Kuenen, F., Liefting, M., Verhoef, H. A., & Ellers, J. (2010). Adapt or disperse: Understanding species persistence in a changing world. *Global Change Biology, 16*, 587–598. https://doi.org/10.1111/j.1365-2486.2009.02014.x

Bernatavichute, Y. V., Zhang, X., Cokus, S., Pellegrini, M., & Jacobsen, S. E. (2008). Genome-wide association of histone H3 lysine nine methylation with CHG DNA methylation in *arabidopsis thaliana. PLoS One, 3*, e3156. https://doi.org/10.1371/journal.pone.0003156

Berr, A., Shafiq, S., & Shen, W. H. (2011). Histone modifications in transcriptional activation during plant development. *Biochimica et Biophysica Acta, 1809*, 567–576. https://doi.org/10.1016/j.bbagrm.2011.07.001

Bohra, A., Saxena, K. B., Varshney, R. K., & Saxena, R. K. (2022). Genomics-assisted breeding for pigeonpea improvement. *Theoretical and Applied Genetics, 135*(11), 3879–3902. https://doi.org/10.1007/s00122-020-03563-7

Bonasio, R., Tu, S., & Reinberg, D. (2010). Molecular signals of epigenetic states. *Science, 330*, 612–616. https://doi.org/10.1126/science.1191078

Budak, H., Khan, Z., & Kantar, M. (2014). History and current status of wheat miRNAs using next-generation sequencing and their roles in development and stress. *Briefings in Functional Genomics, 14*, 189–198. https://doi.org/10.1093/bfgp/elu021

Cao, X., & Jacobsen, S. E. (2002). Role of the *arabidopsis* DRM methyltransferases in de novo DNA methylation and gene silencing. *Current Biology, 12*, 1138–1144. https://doi.org/10.1016/s0960-9822(02)00925-9

Carbonara, N., & Pellegrino, R. (2020). The role of public private partnerships in fostering innovation. *Construction Management and Economics, 38*(2), 140–156. https://doi.org/10.1080/01446193.2019.1610184

Carbonell, I. M. (2016). The ethics of big data in big agriculture. *Internet Policy Review, 5*(1), 1–13. https://doi.org/10.14763/2016.1.405

Chinnusamy, V., Zhu, J., Zhou, T., & Zhu, J. K. (2007). Small RNAs: Big role in abiotic stress tolerance of plants. In M. A. Jenks, P. M. Hasegawa, & S. M. Jain (Eds.), *Advances in molecular breeding toward drought and salt tolerant crops* (pp. 223–260). Springer Netherlands.

Chlingaryan, A., Sukkarieh, S., & Whelan, B. (2018). Machine learning approaches for crop yield prediction and nitrogen status estimation in precision agriculture: A review. *Computers and Electronics in Agriculture, 151*, 61–69. https://doi.org/10.1016/j.compag.2018.05.012

Ding, Y., Avramova, Z., & Fromm, M. (2011). The arabidopsis trithorax-like factor ATX1 functions in dehydration stress responses via ABA-dependent and ABA-independent pathways. *Plant Journal: For Cell and Molecular Biology, 66*(5), 735–744. https://doi.org/10.1111/j.1365-313x.2011.04534.x

Du, J., Zhong, X., Bernatavichute, Y. V., Stroud, H., Feng, S., Caro, E., Vashisht, A. A., Terragni, J., Chin, H. G., Tu, A., & Hetzel, J. (2012). Dual binding of chromomethylase domains to H3K9me2-containing nucleosomes directs DNA methylation in plants. *Cell*, *151*, 167–180. https://doi.org/10.1016/j.cell.2012.07.034

Eckerstorfer, M. F., Engelhard, M., Heissenberger, A., Simon, S., & Teichmann, H. (2019). Plants developed by new genetic modification techniques—comparison of existing regulatory frameworks in the EU and non-EU countries. *Frontiers in Bioengineering and Biotechnology*, *7*, 26. https://doi.org/10.3389/fbioe.2019.00026

Eichten, S. R., & Springer, N. M. (2015). Minimal evidence for consistent changes in maize DNA methylation patterns following environmental stress. *Frontiers in Plant Science*, *6*, 308. https://doi.org/10.3389/fpls.2015.00308

FAO. (2018). *The future of food and agriculture: Alternative pathways to 2050*. Food and Agriculture Organization.

FAO. (2021). *WaPOR: Remote sensing for water productivity*. Food and Agriculture Organization.

Feng, S., & Jacobsen, S. E. (2011). Epigenetic modifications in plants: An evolutionary perspective. *Current Opinion in Plant Biology*, *14*, 179–186. https://doi.org/10.1016/j.pbi.2010.12.002

Fenton, A., & Spencer, M. (2010). Linking population, community and ecosystem ecology within mainstream ecology. In D. Raffaelli & C. L. J. Frid (Eds.), *Ecosystem ecology, a new synthesis* (pp. 19–39). Cambridge University Press.

Ferreira, L. J., Azevedo, V., Maroco, J., Oliveira, M. M., & Santos, A. P. (2015). Salt tolerant and sensitive rice varieties display differential methylome flexibility under salt stress. *PLoS One*, *10*(5), e0124060. https://doi.org/10.1371/journal.pone.0124060

Folsom, J. J., Begcy, K., Hao, X., Wang, D., & Walia, H. (2014). Rice fertilization-independent endosperm1 regulates seed size under heat stress by controlling early endosperm development. *Plant Physiology*, *165*(1), 238–248. https://doi.org/10.1104/pp.113.232413

Friedrich, T., Faivre, L., Bäurle, I., & Schubert, D. (2019). Chromatin-based mechanisms of temperature memory in plants. *Plant, Cell & Environment*, *42*, 762–770. https://doi.org/10.1111/pce.13373

Fu, R., Zhang, M., Zhao, Y., He, X., Ding, C., Wang, S., Feng, Y., Song, X., Li, P., & Wang, B. (2017). Identification of salt tolerance-related microRNAs and their targets in maize (*Zea mays* L.) using high-throughput sequencing and degradome analysis. *Frontiers in Plant Science*, *8*, 864. https://doi.org/10.3389/fpls.2017.00864

Gallego-Bartolomé, J. (2020). DNA methylation in plants: Mechanisms and tools for targeted manipulation. *New Phytologist*, *227*(1), 38–44. https://doi.org/10.1111/nph.16529

Gao, G., Li, J., Li, H., Li, F., Xu, K., Yan, G., Chen, B., Qiao, J., & Wu, X. (2014). Comparison of the heat stress induced variations in DNA methylation between heat-tolerant and heat-sensitive rapeseed seedlings. *Breeding Science*, *64*(2), 125–133. https://doi.org/10.1270/jsbbs.64.125

Gayacharan, A. J., & Joel, A. J. (2013). Epigenetic responses to drought stress in rice (*Oryza sativa* L.). *Physiology and Molecular Biology of Plants*, *19*(3), 379–387. https://doi.org/10.1007/s12298-013-0176-4

Gehring, M., Reik, W., & Henikoff, S. (2009). DNA demethylation by DNA repair. *Trends in Genetics*, *25*, 82–90. https://doi.org/10.1016/j.tig.2008.12.001

Giller, K. E., Hijbeek, R., Andersson, J. A., & Sumberg, J. (2021). Regenerative agriculture: An agronomic perspective. *Outlook on Agriculture*, *50*(1), 13–25. https://doi.org/10.1177/0030727021998063

Guan, Z., Wu, L., Zhao, H., He, M., & Fan, J. (2025). Enhancing collaborative semantics of language model-driven recommendations via graph-aware learning. *IEEE Transactions on Knowledge and Data Engineering*. https://doi.org/10.1109/TKDE.2025.3581606

Guo, W. J., Li, D. W., Xie, S., Yang, L. W., Li, C., Tian, J., Pu, L., & Gu, X. F. (2022). Artificial intelligence accelerates epigenetics and plant breeding. https://doi.org/10.13304/j.nykjdb.2022.1031

Gyarmati, G., & Mizik, T. (2020, June). The present and future of the precision agriculture. *2020 IEEE 15th International Conference of System of Systems Engineering (SoSE)*, 593–596. IEEE. https://doi.org/10.1109/SoSE50414.2020.9130481

Hernández Hernández, G. C., Gómez Gómez, J., & Jiménez-Cabas, J. (2025). Predictive models based on artificial intelligence to estimate crop yield: A literature review. *Agriculture, 15*(23), 2438. https://doi.org/10.3390/agriculture15232438

Hilbeck, A., Binimelis, R., Defarge, N., Steinbrecher, R., Székács, A., Wickson, F., & Wynne, B. (2015). No scientific consensus on GMO safety. *Environmental Sciences Europe, 27*(1), 1–6. https://doi.org/10.1186/s12302-014-0034-1

Horlacher, M., Wagner, N., Moyon, L., Kuret, K., Goedert, N., Salvatore, M., Ule, J., Gagneur, J., Winther, O., & Marsico, A. (2023). Towards in silico CLIP-seq: Predicting protein-RNA interaction via sequence-to-signal learning. *Genome Biology, 24*(1), 180. https://doi.org/10.1186/s13059-023-03015-7

Iwasaki, M., & Paszkowski, J. (2014). Epigenetic memory in plants. *The EMBO Journal, 33*, 1–12. https://doi.org/10.15252/embj.201488883

Jewkes, R., Sikweyiya, Y., Morrell, R., & Dunkle, K. (2011). Gender inequitable masculinity and sexual entitlement in rape perpetration South Africa: Findings of a cross-sectional study. *PLoS One, 6*(12), e29590. https://doi.org/10.1371/journal.pone.0029590

Kamilaris, A., Kartakoullis, A., & Prenafeta-Boldú, F. X. (2017). A review on the practice of big data analysis in agriculture. *Computers and Electronics in Agriculture, 143*, 23–37. https://doi.org/10.1016/j.compag.2017.09.037

Kareiva, P. M., Kingsolver, J. G., & Huey, R. B. (Eds.). (1993). *Biotic interactions and global change*. Sinauer Associates Inc.

Khraiwesh, B., Arif, M. A., Seumel, G. I., Ossowski, S., Weigel, D., Reski, R., & Frank, W. (2010). Transcriptional control of gene expression by microRNAs. *Cell, 140*, 111–122. https://doi.org/10.1016/j.cell.2009.12.023

Khraiwesh, B., Zhu, J. K., & Zhu, J. (2012). Role of miRNAs and siRNAs in biotic and abiotic stress responses of plants. *Biochimica et Biophysica Acta, 1819*, 137–148. https://doi.org/10.1016/j.bbagrm.2011.05.001

Kim, J. M., To, T. K., Ishida, J., Morosawa, T., Kawashima, M., Matsui, A., Toyoda, T., Kimura, H., Shinozaki, K., & Seki, M. (2008). Alterations of lysine modifications on the histone H3 N-tail under drought stress conditions in *arabidopsis thaliana*. *Plant and Cell Physiology, 49*(10), 1580–1588. https://doi.org/10.1093/pcp/pcn133

Kim, J. M., To, T. K., Ishida, J., Matsui, A., Kimura, H., & Seki, M. (2012). Transition of chromatin status during the process of recovery from drought stress in *arabidopsis thaliana*. *Plant and Cell Physiology, 53*(5), 847–856. https://doi.org/10.1093/pcp/pcs053

Kipf, T. N., & Welling, M. (2016). Variational graph auto-encoders. *arXiv preprint arXiv: 1611.07308*. https://doi.org/10.48550/arXiv.1611.07308

van Klompenburg, T., Kassahun, A., & Catal, C. (2020). Crop yield prediction using machine learning: A systematic literature review. *Computers and Electronics in Agriculture, 177*, 105709. https://doi.org/10.1016/j.compag.2020.105709

Kumar, S., Beena, A. S., Awana, M., & Singh, A. (2017). Salt-induced tissue-specific cytosine methylation downregulates expression of HKT genes in contrasting wheat (*Triticum aestivum* L.) genotypes. *DNA and Cell Biology*, *36*(4), 283–294. https://doi.org/10.1089/dna.2016.3505

Kumar, V., Khare, T., Shriram, V., & Wani, S. H. (2018). Plant small RNAs: The essential epigenetic regulators of gene expression for salt-stress responses and tolerance. *Plant Cell Reports*, *37*, 61–75. https://doi.org/10.1007/s00299-017-2210-4

Kussul, N., Lavreniuk, M., Skakun, S., & Shelestov, A. (2017). Deep learning classification of land cover and crop types using remote sensing data. *IEEE Geoscience and Remote Sensing Letters*, *14*(5), 778–782. https://doi.org/10.1109/LGRS.2017.2681128

Lee, J. H., Mazarei, M., Pfotenhauer, A. C., Dorrough, A. B., Poindexter, M. R., Hewezi, T., Lenaghan, S. C., Graham, D. E., & Stewart, C. N., Jr. (2020). Epigenetic footprints of CRISPR/Cas9-mediated genome editing in plants. *Frontiers in Plant Science*, *10*, 1720. https://doi.org/10.3389/fpls.2019.01720

Li, X., Xie, Y., Zhu, Q., & Liu, Y. G. (2020). Targeted genome editing in genes and cis-regulatory regions improves qualitative and quantitative traits in crops. *Molecular Plant*, *13*(6), 811–815. https://doi.org/10.1016/j.molp.2020.04.004

Lobell, D. B., Thau, D., Seifert, C., Engle, E., & Little, B. (2020). A scalable satellite-based crop yield mapper. *Remote Sensing of Environment*, *164*, 324–333. https://doi.org/10.1016/j.rse.2015.04.021

Lowder, S. K., Sánchez, M. V., & Bertini, R. (2021). Which farms feed the world and has farmland become more concentrated? *World Development*, *142*, 105455. https://doi.org/10.1016/j.worlddev.2021.105455

Luo, X., & He, Y. (2020). Experiencing winter for spring flowering: A molecular epigenetic perspective on vernalization. *Journal of Integrative Plant Biology*, *62*, 104–117. https://doi.org/10.1111/jipb.12896

Martella, A., Firth, M., Taylor, B. J., Goppert, A., Cuomo, E. M., Roth, R. G., Dickson, A. J., & Fisher, D. I. (2019). Systematic evaluation of CRISPRa and CRISPRi modalities enables development of a multiplexed, orthogonal gene activation and repression system. *ACS Synthetic Biology*, *8*(9), 1998–2006. https://doi.org/10.1021/acssynbio.8b00527

Mastan, S. G., Rathore, M. S., Bhatt, V. D., Yadav, P., & Chikara, J. (2012). Assessment of changes in DNA methylation by methylation-sensitive amplification polymorphism in *Jatropha curcas* L. subjected to salinity stress. *Gene*, *508*, 125–129. https://doi.org/10.1016/j.gene.2012.07.063

Mathieu, O., Reinders, J., Čaikovski, M., Smathajitt, C., & Paszkowski, J. (2007). Transgenerational stability of the *arabidopsis* epigenome is coordinated by CG methylation. *Cell*, *130*, 851–862. https://doi.org/10.1016/j.cell.2007.07.007

Matzke, M. A., Kanno, T., & Matzke, A. J. (2015). RNA-directed DNA methylation: The evolution of a complex epigenetic pathway in flowering plants. *Annual Review of Plant Biology*, *66*, 243–267. https://doi.org/10.1146/annurev-arplant-043014-114633

Maurício, J., Domingues, I., & Bernardino, J. (2023). Comparing vision transformers and convolutional neural networks for image classification: A literature review. *Applied Sciences*, *13*(9), 5521. https://doi.org/10.3390/app13095521

Meyer, P. (2015). Epigenetic variation and environmental change. *Journal of Experimental Botany*, *66*, 3541–3548. https://doi.org/10.1093/jxb/eru502

Mirouze, M., & Paszkowski, J. (2011). Epigenetic contribution to stress adaptation in plants. *Current Opinion in Plant Biology*, *14*, 267–274. https://doi.org/10.1016/j.pbi.2011.03.004

Montesinos-López, O. A., Montesinos-López, A., Pérez-Rodríguez, P., Barrón-López, J. A., & Martini, J. W. R. (2021). A review of deep learning applications for genomic selection. *BMC Genomics*, *22*(1), 1–23. https://doi.org/10.1186/s12864-020-07319-x

Montoya, J. M., & Raffaelli, D. (2010). Climate change, biotic interactions and ecosystem services. *Philosophical Transactions of the Royal Society, B: Biological Sciences, 365*(1549), 2013–2018. https://doi.org/10.1098/rstb.2010.0114

Montoya, J. M., Pimm, S. L., & Sole´, R. V. (2006). Ecological networks and their fragility. *Nature, 442*, 259–264. https://doi.org/10.1038/nature04927

Naydenov, M., Baev, V., Apostolova, E., Gospodinova, N., Sablok, G., Gozmanova, M., & Yahubyan, G. (2015). High-temperature effect on genes engaged in DNA methylation and affected by DNA methylation in *arabidopsis*. *Plant Physiology and Biochemistry, 87*, 102–108. https://doi.org/10.1016/j.plaphy.2014.12.022

Petchey, O. L., Brose, U., & Rall, B. C. (2010). Predicting the effects of temperature on food web connectance. *Philosophical Transactions of the Royal Society, B: Biological Sciences, 365*, 2081–2091. https://doi.org/10.1098/rstb.2010.0011

Pikaard, C. S., & Scheid, O. M. (2014). Epigenetic regulation in plants. *Cold Spring Harbor Perspectives in Biology, 6*(12), a019315. https://doi.org/10.1101/cshperspect.a019315

Pimm, S. L. (1991). *The balance of nature?* Chicago University Press.

Popova, O. V., Dinh, H. Q., Aufsatz, W., & Jonak, C. (2013). The RdDM pathway is required for basal heat tolerance in arabidopsis. *Molecular Plant, 6*, 396–410. https://doi.org/10.1093/mp/sst023

Pretty, J., Benton, T. G., Bharucha, Z. P., Dicks, L. V., Flora, C. B., Godfray, H. C. J., & Wratten, S. (2018). Global assessment of agricultural system redesign for sustainable intensification. *Nature Sustainability, 1*(8), 441–446. https://doi.org/10.1038/s41893-018-0114-0

Rendina González, A. P., Preite, V., Verhoeven, K. J., & Latzel, V. (2018). Transgenerational effects and epigenetic memory in the clonal plant *Trifolium repens*. *Frontiers in Plant Science, 9*, 1677. https://doi.org/10.3389/fpls.2018.01677

Samek, W., Montavon, G., Vedaldi, A., Hansen, L. K., & Müller, K. R. (2019). *Explainable AI: Interpreting, explaining and visualizing deep learning*. Springer Nature.

Schmitz, O. J., Post, E., Burns, C. E., & Johnston, K. M. (2003). Ecosystem responses to global climate change: Moving beyond color mapping. *BioScience, 53*, 1199–1205. https://doi.org/10.1641/0006-3568(2003)053[1199:ERTGCC]2.0.CO;2

Shah, J. M. (2021). Epimutations and mutations, nurturing phenotypic diversity. *Genetica, 150*(3), 171–181. https://doi.org/10.1007/s10709-021-00124-8

Shanker, A. K., Bhanu, D., & Maheswari, M. (2020). Epigenetics and transgenerational memory in plants under heat stress. *Plant Physiology Reports, 25*(4), 583–593. https://doi.org/10.1007/s40502-020-00557-x

Sharma, A., Jain, A., Gupta, P., & Chowdary, V. (2020). Machine learning applications for precision agriculture: A comprehensive review. *IEEE Access, 9*, 4843–4873. https://doi.org/10.1109/ACCESS.2020.3048415

Singh, A., Ganapathysubramanian, B., Singh, A. K., & Sarkar, S. (2022). Machine learning for high-throughput stress phenotyping in plants. *Trends in Plant Science, 27*(1), 91–101. https://doi.org/10.1016/j.tplants.2015.10.015

Smith, P., Calvin, K., Nkem, J., Campbell, D., Cherubini, F., Grassi, G., Korotkov, V., Le Hoang, A., Lwasa, S., McElwee, P., & Nkonya, E. (2020). Which practices co-deliver food security, climate change mitigation and adaptation, and combat land degradation? *Global Change Biology, 26*(3), 1532–1575. https://doi.org/10.1111/gcb.14878

Song, Y., Ji, D., Li, S., Wang, P., Li, Q., & Xiang, F. (2012). The dynamic changes of DNA methylation and histone modifications of salt responsive transcription factor genes in soybean. *PLoS One, 7*(7), e41274. https://doi.org/10.1371/journal.pone.0041274

Sprink, T., Eriksson, D., Schiemann, J., & Hartung, F. (2022). Regulatory hurdles for genome editing: A case study of the EU. *In Vitro Cellular & Developmental Biology—Plant, 58*(2), 287–295.

Stroud, H., Do, T., Du, J., Zhong, X., Feng, S., Johnson, L., Patel, D. J., & Jacobsen, S. E. (2014). Non-CG methylation patterns shape the epigenetic landscape in *arabidopsis*. *Nature Structural & Molecular Biology, 21*, 64–72. https://doi.org/10.1038/nsmb.2735

Suji, K. K., & Joel, A. J. (2010). An epigenetic change in rice cultivars under water stress conditions. *Electronic Journal of Plant Breeding, 1*(4), 1142–1143.

Tang, K., Lang, Z., Zhang, H., & Zhu, J. K. (2016). The DNA demethylase ROS1 targets genomic regions with distinct chromatin modifications. *Nature Plants, 2*, 16169. https://doi.org/10.1038/nplants.2016.169

Tester, M., & Langridge, P. (2010). Breeding technologies to increase crop production in a changing world. *Science, 327*(5967), 818–822. https://doi.org/10.1126/science.1183700

Thiebaut, F., Hemerly, A. S., & Ferreira, P. C. G. (2019). A role for epigenetic regulation in the adaptation and stress responses of non-model plants. *Frontiers in Plant Science, 10*, 246–253. https://doi.org/10.3389/fpls.2019.00246

To, T. K., & Kim, J. M. (2014). Epigenetic regulation of gene responsiveness in arabidopsis. *Frontiers in Plant Science, 4*, 548. https://doi.org/10.3389/fpls.2013.00548

Tripoli, M., & Schmidhuber, J. (2018). *Emerging opportunities for the application of blockchain in the agri-food industry*. FAO and ICTSD.

Tylianakis, J. M., Didham, R. K., Bascompte, J., & Wardle, D. A. (2008). Global change and species interactions in terrestrial ecosystems. *Ecology Letters, 11*, 1351–1363. https://doi.org/10.1111/j.1461-0248.2008.01250.x

Vaillant, I., & Paszkowski, J. (2007). Role of histone and DNA methylation in gene regulation. *Current Opinion in Plant Biology, 10*, 528–533. https://doi.org/10.1016/j.pbi.2007.06.008

Vanyushin, B. F., & Ashapkin, V. V. (2011). DNA methylation in higher plants: Past, present and future. *Biochimica et Biophysica Acta, 1809*, 360–368. https://doi.org/10.1016/j.bbagrm.2011.04.006

Varshney, R. K., Sinha, P., Singh, V. K., Kumar, A., Zhang, Q., & Bennetzen, J. L. (2021). 5Gs for crop genetic improvement. *Current Opinion in Plant Biology, 56*, 190–196. https://doi.org/10.1016/j.pbi.2019.12.004

Verhoeven, K. J., Jansen, J. J., van Dijk, P. J., & Biere, A. (2010). Stress-induced DNA methylation changes and their heritability in asexual dandelions. *The New Phytologist, 185*, 1108–1118. https://doi.org/10.1111/j.1469-8137.2009.03121.x

Voinnet, O. (2009). Origin, biogenesis, and activity of plant microRNAs. *Cell, 136*, 669–687. https://doi.org/10.1016/j.cell.2009.01.046

Wada, Y., Miyamoto, K., Kusano, T., & Sano, H. (2004). Association between up-regulation of stress-responsive genes and hypomethylation of genomic DNA in tobacco plants. *Molecular Genetics and Genomics, 271*(6), 658–666. https://doi.org/10.1007/s00438-004-1018-4

Walther, G.-R. (2010). Community and ecosystem responses to recent climate change. *Philosophical Transactions of the Royal Society B, 365*, 2019–2024. https://doi.org/10.1098/rstb.2010.0021

Wang, W., Zhao, X., Pan, Y., Zhu, L., Fu, B., & Li, Z. (2011a). DNA methylation changes detected by methylation-sensitive amplified polymorphism in two contrasting rice genotypes under salt stress. *Journal of Genetics and Genomics, 38*(9), 419–424. https://doi.org/10.1016/j.jgg.2011.07.006

Wang, W.-S., Pan, Y.-J., Zhao, X.-Q., Dwivedi, D., Zhu, L.-H., Ali, J., Fu, B.-Y., & Li, Z.-K. (2011b). Drought-induced site-specific DNA methylation and its association with drought tolerance in rice (*Oryza sativa* L.). *Journal of Experimental Botany, 62*, 1951–1960. https://doi.org/10.1093/jxb/erq391

Wang, W., Huang, F., Qin, Q., Zhao, X., Li, Z., & Fu, B. (2015). Comparative analysis of DNA methylation changes in two rice genotypes under salt stress and subsequent recovery. *Biochemical and Biophysical Research Communications, 465*(4), 790–796. https://doi.org/10.1016/j.bbrc.2015.08.089

Wang, X., Xin, C., Cai, J., Zhou, Q., Dai, T., Cao, W., & Jiang, D. (2016). Heat priming induces trans-generational tolerance to high temperature stress in wheat. *Frontiers in Plant Science, 7*. https://doi.org/10.3389/fpls.2016.00501

Wendte, J. M., & Pikaard, C. S. (2017). The RNAs of RNA-directed DNA methylation. *Biochimica et Biophysica Acta, 1860*, 140–148. https://doi.org/10.1016/j.bbagrm.2016.08.004

Wiseman, L., Sanderson, J., Zhang, A., & Jakku, E. (2019). Farmers and their data: An examination of farmers' reluctance to share their data through the lens of the laws impacting smart farming. *NJAS: Wageningen Journal of Life Sciences, 90*, 100301. https://doi.org/10.1016/j.njas.2019.04.007

Woo, H. R., Dittmer, T. A., & Richards, E. J. (2008). Three SRA-domain methylcytosine-binding proteins cooperate to maintain global CpG methylation and epigenetic silencing in *arabidopsis*. *PLoS Genetics, 4*, e1000156. https://doi.org/10.1371/journal.pgen.1000156

Wootton, J. T., & Emmerson, M. E. (2005). Measurement of interaction strength in nature. *Annual Review of Ecology, Evolution, and Systematics, 36*, 419–444. https://doi.org/10.1146/annurev.ecolsys.36.091704.175535

Xu, R., Wang, Y., Zheng, H., Lu, W., Wu, C., Huang, J., Yan, K., Yang, G., & Zheng, C. (2015). Salt-induced transcription factor MYB74 is regulated by the RNA-directed DNA methylation pathway in arabidopsis. *Journal of Experimental Botany, 66*, 5997–6008. https://doi.org/10.1093/jxb/erv312

Zarco-Tejada, P. J., Hubbard, N., & Loudjani, P. (2014). *Precision agriculture: An opportunity for EU farmers—potential support with the CAP 2014-2020*. European Parliament.

Zemach, A., Kim, M. Y., Hsieh, P. H., Coleman-Derr, D., Eshed-Williams, L., Thao, K., Harmer, S. L., & Zilberman, D. (2013). The *arabidopsis* nucleosome remodeler DDM1 allows DNA methyltransferases to access H1-containing heterochromatin. *Cell, 153*, 193–205. https://doi.org/10.1016/j.cell.2013.02.033

Zhang, X., Clarenz, O., Cokus, S., Bernatavichute, Y. V., Pellegrini, M., Goodrich, J., & Jacobsen, S. E. (2007a). Whole-genome analysis of histone H3 lysine 27 trimethylation in *arabidopsis*. *PLoS Biology, 5*, e129. https://doi.org/10.1371/journal.pbio.0050129

Zhang, X., Germann, S., Blus, B. J., Khorasanizadeh, S., Gaudin, V., & Jacobsen, S. E. (2007b). The *arabidopsis* LHP1 protein colocalizes with histone H3 Lys27 trimethylation. *Nature Structural & Molecular Biology, 14*, 869–871. https://doi.org/10.1038/nsmb1283

Zhang, X., Bernatavichute, Y. V., Cokus, S., Pellegrini, M., & Jacobsen, S. E. (2009). Genome-wide analysis of mono-, di- and trimethylation of histone H3 lysine 4 in *arabidopsis thaliana*. *Genome Biology, 10*, R62. https://doi.org/10.1186/gb-2009-10-6-r62

Zhang, H., Lang, Z., & Zhu, J. K. (2018). Dynamics and function of DNA methylation in plants. *Nature Reviews Molecular Cell Biology, 19*(8), 489–506. https://doi.org/10.1038/s41580-018-0016-z

Zhao, C., Liu, B., Piao, S., Wang, X., Lobell, D. B., Huang, Y., & Asseng, S. (2017). Temperature increase reduces global yields of major crops in four independent estimates. *Proceedings of the National Academy of Sciences, 114*(35), 9326–9331. https://doi.org/10.1073/pnas.1701762114

3

AI-Driven Plant Epigenome Engineering for Developing Resilient Crops

Neslihan Turgut Kara and Burcu Arıkan

Faculty of Science, Department of Molecular Biology and Genetics, Istanbul University, Istanbul, Türkiye

3.1 Introduction: Climate Change and the Need for Epigenetic Innovation

The world population is expected to reach 10 billion by 2050 (Lutz & Samir, 2010). The food producer-to-consumer ratio is constantly decreasing, and maintaining higher production of agricultural products using less land and labor is increasing rapidly. Besides this, other severe challenges for agricultural production worldwide involve negative weather conditions and the growing frequency of diseases and pests due to the effects of climate change (Raza et al., 2020). These stress factors disturb all the physiological, biochemical, molecular, and cellular processes of plants, which ultimately reduce overall growth and productivity (Raza et al., 2020, 2021). However, existing plant species and agricultural practices are insufficient to cope with these problems (Voss-Fels et al., 2019).

From a plant breeding perspective, the loss of genetic diversity, the time taken to develop new and superior varieties, and the unsuitability of agricultural management practices being faced highlight the need for new technologies and approaches that can enable rapid crop improvement (Voss-Fels et al., 2019; Williams et al., 2024).

Plant breeding is the science and art of developing improved varieties based on the selection of desired traits of a particular species (Breseghello & Coelho, 2013). This process utilizes the genetic diversity that naturally exists within the species and, traditionally, is achieved through techniques such as artificial pollination, embryo culture, chromosome doubling, and hybridization. Some of the main characteristics for which traits are targeted in breeding studies include yield, quality, durability, resistance to pests, tolerance to abiotic stresses, and ease of processing (Anand et al., 2023; Breseghello & Coelho, 2013). Also, these factors are a major component that captures the strategic relationship between the success of the product and the food supply. However, many such breeding tools that earlier helped accelerate gains in yields are presently inadequate in view of stress factors due to climate change and the growth of the human population. One of the leading methods in this regard is speed breeding, which reduces the plant generation cycle. Thus, it offers rapid generation production, thereby accelerating agricultural research as a whole (Alahmad et al., 2018; Williams et al., 2024).

Nowadays, new-generation methods that utilize omics technologies are adopted for improving agriculture. In light of a growing world population, the latest scientific advances in plant epigenetics must be put to effective use. An increasing number of studies show that epigenetics contributes to the development of important traits in many plant species. The definition of epigenetics includes gene expression changes that are stably passed across generations with no change in the DNA sequence. Plant epigenetics brings together different areas of research and helps us understand how plants can adapt their phenotypes to extreme stress conditions without altering their DNA sequences (Abdulraheem et al., 2024).

Recent insights from genomic, transcriptomic, proteomic, and metabolomic studies are facilitating a deeper understanding of the molecular basis of stress tolerance and adaptation in plants. However, this data explosion also creates significant challenges in terms of computation, data integration, and accessibility.

The key bottlenecks are solved by AI-driven approaches such as machine learning (ML), deep learning (DL), and large language models (LLMs), which allow for real-time pattern recognition, predictive modeling, and multiomics data integration and accelerate the pace of breeding climate-resilient plant varieties (Thingujam et al., 2025). Integration of these digital innovations with plant genomics thus forms the backbone of AI-assisted precision breeding, which makes possible the creation of next-generation resilient plant varieties, faster to produce, more targeted, and ethically safe (Zafar et al., 2025).

The significance of AI-based techniques in determining targets for epigenome engineering is illustrated in this chapter. Increasing access to scientific databases facilitates the collection of data from various studies, improving the diversity and dependability of methods and preventing experimental errors. For tasks like identifying CRISPR targets and predicting possible epigenomic modification sites, AI applications like ML, DL, and LLMs have been developed. Remarkably, the majority of existing tools for epigenomic target discovery were published by 2025. While studies on genetic target identification exist, examples directly focusing on epigenomic target prediction for the development of resilient plant phenotypes are still limited. This section aims to provide an up-to-date overview of advances in the field and future research potential by reviewing current AI-based approaches and the opportunities these technologies offer for enhancing stress tolerance in plant epigenome engineering.

3.2 Epigenetic Mechanisms in Plant Stress Response

Plants respond to environmental stress factors not only through genetic regulation but also through epigenetic mechanisms. Epigenetics denotes the change in gene expression and cellular functions through DNA and histone modifications and the synthesis of noncoding RNAs without altering the DNA sequence. If these modifications occur across the whole genome, they are referred to as epigenomics. These modifications are crucial in plant development and in the responses of plants to environmental stresses (Kumar, 2019).

DNA methylation in plants functions as a key epigenetic mechanism for regulating responses to environmental stress conditions and in gene silencing. DNA methylation typically occurs through the addition of a methyl group to cytosine bases in CG (or CpG), CHG, or CHH contexts (where H corresponds to A, T, or C), with 5-methylcytosine (5mC) being the predominant form. Besides, N6-methyladenine (6mA) is another DNA modification that plays a critical role in genome stability, DNA repair, and stress response regulation. Methylation events are basically maintained by methyltransferases MET1, CMT2, and CMT3, while active demethylation occurs with the help of the enzymes Repressor of Silencing 1 (ROS1), Demeter (DME), Demeter-like 2 (DML2), and Demeter-like 3 (DML3) (Choi et al., 2002).

These methylation changes have a bidirectional effect on the regulation of gene expression. Whereas high methylation in promoter regions silences genes, methylation within gene bodies can often lead to higher expression. For instance, according to Xue et al. (2022), CG methylation in the gene body exerted a gene expression-enhancing effect. Conversely, CG methylation within upstream and downstream regions showed consistent gene expression suppression. The general effects of CHG and CHH methylation were the suppression of gene expression, except for CHH methylation in upstream regions. For instance, according to Xue et al. (2022), CG methylation in the gene body exhibited a gene expression-enhancing effect. Conversely, CG methylation observed in both upstream (promoter) and downstream regions consistently repressed gene expression. The general effects of CHG and CHH methylation were the suppression of gene expression, except for CHH methylation in upstream regions. Generally, such a relationship between DNA methylation and gene expression is conserved across most studied plant species. These findings indicate that DNA methylation plays a fundamental role in regulating gene expression and can perform different functions depending on gene regions and sequence contexts. This validates the general view that DNA methylation is one of the dynamic regulators of gene expression levels in plants. Furthermore, under stress conditions, reactivation of ABA, JA, and auxin-responsive genes through active DNA demethylation is an important process for the development of stress memory in plants. For example, in *Oryza sativa*, mild drought stress caused hypomethylation in the promoter region of *OsNCED*, a gene involved in ABA biosynthesis; these epigenetic changes were maintained for at least two generations, enhancing the plant's drought tolerance and water-use efficiency (Ghosh, 2019).

Methylation occurs through chemical modifications of cytosine and adenine bases, affecting DNA accessibility and thereby influencing the binding ability of transcription factors. This process plays a critical role both in gene silencing and in maintaining genome integrity. In *Tartary buckwheat*, a general decrease in DNA methylation levels was observed under cold stress, with this change being particularly pronounced in intergenic regions (Song et al., 2020). Similarly, in *Arabidopsis thaliana*, abiotic stresses such as salt, cold, and water deficiency led to demethylation in the promoter regions of *ACD6*, *ACO3*, and *GSTF14* genes, resulting in increased expression of these genes. Interestingly, priming treatments with BABA (β-aminobutyric acid) altered DNA methylation patterns, and these changes were found to be transmitted across generations (Yang et al., 2022).

Stress responses in plants are modulated not only through DNA methylation but also through histone modification and chromatin remodeling. Histones are subject to various chemical modifications: acetylation, methylation, ubiquitination, phosphorylation, and sumoylation. These modifications alter the compactness of chromatin; thus, they determine the activity of genes. This increases or decreases the accessibility of DNA to transcription factors (Sharma et al., 2023).

Chromatin remodeling is associated with these modifications. In plant cells, the chromatin-remodeling complexes such as SWItch/Sucrose Non-Fermentable (SWI/SNF), CHD, and ISWI consume ATP energy to reposition nucleosomes, thereby controlling gene accessibility. Notably, studies in *Arabidopsis* have shown that the *BRM (BRAHMA)* gene, a component of the SWI/SNF complex, enhances chromatin relaxation under stress, allowing genes encoding heat shock proteins to be activated more rapidly. This mechanism contributes to the plant's ability to respond faster to subsequent stresses by epigenetically encoding the information learned from its first stress experience (Nishio et al., 2024).

Arıkan et al. (2026) point out that histone modifications are one of the decisive factors in the process of stress priming in plants. The active marks like H3K4me3, and repressive marks like H3K27me3, regulate the opening and closing cycles of genes upon environmental stresses (Yu et al., 2025). For instance, repeated heat stress in *A. thaliana* resulted in the eviction of H3K27me3 marks, which allowed the rapid activation of heat-tolerance genes such as *HSP22* and *HSP17.6C*. Similarly, increases in H3K9ac and H3K4me3 in rice and barley enhanced the transcriptional activation of genes responsive to drought and salinity stress (Yamaguchi et al., 2021).

Histone modifications also interact with ncRNAs. Small RNAs, siRNAs, and miRNAs guide the structure of chromatin through recruiting histone methyltransferases and demethylases at their specific target loci. This coordination supports the persistence of epigenetic marks and the transgenerational inheritance of gene silencing patterns (Wilkinson et al., 2023).

Another approach in epigenetic engineering involves RNA interference, RNAi. RNAi represses gene expression via a dsRNA molecule. It has also been shown that dsRNA can direct DNA methylation (Mette et al., 2000). During this process, small noncoding RNAs, microRNAs, and siRNAs, which are generated by Dicer, become active, and together with proteins like Argonaute in the RISC complex, degrade target messenger RNAs, thus accomplishing gene silencing. RNAi systems can be transiently applied to plants through classic approaches like *Agrobacterium*-mediated transformation, virus-mediated delivery, and particle bombardment (Collins et al., 2011).

In conclusion, epigenetic modifications provide a flexible yet stable layer of regulation that enables plants to adapt to stress conditions. These mechanisms allow gene expression to be controlled not just transiently, but in a memory-based and heritable manner, contributing to the development of long-term resilience in plants against environmental challenges.

3.2.1 Epigenetic Memory in Environmental Stress Adaptation

Plants, throughout their life cycles, experience variable environmental conditions. However, they do not merely develop short-term stress responses; instead, they record such experiences as epigenetic memory to be more resilient in case of further events of stress. Epigenetic memory involves reprogramming gene expression through DNA methylation, histone modifications, small RNAs, and chromatin remodeling with no changes in the DNA sequence itself. This allows plants to "remember" past stress experiences and respond more quickly and strongly upon subsequent stress (Liu et al., 2022; Turgut Kara et al., 2025).

Epigenetic memory can function at two levels: somatic memory that enables the plant to adapt to new stresses within the same life cycle, and intergenerational or transgenerational memory, which allows the transfer of stress information to subsequent generations (Lämke & Bäurle, 2017). Epigenetic marks in genes activated during abiotic stresses such as drought, salinity, and temperature are particularly associated with DNA demethylation and the increase of active histone marks (H3K4me3, H3K9ac). For example, in *A. thaliana*, the loss of the H3K27me3 mark in the promoter regions of *HSP22* and *HSP17.6C* enabled these genes to be activated more rapidly during subsequent heat stress (Yamaguchi et al., 2021).

Similarly, Arıkan et al. (2026) also relate epigenetic memory to the "learning" potential of plants. They describe how the process of preconditioning results in a mild stress that prepares the plant against future severe stresses. During this process, epigenetic modifications ensure that genes for stress tolerance remain in an "open" state or the repressive epigenetic marks, such as H3K27me3, are removed. A striking feature of epigenetic memory after priming is that certain epigenetic marks are retained and stabilized even after the stress has subsided. Thus, plants are provided not only with temporary but also with long-lasting resilience.

Sometimes, this epigenetic memory is transmitted across generations. For instance, parent plants subjected to either salt or drought stress showed changes in methylation that were passed on to seeds, with the resultant similar expression of genes in the offspring. This demonstrates that plants can convey information on environmental signals through epigenetic inheritance without the need for classical genetic mutations (Karalija et al., 2025).

In summary, epigenetic memory represents a flexible, stable, and heritable regulatory mechanism for plant adaptation to environmental stress conditions. This system, shaped through DNA methylation, histone modifications, and small RNA interactions, confers plants with acquired resistance both at the individual level and trans-generationally. The accrued knowledge would, therefore, provide a scientific basis for developing plant species resilient to future climate change.

3.3 Role of AI in Plant Epigenomic Decoding

AI plays a pivotal role in interpreting multilayered biological data in plant epigenetics and breeding. The most valuable contribution of AI in plant breeding is its ability to model the complex relationships between genetic, epigenetic, and environmental variables beyond the capabilities of traditional methods of data analysis. For instance, ML algorithms have been applied to predict epigenetic marks associated with complex traits in plants like stress tolerance, yield, and disease resistance. Further, these approaches enable the resolution of promoter methylation patterns, histone modification regions, and transcriptional networks in plant genomes, allowing the prediction of gene activation under stress conditions (Farooq et al., 2024).

Integrated analyses assessing genomic, transcriptomic, proteomic, metabolomic, and epigenomic data together have achieved high accuracy in identifying epigenetic defense marks and stress-responsive genes within the AI-assisted omics framework. Thus, AI becomes the major tool for unraveling the interaction between plant defense mechanisms and epigenetic regulation (Murmu et al., 2024). Among others, DL models can automatically identify the regulatory motifs based on DNA methylation and histone modifications to enable a rapid detection of stress-associated epigenetic signatures. All these approaches strongly support AI not only as a predictive but also as a prescriptive tool in engineering plant epigenomes (Murmu et al., 2024).

In recent years, these technologies have become widely used for target discovery processes at the epigenomic level (Farooq et al., 2024). Methods such as artificial neural networks (ANN), support vector machines (SVM), and decision trees enable the successful analysis of DNA methylation maps and the classification of epigenetic markers. These systems allow for faster and more precise identification of epigenetic targets compared to classical gene-editing approaches. Furthermore, the integration of AI-based models with technologies such as CRISPR/dCas9 makes it possible to perform targeted and reversible modifications at the epigenome level. Thus, epigenetic editing transitions from being merely an analytical tool to a design-based modern breeding strategy (Murmu et al., 2024).

In summary, AI is not only a platform for plant epigenomic data interpretation but also a cognitive one that can handle predictions and design. Artificial intelligence, coupled with the latest gene-editing technology and used for epigenetic analysis, opens new avenues toward developing high-yielding, stress-tolerant, and sustainable crop varieties.

3.3.1 AI-Driven Integration of Multiomics and Epigenetic Data

The exponential growth in genomic, transcriptomic, proteomic, and epigenomic datasets poses an enormous analytical challenge but concurrently offers a transformative opportunity. Therefore, epigenetic research has transitioned into a data-driven era. Traditional statistical approaches are often limited due to the high dimensionality and multicomponent nature of omics data, but these limitations are increasingly being addressed through the support of AI methods.

AI, particularly through ML and DL models, enables the extraction of meaningful biological patterns from complex and large-scale datasets. These models can provide significant biological insights even in complex relationships, such as gene–environment (GxE) interactions (Boye et al., 2024; Butera et al., 2023). While traditional models typically detect only single-gene expression differences or simple correlations, AI systems can integrate multiple biological layers. This allows for the analysis of genomic, transcriptomic, epigenomic, and proteomic data within a single framework. Such an integrated approach provides the most powerful targets for predicting phenotypic outcomes and discovering novel biological insights (Sun et al., 2021; Yu et al., 2022).

Supervised ML models are at the core of predictive epigenetic analyses. Such as SVMs, a ML technique used for classification and regression (Cortes & Vapnik, 1995), random forests (Breiman, 2001) are applied to rank genomic features according to their contributions to phenotypic variation, and Extreme Gradient Boosting (XGBoost) (Chen & Guestrin, 2016) provides high-accuracy classification. Additionally, LASSO (Least Absolute Shrinkage and Selection Operator) regression (Zemlianskaia et al., 2022) is widely employed in epigenetic modeling to identify the most relevant methylation regions or chromatin states associated with developmental stages.

Moreover, DL approaches such as PlantDeepMeth, developed to predict DNA methylation states in plants (Guo et al., 2025), have emerged as powerful tools for identifying patterns between epigenetic signatures and phenotypic traits. The first multiomics database in plant epigenetics, PlantEMS, is a collection of genomic, transcriptomic, and proteomic modifications from 54 plant species, capable of recognizing over 13 million modification regions across these disciplines (Dao et al., 2025).

The process starts with the definition of the research question and the gathering of the corresponding data, such as DNA methylation profiles, from various sources. Data preprocessing is an important stage that ensures consistency, suitability for analysis, and involves tasks like normalization, missing value handling, and encoding of categorical variables. Feature selection and feature engineering steps are then used to highlight the most influential variables in the dataset. At the heart of the process, the ML model most suitable for the data type and problem will be chosen and trained. Validation and testing phases will check the generalization ability of the model using different performance metrics such as accuracy and precision. Ensuring biological coherence of the results makes the model reliable and applicable. Finally, continuous monitoring and updating of the model are crucial to ensure accuracy and relevance, considering new data and scientific evolution (Aref-Eshghi et al., 2025).

3.3.2 Deep Learning and Generative Network Approaches in Epigenetic Systems

DL is an artificial intelligence approach that teaches models with multiple layers to extract representations from data at different levels of abstraction. This method has led to significant advances in fields such as image recognition, speech processing, object detection, drug development, and genomics. The learning process occurs through the backpropagation algorithm, which adjusts the parameters in each layer based on the information received from the previous layer (LeCun et al., 2015). In short, DL is a branch of machine learning that can extract meaningful features from raw data without human intervention. It has emerged as one of the powerful tools developed in recent times for bioinformatics and epigenomic analyses. DL allows holistic integration of large-scale omics data, enabling deeper insight into the complex, multilayered structure characteristic of biological systems (Ballard et al., 2024).

Methods such as DNA bisulfite sequencing (BS-seq) for 5mC, single-molecule real-time sequencing (SMRT-seq) for 6mA, methylated RNA immunoprecipitation sequencing (MeRIP-seq) for m6A, and chromatin immunoprecipitation sequencing (ChIP-seq) for histone modifications are employed to quantify epigenetic marks genome-wide (Chen et al., 2025a). However, these datasets underrepresent the epigenomic diversity in plants because of limitations on tissue types, genotypes, and environmental conditions. As such, computational approaches able to predict the locations of epigenomic modifications in plants are called for (Wang et al., 2021a).

Convolutional neural networks (CNNs) to extract new features from input data have emerged as a promising approach to develop predictive tools in plant epigenetics. Several studies have demonstrated that CNN-based models are successful at the prediction of genomic and epigenomic marks; however, most were developed for non-plant species and usually predict a single epigenetic mark (LeCun et al., 2015; Wang et al., 2021a).

In the study by Wang et al. (2021a), a DL model called SMEP (Smart Model for Epigenetics in Plants) was presented, capable of predicting six different epigenetic modifications (5mC, 6mA, m6A, H3K27me3, H3K4me3, and H3K9ac) genome-wide in plants. In this study, sequences containing epigenetic modifications were used as input for DeepLIFT. DeepLIFT (Shrikumar et al., 2017) is a feature attribution method that calculates the contribution of each base to a specific output prediction of a CNN model. The method decomposes the difference between the prediction for the actual input and that for a neutral reference input into contribution scores for all bases in the sequence. As a result, SMEP trained on Japonica rice Nipponbare data achieved high accuracy for DNA, RNA, and histone modifications and correctly predicted 6mA dynamics under heat stress. The model was also successfully applied to other species, including Indica rice 93-11 and maize B73. Additionally, an online SMEP prediction tool was developed to facilitate researchers in examining plant epigenomic marks. SMEP provides a powerful computational resource for understanding the dynamic nature of epigenetic regulation in plants.

With access to larger and contextually relevant epigenomic datasets, DL models can reach even higher accuracy. Such models will be able to be combined with transcriptomic data from different stages of development and conditions of environmental stress, therefore accelerating biological discovery and making the processes of genome engineering more efficient for researchers and breeders. While experimental techniques remain far behind in fully capturing all epigenomic marks, deep learning-based approaches allow one to derive more comprehensive and functional insights from already existing data (Wang et al., 2021a).

Wen et al. (2023) conducted a comprehensive survey on state-of-the-art DL-based multiomics data integration methods in the biomedical field. They classified these approaches into six categories: fully connected neural networks (FCNN), convolutional neural networks (CNNs), autoencoders (AE), graph neural networks (GNN), capsule networks (CapsNet), and generative adversarial networks (GAN). These models can capture nonlinear relationships between different data layers, thereby contributing to a holistic understanding of gene regulatory mechanisms and phenotypic diversity (Ballard et al., 2024). In particular, their ability for automatic feature extraction allows meaningful patterns to be identified in high-dimensional data without human intervention (Chen et al., 2022). Within this framework, DL-based approaches are successfully used to generate biologically meaningful representations from multiomics data and to resolve gene regulatory networks, classify disease subtypes, and identify potential biomarkers. Therefore, deep learning-based multiomics integration plays a critical role in elucidating the mechanistic complexity of biological systems (Ballard et al., 2024; Wang et al., 2021b).

3.3.3 Large Language Models in Plant Epigenomics

The large datasets generated by omics technologies challenge classical analytical approaches, particularly due to the "small n, large p" problem (Kirpich et al., 2018). Consequently, researchers have turned to ML and DL methods capable of handling high-volume data. Ensemble ML models improve analysis accuracy, while DL architectures offer a significant advantage with their ability to automatically learn complex biological patterns (Farooq et al., 2024; Sartori et al., 2025).

In recent years, DL methods have emerged as a powerful tool to overcome these limitations by reducing reliance on genome annotations. DL models can learn complex biological patterns directly from raw data. For example, the DeepCpG model predicted methylation states with high accuracy in human and animal cells using only DNA sequences and neighboring methylation information, outperforming traditional methods (Angermueller et al., 2016). However, plant genomes contain three types of methylation—CpG, CHG, and CHH (H = A, C, or T), so models designed for animal genomes do not fully capture the structural complexity of plant epigenomes (Guo et al., 2025).

To overcome such challenges, increasing attention has been paid to transfer learning. Transfer learning enables knowledge learned in one domain (e.g., animal genomes) to be transferred into a related domain (e.g., plant genomes). For example, the DeepSEA model, initially developed for the human genome, has been successfully used to predict the effects of regulatory variants in plant genomes (Guo et al., 2025; Zhao et al., 2021; Zhou & Troyanskaya, 2015). Similarly, the DeepSignal-Plant and DeepPlant models have been effective in predicting cytosine methylation in plants using third-generation sequencing data (Chen et al., 2025b). During the development of DeepPlant, a triple-encoder architecture, similar to that used in previous tools such as DeepSignal and DeepBam, was employed (Ni et al., 2019).

General LLMs are not suitable for analyzing DNA sequences because they are designed for human language and operate at the character level. Consequently, DNA-specific LLMs such as DNABERT, DNABERT 2 (Ji et al., 2021; Zhou et al., 2023), and Nucleotide Transformer (Dalla Torre et al., 2025), which work with k-mer-based vocabularies, have been developed. However, these models were not specifically designed for plant genomes. While InstaDeep's AgroNT model (Mendoza Revilla et al., 2024) partially addresses this gap for plant species, its large size makes training on personal computers difficult. To meet these requirements, researchers developed PDLLMs models based on five different architectures adapted to the unique structure of plant genomes, fine-tuning 198 models across ten plant datasets (Liu et al., 2025).

Overall, in the past two years, LLMs have witnessed remarkable success in the field of AI-generated content. At this point, LLMs represent a new era in plant science (Liu et al., 2025). With their capacity to recognize patterns, uncover relationships, and make human-like inferences from high-volume and heterogeneous data, LLMs have the potential to revolutionize the breeding process. In plant breeding, LLMs can analyze complex datasets comprising high-throughput omics, phenotypic, and environmental data, thereby improving predictive accuracy and revealing novel genetic and epigenetic relationships. These models provide breeders not only with the ability to leverage existing knowledge but also to generate new hypotheses and accelerate decision-making processes (Yoosefzadeh-Najafabadi, 2025).

3.4 AI-Enabled Targeted Epigenome Editing

Epigenetic regulation encompasses heritable changes that modulate gene expression without altering the underlying DNA sequence, transmitted through both meiotic and mitotic divisions (Pulecio et al., 2017). Epigenome editing has emerged as a powerful strategy for manipulating epigenetic marks at defined genomic loci. Modern genome editing technologies derived from microbial CRISPR–Cas systems—originally components of prokaryotic adaptive immunity—are now central to applications in medicine and agriculture due to their precision, flexibility, and transformative potential (Fadul et al., 2023).

AI, particularly ML and DL algorithms, plays a critical role in enhancing the efficiency and accuracy of these editing systems. By extracting complex patterns from high-dimensional biological datasets, AI can optimize experimental workflows, reduce failure rates, and assist in crucial design steps such as guide RNA (gRNA) selection and off-target prediction (Aghziel et al., 2025; Fadul et al., 2023; Tahir et al., 2024).

In plant breeding, multiple strategies aim to exploit epigenetic diversity to improve traits such as stress tolerance. Population-level epigenetic variation, including heritable DNA methylation patterns, histone modifications, and regulatory chromatin changes, can give rise to stable or semi-stable "epialleles," which can be subjected to artificial selection similarly to genetic alleles (Dobránszki et al., 2025; Lieberman-Lazarovich et al., 2022; Singh et al., 2024).

Epigenomes can be modified either globally or in a targeted manner. Chemical agents such as 5-Azacytidine (5-AzaC) induce genome-wide hypomethylation by inhibiting cytosine methyltransferases, generating new epigenetic variation (Osorio-Montalvo et al., 2018). Such "epimutagenesis" has been applied in crops; for example, in sugarcane, 5-AzaC treatment followed by selection under aluminum stress produced a line (Mut 2) showing distinct epigenetic signatures and enhanced stress tolerance. However, chemical treatments lack genomic specificity (Koetle et al., 2023).

In contrast, CRISPR, TALEN, and ZFN platforms enable precise, locus-specific epigenetic modifications. Notable examples include targeted alteration of DNA methylation to improve cassava resistance to bacterial blight (Veley et al., 2023), reactivation of the FWA gene via locus-specific DNA demethylation leading to late flowering in Arabidopsis (Gallego-Bartolomé et al., 2018), removal of repressive histone marks such as H3K27me3 at the CUC3 locus (Fal et al., 2025), and activation of drought-responsive genes via histone acetylation (Roca Paixão et al., 2019). Additionally, dCas9-ROS1 fusions have demonstrated targeted demethylation of specific methylated DNA regions, providing further proof-of-concept for precise "epigenomic surgery" in plants (Devesa-Guerra et al., 2020).

Although CRISPR/Cas systems are powerful, practical limitations like off-target effects, PAM constraints, and delivery challenges their widespread use (Cardi et al., 2023). A major conceptual bottleneck is the incomplete understanding of epigenome–phenotype relationships across species. Integrating AI with high-throughput phenotyping and multiomics data offers new opportunities to model these relationships and accelerate the development of stress-tolerant crops (Dobránszki et al., 2025).

3.4.1 CRISPR/dCas9 Systems for Epigenetic Modifications: Designing Target Sites Through AI Prediction Models

Originally derived from the adaptive immune defense of prokaryotes, bacteria, and archaea, CRISPR–Cas systems have become popular tools in genome engineering. The canonical system includes Clustered Regularly Interspaced Short Palindromic Repeats (CRISPR) together with associated Cas proteins, enabling correct genomic modifications. Whereas the original CRISPR-Cas9 executes targeted DNA cleavage, the catalytically dead version, dCas9, provides programmable DNA binding without inducing double-strand breaks. Therefore, dCas9 is an excellent scaffold for targeting transcriptional and epigenetic effectors to specific genomic positions to achieve targeted gene regulation or epigenome modulation (Chen et al., 2019; Moradpour & Abdulah, 2020).

Researchers can induce locus-specific modifications by fusing dCas9 with various epigenetic effector domains, including DNA methyltransferases, the dioxygenase enzyme Ten-Eleven Translocation 1 (TET1) that initiates the DNA demethylation cycle, histone acetyltransferases such as E1A-binding protein p300, and chromatin-modifying enzymes like lysine-specific histone demethylase 1A (LSD1). Such modifications include DNA methylation and demethylation, histone posttranslational modifications, and remodeling of chromatin architecture. This approach, in general, can be referred to as Epi-CRISPR, which allows for precise transcriptional reprogramming and targeted epigenetic editing. It has been demonstrated in Arabidopsis that the dCas9-TET1 fusion can efficiently demethylate the FWA promoter, showing highly effective locus-specific epigenome manipulation (Gallego-Bartolomé et al., 2018).

While offering tremendous transformative potential, however, editing approaches based on CRISPR/dCas9 are dependent on the correct design of single-guide RNAs to guarantee effective targeting. The prediction of sgRNA activity remains one of the key bottlenecks in this respect, since small differences in sequence are able to drastically alter binding efficiency and off-target effects. The more recent progress made in AI—particularly deep learning—has given rise to a host of tools capable of overcoming this issue. DL algorithms can consider complex features of the sequence, understand intricate dependencies within nucleotides, and predict sgRNA efficacy accurately. Integrating AI-driven models with the design pipeline for CRISPR/dCas9 accelerates the identification of optimal target sites, thus favorably affecting both the precision and efficiency of epigenetic editing workflows (Chen et al., 2019; LeCun et al., 2015).

Taken together, the synergy between AI-driven sgRNA design and CRISPR/dCas9-based epigenetic tools establishes a powerful platform for programmable, locus-specific remodeling of plant and crop genomes. This integrative platform bears considerable potential for deconstructing regulatory networks, reprogramming gene expression, and engineering stress-tolerant phenotypes via specific epigenomic interventions.

3.5 Case Studies: Resilient Crop Development

AI, especially ML and DL, has been increasingly transforming plant science by providing powerful tools to address complex challenges in crop improvement and stress resilience. It facilitates data-driven decision-making, predictive modeling, and high-throughput analysis in agriculture, hereby enhancing precision farming, crop monitoring, and resource optimization while addressing critical issues such as food security and environmental sustainability (Gerakari et al., 2025).

In plant breeding, AI accelerates genomic selection by integrating all genetic markers, including those with minor effects or high correlations, into predictive models. ML and DL can capture complex, nonadditive interactions, such as dominance and epistasis, without requiring predefined assumptions about predictor distributions (Li et al., 2025). In addition, AI-driven approaches complement gene-editing technologies like CRISPR-Cas9 by enabling efficient identification and validation of target genes that improve the precision and efficiency of modifying key traits related to salinity tolerance, drought resilience, and pest resistance. AI applied to bioinformatics extends this further by analyzing multiomics data for the identification of regulatory networks, stress-responsive genes and molecular markers, and predictive models related to plant resilience. Applications involve the discovery of genes tolerant to drought and salt, real-time monitoring of plant health, and the management of resources in changing climates. Furthermore, phenomics platforms equipped with AI support high-throughput automation, cutting labor needs and improving efficiency (Zhang et al., 2024).

Standardization of datasets, model interpretability, overfitting, and accessibility are some challenges, despite these advances. To fully harness the power of these AI and bioinformatics tools, education as well as collaboration across disciplines and training targeted at different skill levels, is required. Together, they provide an entire framework for engineering stress-resilient plant varieties that support sustainable agriculture and contribute toward global food security with climate change (Gerakari et al., 2025; Zhang et al., 2024).

Moreover, various applications of AI—including ML, DL, and LLMs—have recently been developed for predicting potential epigenomic modification sites and identifying CRISPR targets. Notably, all currently available tools specifically dedicated to epigenomic target discovery were published in 2025. Thus, although numerous studies exist on genomic target-gene identification, examples that directly address epigenomic target prediction for developing resilient plant phenotypes remain limited. Therefore, this chapter further explores existing AI-driven research efforts supporting the development of stress-resilient plants, highlighting the current capabilities and future prospects of these technologies within the emerging field of plant epigenome engineering.

3.5.1 Drought and Heat Tolerance Through Epigenetic Regulation

Plants possess complex systems for sensing, storing, and responding to environmental cues through cellular, molecular, and epigenetic networks that adjust dynamically under stresses like drought and heat. For crop improvement, understanding these regulatory mechanisms is essential. Recent advances in AI, ML, and DL have provided powerful tools to study plant responses to abiotic stress by enabling high-throughput phenotyping, genomic analysis, and accurate prediction of stress-tolerant traits (Dobránszki et al., 2025; Han & Wang, 2025).

Various DL models, including CNNs, Recurrent Neural Networks (RNNs), Long Short-Term Memory networks (LSTMs), and U-Net Convolutional Neural Network Architecture (U-Nets), have demonstrated very remarkable performance in the extraction of complex spatiotemporal patterns from multimodal datasets comprising RGB, thermal, fluorescence, multispectral, and hyperspectral imagery. These models enable the accurate characterization of physiological and phenotypic changes under stress and have achieved higher than 95% predictive accuracy in applications like leaf area estimation, canopy segmentation, pest and disease detection, and drought assessment. Attention-based networks and pretrained models have enabled automated feature extraction, transfer learning, and adaptation across species and environmental conditions through Residual Neural Network (ResNet), Visual Geometry Group Network (VGGNet), and Inception Neural Network Architecture (InceptionNet) (Han & Wang, 2025).

High-throughput phenotyping platforms combined with DL algorithms have been successfully applied in real-world studies. For example, Boguszewska-Mańkowska et al. (2022) evaluated drought tolerance in 50 potato cultivars over 11 years based on morphological traits, tuber yield, and climatic water balance. ML classifiers, such as random forest, Extra Trees, AdaBoost, and XGBoost, achieved up to 96.7% accuracy in categorizing cultivars by drought tolerance. In cotton, VisStressPredict and HyperStressPropagateNet for drought stress modeling were developed. Time-lapse RGB and hyperspectral images were analyzed via dynamic time warping and deep CNNs, allowing early stress detection and visualization of stress propagation over time (Das Choudhury et al., 2023).

AI-driven approaches have identified key stress-responsive genes and regulatory networks in crops such as wheat, rice, and tomato. ML and DL frameworks enable large-scale analyses of genomic data, variant detection, functional annotation, and construction of gene co-expression and regulatory networks to predict candidate genes underlying drought and heat tolerance (Lin et al., 2025). In bread wheat (*Triticum aestivum*), ML models examining DNA methylation in promoters, coding sequences, and exons predicted tissue-specific differentially expressed genes with 81% accuracy, pointing to the critical role of CG-context methylation in their regulation (N'Diaye et al., 2020). Similarly, the SMEP predicted six epigenomic marks in rice: DNA 5-methylcytosine (5mC), DNA 6-methyladenosine (6mA), RNA m6A, and histone modifications. Consequently, up to 95% accuracy was reached for 6mA and 80% for the rest. Moreover, the validation after heat shock confirmed the ability of the model to detect changes in methylation induced by stresses (Wang et al., 2021a).

Traditional ML models, such as random forests, and DL architectures have tracked DNA methylation dynamics under various intensities and durations of stress, thereby enabling the identification of stress-responsive epigenetic marks in rice and wheat (Smet et al., 2023). Furthermore, studies integrating chromatin modifications and transcription factor (TF) binding data in Arabidopsis and rice showed that only a subset of histone marks—such as H3K4me3 and H3K36me3—predicted gene expression patterns, with emphasis on the combinatorial regulation of transcription by epigenetic states and TF activity (Wu et al., 2019). Hence, AI has improved the precision of genomic editing, enabling more efficient selection of targets, reduction in costs, and better prediction of the effects of stress responses. This contributes to more efficient strategies of crop improvement (Lin et al., 2025).

ML and DL approaches have also been applied to transcriptomic and miRNA data for the prediction of stress tolerance. Drought-responsive genes in tomato were identified using gradient boosting models (XGBoost), illuminating key pathways for genetic improvement (Chowdhury et al., 2023). The TransGeneSelector DL framework successfully extracted regulatory genes from limited RNA-seq datasets in Arabidopsis under heat stress (Huang et al., 2025). Another study from cucumber combined miRNA profiling with an SVM optimized via a genetic algorithm and particle swarm optimization, accurately predicting drought and salinity stresses and identifying miRNA-477b as a crucial predictor (Mohammadi & Asefpour Vakilian, 2023).

High-throughput phenotyping and remote sensing technologies, along with AI, make it possible to monitor for early signs of stress tolerance. With the help of hyperspectral imaging and attention-based DL, visible drought stress has been detected in potato cultivars before the appearance of symptoms (Lapajne et al., 2024). The usage of transformer-based architectures like LWC (leaf water content)-former has allowed accurate drought phenotyping using hyperspectral reflectance to estimate leaf water content in blueberries and, most importantly, was able to outperform all earlier models based on ML and DL (Rahman et al., 2025). Among many applications are Least Squares Support Vector Machine (LS-SVM) in heat-stress detection of garlic plants (Ryu et al., 2023) and CNN-LSTM-attention models that have identified high accuracy of drought and heat stress in Chinese fir seedlings (Xing et al., 2023). In wheat, elastic-net, Gaussian processes, and XGBoost models were found to predict germination-stage drought tolerance reliably from key traits as a rapid and low-resource early screening strategy (Benlioğlu et al., 2024).

The integration of AI with phenomic and genomic data enhances breeding efficiency and accuracy. For instance, Fumia et al. (2023) assessed the tolerance of 300 *Capsicum annuum* accessions to high temperatures using phenotypic clustering, random forest models, and genomic prediction with Weighted Root Sum of Squares Index (WRSSI), which selected 33 promising lines for breeding. Ali et al. (2024) proposed multimodal LSTM models that integrated molecular and phenotypic data across 101 plant genera to reach 97% predictive accuracy for drought stress; this underlines the contribution of AI in enabling early detection of stress and precision agriculture.

Most of the epigenetic analyses are time-consuming and costly when performed in the laboratory. AI and ML provide time-saving alternatives for mapping phenotypic manifestations of epigenetic modifications and predicting the stress responses of plants (Tahir et al., 2024). This genomic editing through AI provides far better target selection to increase the efficiency and reduce the cost, thereby accelerating the process of developing stress-resilient varieties (Ali et al., 2024; Lin et al., 2025) (Table 3.1). These integrative approaches bridge the molecular regulation of key traits with phenotyping to accelerate the identification of stress-tolerant genotypes, guide precision breeding strategies, and underpin sustainable crop production in an increasingly challenging environment.

3.5.2 Salinity Resistance and Metabolic Adaptation

Salinity stress is one of the major abiotic stresses that limit crop productivity worldwide. New approaches are needed to identify tolerant genotypes and understand physiological and metabolic mechanisms underlying salinity tolerance. Salinity resistance study has recently taken a step forward with the integration of high-throughput phenotyping platforms, machine learning, and artificial intelligence. These integrated approaches enable large-scale phenotypic data processing, allowing the identification of complex trait associations with salt tolerance (Morton et al., 2024). AI-driven analysis has also pointed out candidate genes controlling salinity responses that inform trait-focused breeding programs (Aamir et al., 2025; Fatima et al., 2023).

DL architectures, including CNNs, RNNs, Transformers, and GANs, have overcome limitations inherent in traditional ML methods by enabling the direct extraction of salient features from raw data and robust generalization across diverse conditions. Applications

Table 3.1 AI-based models for detecting drought and heat-stress-related traits regulation across diverse plant species.

Plant	Method	Results	Reference
Rice	Machine learning models analyzing DNA methylation patterns	Unique DNA methylation patterns associated with drought tolerance were identified	Smet et al. (2023)
Triticum aestivum (bread wheat)	Machine learning models analyzing DNA methylation patterns in promoters, coding sequences, and exons	Predicted differentially expressed genes with 81% accuracy; highlighted CG-context methylation's role in tissue-specific gene regulation	N'Diaye et al. (2020)
Oryza sativa (japonica rice, Nipponbare)	Deep learning (SMEP) to predict six types of epigenomic marks	Achieved 95% accuracy for 6mA and ~80% for 5mC, m6A, and histone marks; accurately predicted stress-induced epigenetic changes under heat stress	Wang et al. (2021a)
Arabidopsis thaliana and rice	Machine learning models integrating multiple histone modification datasets	Identified minimal sets of histone marks (H3K4me3 and H3K36me3) and transcription factors sufficient to predict gene expression; provided quantitative insight into epigenetic regulation	Wu et al. (2019)
Tomato	XGBoost machine learning of RNA-seq data	Six candidate genes were identified involved in drought adaptation	Chowdhury et al. (2023)
Arabidopsis thaliana	TransGeneSelector: Wasserstein GAN + Transformer on RNA-seq data	Identified key regulatory genes under germination and heat stress; high classification accuracy; validated by RT-qPCR	Huang et al. (2025)
Potato (*Solanum tuberosum*)	QDA, random forest, Extra Trees, AdaBoost, XGBoost on morphological + yield traits	XGBoost achieved 96.7% accuracy in classifying drought tolerance among 50 cultivars	Boguszewska-Mańkowska et al. (2022)
Potato (*Solanum tuberosum*)	Hyperspectral imaging + attention-based deep learning (VNIR + SWIR)	Early detection of drought stress; precision irrigation; supports breeding of drought-tolerant cultivars	Lapajne et al. (2024)
Pepper (*Capsicum annuum*)	Phenomic (k-means + RF) + Genomic (WRSSI) selection	33 heat-tolerant lines identified; 13 overlapping across methods; integrating phenomics + genomics improved selection efficiency	Fumia et al. (2023)

(*Continued*)

Table 3.1 (Continued)

Plant	Method	Results	Reference
Cotton	VisStressPredict (RGB + DTW) and HyperStressPropagateNet (CNN on hyperspectral images)	RGB- and hyperspectral-based models accurately detected stress onset and mapped stress propagation, correlating strongly with soil moisture	Das Choudhury et al. (2023)
Cucumber	SVM models optimized by GA and PSO using morphological, physiological/biochemical, and miRNA traits	miRNA profiles (esp. miRNA-477b) gave highest accuracy ($R^2 = 0.99$); superior to morphological and biochemical traits; enables early, precise stress detection	Mohammadi and Asefpour Vakilian (2023)
101 plant genera	ML/DL (SVM, Gradient Boosting, RNN, LSTM) + protein domain data	Multimodal LSTM combining molecular + phenotypic data reached 97% accuracy; superior to RNN (94%), GB (96%), SVM (82%); shows strong capability for early stress detection and resilience improvement	Ali et al. (2024)
Blueberry (*Vaccinium spp.*)	Transformer-based LWC-former using hyperspectral reflectance + patch embeddings	Achieved $R^2 = 0.81$ across cultivars; outperformed TabTransformer, DeepRWC, MLP, PLSR, SVR, RF; enables fast, scalable drought screening using high-throughput hyperspectral imaging	Rahman et al. (2025)
Garlic	Multispectral snapshot imaging + ML models (PLS-DA, LS-SVM, DNN, RP-CNN)	LS-SVM performed best; model ranking LS-SVM > RP-CNN > DNN > PLS-DA; multispectral imaging effective for heat-stress monitoring	Ryu et al. (2023)
Chinese fir	CNN-LSTM-attention model (ResNet50-LSTM-att + Grad-CAM)	Achieved 96.91% accuracy for heat stress and 96.05% for drought; high R^2 (0.957 heat, 0.944 drought); effective tool for stress phenotyping and applicable to other tasks	Xing et al. (2023)
Wheat (*Triticum aestivum*)	PCA + biplots + ML models (Elastic-net, Gaussian Processes, XGBoost)	Elastic-net best predicted germination traits; Gaussian processes excelled for root length; and XGBoost most accurately estimated shoot and biomass, enabling fast drought-tolerance screening.	Benlioğlu et al. (2024)

include yield prediction, stress detection, pest and disease identification, and phenotypic forecasting (Han & Wang, 2025). For example, Johansen et al. (2020) illustrated that UAV-acquired RGB and multispectral imagery, in conjunction with random forest algorithms, can be used to predict yield and biomass with high accuracy for wild tomato under both normal and salt-stressed conditions, thus providing early and accurate forecasting in precision agriculture.

Image-based phenotyping coupled with multivariate analysis has been effective in distinguishing salinity tolerance in halophytes. In *Salicornia europaea* L., for instance, computer vision systems, coupled with MDA, classified plants based on morphometric and color traits with 90–100% accuracy in salinity tolerance classification. The models predicted biomass and the optimal salinity level with near-perfect accuracy, demonstrating that non-destructive, AI-assisted phenotyping is highly valuable for breeding, ecological research, and bio-agriculture applications (Cárdenas-Pérez et al., 2024).

Time-resolved physiological measurements, including electrical signaling, have also benefited from DL-based modeling. Yao et al. (2021) proposed an SLSTM-TCNN framework to model wheat leaf electrical signals under different NaCl concentrations. Combining this framework with a NaCl stress detection module and a salt tolerance-grading system facilitates rapid and accurate identification of salt-tolerant wheat genotypes, providing for early-stage screening in a scalable, data-driven manner.

Hyperspectral imaging adds to salinity resistance assessment by recording stress-induced physiological and metabolic changes. Feng et al. (2020) showed that hyperspectral imaging, coupled with deep-learning-driven image segmentation, can nondestructively quantify changes in chlorophyll content, photosynthetic capability, and ion accumulation in okra leaves under salt stress. Predictive AI models derived from spectral data were found to reliably reflect key stress-related traits, hence providing a robust pipeline for high-throughput screening in breeding programs. Similarly, other DL models, using attention mechanisms, have been implemented to enhance the classification of cereal stress. Khotimah et al. (2022) presented SC-CAN, a network that integrates spectral convolution and a channel-attention mechanism to emphasize informative spectral features while offering sensitive early-stage detection of salinity stress in wheat. The Spatial-Channel Convolutional Attention Network (SC-CAN) showed 82–89% accuracy for various wheat datasets, demonstrating its reliability for both abiotic and biotic stress classification.

In legumes, Deng et al. (2024) combined CNN architectures with chlorophyll fluorescence imaging to detect soybean seedlings from salinity stress. By fusing features from key chlorophyll fluorescence parameters (ChlF), Fv/Fm, Quantum Yield of Non-regulated Energy Dissipation [Y(NO)], and Inh (Inhibition of Photosystem II) within the Residual Neural Network 50-layer architecture (ResNet50) network and using SVM for classification, the approach achieved 98.61% accuracy. Uniform Manifold Approximation and Projection (UMAP)-based visualization confirmed that the fused high-dimensional feature space has strong discrimination and provided a rapid, sample-preserving, and scalable tool for precision agriculture.

These studies collectively demonstrate that the combination of high-throughput phenotyping, spectral analysis, and DL may bring about a paradigm shift in the understanding of the mechanisms that underpin salinity resistance. Precise, early detection of the stress responses via this approach, while integrating metabolic and physiological data, underpins

Table 3.2 AI-based models for detecting salt-stress-related traits across diverse plant species.

Plant	Method	Results	Reference
Wild tomato (*Solanum pimpinellifolium*)	UAV-based RGB and multispectral imaging combined with machine learning (random forest)	The approach enabled early and precise yield forecasting for both healthy and salt-stressed tomato plants	Johansen et al. (2020)
Wheat	Electrical signal time-series + Deep Learning (DL)	Long-term prediction of electrical signals under salt stress to identify salt-tolerant lines	Yao et al. (2021)
Cotton (*Gossypium hirsutum*)	Hyperspectral imaging + machine learning	Enabled high-throughput phenotyping for salt-stress tolerance and discrimination of salt-tolerant genotypes	Feng et al. (2020)
Wheat (*Triticum aestivum*)	Hyperspectral imaging + SC-CAN (spectral convolution + channel-attention network)	Enabled early and accurate classification of salt and biotic stress in wheat	Khotimah et al. (2022)
Soybean (*Glycine max*)	Chlorophyll fluorescence imaging + deep learning (CNNs, feature fusion with SVM)	Achieved highly accurate detection of salt-stress levels in soybean seedlings (up to 98.61%)	Deng et al. (2024)
Salicornia europaea L.	Computer vision system + Multivariate analysis	Predicted biomass and classified different salt-tolerant populations with high accuracy	Cárdenas-Pérez et al. (2024)

the generation of salt-tolerant varieties of crops and furthers breeding programs for enhanced global food security (Table 3.2).

3.6 Conclusion and Future Perspectives

The transition into the Breeding 5.0 era is foreseen to lie in the combination of AI, CRISPR, and Synthetic Biology (SynBio). This approach intends to generate faster, more precise, and sustainable breeding strategies through the integration of multiomics data (Goell et al., 2024). Speed breeding techniques significantly reduce time cycles of breeding (Watson et al., 2018), whereas AI-assisted models enable the more efficient optimization of complex biological traits. Additionally, Explainable AI (XAI) approaches enhance interpretability within biological processes. This process subsequently strengthens the reliability of AI-based frameworks that adopt decisions derived from the same procedures (Koh et al., 2024). Thus, the concept of sustainable future breeding depends on an AI–CRISPR–SynBio alliance, which would take shape within an ethical, transparent, and interdisciplinary framework (Ali et al., 2024).

This transformation has great potential at both the genetic level and epigenomic editing level. With the increasing impacts of global climate change, developing plant varieties that are resistant to environmental stress factors becomes crucial. Targeted epigenomic modifications offer the advantage of control in changing the gene expression without actually changing the DNA sequence, hence offering a new paradigm in sustainable agriculture. At this point, AI takes center stage in this transformation by learning from large biological datasets and deciphering epigenetic patterns.

The evolution of AI-based systems has fundamentally changed how we understand the building blocks of biology. The Evo model is one example of such a change. The Evo model, developed by Nguyen et al. (2024), is a long-context foundational biological model based on the architecture of StripedHyena to comprehend the languages of biology: DNA, RNA, and proteins. Following training on 2.7 million prokaryotic genomes, Evo carries out both predictive analyses and molecular redesign while sustaining contextual understanding for longer sequences than 650,000 tokens.

The integration of these large-scale biological language models in plant epigenome engineering ushers in a new era in predicting how epigenetic regulatory mechanisms, such as DNA methylation, histone modifications, and chromatin accessibility affect gene expression and stress responses. When combined with CRISPR-based editing technologies, Evo-like models allow for the design of targeted epigenetic interventions to enhance resilience against environmental stresses such as drought, salinity, or extreme temperatures.

In conclusion, the integration of AI and epigenome engineering represents a profound paradigm shift in the field of biology and plant science. This integrated approach moves beyond genetic information to enable design based on epigenetic knowledge, facilitating the development of climate-resilient, high-yield, and robust plant varieties in a faster, ethical, and sustainable way.

References

Aamir, M., Kashyap, S. P., Krishna, R., Mishra, P., Singh, P., Ansari, W. A., Moharana, D. P., Tiwari, S. K., Singh, P. M., Rai, N., & Singh, A. K. (2025). Advances in salinity tolerance breeding in tomato: From conventional breeding tools to genome editing. *Environmental Sustainability*, *8*(2), 139–157. https://doi.org/10.1007/s42398-025-00354-y

Abdulraheem, M. I., Xiong, Y., Moshood, A. Y., Cadenas-Pliego, G., Zhang, H., & Hu, J. (2024). Mechanisms of plant epigenetic regulation in response to plant stress: Recent discoveries and implications. *Plants (Basel)*, *13*(2), 163. https://doi.org/10.3390/plants13020163

Aghziel, A., Mahraz, M. A., Tairi, H., & Aherrahrou, N. (2025). Artificial intelligence for comprehensive DNA methylation analysis: Overview, challenges, and future directions. *Briefings in Bioinformatics*, *26*(5). https://doi.org/10.1093/bib/bbaf468

Alahmad, S., Dinglasan, E., Leung, K. M., Riaz, A., Derbal, N., Voss-Fels, K. P., Able, J. A., Bassi, F. M., Christopher, J., & Hickey, L. T. (2018). Speed breeding for multiple quantitative traits in durum wheat. *Plant Methods*, *14*, 36. https://doi.org/10.1186/s13007-018-0302-y

Ali, T., Rehman, S. U., Ali, S., Mahmood, K., Obregon, S. A., Iglesias, R. C., Khurshaid, T., & Ashraf, I. (2024). Smart agriculture: Utilizing machine learning and deep learning for drought stress identification in crops. *Scientific Reports*, *14*(1), 30062. https://doi.org/10.1038/s41598-024-74127-8

Anand, A., Subramanian, M., & Kar, D. (2023). Breeding techniques to dispense higher genetic gains. *Frontiers in Plant Science*, *13*, 1076094. https://doi.org/10.3389/fpls.2022.1076094

Angermueller, C., Pärnamaa, T., Parts, L., & Stegle, O. (2016). Deep learning for computational biology. *Molecular Systems Biology, 12*(7), 878. https://doi.org/10.15252/msb.20156651

Aref-Eshghi, E., Abadi, A. B., Farhadieh, M.-E., Hooshmand, A., Ghasemi, F., Youssefian, L., Vahidnezhad, H., Kerrins, T. M., Zhao, X., Akbarzadeh, M., Hakonarson, H., & Saeidian, A. H. (2025). DNA methylation and machine learning: Challenges and perspective toward enhanced clinical diagnostics. *Clinical Epigenetics, 17*(1), 1–43. https://doi.org/10.1186/s13148-025-01967-0

Arıkan, B., Çakır, Ö., & Turgut Kara, N. (2026). Epigenetic and transcriptomic features of stress memory during salt-stress priming in arabidopsis. *Plant Science, 362*, 112775. https://doi.org/10.1016/j.plantsci.2025.112775

Ballard, J. l., Wang, Z., Li, W., Shen, L., & Long, Q. (2024). Deep learning-based approaches for multi-omics data integration and analysis. *Biodata Mining, 17*(1), 38. https://doi.org/10.1186/s13040-024-00391-z

Benlioğlu, B., Demirel, F., Türkoğlu, A., Haliloğlu, K., Özaktan, H., Kujawa, S., Piekutowska, M., Wojciechowski, T., & Niedbała, G. (2024). Insights into drought tolerance of tetraploid wheat genotypes in the germination stage using machine learning algorithms. *Agriculture, 14*(2), 206. https://doi.org/10.3390/agriculture14020206

Boguszewska-Mańkowska, D., Ruszczak, B., & Zarzyńska, K. (2022). Classification of potato varieties drought stress tolerance using supervised learning. *Applied Sciences, 12*(4), 1939. https://doi.org/10.3390/app12041939

Boye, C., Nirmalan, S., Ranjbaran, A., & Luca, F. (2024). Genotype × environment interactions in gene regulation and complex traits. *Nature Genetics, 56*(6), 1057–1068. https://doi.org/10.1038/s41588-024-01776-w

Breiman, L. (2001). Random forests. *Machine Learning, 45*(1), 5–32. https://doi.org/10.1023/A:1010933404324

Breseghello, F. and Coelho, A. S. (2013). Traditional and modern plant breeding methods with examples in rice (Oryza sativa L.). *Journal of Agricultural and Food Chemistry* 61(35): 8277–86. https://doi.org/10.1021/jf305531j

Butera, A., Smirnova, L., Ferrando-May, E., Hartung, T., Brunner, T., Leist, M., & Amelio, I. (2023). Deconvoluting gene and environment interactions to develop an "epigenetic score meter" of disease. *EMBO Molecular Medicine, 15*(9), e18208. https://doi.org/10.15252/emmm.202318208

Cárdenas-Pérez, S., Grigore, M. N., & Piernik, A. (2024). Prediction of *Salicornia europaea* L. biomass using a computer vision system to distinguish different salt-tolerant populations. *BMC Plant Biology, 24*(1), 1086. https://doi.org/10.1186/s12870-024-05743-9

Cardi, T., Murovec, J., Bakhsh, A., Boniecka, J., Bruegmann, T., Bull, S. E., Eeckhaut, T., Fladung, M., Galovic, V., Linkiewicz, A., Lukan, T., Mafra, I., Michalski, K., Kavas, M., Nicolia, A., Nowakowska, J., Sági, L., Sarmiento, C., Yıldırım, K., ... van Laere, K. (2023). CRISPR/Cas-mediated plant genome editing: Outstanding challenges a decade after implementation. *Trends in Plant Science, 28*(10), 1144–1165. https://doi.org/10.1016/j.tplants.2023.05.012

Chen, T., & Guestrin, C. (2016, August 13–17). XGBoost: A scalable tree boosting system. *Proceedings of the 22nd ACM SIGKDD International Conference on Knowledge Discovery and Data Mining*. San Francisco, California, USA: Association for Computing Machinery.

Chen, K., Wang, Y., Zhang, R., Zhang, H., & Gao, C. (2019). CRISPR/Cas genome editing and precision plant breeding in agriculture. *Annual Review of Plant Biology, 70*, 667–697. https://doi.org/10.1146/annurev-arplant-050718-100049

Chen, W., Qiao, X., Yin, S., Zhang, X., & Xu, X. (2022). Integrating radiomics with genomics for non-small cell lung cancer survival analysis. *Journal of Oncology, 2022*, 5131170. https://doi.org/10.1155/2022/5131170

Chen, X., Xu, H., Shu, X., & Song, C. X. (2025a). Mapping epigenetic modifications by sequencing technologies. *Cell Death and Differentiation, 32*(1), 56–65. https://doi.org/10.1038/s41418-023-01213-1

Chen, H. X., Liu, Z. D., Bai, X., Wu, B., Song, R., Yao, H. C., Chen, Y., Chi, W., Hua, Q., Cheng, L., & Xiao, C. L. (2025b). Accurate cross-species 5mC detection for Oxford Nanopore sequencing in plants with DeepPlant. *Nature Communications, 16*(1), 3227. https://doi.org/10.1038/s41467-025-58576-x

Choi, Y., Gehring, M., Johnson, L., Hannon, M., Harada, J. J., Goldberg, R. B., Jacobsen, S. E., & Fischer, R. L. (2002). DEMETER, a DNA glycosylase domain protein, is required for endosperm gene imprinting and seed viability in arabidopsis. *Cell, 110*(1), 33–42. https://doi.org/10.1016/S0092-8674(02)00807-3

Chowdhury, R. H., Eti, F. S., Ahmed, R., Gupta, S. D., Jhan, P. K., Islam, T., Bhuiyan, M. A. R., Rubel, M. H., & Khayer, A. (2023). Drought-responsive genes in tomato: Meta-analysis of gene expression using machine learning. *Scientific Reports, 13*(1), 19374. https://doi.org/10.1038/s41598-023-45942-2

Collins, L. J., Schönfeld, B., & Chen, X. S. (2011). Chapter 4—the epigenetics of non-coding RNA. In T. Tollefsbol (Ed.), *Handbook of epigenetics, 49-61*. Academic Press. https://doi.org/10.1016/B978.0-12-375709-8.00004-6

Cortes, C., & Vapnik, V. (1995). Support-vector networks. *Machine Learning, 20*(3), 273–297. https://doi.org/10.1007/BF00994018

Dalla-Torre, H., Gonzalez, L., Mendoza-Revilla, J., Lopez Carranza, N., Grzywaczewski, A. H., Oteri, F., Dallago, C., Trop, E., Almeida, B. P., Sirelkhatim, H., Richard, G., Skwark, M., Beguir, K., Lopez, M., & Pierrot, T. (2025). Nucleotide transformer: Building and evaluating robust foundation models for human genomics. *Nature Methods, 22*(2), 287–297. https://doi.org/10.1038/s41592-024-02523-z

Dao, F., Xie, X., Zhang, H., Guan, Z., Wu, C., Su, W., Wei, Y., Hong, F., Luo, X., Xie, S., Lai, H., Gao, D., Yang, Y., Zhang, Y., Ning, L., Li, S., Hao, Y., Lebeau, B., Ling, C. C. Y., ... Lv, H. (2025). PlantEMS: A comprehensive database of epigenetic modification sites across multiple plant species. *Plant Communications, 6*(4), 101228. https://doi.org/10.1016/j.xplc.2024.101228

Das Choudhury, S., Saha, S., Samal, A., Mazis, A., & Awada, T. (2023). Drought stress prediction and propagation using time series modeling on multimodal plant image sequences. *Frontiers in Plant Science, 14*, 1003150. https://doi.org/10.3389/fpls.2023.10.3150

Deng, Y., Xin, N., Zhao, L., Shi, H., Deng, L., Han, Z., & Wu, G. (2024). Precision detection of salt stress in soybean seedlings based on deep learning and chlorophyll fluorescence imaging. *Plants, 13*(15), 2089. https://doi.org/10.3390/plants13152089

Devesa-Guerra, I., Morales-Ruiz, T., Pérez-Roldán, J., Parrilla-Doblas, J. T., Dorado-León, M., García-Ortiz, M. V., Ariza, R. R., & Roldán-Arjona, T. (2020). DNA methylation editing by

CRISPR-guided excision of 5-Methylcytosine. *Journal of Molecular Biology*, *432*(7), 2204–2216. https://doi.org/10.1016/j.jmb.2020.02.007

Dobránszki, J., Vassileva, V., Agius, D. R., Moschou, P. N., Gallusci, P., Berger, M. M. J., Farkas, D., Basso, M. F., & Martinelli, F. (2025). Gaining insights into epigenetic memories through artificial intelligence and omics science in plants. *Journal of Integrative Plant Biology*, *67*(9), 2320–2349. https://doi.org/10.1111/jipb.13953

Fadul, S. M., Arshad, A., & Mehmood, R. (2023). CRISPR-based epigenome editing: Mechanisms and applications. *Epigenomics*, *15*(21), 1137–1155. https://doi.org/10.2217/epi-2023-0281

Fal, K., El Khoury, S., Le Masson, M., Berr, A., & Carles, C. C. (2025). CRISPR/dCas9-targeted H3K27me3 demethylation at the CUC3 boundary gene triggers ectopic transcription and impacts plant development. *Iscience*, *28*(5), 112475. https://doi.org/10.1016/j.isci.2025.112475

Farooq, M. A., Gao, S., Hassan, M. A., Huang, Z., Rasheed, A., Hearne, S., Prasanna, B., Li, X., & Li, H. (2024). Artificial intelligence in plant breeding. *Trends in Genetics*, *40*(10), 891–908. https://doi.org/10.1016/j.tig.2024.07.001

Fatima, K., Sadaqat, M., Azeem, F., Rao, M. J., Albekairi, N. A., Alshammari, A., & Tahir Ul Qamar, M. (2023). Integrated omics and machine learning-assisted profiling of cysteine-rich-receptor-like kinases from three peanut spp. revealed their role in multiple stresses. *Frontiers in Genetics*, *14*, 1252020. https://doi.org/10.3389/fgene.2023.1252020

Feng, X., Zhan, Y., Wang, Q., Yang, X., Yu, C., Wang, H., Tang, Z., Jiang, D., Peng, C., & He, Y. (2020). Hyperspectral imaging combined with machine learning as a tool to obtain high-throughput plant salt-stress phenotyping. *The Plant Journal*, *101*(6), 1448–1461. https://doi.org/10.1111/tpj.14597

Fumia, N., Kantar, M., Lin, Y.-P., Schafleitner, R., Lefebvre, V., Paran, I., Börner, A., Diez, M. J., Prohens, J., Bovy, A., & Boyaci, F. (2023). Exploration of high-throughput data for heat tolerance selection in *Capsicum annuum*. *The Plant Phenome Journal*, *6*(1), e20071. https://doi.org/10.1002/ppj2.20071

Gallego-Bartolomé, J., Gardiner, J., Liu, W., Papikian, A., Ghoshal, B., Kuo, H. Y., Zhao, J. M., Segal, D. J., & Jacobsen, S. E. (2018). Targeted DNA demethylation of the arabidopsis genome using the human TET1 catalytic domain. *Proceedings of the National Academy of Sciences of the United States of America*, *115*(9), E2125–e2134. https://doi.org/10.1073/pnas.1716945115

Gerakari, M., Katsileros, A., Kleftogianni, K., Tani, E., Bebeli, P. J., & Papasotiropoulos, V. (2025). Breeding of solanaceous crops using AI: Machine learning and deep learning approaches—a critical review. *Agronomy*, *15*(3), 757. https://doi.org/10.3390/agronomy15030757

Ghosh, M. (2019). Climate-smart agriculture, productivity and food security in India. *Journal of Development Policy and Practice*, *4*(2), 166–187. https://doi.org/10.1177/24551333198624

Goell, J., Li, J., Mahata, B., Ma, A. J., Kim, S., Shah, S., Contreras, M., Misra, S., Reed, D., Bedford, G. C., Escobar, M., & Hilton, I. B. (2024). Tailoring a CRISPR/Cas-based epigenome editor for programmable chromatin acylation and decreased cytotoxicity. *bioRxiv*. doi: https://doi.org/10.1101/2024.09.22.611000

Guo, Z., Fan, W., Cai, C., Zhang, K., Hou, X., Li, Y., & Cheng, F. (2025). Plantdeepmeth: A deep learning model for predicting DNA methylation states in plants. *Plants*, *14*(11), 1724. https://doi.org/10.3390/plants14111724

Han, S., & Wang, H. (2025). Application of deep learning technology in monitoring plant attribute changes. *Sustainability*, *17*(17), 7602. https://doi.org/10.3390/su17177602

Huang, K., Tian, J., Sun, L., Hu, H., Huang, X., Zhou, S., Deng, A., Zhou, Z., Jiang, M., Li, G., Xie, P., Wang, Y., & Jiang, X. (2025). TransGeneSelector: Using a transformer approach to mine key genes from small transcriptomic datasets in plant responses to various environments. *BMC Genomics*, *26*(1), 259. https://doi.org/10.1186/s12864-025-11434-y

Ji, Y., Zhou, Z., Liu, H., & Davuluri, R. V. (2021). DNABERT: Pre-trained bidirectional encoder representations from transformers model for DNA-language in genome. *Bioinformatics*, *37*(15), 2112–2120. https://doi.org/10.1093/bioinformatics/btab083

Johansen, K., Morton, M. J. L., Malbeteau, Y., Aragon, B., al-Mashharawi, S., Ziliani, M. G., Angel, Y., Fiene, G., Negrão, S., Mousa, M. A. A., Tester, M. A., & McCabe, M. (2020). Predicting biomass and yield in a tomato phenotyping experiment using UAV imagery and random forest. *Frontiers in Artificial Intelligence*, *3*, 28. https://doi.org/10.3389/frai.2020.00028

Karalija, E., Ibragić, S., Dahija, S., & Šamec, D. (2025). Transgenerational memory of phenotypic traits in plants: Epigenetic regulation of growth, hormonal balance, and stress adaptation. *Current Issues in Molecular Biology*, *47*(6), 404. https://doi.org/10.3390/cimb47060404

Khotimah, W. N., Boussaid, F., Sohel, F., Xu, L., Edwards, D., Jin, X., & Bennamoun, M. (2022). SC-can: Spectral convolution and channel attention network for wheat stress classification. *Remote Sensing*, *14*(17), 4288. https://doi.org/10.3390/rs14174288

Kirpich, A., Ainsworth, E. A., Wedow, J. M., Newman, J. R. B., Michailidis, G., & McIntyre, L. (2018). Variable selection in omics data: A practical evaluation of small sample sizes. *PLoS One*, *13*(6), e0197910. https://doi.org/10.1371/journal.pone.0197910

Koetle, M. J., Osborn, C., Snyman, S. J., & Rutherford, R. S. (2023). The 5-azacytidine-induced epimutagenesis of sugarcane (Saccharum spp. hybrids) for aluminium tolerance. *Biochimica et Biophysica Acta, General Subjects*, *1867*(12), 130491. https://doi.org/10.1016/j.bbagen.2023.130491

Koh, E., Sunil, R. S., Lam, H. Y. I., & Mutwil, M. (2024). Confronting the data deluge: How artificial intelligence can be used in the study of plant stress. *Computational and Structural Biotechnology Journal*, *23*, 3454–3466. https://doi.org/10.1016/j.csbj.2024.09.010

Kumar, S. (2019). Epigenomics for crop improvement: Current status and future perspectives. *Journal of Genetics and Cell Biology*, *3(1)*, 128–134.

Lämke, J., & Bäurle, I. (2017). Epigenetic and chromatin-based mechanisms in environmental stress adaptation and stress memory in plants. *Genome Biology*, *18*(1), 124. https://doi.org/10.1186/s13059-017-1263-6

Lapajne, J., Vojnović, A., Vončina, A., & Žibrat, U. (2024). Enhancing water-deficient potato plant identification: Assessing realistic performance of attention-based deep neural networks and hyperspectral imaging for agricultural applications. *Plants*, *13*(14), 1918. https://doi.org/10.3390/plants13141918

Lecun, Y., Bengio, Y., & Hinton, G. (2015). Deep learning. *Nature*, *521*(7553), 436–444.

Li, M., Hall, T., MacHugh, D. E., Chen, L., Garrick, D., Wang, L., & Zhao, F. (2025). KPRR: A novel machine learning approach for effectively capturing nonadditive effects in genomic prediction. *Briefings in Bioinformatics*, *26*(1), bbae683. https://doi.org/10.1093/bib/bbae683

Lieberman-Lazarovich, M., Kaiserli, E., Bucher, E., & Mladenov, V. (2022). Natural and induced epigenetic variation for crop improvement. *Current Opinion in Plant Biology*, *70*, 102297. https://doi.org/10.1016/j.pbi.2022.102297

Lin, J., Zhao, Y., Tang, J., & Wang, J. (2025). The application and progress of artificial intelligence in improving drought and salinity tolerance in wheat. *Advances in Resources Research*, *5*(2), 736–750. https://doi.org/10.50908/arr.5.2_736

Liu, H., Able, A. J., & Able, J. A. (2022). Priming crops for the future: Rewiring stress memory. *Trends in Plant Science*, *27*(7), 699–716. https://doi.org/10.1016/j.tplants.2021.11.015

Liu, G., Chen, L., Wu, Y., Han, Y., Bao, Y., & Zhang, T. (2025). PDLLMs: A group of tailored DNA large language models for analyzing plant genomes. *Molecular Plant*, *18*(2), 175–178. https://doi.org/10.1016/j.molp.2024.12.006

Lutz, W., & Samir, K. C. (2010). Dimensions of global population projections: What do we know about future population trends and structures? *Philosophical Transactions of the Royal Society, B: Biological Sciences*, *365*(1554), 2779–2791. https://doi.org/10.1098/rstb.2010.0133

Mendoza-Revilla, J., Trop, E., Gonzalez, L., Roller, M., Dalla-Torre, H., de Almeida, B. P., Richard, G., Caton, J., Lopez Carranza, N., Skwark, M., Laterre, A., Beguir, K., Pierrot, T., & Lopez, M. (2024). A foundational large language model for edible plant genomes. *Communications Biology*, *7*(1), 835. https://doi.org/10.1038/s42003-024-06465-2

Mette, M. F., Aufsatz, W., Van Der Winden, J., Matzke, M. A., & Matzke, A. J. (2000). Transcriptional silencing and promoter methylation triggered by double-stranded RNA. *The EMBO Journal*, *19*(19), 5194–5201. https://doi.org/10.1093/emboj/19.19.5194

Mohammadi, P., & Asefpour Vakilian, K. (2023). Machine learning provides specific detection of salt and drought stresses in cucumber based on miRNA characteristics. *Plant Methods*, *19*(1), 123. https://doi.org/10.1186/s13007-023-01095-x

Moradpour, M., & Abdulah, S. N. A. (2020). CRISPR/dCas9 platforms in plants: Strategies and applications beyond genome editing. *Plant Biotechnology Journal*, *18*(1), 32–44. https://doi.org/10.1111/pbi.13232

Morton, M., Fiene, G., Ahmed, H. I., Rey, E., Abrouk, M., Angel, Y., Johansen, K., Saber, N. O., Malbeteau, Y., al-Mashharawi, S., Ziliani, M. G., Aragon, B., Oakey, H., Berger, B., Brien, C., Krattinger, S. G., Mousa, M. A. A., McCabe, M., Negrão, S., ... Julkowska, M. M. (2024). Deciphering salt stress responses in Solanum pimpinellifolium through high-throughput phenotyping. *The Plant Journal*, *119*(5), 2514–2537. https://doi.org/10.1111/tpj.16894

Murmu, S., Sinha, D., Chaurasia, H., Sharma, S., das, R., Jha, G. K., & Archak, S. (2024). A review of artificial intelligence-assisted omics techniques in plant defense: Current trends and future directions. *Frontiers in Plant Science*, *15*, 1292054. https://doi.org/10.3389/fpls.2024.1292054

N'diaye, A., Byrns, B., Cory, A. T., Nilsen, K. T., Walkowiak, S., Sharpe, A., Robinson, S. J., & Pozniak, C. J. (2020). Machine learning analyses of methylation profiles uncovers tissue-specific gene expression patterns in wheat. *Plant Genome*, *13*(2), e20027. https://doi.org/10.1002/tpg2.20027

Nguyen, E., Poli, M., Durrant, M. G., Kang, B., Katrekar, D., Li, D. B., Bartie, L. J., Thomas, A. W., King, S. H., Brixi, G., Sullivan, J., Ng, M. Y., Lewis, A., Lou, A., Ermon, S., Baccus, S. A., Hernandez-Boussard, T., Ré, C., Hsu, P. D., & Hie, B. L. (2024). Sequence modeling and design from molecular to genome scale with Evo. *Science*, *386*(6723), eado9336. https://doi.org/10.1126/science.ado9336

Ni, P., Huang, N., Zhang, Z., Wang, D. P., Liang, F., Miao, Y., Xiao, C. L., Luo, F., & Wang, J. (2019). DeepSignal: Detecting DNA methylation state from Nanopore sequencing reads using deep-learning. *Bioinformatics*, *35*(22), 4586–4595. https://doi.org/10.1093/bioinformatics/btz276

Nishio, H., Kawakatsu, T., & Yamaguchi, N. (2024). Beyond heat waves: Unlocking epigenetic heat stress memory in *arabidopsis*. *Plant Physiology*, *194*(4), 1934–1951. https://doi.org/10.1093/plphys/kiad558

Osorio-Montalvo, P., Sáenz-Carbonell, L., & De-La-Peña, C. (2018). 5-azacytidine: A promoter of epigenetic changes in the quest to improve plant somatic embryogenesis. *International Journal of Molecular Sciences*, *19*(10), 3182. https://doi.org/10.3390/ijms19103182

Pulecio, J., Verma, N., Mejía-Ramírez, E., Huangfu, D., & Raya, A. (2017). CRISPR/Cas9-based engineering of the epigenome. *Cell Stem Cell*, *21*(4), 431–447. https://doi.org/10.1016/j.stem.2017.09.006

Rahman, M. H., Busby, S., Ru, S., Hanif, S., Sanz-Saez, A., Zheng, J., & Rehman, T. U. (2025). Transformer-based hyperspectral image analysis for phenotyping drought tolerance in blueberries. *Computers and Electronics in Agriculture*, *228*, 109684. https://doi.org/10.1016/j.compag.2024.109684

Raza, A., Ashraf, F., Zou, X., Zhang, X., & Tosif, H. (2020). Plant adaptation and tolerance to environmental stresses: Mechanisms and perspectives. In M. Hasanuzzaman (Ed.), *Plant ecophysiology and adaptation under climate change: Mechanisms and perspectives I: General consequences and plant responses* (pp. 117–145). Springer Singapore. https://doi.org/10.1007/978-981-15-2156-0_5

Raza, A., Tabassum, J., Kudapa, H., & Varshney, R. K. (2021). Can omics deliver temperature resilient ready-to-grow crops? *Critical Reviews in Biotechnology*, *41*(8), 1209–1232. https://doi.org/10.1080/07388551.2021.1898332

Roca Paixão, J. F., Gillet, F. X., Ribeiro, T. P., Bournaud, C., Lourenço-Tessutti, I. T., Noriega, D. D., Melo, B. P., de Almeida-Engler, J., & Grossi-de-Sa, M. F. (2019). Improved drought stress tolerance in arabidopsis by CRISPR/dCas9 fusion with a histone AcetylTransferase. *Scientific Reports*, *9*(1), 8080. https://doi.org/10.1038/s41598-019-44571-y

Ryu, J., Wi, S., & Lee, H. (2023). Snapshot-based multispectral imaging for heat stress detection in Southern-type garlic. *Applied Sciences*, *13*(14), 8133. https://doi.org/10.3390/app13148133

Sartori, F., Codicè, F., Caranzano, I., Rollo, C., Birolo, G., Fariselli, P., & Pancotti, C. (2025). A comprehensive review of deep learning applications with multi-omics data in cancer research. *Genes*, *16*(6), 648. https://doi.org/10.3390/genes16060648

Sharma, M., Sidhu, A. K., Samota, M. K., Gupta, M., Koli, P., & Choudhary, M. (2023). Post-translational modifications in histones and their role in abiotic stress tolerance in plants. *Proteomes*, *11*(4), 38. https://doi.org/10.3390/proteomes11040038

Shrikumar, A., Greenside, P., & Kundaje, A. (2017, August 6–11). Learning important features through propagating activation differences. In: P. Doina, & T. Yee Whye (Eds.), *Proceedings of the 34th International Conference on Machine Learning, ICML 2017, Sydney, NSW, Australia. Proceedings of Machine Learning Research*. PMLR.

Singh, V. K., Ahmed, S., Saini, D. K., Gahlaut, V., Chauhan, S., Khandare, K., Kumar, A., Sharma, P. K., & Kumar, J. (2024). Manipulating epigenetic diversity in crop plants: Techniques, challenges and opportunities. *Biochimica et Biophysica Acta (BBA)—General Subjects*, *1868*(2), 130544. https://doi.org/10.1016/j.bbagen.2023.130544

Smet, D., Opdebeeck, H., & Vandepoele, K. (2023). Predicting transcriptional responses to heat and drought stress from genomic features using a machine learning approach in rice. *Frontiers in Plant Science*, *14*, 1212073. https://doi.org/10.3389/fpls.2023.1212073

Song, Y., Jia, Z., Hou, Y., Ma, X., Li, L., Jin, X., & An, L. (2020). Roles of DNA methylation in cold priming in tartary buckwheat. *Frontiers in Plant Science, 11*, 608540. https://doi.org/10.3389/fpls.2020.608540

Sun, L., Xu, K., Huang, W., Yang, Y. T., Li, P., Tang, L., Xiong, T., & Zhang, Q. C. (2021). Predicting dynamic cellular protein–RNA interactions by deep learning using in vivo RNA structures. *Cell Research, 31*(5), 495–516. https://doi.org/10.1038/s41422-021-00476-y

Tahir, M., Norouzi, M., Khan, S. S., Davie, J. R., Yamanaka, S., & Ashraf, A. (2024). Artificial intelligence and deep learning algorithms for epigenetic sequence analysis: A review for epigeneticists and AI experts. *Computers in Biology and Medicine, 183*, 109302. https://doi.org/10.1016/j.compbiomed.2024.109302

Thingujam, D., Gouli, S., Cooray, S. P., Chandran, K. B., Givens, S. B., Gandhimeyyan, R. V., Tan, Z., Wang, Y., Patam, K., Greer, S. A., Acharya, R., Moseley, D. O., Osman, N., Zhang, X., Brooker, M. E., Tagert, M. L., Schafer, M. J., Jeong, C., Hoffseth, K. F., ... Mukhtar, M. S. (2025). Climate-resilient crops: Integrating AI, multi-omics, and advanced phenotyping to address global agricultural and societal challenges. *Plants, 14*(17), 2699. https://doi.org/10.3390/plants14172699

Turgut Kara, N., Arıkan, B., Pulat, E., & Çakır, Ö. (2025). The role of plant epigenetic memory mechanisms in abiotic stress tolerance. Epigenetics for Climate-Smart and Sustainable Agriculture. CABI GB. https://doi.org/10.1079/9781800626102.000

Veley, K. M., Elliott, K., Jensen, G., Zhong, Z., Feng, S., Yoder, M., Gilbert, K. B., Berry, J. C., Lin, Z. D., Ghoshal, B., Gallego-Bartolomé, J., Norton, J., Motomura-Wages, S., Carrington, J. C., Jacobsen, S. E., & Bart, R. S. (2023). Improving cassava bacterial blight resistance by editing the epigenome. *Nature Communications, 14*(1), 85. https://doi.org/10.1038/s41467-022-35675-7

Voss-Fels, K. P., Stahl, A., & Hickey, L. T. (2019). Q&a: Modern crop breeding for future food security. *BMC Biology, 17*(1), 18. https://doi.org/10.1186/s12915-019-0638-4

Wang, Y., Zhang, P., Guo, W., Liu, H., Li, X., Zhang, Q., du, Z., Hu, G., Han, X., Pu, L., Tian, J., & Gu, X. (2021a). A deep learning approach to automate whole-genome prediction of diverse epigenomic modifications in plants. *New Phytologist, 232*(2), 880–897. https://doi.org/10.1111/nph.17630

Wang, T., Shao, W., Huang, Z., Tang, H., Zhang, J., Ding, Z., & Huang, K. (2021b). MOGONET integrates multi-omics data using graph convolutional networks allowing patient classification and biomarker identification. *Nature Communications, 12*(1), 3445. https://doi.org/10.1038/s41467-021-23774-w

Watson, A., Ghosh, S., Williams, M. J., Cuddy, W. S., Simmonds, J., Rey, M. D., Asyraf Md Hatta, M., Hinchliffe, A., Steed, A., Reynolds, D., Adamski, N. M., Breakspear, A., Korolev, A., Rayner, T., Dixon, L. E., Riaz, A., Martin, W., Ryan, M., Edwards, D., ... Hickey, L. T. (2018). Speed breeding is a powerful tool to accelerate crop research and breeding. *Nature Plants, 4*(1), 23–29. https://doi.org/10.1038/s41477-017-0083-8

Wen, Y., Zheng, L., Leng, D., Dai, C., Lu, J., Zhang, Z., He, S., & Bo, X. (2023). Deep learning-based multiomics data integration methods for biomedical application. *Advanced Intelligent Systems, 5*(5), 2200247. https://doi.org/10.1002/aisy.202200247

Wilkinson, S., Hannan Parker, A., Muench, A., Wilson, R. S., Hooshmand, K., Henderson, M. A., Moffat, E. K., Rocha, P. S. C. F., Hipperson, H., Stassen, J. H. M., López Sánchez, A., Fomsgaard, I. S., Krokene, P., Mageroy, M. H., & Ton, J. (2023). Long-lasting memory of jasmonic acid-dependent immunity requires DNA demethylation and ARGONAUTE1. *Nature Plants*, *9*(1), 81–95. https://doi.org/10.1038/s41477-022-01313-9

Williams, K., Subramani, M., Lofton, L. W., Penney, M., Todd, A., & Ozbay, G. (2024). Tools and techniques to accelerate crop breeding. *Plants*, *13*(11), 1520. https://doi.org/10.3390/plants13111520

Wu, Z., Tang, J., Zhuo, J., Tian, Y., Zhao, F., Li, Z., Yan, Y., & Yang, R. (2019). Chromatin signature and transcription factor binding provide a predictive basis for understanding plant gene expression. *Plant & Cell Physiology*, *60*(7), 1471–1486. https://doi.org/10.1093/pcp/pcz051

Xing, D., Wang, Y., Sun, P., Huang, H., & Lin, E. (2023). A CNN-LSTM-att hybrid model for classification and evaluation of growth status under drought and heat stress in chinese fir (Cunninghamia lanceolata). *Plant Methods*, *19*(1), 66. https://doi.org/10.1186/s13007-023-01044-8

Xue, Y., Zou, C., Zhang, C., Yu, H., Chen, B., & Wang, H. (2022). Dynamic DNA methylation changes reveal tissue-specific gene expression in sugarcane. *Frontiers in Plant Science*, *13*, 1036764. https://doi.org/10.3389/fpls.2022.1036764

Yamaguchi, N., Matsubara, S., Yoshimizu, K., Seki, M., Hamada, K., Kamitani, M., Kurita, Y., Nomura, Y., Nagashima, K., Inagaki, S., Suzuki, T., Gan, E. S., To, T., Kakutani, T., Nagano, A. J., Satake, A., & Ito, T. (2021). H3K27me3 demethylases alter HSP22 and HSP17. 6C expression in response to recurring heat in arabidopsis. *Nature Communications*, *12*(1), 3480. https://doi.org/10.1038/s41467-021-23766-w

Yang, L., Lang, C., Wu, Y., Meng, D., Yang, T., Li, D., Jin, T., & Zhou, X. (2022). ROS1-mediated decrease in DNA methylation and increase in expression of defense genes and stress response genes in arabidopsis thaliana due to abiotic stresses. *BMC Plant Biology*, *22*(1), 104. https://doi.org/10.1186/s12870-022-03473-4

Yao, J.-P., Wang, Z.-Y., De Oliveira, R. F., Wang, Z. Y., & Huang, L. (2021). A deep learning method for the long-term prediction of plant electrical signals under salt stress to identify salt tolerance. *Computers and Electronics in Agriculture*, *190*, 106435. https://doi.org/10.1016/j.compag.2021.106435

Yoosefzadeh-Najafabadi, M. (2025). From text to traits: Exploring the role of large language models in plant breeding. *Frontiers in Plant Science*, *16*, 1583344. https://doi.org/10.3389/fpls.2025.1583344

Yu, X., Jiang, L., Jin, S., Zeng, X., & Liu, X. (2022). PreMLI: A pre-trained method to uncover microRNA-lncRNA potential interactions. *Briefings in Bioinformatics*, *23*(1), bbab470. https://doi.org/10.1093/bib/bbab470

Yu, M. H., Liao, W. C., & Wu, K. (2025). Histone methylation in plant responses to abiotic stresses. *Journal of Experimental Botany*, *76*(17), 4771–4786. https://doi.org/10.1093/jxb/eraf058

Zafar, I., Unar, A., Khan, N. U., Abalkhail, A., & Jamal, A. (2025). Molecular biology in the exabyte era: Taming the data deluge for biological revelation and clinical transformation. *Computational Biology and Chemistry*, *119*, 108535. https://doi.org/10.1016/j.compbiolchem.2025.108535

Zemlianskaia, N., Gauderman, W. J., & Lewinger, J. P. (2022). A scalable hierarchical lasso for gene-environment interactions. *Journal of Computational and Graphical Statistics*, *31*(4), 1091–1103. https://doi.org/10.1080/10.18600.2022.2039161

Zhang, X., Ibrahim, Z., Khaskheli, M. B., Raza, H., Zhou, F., & Shamsi, I. H. (2024). Integrative approaches to abiotic stress management in crops: Combining bioinformatics educational tools and artificial intelligence applications. *Sustainability*, *16*(17), 7651. https://doi.org/10.3390/su16177651

Zhao, H., Tu, Z., Liu, Y., Zong, Z., Li, J., Liu, H., Xiong, F., Zhan, J., Hu, X., & Xie, W. (2021). PlantDeepSEA, a deep learning-based web service to predict the regulatory effects of genomic variants in plants. *Nucleic Acids Research*, *49*(W1), W523–w529. https://doi.org/10.1093/nar/gkab383

Zhou, J., & Troyanskaya, O. G. (2015). Predicting effects of noncoding variants with deep learning-based sequence model. *Nature Methods*, *12*(10), 931–934. https://doi.org/10.1038/nmeth.3547

Zhou, Z., Ji, Y., Li, W., Dutta, P., Davuluri, R., & Liu, H. (2023). Dnabert-2: Efficient foundation model and benchmark for multi-species genome. *arXiv preprint arXiv:2306.15006*. https://doi.org/10.48550/arXiv.2306.15006

4

AI-Based Studies on Epigenetic Mechanisms: Highlighting Plant Adaptation and Domestication

Prioty Bandhan Roop and Sandip Debnath

Department of Genetics and Plant Breeding, Institute of Agriculture, Visva-Bharati University, Sriniketan, West Bengal, India

4.1 Introduction

Plant adaptation and domestication are the two preliminary evolutionary processes of diversity and resilience in plants. Adaptation is the acquired capacity of the plant to survive in any dynamic new environmental condition. Domestication is a drive toward desirable agronomic phenotypic changes under human selection. Traditionally, plant adaptation and domestication are thought to be driven by sheer genetic phenomena like mutations, recombination, and selection pressure, shaping phenotypic diversity that supports survival and the transition of wild progenitors into cultivated crops (Meyer & Purugganan, 2013). But now evidence says that genetic variation alone cannot explain the rapid phenotypic diversification during domestication and environmental adaptation (Vogt, 2017). These findings challenge classical genetic approaches and suggest rapid epigenetic responses through epigenetic mechanisms as a guide to genetic evolution (Abdolmaleky et al., 2025). It has been seen in maize and rice studies that changes in DNA methylation patterns near promoter regions can influence the yield and stress resistance of these plants (Li et al., 2019). A shift of growth rate and immunity within a single generation through methylation of genes in the case of animal domestication is also to be noticed (Podgorniak et al., 2022). Epigenetic mechanisms are nothing but another complementary layer of heritable variation to regulate gene expression without any alteration in DNA sequence (Miryeganeh, 2025).

Cytosine DNA methylation (5mCs) is the most studied epigenetic mechanism in relation to regulating biological processes of plants (Feng et al., 2010; Henderson & Jacobsen 2007). This methylation (5mCs) can happen in three sequence contexts, viz. CpG, CHG, and CHH (where H = A, C, OR T) resulting in gene expression regulation (Ni et al., 2021). High-resolution mapping through whole-genome bisulfite sequencing (WGBS) of the high altitude Tibetan population shows a specific methylation pattern that is correlated with high altitude adaptive phenotypes (Basak et al., 2017). Histone modification by methylation and acetylation of the histone tail can change chromosome accessibility and transcriptional potential. Recently, it has been found that heterochromatin-defining H3K9 methylation during stress can be redistributed across the whole genome to establish new adaptive phenotypes (Larkin et al., 2023). This kind of chromatin remodeling of nucleosomes and 3D genomic architecture affects long-range regulatory interactions (García-Pérez et al., 2021). The next layer of epigenetic control is small RNAs (sRNAs) that have been documented across several plant species in stress resilience and transgenerational inheritance. Recent studies combining WGBS, chromatin immunoprecipitation sequencing (ChIP-seq), accessible chromatin sequencing (ATAC-seq), and high-throughput chromosome conformation capture sequencing (Hi-C-seq) with multiomics frameworks unravel chromatin state transition during development and stress response (Cheng & McCarrey, 2023a). ATAC-seq and WGBS have been combined to identify regulatory hotspots associated with growth and stress resistance, such as the PToRAP2.12-PtoGntK axis in poplar (He et al., 2024). Now in the study of epigenetic regulation, the standard techniques are these WGBS, ChIP-seq, ATAC-seq, and Hi-C in the era of quantitative epigenomics (Qin et al., 2024).

High-throughput sequencing technologies have generated massive nonlinear, multilayered data that cannot be analyzed through traditional biostatistical tools. As models like linear regression, analysis of variance [ANOVA], and Principal component analysis (PCA) assume linearity and independence, they cannot comprehend complex regulatory architectures of the plant genome (Cheng & McCarrey, 2023b). Here comes the artificial intelligence (AI) and deep learning (DL), which can read and interpret these complex, nonlinear dependent datasets (Yassi et al., 2023). AI-backed approaches have become an inseparable part to uncover regulatory motifs and interactions by processing genomic, methylation, histone, and transcriptomic layers simultaneously (Nguyen et al., 2021). The Smart Model for Epigenetics in Plants and DeepSignal-plant can interpret multicontext methylation and histone modification data with more than 90% accuracy, showing the real power of DL (Ni et al., 2021; Wang et al., 2021). Similarly, AtacWorks DL frameworks can detect regulatory peaks from low-quality ATAC-seq data (Lal et al., 2021). Deep signal-plant model can analyze methylation profiling by learning contextual methylation signatures from nano long reads (Ni et al., 2021). A generative AI model named DMRU has more than 90% accuracy in predicting condition-specific cytosine methylation by learning the relationship between methylation pattern and transcriptomic status (Gupta et al., 2025). AI frameworks like DeepATAC and AtacWork can identify transcription factor binding and chromatin accessibility even from low-resolution ATAC-seq datasets (Hiranuma et al., 2017; Lal et al., 2021). These models can annotate cis-regulatory elements in the plant genome related to domestication and stress adaptation. Further Transcription Regulator Activity Prediction Tool (TRAPT) is capable of predicting functional transcriptional regulators from a queried genome by integrating it with histone modification, ATAC-seq, and TR-ChIP-seq data (Zhang et al., 2024a). These approaches are interpreting nonlinear gene regulatory hierarchies to predict the underlying traits like drought tolerance and flowering time. These innovations can efficiently perform cross-species generalization, predictive modeling of adaptation, and generate hypotheses for crop improvement and many more, which can be marked as the AI renaissance in plant epigenetics (Abdolmaleky et al., 2025; Miryeganeh, 2025).

Even after tremendous advancement, two basic questions float in our mind are "How can AI reveal the hidden epigenetic mechanisms governing adaptation and domestication?" And "What predictive insights can it offer for breeding and climate resilience?" In layman's language, first, epigenetic regulation operates in interdependent multiple layers which vary spatially and temporally, thus AI models must go beyond classification toward causal inference (Choudhary et al., 2022). Second, we need interpretable AI frameworks that can explain complex biological data rather than making predictions. Models like MHCpG are the hope toward merging statistical interpretability with DL accuracy (Fu et al., 2020). Finally, the success of epigenetically engineered crops for climate resilience lies in the hands of translational applications. Integration of epigenomic states with phenotypic outcomes and environmental variables can change the fundamentals of plant breeding in light of DL systems (Zhu et al., 2025).

4.2 Epigenomic Data to Predictive Models: From Data to Discovery

As technology evolved, computational epigenetics shifted from descriptive statistical analysis to AI-driven predictive models. With time, high-throughput techniques become widespread, and data complexities quickly exceed the traditional approaches (Xue et al., 2022). The introduction of DL and network-based models allowed nonlinear mapping of high-dimensional epigenomic interactions (Kim, 2016). As we go further, we realize that only prediction is not enough; we need to unwind the causality behind the complex biological activities. Here comes the explainable AI (XAI), which can reveal causal patterns within multilayered biological systems. OmiXAI are such models that combine convolutional and graph neural networks (GNNs) to explain transparent feature attribution in epigenomic data (Alaeva et al., 2025). Another XAI framework, LDEncoder, can use latent spaces for cross-omics transfer learning to generalize models across species (Park & Bryson, 2021). Epigenomic science is fully aware of the fact that no single molecular layer can explain adaptation and domestication fully; that is why multiomics integration for biological inference is absolutely necessary (Ruprecht et al., 2024). Multiomics factor analysis (MOFA) is the pioneer of unsupervised data fusion, leading to the discoveries of shared and unique variation across transcriptomic and methylomic layers (Argelaguet et al., 2018). Models like multiomics transformer (MOT) and MAPLE are capable of integrating heterogeneous data as a unified representation and explaining system-level gene regulation insights. MOT uses hierarchical representations of transcriptomic and epigenomic data for phenotypic prediction with a 98% accuracy level (Osseni et al., 2022). MAPLE works on methylation expression linkages at the cell level to construct regulatory networks (Uzun et al., 2021). Another model, Garfield (graph-based contrastive learning enables fast single-cell embedding), uses geometric DL to reveal spatial regulatory niches and long-range interactions (Zhou et al., 2025). These AI-based models are forming the foundation of the new era of multiomics system biology and showing immense applications from cancer biology to agriculture and ecological genomics.

Research in neuro-epigenomics and aging shows how AI models are capable of finding regulatory signatures behind complex traits. For instance, DL-XAI can provide insights into the inflammatory gene subnetworks associated with aging (Li et al., 2025). In the mouse brain, using single-cell methylome atlases presented methylation diversity maps to functional specialization, which is applicable directly as a cell type-specific epigenetic variation in plant tissues (Liu et al., 2024). Comparative studies across species show integration of chromatin conformation, methylation, and transcriptome data is the key to identifying conserved regulatory modules controlling cell identity (Seal et al., 2020). It is mentioned that human and animal studies can be aligned with the algorithm translation of plants and show the universality of computational epigenomic pipelines.

Machine learning (ML) models that can predict methylation states and chromatin accessibility under stress conditions could be the most important guide to make epigenetically optimized genotypes (Shittu et al. 2024). Multiomics field data, along with environmental parameters, can guide us toward predictive modeling of yield, drought, and disease resistance (Zhang et al., 2025), which is directly connected to plant adaptation and domestication.

4.3 AI/ML/DL Frameworks for Epigenetic Analysis in Plants

4.3.1 Machine Learning Approaches

ML can be regarded as the foundation of modern plant epigenomic analysis. There are several ML algorithms, among which the following are the most widely used, viz. support vector machines (SVMs), random forests (RFs), naive Bayes classifier, and k-means clustering. These algorithms can efficiently locate patterns in DNA methylation and histone modification datasets. Among them, some are supervised and some unsupervised; both methods are highly efficient in correlating features like CpG methylation sites and differential methylation regions (DMRs) with environmental stresses (Wu et al., 2022). SVMs are supervised learning algorithms that work well with labeled datasets (e.g., methylated vs. unmethylated sites). It is useful to analyze feature-rich genomic datasets for higher precision. RF algorithms also fall under supervised algorithms, but work well with even noisy datasets. It could integrate methylation intensity, chromatin accessibility, and transcriptional data from multiple types of cells (Mienye & Swart, 2024). Unsupervised algorithms are used when data labels are absent. This technique consists of k-means clustering, hierarchical clustering, and Gaussian mixture models can reveal latent regulatory subgroups linked to epigenetic reprogramming through pattern recognition. PCA and t-distribution stochastic neighbor embedding (t-SNE) can be used for the reduction of dimensionality that converts complex methylation data into low-dimensional spaces helpful for visualization of data (Chauhan & Singh, 2018).

A simple ML workflow in plant epigenomics consists of the following: (1) Data preprocessing (e.g., bisulfite-seq read alignment and methylation calling); (2) feature engineering, in this process, statistical features like methylation percentage, GC content, and sequence motifs are extracted; (3) feature selection includes algorithms like recursive feature elimination to retain the most relevant biological variables in the study; (4) model training and cross-validation, use stratified folds to avoid any overfitting; (5) performance evaluation, using different metrics like accuracy and the Matthews correlation coefficient.

Tree-based models like RFs and eXtreme gradient boosting (XGBoost) had unwound the stress-responsive methylation signatures in *Arabidopsis* and *Oryza sativa* and pinpointed epigenetic markers to phenotypic resilience (Shiri et al., 2024). These applications of ML in histone and methylation marks prediction with a high level of accuracy lead us to the path of ML-assisted stress signature discovery.

4.3.2 Deep Learning for Epigenetic Pattern Recognition

DL is based on artificial neural networks (ANNs), is a subfield of ML having multiple hidden layers that can be trained to learn different hierarchical features from complex biological datasets. Nowadays, DL is an essential part of inferring chromatin states, methylation landscapes, and stress-induced epigenetic plasticity from multiomics data (Beknazarov & Poptsova, 2023). Here, four types of ANNs have been discussed, viz. convolutional neural networks (CNNs) and recurrent neural networks (RNNs), GNNs, and Transformers.

4.3.2.1 CNNs for Spatial Epigenetic Patterns

CNNs are efficient in detecting spatially localized features thus are a central pillar in epigenomic pattern recognition (Figure 4.1). It uses convolutional filters across the input matrices to find positional dependencies among CpG sites. CNNs have shown

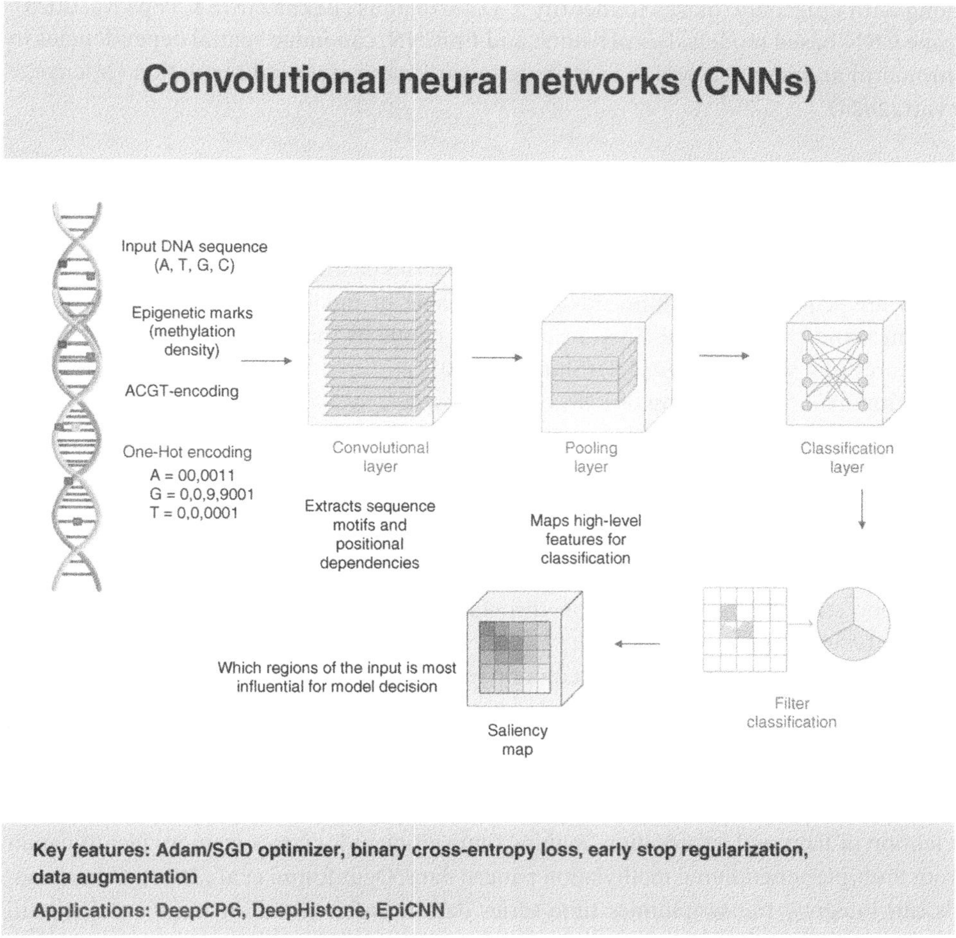

Figure 4.1 Convolutional neural networks (CNNs) for spatial epigenetic patterns in plants.

significant results for methylation motif detection and chromatin state classification in plant systems. CNNs for plant epigenomics use Adam optimizers or stochastic gradient descent with a binary cross-entropy loss. It prevents overfitting to limited labeled data by early stopping and dropout regularization. Two techniques convert bases (A, T, G, C) into binary data and introduce noise for robustness of the model, viz. one-hot encoding and data augmentation. There is a CNN framework called DeepCpG, which, for mammalian data, has been successfully applied to *Arabidopsis thaliana* to explain single-cell methylation data. The basic structure of CNNs is as follows: (1) Convolutional layers, for the extraction of sequence motifs and methylation patterns; (2) pooling layers help to reduce dimensionality with sufficient feature retention; (3) fully connected layers, which map final convolutional outputs. CNN-based classifiers can explain methylation probability for each cytosine site using a flanking sequence window. Further saliency maps and filter visualization can pinpoint which sequence motifs drive the model prediction (Wu et al., 2022). There are even more advanced CNNs that can explain multi-channel inputs from ChIP-seq and ATAC-seq assays. DeepZ is based on one of the most advanced CNNs that analyzes beyond methylation patterns by sequence-based features along with epigenetic marks to identify Z-DNA regions (Beknazarov & Popsova, 2023). Some CNN-based models, DeepHistone and EpiCNN, can judge spatial dependencies in chromatin architecture, which contribute toward transcriptional regulation (Mienye & Swart, 2024).

4.3.2.2 RNNs for Temporal Epigenetic Dynamics

On the other hand, RNNs are an algorithm based on temporal modeling. It is capable of interpreting dependencies across epigenetic time series or stress-induced memory datasets, which are sequential in nature. RNNs are capable of modeling epigenetic inheritance dynamics and chronological stress responses in plants by maintaining a hidden memory that stores past inputs (Figure 4.2) (Shin et al., 2017).

The mathematical equation of RNNs is as follows:

$$h_t = f\left(W_h\, h_{t-1} + W_x\, x_t + b\right)$$

where h_t = hidden state at time t, x_t = the input (e.g., methylation at time t), and W_h, W_x are the weight matrices. Parameters are updated based on temporal dependencies between epigenetic states by backpropagation through time (BPTT) (Lillicrap & Santoro, 2019; Pineda, 1989).

RNNs are very reliable for taking temporal methylome snapshots (e.g., drought response across developmental stages), which help us to identify persistent regulatory signatures. In the case of RNNs, there is a problem of vanishing gradients, which limits their capacity for long-range use. To overcome this limitation, long short-term memory and gated recurrent unit have been introduced. These new models use input, output, and forget gates to aid retention of temporal information, making them efficient for stress memory identification from multiple generational methylation pattern data (Ogunfunmi et al., 2019). Other models can integrate transcriptomics time series data to infer causal environmental stimuli

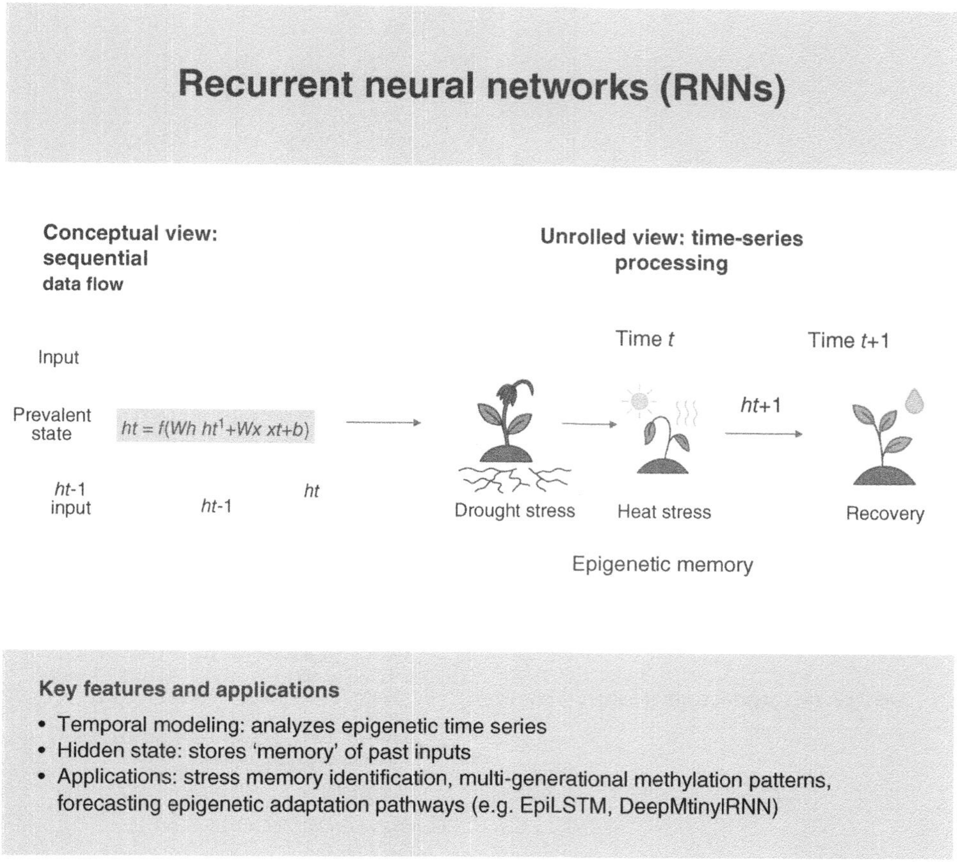

Figure 4.2 An overview of recurrent neural networks (RNNs) for modeling temporal epigenetic dynamics, sequential stress responses and epigenetic memory in plants.

toward epigenetic adaptation pathways. EpiLSTM and DeepMethylRNN are among those models, capable of utilizing time series biological data to forecast changes in response to sequential stimuli (e.g., heat stress———> drought———>recovery). RNNs are a kind of algorithm that enables us to peek into the epigenetic memory and adaptation from the dynamic methylation pattern over time and stress cycles (Voytetskiy et al., 2022).

4.3.2.3 Graph Neural Networks for 3D Chromatin Architecture

Next is the GNNs, most used in modeling 3D chromatin topology and enhancer promoter interaction from Hi-C and ChIA-PET assays (Figure 4.3). Unlike RNNs and CNNs, this model works with relational data dependencies. In this model, chromatin conformational data are presented in a structured graph where nodes are genomic loci (e.g., promoters and enhancers), and edges are physical interactions between the regions. GNNs can capture higher-order genome foldings and long-range regulatory communication by simply relating to aggregating feature information from neighboring nodes.

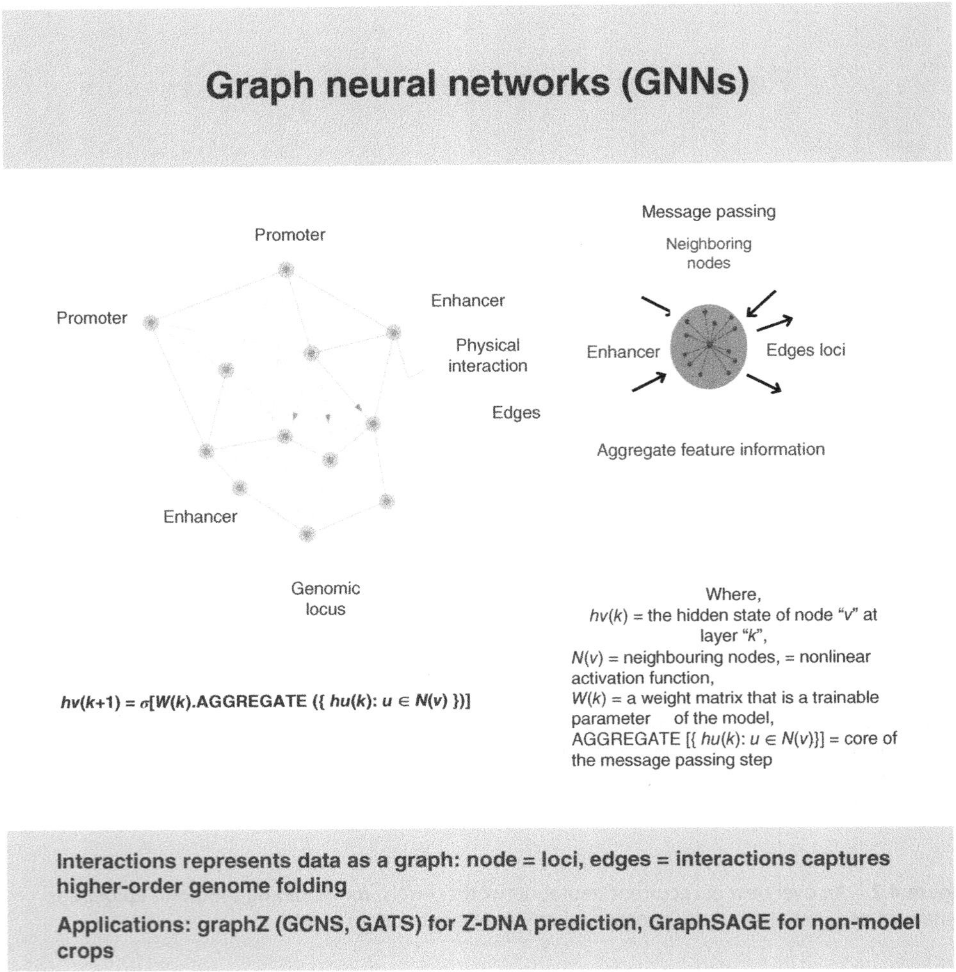

Figure 4.3 Graph neural networks (GNNs) for 3D chromatin architecture in plants.

The general message passing update rule can be expressed by the following mathematical equation:

$$h_v^{(k+1)} = \sigma\left[W^{(k)}.\text{AGGREGATE}\left(\left\{h_u^{(k)} : u \in N(v)\right\}\right)\right]$$

where $h_v^{(k)}$ = the hidden state of node "v" at layer "k," and $N(v)$ is the neighboring nodes, σ = nonlinear activation function, = $W^{(k)}$ a weight matrix that is a trainable parameter of the model, AGGREGATE ($\{h_u^{(k)}$: $u \in N(v)\}$ = core of the message passing step (Fan et al., 2023; Jørgensen et al., 2018; Zhang et al., 2020).

Graph-Z is a model that utilizes graph convolutional networks and graph attention networks (GATs) have been seen to outperform CNN–RNN models in predicting Z-DNA and chromatin domains. Graph-Z uses a graph-based cluster of sequence, methylation, and

Hi-C data (Voytetskiy et al., 2022). GNNs learn regulatory networks that capture 3D chromatin folding by simply placing each genomic segment as nodes with methylation density and transcription factor occupancy features. GNNs are inferring enhancer promoter connectivity, which is the driving force of gene expression under environmental stress, by utilizing Hi-C, ATAC-seq, and RNA-seq data. For nonmodel crops, GraphSAGE has been a game-changer as it uses representation learning, which allows generalization of unseen loci or species (Beknazarov & Poptsova, 2023).

4.3.2.4 Transformers for Multiomics Integration

Transformers is another AI model that is based on query-key-value attention matrices to learn interspecies relationships within genomic or omics data (Figure 4.4). It has been designed upon a self-attention mechanism that can capture contextual dependencies between all possible positions in an input sequence simultaneously.

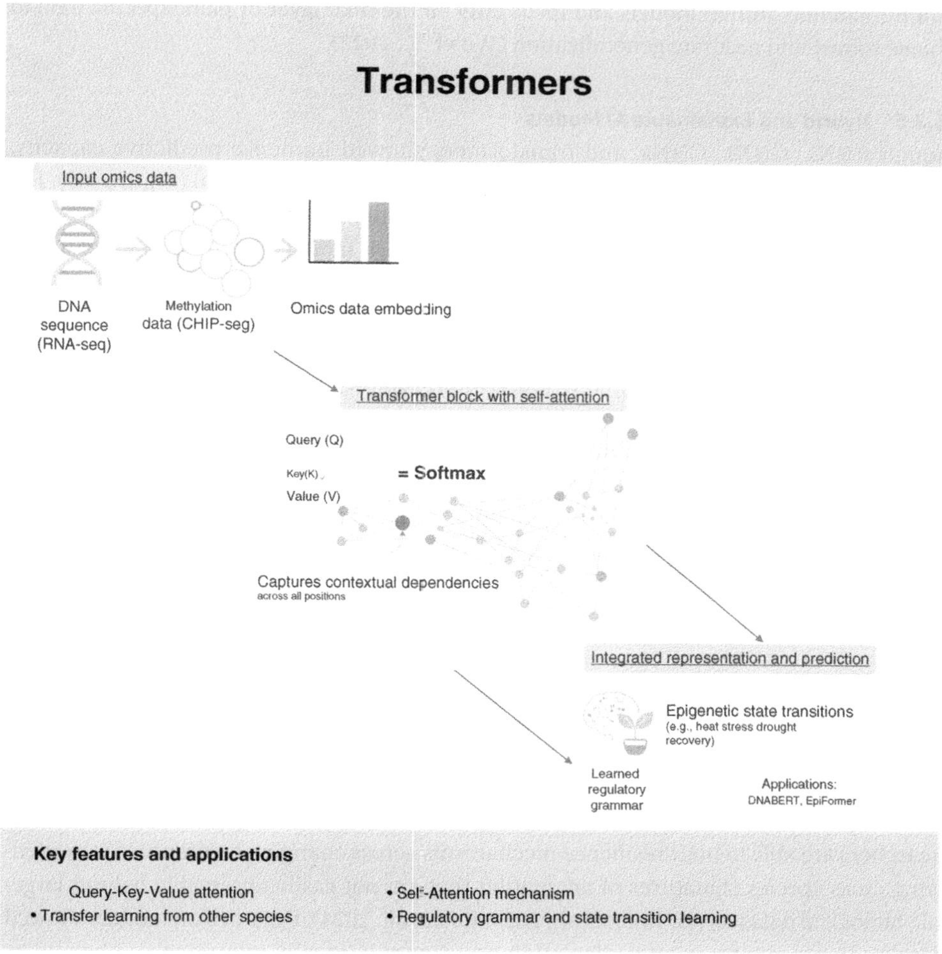

Figure 4.4 Transformers for multiomics integration in plants, including learning regulatory grammar and state transitions from diverse data.

Mathematically, it is represented as follows:

$$\text{Attention}\left(Q, K, V\right) = \text{softmax}\left(QK^{T} / \sqrt{d_k}\right)V$$

where Q, K, V are the query, key, and value matrices representing sequence features. K^T = transpose of key matrix and d_k = dimension of key vector (Guo et al., 2023; Han et al., 2024; Zhang et al., 2024c).

DNA Bidirectional Encoder Representations from Transformers [DNABERT] learns regulatory grammar from billions of base pairs by applying a transformer architecture to a 6-mer tokenized DNA sequence (Mienye & Swart, 2024). Another transformer-based model, EpiFormer, trained on *O. sativa* datasets, can infer epigenetic state transitions under heat stress by analyzing methylation and gene expression time series jointly (Zhou et al., 2025). But there is a catch that transformer-based models require extensive amounts of data with immense computational power. That is why researchers use transfer learning from human and animal models and focus only on the final layer of plant-specific data to achieve robust and accurate generalization (Wu et al., 2022).

4.3.2.5 Hybrid and Explainable AI Models

Though RNNs, GNNs, CNNs, and transformers showed immense predictive capacity, decision transparency and interpretability of AI models remain critical (Figure 4.5). For that hybrid architecture came into the picture, combining GNNs, RNNs, CNNs and transformers into one, leveraging the strength of each model. The hybrid approach of the CNN–RNN fusion model gives higher accuracy in methylation site classification and pattern generalization (Wu et al., 2022). DeepZHybrid models utilizing CNN and GNN exploit both spatial and relational chromatin features (Voytetskiy et al., 2022). SHapley Additive exPlanations (SHAP) and local interpretable model agnostic explanations (LIME) are two XAI based models that further minimize the gap between prediction and interpretability. SHAP quantifies each feature's contribution to the model output whereas LIME alters input features locally to observe the model outcome, offering instance-level interpretability. SHAP outcomes coupled with CNN filters can highlight which DNA motifs are most influential, which can be used for hypothesis formation for experimental validation (Huang, 2025).

4.4 AI Insights into Epigenetic Mechanisms of Plant Adaptation

Plant adaptation is a mechanism, one of the main reasons for plant resilience. Now, researchers are able to infer resilience mechanisms across changing environments by identifying cross-species signatures of adaptation that are not easily observable behind large-scale biological datasets (Le Roux et al., 2024; Rao et al., 2024). Data-driven models allowed scientists to see environmental adaptation as a computationally interpretable phenotype. These techniques made them capable of not just measuring genomic and phenomic data but also predicting them under changing climatic conditions (Ali et al., 2024). As we

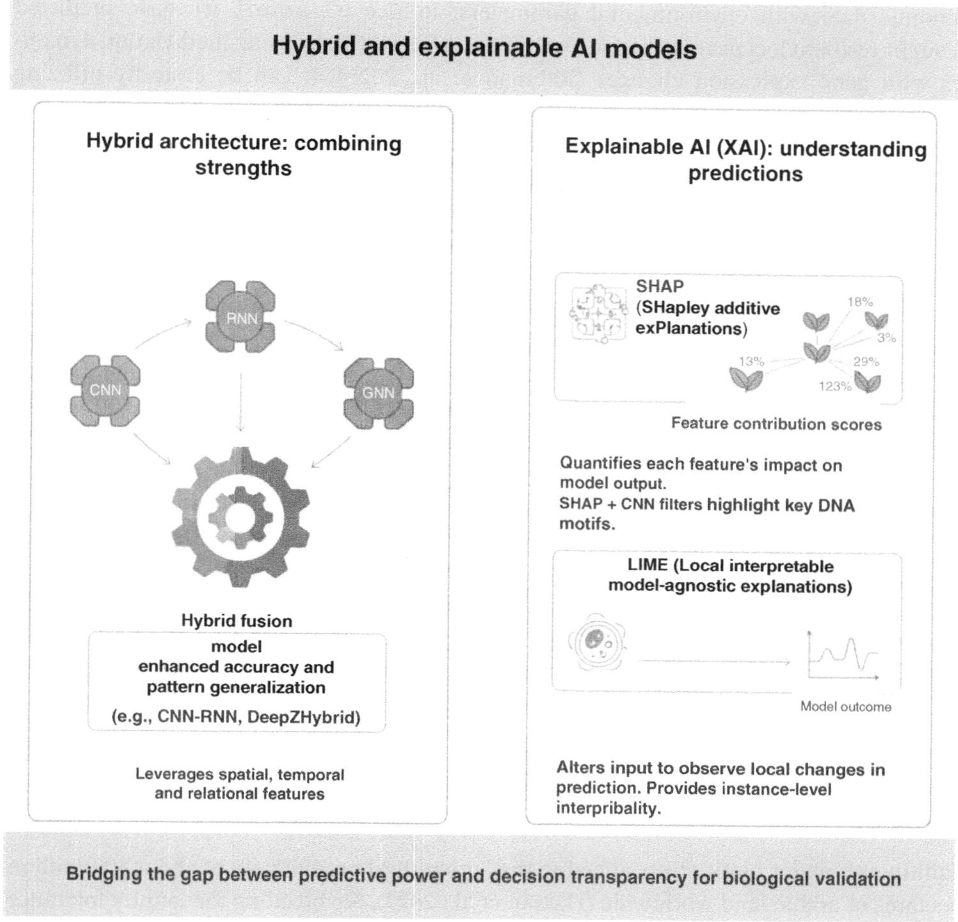

Figure 4.5 Hybrid and explainable AI models for plant epigenetics.

discussed in the previous section, CNNs, GNNs, RNNs, and transformers, along with the hybrid models, are capable of merging complex datasets of WGBS), transcriptomics, and chromatin accessibility data. These integrations made it possible to predict nonlinear regulatory hierarchies and latent molecular interactions that drive the plant resilience to a multistress environment (Guru et al., 2021). This section focuses on the stress-specific AI models and how they help to infer crop environment interaction during drought, salinity, heat, cold, flooding, and UV/multistress conditions. And it compares adaptation mechanisms of major cereal, legume, and orphan crops in the light of AI interpretability.

4.4.1 Drought Adaptation: Predictive Epigenomics and Transcriptomic Modeling

Drought is a highly destructive stress which accounts for affecting 60% of the land globally every year (Nouri et al., 2020). So, prediction of drought tolerance across crop species like rice, wheat, pearl millet, barley, and chickpea is a major part of AI-assisted bridging of genomic data with environmental parameters. In rice (*O. sativa*), we have predicted drought-resilient loci near DREB1A, OsHKT1.5, AND HSP70 by using methylation dynamics with gene expression changes (Rahman et al., 2025). It can be done by utilizing transformer-based models. Likewise, CNN-based multiomics fusion models have been utilized in the case of wheat (*Triticum aestivum*) (Le Roux et al., 2024) and barley (*Hordeum vulgare*) (Falahi et al., 2022) to merge epigenomic and physiological data (e.g., leaf temperature, root architecture, and yield potential). These models identified regulatory drought-associated promoters and signal networks from stress-correlated DNA methylation. In chickpea (*Cicer arietinum*), we have noticed DREB-linked modules as important drought endurance regulators using RF classifiers (Rai & Rai, 2020). In barley and rice, RNNs unfold the DNA methylation states that return to the previous state after rehydration but leave epigenetic marks on certain loci (Falahi et al., 2022). These findings indicated that drought resilience is not a singular genetic trait but a temporal epigenetic process (Rao et al., 2024).

Meta learning models are equipped beyond single crop studies with the tools to build "drought resilience indices" leading to yield forecasting at variable water stress levels (Zhang et al., 2021a). These indices are formed based on ensemble learning by comparing methylation, transcriptomic, and phenotypic data from rice, wheat, and legumes, and are capable of predicting adaptive performance in untested genotypes (Ali et al., 2024). Now plant biologists can simulate adaptation trajectories and forecast the evolution of crops across drought cycles (Le Roux et al., 2024). Precision breeding for drought tolerance and translatable environmental adaptation models largely depends on AI-based prediction with multiomics datasets to find stable biomarkers (Le Roux et al., 2024; Rao et al., 2024).

4.4.2 Salinity Stress: AI-Revealed Chromatin and Ion Homeostasis Networks

Salinity is a major biotic stress affecting the agricultural productivity of about 800 million hectares of arable land worldwide (Hassan et al., 2025). So, breeding for salinity tolerance is an important part of resilient crop breeding. DL, GNNs, and probabilistic inference frameworks have become great tools to uncover layers of chromatin topology and ion homeostasis networks in crops. GNNs have been used to model salinity-induced rearrangement of chromatin topology in *Zea mays* and *Solanum lycopersicum*. This model successfully identified regulatory interactions between ion transporter genes like SOS1, NHX1, and HKT1 (Basu et al., 2020). AI systems trained on ATAC-seq and RNA-seq data of *Z. mays* and *Manihot esculenta* are capable of linking transcriptional and post-transcriptional datasets with physiological traits. It eventually found cis-regulatory regions that become epigenetically active during salinity stress. And these regions correspond to the loci of vacuolar antiporters and calcium transport proteins that balance intracellular ion concentration (Kumar et al., 2021). DL models trained on Hi-C and methylome datasets can find chromatin condensation near stress-regulated promoters, which maintain key stress response activity (Singhal et al., 2021). 3D chromatin simulation is a way of predicting "salinity resilient hubs" of genes encoding vacuolar H+-ATPases and Na+/H+ exchangers. These patterns completely align with the discovery that tolerance cultivars of *Z. mays* and *O. sativa* maintain a certain level of K+/Na+ ratio via dynamic transcriptional regulation of ion channels (Basu et al., 2020). Chromatin modeling of *M. esculenta* showed epigenetic crosswalk between methylation and histone marks under salinity stress, which control ion channel expression and osmotic signaling pathways (Hassan et al., 2025). AI-assisted integration of multiomics layers can provide a system-level, unprecedented perspective of ionic homeostasis of plants during salt stress (Singhal et al., 2021). Multitask DL has combined histone modification profiles with expression data and revealed histone marks H3K4me3 near SOS1 and NHX1 loci, correlating strongly with transcriptional stability during salt exposure in the case of *S. lycopersicum* (Kumar & Srivastava). Ensemble AI models in quinoa (*Chenopodium quinoa)* found methylation patterns and transcriptional regulators that maintain antioxidant balance during salt stress (Zamanzadeh-Nasrabadi et al., 2023). Deep reinforcement learning models using simulations to predict metabolic and ionic balance for sustenance of long-term salt exposure and finding adaptive dynamics between osmotic regulation and energy expenditure (Hassan et al., 2025). It has been found that genes regulating Na+ compartmentalization (NHX1, HKT1), ROS detoxification (GSTU4), and methylation dynamics (MET1, CMT3) have key roles in AI-driven tolerance models (Basu et al., 2020). AI models have built learnable molecular networks for plant salinity tolerance to improve accuracy with emerging immense omics data (Singhal et al., 2021).

4.4.3 Heat Stress: AI-Modeled Epigenetic Dynamics and Thermotolerance

As global temperatures are rising, plant thermotolerance has become critically important for resilient, sustainable agriculture (Kourani et al., 2022). AI-driven architectures like CNNs and GNNs have been a great help in achieving our target. CNN-based heat phenotyping systems have quantified stress levels with 90% accuracy in *H. vulgare* and *T. aestivum*. They are using thermal imaging and environmental data for this purpose (Guidara et al., 2024). These AI-based phenomics models have proven valuable to identify early thermal injury signals at the canopy level from molecular and transcriptomic patterns (He et al., 2025). Heat stress triggers large-scale transcriptomic remodeling, histone modification, and DNA methylation changes, which in turn regulate stress-responsive genes (e.g., HSP70, HSP90, AND DREB2A) (Chatterjee et al., 2023). OmiXAI models combine transcriptome, methylome, and proteome data to identify causal driving features of heat tolerance (Alaeva et al., 2025). Variational autoencoders (VAEs) and transformer-based networks have been successfully applied to *Sorghum bicolor* and *G. hirsutum* to identify "latent adaptive manifolds." And we observed dynamic change in methylation pattern near promoters of HSP and ROS scavenging genes, showing DNA methylation configuration during extreme heat stress. In *Brassica napus* oscillatory genes have shown linkage to thermotolerance during reproductive stages. Heat shock proteins and splicing regulators are the preservers of floral fertility during temperature fluctuations (Kourani et al., 2022). AI models have the ability to map temporal methylation-demethylation cycles, pinpointing molecular "memory." That supports adaptive thermotolerance. Observations in *A. thaliana* through cis-regulatory modeling of heat and drought responses lead us to similar results (Azodi et al., 2020). Platforms like mosGraphGen can produce multiomics signaling graphs by linking transcriptomic reprogramming with cellular homeostasis under heat and create multiomics signaling graphs (Zhang et al., 2024b). Foundation models (large transformer-based systems) are capable of reading hidden regulatory dynamics that are related to tissue-specific thermotolerance by combining single-cell transcriptomics and proteomics data (Chau et al., 2024). CNNs and graph-based learning can establish a correlation between leaf surface temperature, epigenetic reprogramming, and HSP expression stability. These are helpful for the identification of genomic markers for heat-tolerant lines (He et al., 2025). Also, deep reinforcement learning and Bayesian inference can identify persistent methylation signatures that act as "epigenetic thermomemory" by cyclic methylation pattern simulation across heat cycles (Kourani et al., 2022). XAI has enabled researchers to trace predictive accuracy back to key genomic and metabolomic factors (Alaeva et al., 2025). Through these systems, methylation near HSP70 and HSP90 promoters, histone acetylation patterns, and ncRNA networks have been recognized as dominant features of thermotolerance (Aliyu, 2024). Further cross-species insights from wheat, sorghum, cotton, and canola lead us to next-generation thermotolerant breeding (He et al., 2025).

4.4.4 Cold Stress and Flooding: Temporal AI Modeling of Stress Memory

Cold stress and flooding-induced hypoxia are the two most complex environmental pressures to plant metabolism. Cold tolerance includes transcriptional and epigenetic changes that induce CBF/ICE regulatory cascades. RNN models have the ability to simulate these transitions, discovering a "molecular memory circuit" that determines the cold adaptation persistence (Banerjee et al., 2017). In *A. thaliana,* cycles of DNA methylation-demethylation around CBF and COR genes show aligned changes with cold acclimation responses (Ouellet & Charron, 2013). Studies in *B. napus* and *A. thaliana* showed that cold stress induces activation of ATBS1 interacting factor 2 (AIF2), which controls phosphorylation feedback loops to regulate ICE1/CBF stability. DL based causal networks link AIF2-MAPK3/6 interaction with freezing tolerance in a time-dependent manner (Kim et al., 2024). Cold-induced histone acetylation patterns and sRNA expression waves are the short-term "epigenetic memory units" that help plants to remember stress exposure. This kind of relationship can be unfolded by the AI integration to find protective genes during recurrent chilling or flooding events (Satyakam et al., 2022). By combining flooding and temperature data of *Festuca pratensis,* deep regression models found a correlation of enhanced sugar metabolism with CBF6 expression (Jurczyk et al., 2013). Transformer-based systems found promoter hypomethylation of ADH1 and GSTU4 genes for energy metabolism during hypoxic conditions in rice and soybean (Charrier et al., 2023). Graph-based models have linked chromatin topology with ROS scavenging and carbohydrate mobilization pathways during post-flood situations. These models inferred hypoxic reprogramming as a result of periodic demethylation-acetylation cycles (Ambroise et al., 2019). AI meta models trained on transcriptomics datasets from eucalyptus, wheat, potato, and maize have had the capacity to find conserved cold resilience genes and efficiently found the shared CBF-mediated regulatory architectures in several species (Fernandez et al., 2015). Genome-wide AI selection frameworks have located loci controlling frost tolerance on chromosome 5B in wheat, which is out of the known Fr-B1/B2 regions (Zhao et al., 2013).

4.4.5 UV and Multistress Environments: Hybrid Adaptation Signatures

High altitude with multistress environments (ultraviolet radiation, low temperature, and oxidative stress) is one of the most challenging adaptive landmarks for plants. Multistress responses integrate photoprotection, osmotic regulation, and antioxidant metabolism to form cross-regulatory responses. AI-based omics integration made it possible to understand shared molecular signatures by making hybrid adaptive clusters (Perrone & Martinelli, 2020). CNNs and transformer-based models have been very helpful to assess UV-induced gene expression patterns and epigenetic changes simultaneously and infer antioxidant defense networks (Chen et al., 2022). These frameworks can also infer cross-stress co-regulation. *C. quinoa* and *Amaranthus hypochondriacus* showed coordinated regulation of flavonoid and ROS detoxification pathways between UV rays and oxidative tolerance with the help of deep reinforcement learning (Thakur et al., 2023). *Rheum nobile* and *Potentilla saundersiana,* like high altitude plants, face combined effects of UV exposure, hypoxia, and low temperature; in these plants, multiomics models uncovered methylation-driven adjustments to flavonoid biosynthesis and antioxidant metabolism (Li et al., 2023; Ma et al., 2015). An AI model specifically developed for Qinghai-Tibet Plateau flora identified high altitude adaptive modules related to UV screening, ROS scavenging, and energy metabolism (Sun et al., 2025). Recurrent UV exposure triggers stable DNA methylation imprints (Zhang et al., 2021b). Transformers models used in *Arabidopsis* and *Populus species* show UV exposure increases the activity of antioxidant genes in subsequent drought and heat cycles (Muller-Xing et al., 2014) (Table 4.1).

4.5 AI Insights into Plant Domestication Epigenomics

4.5.1 Epigenetic Footprints of Domestication in Crops

AI-assisted methylome and chromatin analysis are capable of tracking system-level domestication footprints. It transformed the grounds of plant evolution and crop improvement studies (Álvarez-Venegas & De-la-Peña, 2016). MLACNN is a model for the integration of attention mechanisms to methylation prediction (Bai et al., 2023) to find DMRs and long-range enhancer promoter loops.

CNN-based classifiers trained on wild and cultivated methylomes of *S. lycopersicum* found epialleles near FW2.2 and RIN, which control fruit size and ripening. Other studies showed a connection between stress adaptation and domestication through the CNN methylation pattern of drought-responsive genes (Asr 1) (González et al., 2011). These results support tissue-specific gene repression and epigenetic memory through CNN methylation (Diéguez et al., 1997). Bisulfite and histone data interpretation through autoencoder networks show loss of methylation variance in stress-responsive genes (HvCBF2). Fe deficiency and cold exposure enhance stable methylation states in barley (*H. vulgare*), which can be canalized through breeding (Álvarez-Venegas & De-la-Peña, 2016). Methylation profiling of 95 rice accessions revealed that domestication leads to a decrease in methylation diversity and fixation of adaptive chromatin loops (Cao et al., 2023). Persistent small-RNA clusters with panicle architecture and yield analyzed with transformer-based multiomics integration clearly state the convergence between selection and stress-induced methylation networks. Reconstructed 3D chromatin topology of *Z. mays* domestication genes analyzed by Hi-C-based GNNs to find compaction near kernel shape QTLs. RNA-based silencing and chromatin remodeling during maize embryogenesis showed stress-related miRNA

Table 4.1 Crop-specific stress adaptations via AI-driven models.

Crop	Stress type	AI framework	Epigenetic layer	Adaptive insight
Rice (*Oryza sativa*)	Drought and flooding	Transformer-based WGBS transcriptome model	DNA methylation and promoter site differential methylation regions	Water-deficit endurance by DREB1A and OsHKT1.5 activation QTLs (Jaya et al., 2021).
Maize (*Zea mays*)	Salinity	Deep learning (Hi-C + ATAC-seq)	3D chromatin topology and histone marks	Salt-responsive chromatin hubs controlling SOS1/NHX1 stability (Khan & Khan, 2019).
Wheat (*Triticum aestivum*)	Heat and drought	CNN and phenome-omics fusion	Methylome and leaf thermal image	Thermotolerance loci; omics-AI synergy enhances drought resilience (Le Roux et al., 2024).
Sorghum (*Sorghum bicolor*)	Heat	Variational autoencoder (VAE)	DNA methylation and RNA-seq	HSP70–HSP90 regulatory networks predicting thermotolerance (Ali, 2023).
Pearl Millet (*Pennisetum glaucum*)	Drought	Reinforcement learning	ncRNA-metabolite interactions	miRNA-osmolyte feedback loops enhancing osmotic regulation under dehydration (Abdelrahman et al., 2018).
Cassava (*Manihot esculenta*)	Salinity	Graph neural networks	Methylation chromatin interactions	Salt-induced methylation–ion homeostasis coordination (Khan & Khan, 2019).
Soybean (*Glycine max*)	Flooding and hypoxia	Transformer-based multiomics	Promoter methylation and transcriptome	Promoter hypomethylation sustaining anaerobic metabolism (Wu & Xie, 2024).
Canola (*Brassica napus*)	Cold and heat	Deep neural networks	Time series epigenomics	Cold and heat co-tolerance by oscillatory methylation enhancing (Ali, 2023).
Quinoa (*Chenopodium quinoa*)	UV and salinity	Ensemble deep learning	Methylation and antioxidant genes	UV protection and ion regulation (Ali, 2023).
Amaranth (*Amaranthus hypochondriacus*)	UV and drought	Transfer learning method	Transcriptome and methylome	UV antioxidant resilience with quinoa (Wu & Xie, 2024).
Tomato (*Solanum lycopersicum*)	Salinity and heat	Multitask deep learning	Histone marks and transcriptome	H3K4me3 and H3K27me3 near stress loci (Ali, 2023).
Cotton (*Gossypium hirsutum*)	Heat and salinity	Graph neural networks	DNA methylation and ncRNA	Conserved ncRNA clusters regulated fiber quality (Wu & Xie, 2024).
Teff (*Eragrostis tef*)	Heat	Bayesian probabilistic models	Methylation and transcriptome	Thermotolerance by methylation yield correlations (Ali, 2023).
Arabidopsis thaliana	Cold and UV	Recurrent AI models	Dynamic methylation and CBF/DREB regulation	Oscillatory methylation in cold and UV memory circuits (Wu & Xie, 2024).
Rheum nobile	UV and high altitude	Multiomics deep learning	Transcriptome and metabolome	Flavonoid-ROS detoxification (Ali, 2023).

pathways are epigenetically regulated (Álvarez-Venegas & De-la-Peña, 2016). Bayesian models can predict silencing of dormancy genes based on sRNA and methylation for synchronized germination and seed uniformity. Population-wide methylome analysis in wild and cultivated accessions of soybean (*Glycine max*) predicted over 5,000 differentially methylated regions having only 23% correlation to SNPs, explaining the strong epigenetic basis of domestication (Shen et al., 2018). CNN–GNN hybrid models found selection-driven fixation of carbohydrate metabolism circuits in cassava (*M. esculenta*) through analyzing consistent methylation near GBSS and AGPase regions. Epigenomic fingerprinting in cassava has perfected differential methylation between clonally propagated and meristem-cultured plants, leading to stable epimarks at starch biosynthesis loci (Álvarez-Venegas & De-la-Peña, 2016). Methylation of heat response genes expresses itself recurrently under prolonged stress, ultimately toward aridity adaptation in sorghum (*S. bicolor*) (Álvarez-Venegas & De-la-Peña, 2016). Methylome sequencing reveals incomplete domestication and polyploidization-driven adaptation in *Echinochloa crus-galli*, showing interplay between genomic complexity and methylation stability (Chu et al., 2025).

4.5.2 Cross-Species Generalization

CNN-vision transformer (ViT) and transformer-Bayesian pipelines are the hybrid architectures performing excellently in cross-species generalization. These models trained on *Arabidopsis* or *Sorghum* can infer methylation and CRE activity in tomato and rice (Padshetty & Umashetty, 2024). These architectures use feature attention modules and multilevel fusion layers to combine methylation motifs with regulatory landscapes. This kind of capacity of epigenomic transfer learning not only infers existing domestication signals but also predicts emergent domestication trajectories under stress and future breeding scenarios (Mienye & Swart, 2024). Current XAI are capable of feature attribution in epigenetic selection indices to explain which DMRs or histone marks are mostly influencing predicted traits (Amin et al., 2024). Further, CRISPR/dCas9-TET1 and dCas9-DNMT3A could directly utilize AI-predicted ESI hotspots for crop improvement (Shaikh et al., 2022). Spike-driven transformers are working well to emulate adaptive efficiency by taking inspiration from biological networks to achieve energy-aware plasticity modeling (Yao et al., 2024). Epigenetic domestication is not only bound to plants; early generation fish domestication experiments found shifts in genome-wide methylation in a single generation due to environmental confinement and diet (Podgorniak et al., 2022). Also, pig behavioral data modeled with transformer–CNN hybrids achieved more than 96% accuracy in vocal stress signatures (Liao et al., 2023).

4.6 Future Directions in AI-Driven Epigenomics

Epigenomic datasets are by default heterogeneous in format and biological context arising from WGBS, ATAC-seq, ChIP-seq, and Hi-C. The heterogeneity of datasets complicates model training and feature harmonization. And it leads to a species bias toward model organisms. Addressing these challenges needs standardized ontologies and community-level initiatives for multiomics data annotation and fair data sharing (findable, accessible, interoperable, reusable) (Liu et al., 2024). The scarcity of publicly accessible plant epigenomes is limiting AI generalization. Taking lessons from human and animal epigenome consortia (e.g., ENCODE and NIH Roadmap), we can establish open data deposition under the federal learning architectures model.

A basic technical limitation of current AI integration lies in model generalizability, explainability, and computational scalability. Though deep models can achieve high predictive accuracy, they are not very transferable between plant taxa. We have seen the use of domain adaptation strategies in biomedical AI (Erimus et al., 2024), to calibrate models trained on one species for application on another. Explainability always remained a critical question in plant epigenomics. To address this problem, Bayesian graph-based XAI proved to be efficient. It is capable of finding regulatory causality instead of correlations (Ababio et al., 2025). Even after attention maps and salient visualization aid interpretation, it is the black box problem that persists. Computational cost also remained nontrivial. Quantum federated learning (FL) algorithms (Qu et al., 2023) are a promising system to reduce overhead training. It leverages distributed quantum computing nodes for epigenetic pattern recognition.

Epigenomic data can reveal lineage-specific adaptation and domestication histories. Improper commercialization can lead to biocolonial exploitation. Application of federated and blockchain learning systems could ensure data privacy for cross-institutional collaboration without sharing raw data (Liu et al., 2024; Panda et al., 2025). By which geographically diverse datasets within FL models can be utilized for adaptive fairness optimization (Ramu et al., 2022). FL is a decentralized model training technique where only learned parameters are aggregated globally. FL can be combined with blockchain technologies for tamper-proof model auditing and traceability (Sheng et al., 2024). In this method, regional institutes can collaborate to train a methylation predictor for drought resilience while having sovereign data ownership. Asynchronous FL models (Qu et al., 2024) can improve the scalability of heterogeneous datasets for large-scale plant biodiversity modeling.

Next is the digital twins (DTs), virtual replicas of living systems are the future of integrative biology. Human biomedical DT systems are already predicting disease progression with multimodal AI integration (Okegbile et al., 2025). Similarly, in plants, plant epigenetic DT (PEDT) could simulate stress response trajectories using dynamic equations.

4.7 Conclusion

AI has embedded itself in the modern research workflow as it is serving as a powerful catalyst for engineering epigenetic mechanisms behind plant adaptation and domestication. As we can see, it is serving as a linking framework or computational bridge that enables integration of stress-induced plasticity and the heritable epigenomic signatures associated with domestication stability traits. The convergence of DTs, FL, and AI-guided epigenome editing has the descriptive genomics have become equipped to intervene in agriculture. These innovations collectively can reconstruct the shape of crop improvement research.

New technologies are going beyond technological sophistication and will go further. Now, ethical governance, algorithmic transparency, and equitable data access must be the main concerns for equitable sharing of these breakthroughs of science and technology. These principles are unavoidable to ensure the benefits of the scientific quantum leap are shared fairly across communities. Assisted epigenome study of plant adaptation and domestication is shaping the resilient agriculture of the dawn, and ultimately toward a new era of AI-driven intelligent breeding.

References

Ababio, I. B., Bieniek, J., Rahouti, M., Hayajneh, T., Aledhari, M., Verma, D. C., & Chehri, A. (2025). A blockchain-assisted federated learning framework for secure and self-optimizing digital twins in industrial IoT. *Future Internet*, *17*(1), 13. https://doi.org/10.3390/fi17010013

Abdelrahman, M., Burritt, D. J., & Phan Tran, L.-S. (2018). The use of metabolomic quantitative trait locus mapping and osmotic adjustment traits for the improvement of crop yields under environmental stresses. *Seminars in Cell & Developmental Biology*, *83*, 86–94. https://doi.org/10.1016/j.semcdb.2017.06.020

Abdolmaleky, H. M., Nohesara, S., Zhou, J. R., & Thiagalingam, S. (2025). Epigenetics in evolution and adaptation to environmental challenges: Pathways for disease prevention and treatment. *Epigenomics*, *17*(5), 317–333. https://doi.org/10.1080/17501911.2025.2464529

Alaeva, A., Lapteva, A., Mikhaylovskaya, N., Malkov, V., Herbert, A., Borevskiy, A., & Poptsova, M. (2025). OmiXAI: An ensemble XAI pipeline for interpretable deep learning in omics data. *bioRxiv*. https://doi.org/10.1101/2025.04.28.651097

Ali, H. (2023). Artificial intelligence in multi-omics data integration: Advancing precision medicine, biomarker discovery and genomic-driven disease interventions. *International Journal of Science and Research Archive*, *8*(1), 1012–1030. https://doi.org/10.30574/ijsra.2023.8.1.0189

Ali, T., Rehman, S. U., & Ali, S. (2024). Smart agriculture: Utilizing machine learning and deep learning for drought stress identification in crops. *Scientific Reports*, *14*, 30062. https://doi.org/10.1038/s41598-024-74127-8

Aliyu, B. S. (2024). Artificial intelligence (AI): A powerful tool for advancing plant study and research. *Nigerian Journal of Botany*, *37*(1), 61–68. https://doi.org/10.4314/njbot.v37i1.5

Álvarez-Venegas, R., & De-la-Peña, C. (2016). Editorial: Recent advances of epigenetics in crop biotechnology. *Frontiers in Plant Science*, *7*, 413. https://doi.org/10.3389/fpls.2016.00413

Ambroise, V., Legay, S., Guerriero, G., Hausman, J. F., Cuypers, A., & Sergeant, K. (2019). The roots of plant frost hardiness and tolerance. *Plant Cell Physiology*, *61*(1), 3–20. https://doi.org/10.1093/pcp/pcz196

Amin, M. N., Memon, S., Ali, A., Khan, H., Joshi, R., Rana, M., & Alsaawy, Y. (2024). Enhanced deep learning based precision agriculture: A decision support system for enhancing crop recommendation accuracy using convolutional neural networks (CNN). *Journal of Mechanics of Continua and Mathematical Sciences*, *19*, 166–187. https://doi.org/10.26782/jmcms.2024.09.00014

Argelaguet, R., Velten, B., Arnol, D., Dietrich, S., Zenz, T., Marioni, J. C., Buettner, F., Huber, W., & Stegle, O. (2018). Multi-Omics factor analysis—a framework for unsupervised integration of multi-omics data sets. *Molecular Systems Biology*, *14*(6), e8124. https://doi.org/10.15252/msb.20178124

Azodi, C. B., Lloyd, J. P., & Shiu, S.-H. (2020). The cis-regulatory codes of response to combined heat and drought stress in *arabidopsis thaliana*. *NAR Genomics and Bioinformatics*, *2*(3), lqaa049. https://doi.org/10.1093/nargab/lqaa049

Bai, J., Yang, H., & Wu, C. (2023). MLACNN: An attention mechanism-based CNN architecture for predicting genome-wide DNA methylation. *Theory in Biosciences*, *142*, 359–370. https://doi.org/10.1007/s12064-023-00402-3

Banerjee, A., Wani, S. H., & Roychoudhury, A. (2017). Epigenetic control of plant cold responses. *Frontiers in Plant Science*, *8*, 1643. https://doi.org/10.3389/fpls.2017.01643

Basak, N., Nizamuddin, S., & Thangaraj, K. (2017, October 2–4). Epigenetic signatures of high altitude adaptation in Tibetan population [Abstract]. *Abstracts of the NGBT conference* (Vol. 1, Special Issue, p. 113). Bhubaneswar, Odisha, India. *Canadian Journal of Biotechnology*. https://doi.org/10.24870/cjb.2017-a99

Basu, S., Kumar, A., Benazir, I., & Kumar, G. (2020). Reassessing the role of ion homeostasis for improving salinity tolerance in crop plants. *Physiologia Plantarum*, *171*(4), 502–519. https://doi.org/10.1111/ppl.13112

Beknazarov, N., & Poptsova, M. (2023). DeepZ: A deep learning approach for Z-DNA prediction. In K. K. Kim & V. K. Subramani (Eds.), *Z-DNA* (Vol. 2651, pp. 217–226). Humana. https://doi.org/10.1007/978-1-0716-3084-6_15

Cao, S., Chen, K., Lu, K., Chen, S., Zhang, X., Shen, C., Zhu, S., Niu, Y., Fan, L., Chen, Z. J., Xu, J., & Song, Q. (2023). Asymmetric variation in DNA methylation during domestication and de-domestication of rice. *The Plant Cell*, *35*(9), 3429–3443. https://doi.org/10.1093/plcell/koad160

Charrier, G., Willick, I. R., & Takahashi, D. (2023). Cross-disciplinary insights into the mechanisms of plant cold hardiness: From molecules to ecosystems. *Physiologia Plantarum*, *175*(2), e13901. https://doi.org/10.1111/ppl.13901

Chatterjee, A., Sen, P., Chattopadhyay, T., Maji, A., Gorai, S., Sarkar, A., & Ali, M. N. (2023). Omics for elucidating heat stress tolerance mechanisms in wheat: A review update. *Current Agriculture Research Journal*, *11*(2) https://doi.org/10.12944/CARJ.11.2.02, 362–379.

Chau, T. N., Wang, X., McDowell, J. M., & Li, S. (2024). Advancing plant single-cell genomics with foundation models. *Current Opinion in Plant Biology*, *82*, 102666. https://doi.org/10.1016/j.pbi.2024.102666

Chaudhary, N., Zhang, Y., Kesavan, N., Hall, J., Arpaci-Dusseau, A., Arpaci-Dusseau, R., Wright, S., & Mutlu, O. (2022, May 30–June 3). Accelerating deep learning-based identification of chromatin accessibility from noisy ATAC-seq data. *2022 IEEE International Parallel and Distributed Processing Symposium Workshops (IPDPSW)* (pp. 176–185). IEEE. https://doi.org/10.1109/IPDPSW55747.2022.00042

Chauhan, N. K., & Singh, K. (2018, September 28–29). A review on conventional machine learning vs deep learning. *2018 International Conference on Computing, Power and Communication Technologies (GUCON)* (pp. 347–352). IEEE. https://doi.org/10.1109/GUCON.2018.8675097

Chen, Z., Dong, Y., & Huang, X. (2022).). Plant responses to UV-B radiation: Signaling, acclimation and stress tolerance. *Stress Biology*, *2*, 51. https://doi.org/10.1007/s44154-022-00076-9

Cheng, K., & McCarrey, J. R. (2023a). Profiling the epigenetic landscape of the spermatogonial stem cell—Part 1: Epigenomics assays. In M. Oatley & B. P. Hermann (Eds.), *Spermatogonial stem cells* (Vol. 2656, pp. 78–108). Humana. https://doi.org/10.1007/978-1-0716-3139-3_5

Cheng, K., & McCarrey, J. R. (2023b). Profiling the epigenetic landscape of the spermatogonial stem cell: Part 2—computational analysis of epigenomics data. In M. Oatley & B. P. Hermann (Eds.), *Spermatogonial stem cells* (Vol. 2656, pp. 75–89). Humana. https://doi.org/10.1007/978-1-0716-3139-3_6

Chu, Q., Gong, X., Hu, Y., Zhu, Q.-H., Fan, L., & Ye, C.-Y. (2025). The DNA methylomes of *echinochloa* species provide insights into polyploidization-driven adaptation and orphan crop domestication. *The Plant Journal*, *121*(3), e70033. https://doi.org/10.1111/tpj.70033

Diéguez, M., Bellotto, M., Afsar, K., & Others. (1997). Methylation of cytosines in nonconventional methylation acceptor sites can contribute to reduced gene expression. *Molecular and General Genetics*, *253*, 581–588. https://doi.org/10.1007/s004380050360

Erimus, Ł., Borowska, A., Jaromin, A., Lewko, A., & Ruminski, J. (2024, July 8–11). Data domain adaptation in federated learning in the breast mammography image classification problem. *Proceedings of the 2024 16th International Conference on Human System Interaction (HSI)* (pp. 1–6). IEEE. https://doi.org/10.1109/HSI61632.2024.10.13534

Falahi, A., Zarei, L., & Cheghamirza, K. (2022). Most drought-induced DNA methylation changes switched to pre-stress state after re-irrigation in barley (*Hordeum vulgare* L.) cultivars. *Cereal Research Communications*, *50*, 429–438. https://doi.org/10.1007/s42976-021-00189-3

Fan, X., Gong, M., Wu, Y., Qin, A. K., & Xie, Y. (2023). Propagation enhanced neural message passing for graph representation learning. *IEEE Transactions on Knowledge and Data Engineering*, *35*(2), 1952–1964. https://doi.org/10.1109/TKDE.2021.3102964

Feng, S., Jacobsen, S. E., & Reik, W. (2010). Epigenetic reprogramming in plant and animal development. *Science*, *330*(6004), 622–627. https://doi.org/10.1126/science.1190614

Fernández, M., Troncoso, V., & Valenzuela, S. (2015). Transcriptome profile in response to frost tolerance in *Eucalyptus globulus*. *Plant Molecular Biology Reporter*, *33*(6), 1472–1485. https://doi.org/10.1007/s11105-014-0845-7

Fu, L., Peng, Q., & Chai, L. (2020). Predicting DNA methylation states with hybrid information based deep-learning model. *IEEE/ACM Transactions on Computational Biology and Bioinformatics*, *17*(5), 1721–1728. https://doi.org/10.1109/TCBB.2019.2909237

García-Pérez, R., Esteller-Cucala, P., Mas, G., Lobón, I., di Carlo, V., Riera, M., Kuhlwilm, M., Navarro, A., Blancher, A., di Croce, L., Gómez-Skarmeta, J. L., Juan, D., & Marquès-Bonet, T. (2021). Epigenomic profiling of primate lymphoblastoid cell lines reveals the evolutionary patterns of epigenetic activities in gene regulatory architectures. *Nature Communications*, *12*, 3116. https://doi.org/10.1038/s41467-021-23397-1

González, R. M., Ricardi, M. M., & Iusem, N. D. (2011). Atypical epigenetic mark in an atypical location: Cytosine methylation at asymmetric (CNN) sites within the body of a non-repetitive tomato gene. *BMC Plant Biology*, *11*, 94. https://doi.org/10.1186/1471-2229-11-94

Guidara, A., Riheb, B., Ebel, C., & Nadia, S. (2024, October 31–November 3). Deep learning for stress plant phenotyping. *2024 IEEE International Multi-Conference on Smart Systems & Green Process (IMC-SSGP)* (pp. 1–9). IEEE. https://doi.org/10.1109/IMC-SSGP63352.2024.10.19552

Guo, R., Zhang, Y., Chen, J., Li, H., Wang, L., Liu, Q., & Zhao, Y. (2023, November 5–8). CIMFormer: A 38.9TOPS/W-8b systolic CIM-array based Transformer processor with token-slimmed attention reformulating and principal possibility gathering. *Proceedings of the 2023 IEEE Asian Solid-State Circuits Conference (A-SSCC)* (pp. 1–3). IEEE. https://doi.org/10.1109/A-SSCC58667.2023.10347930

Gupta, S., Kumar, A., Kesarwani, V., & Shankar, R. (2025). DMRU: Generative deep-learning to unravel condition-specific cytosine methylation in plants. *bioRxiv*. https://doi.org/10.1101/2025.02.06.635186

Guru, A., Sahoo, S. K., Dash, G. K., Jena, J., & Dwivedi, P. (2021). Recent advances in multi-omics and breeding approaches towards drought tolerance in crops. In M. K. Gupta & L. Behera (Eds.), *Applications of bioinformatics in rice research* (pp. 333–359). Springer. https://doi.org/10.1007/978-981-16-3997-5_16

Han, D., Ye, T., Han, Y., Xia, Z., Pan, S., Wan, P., Song, S., & Huang, G. (2024). Agent attention: On the integration of softmax and linear attention. *arXiv*. https://arxiv.org/abs/2312.08874

Hassan, M. U., Nawaz, M., Barbanti, L., & Masood, S. (2025). Editorial: Enhancing salinity tolerance in crop plants through agronomic, genetic, molecular, and physiological approaches. *Frontiers in Plant Science, 16*, 1554509. https://doi.org/10.3389/fpls.2025.1554509

He, Y., Zhou, J., Lv, C., Zhang, J., Zhong, L., Zhang, D., Li, P., Xiao, L., Quan, M., Wang, D., Zhang, D., & Du, Q. (2024). Binding of PtoRAP2.12 to demethylated and accessible chromatin regions in the PtoGntK promoter stimulates growth of poplar. *New Phytologist, 245*(1), 232–248. https://doi.org/10.1111/nph.20228

He, J., Ning, K., Naznin, A., Wang, Y., Chen, C., Zuo, Y., Zhou, M., Li, C., Varshney, R., & Chen, Z.-H. (2025). Technological advances in imaging and modelling of leaf structural traits: A review of heat stress in wheat. *Journal of Experimental Botany*. https://doi.org/10.1093/jxb/eraf070

Henderson, I., & Jacobsen, S. (2007). Epigenetic inheritance in plants. *Nature, 447*(7143), 418–424. https://doi.org/10.1038/nature05917

Hiranuma, N., Lundberg, S., & Lee, S.-I. (2017). DeepATAC: A deep-learning method to predict regulatory factor binding activity from ATAC-seq signals. *bioRxiv*.https://doi.org/10.1101/172767

Huang, Z. (2025). Prediction of mental problem based on deep learning models. *Applied and Computational Engineering, 138.*, 56–63. https://doi.org/10.54254/2755-2721/2025.21185

Jaya, P., Ranjan, A., Afsana, A. N., Srivastava, A. K., & Mishra, L. N. (2021). Omics to understand drought tolerance in plants: An update. In A. Kumar, R. Kumar, P. Shukla, & H. K. Patel (Eds.), *Omics technologies for sustainable agriculture and global food security* (Vol. II, pp. 73–100). Springer. https://doi.org/10.1007/978-981-16-2956-3_4

Jørgensen, P. B., Jacobsen, K. W., & Schmidt, M. N. (2018). Neural message passing with edge updates for predicting properties of molecules and materials. *arXiv*. https://arxiv.org/abs/1806.03146

Jurczyk, B., Krępski, T., Kosmala, A., & Rapacz, M. (2013). Different mechanisms trigger an increase in freezing tolerance in *Festuca pratensis* exposed to flooding stress. *Environmental and Experimental Botany, 93*, 45–54. https://doi.org/10.1016/j.envexpbot.2013.06.003

Khan, M. S., & Khan, M. N. (2019). *Understanding plant responses to drought and salt stresses: Advances and challenges in "omics" approaches*. IntechOpen. https://doi.org/10.5772/intechopen.81041

Kim, S. (2016, December 15–18). Networks and models for the integrated analysis of multi-omics data. *2016 IEEE International Conference on Bioinformatics and Biomedicine (BIBM)* (p. 6). IEEE. https://doi.org/10.1109/BIBM.2016.7822479

Kim, Y., Kim, S.-H., Lim, J., & Kim, S.-H. (2024). ATBS1-interacting factor 2 positively regulates freezing tolerance via inducer of CBF expression 1/C-repeat binding factor-induced cold acclimation pathway. *Plant and Cell Physiology*, *65*(9), 1363–1376. https://doi.org/10.1093/pcp/pcae072

Kourani, M., Mohareb, F., Rezwan, F. I., Anastasiadi, M., & Hammond, J. P. (2022). Genetic and physiological responses to heat stress in *Brassica napus*. *Frontiers in Plant Science*, *13*, 832147. https://doi.org/10.3389/fpls.2022.832147

Kumar, V., Srivastava, A. K., Wani, S. H., Shriram, V., & Penna, S. (2021). Transcriptional and post-transcriptional mechanisms regulating salt tolerance in plants. *Physiologia Plantarum*, *173*(4), 1291–1294. https://doi.org/10.1111/ppl.13592

Lal, A., Chiang, Z. D., Yakovenko, N., Duarte, F. M., Israeli, J., & Buenrostro, J. D. (2021). Deep learning-based enhancement of epigenomics data with AtacWorks. *Nature Communications*, *12*, 1507. https://doi.org/10.1038/s41467-021-21765-5

Larkin, A., Kunze, C., Seman, M., Levashkevich, A., Curran, J., Morris-Evans, D., Lemieux, S., Khalil, A. S., & Ragunathan, K. (2023). Mapping the dynamics of epigenetic adaptation during heterochromatin misregulation. *bioRxiv*. https://doi.org/10.1101/2023.07.10.548368

Le Roux, M.-S., Kunert, K. J., Cullis, C. A., & Botha, A.-M. (2024). Unlocking wheat drought tolerance: The synergy of omics data and computational intelligence. *Food and Energy Security*, *13*(6), e70024. https://doi.org/10.1002/fes3.70024

Li, J., Chen, G. B., Rasheed, A., Li, D., Sonder, K., Zavala Espinosa, C., Wang, J., Costich, D. E., Schnable, P. S., Hearne, S. J., & Li, H. (2019). Identifying loci with breeding potential across temperate and tropical adaptation via EigenGWAS and EnvGWAS. *Molecular Ecology*, *28*(15), 3544–3560. https://doi.org/10.1111/mec.15169

Li, Y., Niu, Z., Zhu, M., Wang, Y., Gao, X., Ma, X., Li, H., Chen, L., & Sun, H. (2023). Multi-omics data provide insight into the adaptation of the glasshouse plant *Rheum nobile* to the alpine subnival zone. *Communications Biology*, *6*, 906. https://doi.org/10.1038/s42003-023-05271-6

Li, Z. P., Du, Z., Huang, D. S., & Teschendorff, A. E. (2025). Interpretable deep learning of single-cell and epigenetic data reveals novel molecular insights in aging. *Scientific Reports*, *15*, 5048. https://doi.org/10.1038/s41598-025-89646-1

Liao, J., Li, H., Feng, A., & Zhang, Z. (2023). Domestic pig sound classification based on TransformerCNN. *Applied Intelligence*, *53*, 4907–4923. https://doi.org/10.1007/s10489-022-03581-6

Lillicrap, T. P., & Santoro, A. (2019). Backpropagation through time and the brain. *Current Opinion in Neurobiology*, *55*, 82–89. https://doi.org/10.1016/j.conb.2019.01.011

Liu, K., Yan, Z., Liang, X., Kantola, R., & Hu, C. (2024). A survey on blockchain-enabled federated learning and its prospects with digital twin. *Digital Communications and Networks*, *10*(2), 248–264. https://doi.org/10.1016/j.dcan.2022.08.001

Ma, L., Sun, X., Kong, X., Valero Galvan, J., Li, X., Yang, S., Yang, Y., Yang, Y., & Hu, X. (2015). Physiological, biochemical and proteomics analysis reveals the adaptation strategies of the alpine plant *Potentilla saundersiana* at altitude gradient of the Northwestern Tibetan Plateau. *Journal of Proteomics*, *112*, 63–82. https://doi.org/10.1016/j.jprot.2014.08.009

Meyer, R. S., & Purugganan, M. D. (2013). Evolution of crop species: Genetics of domestication and diversification. *Nature Reviews Genetics, 14*(12), 840–852. https://doi.org/10.1038/nrg3605

Mienye, I. D., & Swart, T. G. (2024). A comprehensive review of deep learning: Architectures, recent advances, and applications. *Information (Basel), 15*(12), 755. https://doi.org/10.3390/info15120755

Miryeganeh, M. (2025). Epigenetic mechanisms driving adaptation in tropical and subtropical plants: Insights and future directions. *Plant, Cell & Environment, 48*(5), 3487–3499. https://doi.org/10.1111/pce.15370

Mortezapour Shiri, F., Perumal, T., Mustapha, N., & Mohamed, R. (2024). A comprehensive overview and comparative analysis on deep learning models. *Journal on Artificial Intelligence, 6*, 301–360. https://doi.org/10.32604/jai.2024.054314

Müller-Xing, R., Xing, Q., & Goodrich, J. (2014). Footprints of the sun: Memory of UV and light stress in plants. *Frontiers in Plant Science, 5*, 474. https://doi.org/10.3389/fpls.2014.00474

Nguyen, T. M., Kim, N., Kim, D. H., Le, H. L., Piran, M. J., Um, S.-J., & Kim, J. H. (2021). Deep learning for human disease detection, subtype classification, and treatment response prediction using epigenomic data. *Biomedicine, 9*(11), 1733. https://doi.org/10.3390/biomedicines9111733

Ni, P., Huang, N., Nie, F., Zhang, J., Zhang, Z., Wu, B., Bai, L., Liu, W., Xiao, C. L., Luo, F., & Wang, J. (2021). Genome-wide detection of cytosine methylations in plant from Nanopore data using deep learning. *Nature Communications, 12*, 5976. https://doi.org/10.1038/s41467-021-26278-9

Nouri, H., Eyshi Rezaei, E., & Siebert, S. (2020, May 4–8). Drought impact assessments for crop production need to be crop-specific (EGU2020-5453). EGU General Assembly 2020, Online. https://doi.org/10.5194/egusphere-egu2020-5453

Ogunfunmi, T., Ramachandran, R. P., & Togneri, R.& others. (2019). A primer on deep learning architectures and applications in speech processing. *Circuits, Systems, and Signal Processing, 38*, 3406–3432. https://doi.org/10.1007/s00034-019-01157-3

Okegbile, S. D., Zhang, X., Wang, J., Liu, Y., Zhou, J., & Chen, M. (2025). FLeS: A federated learning-enhanced semantic communication framework for mobile AIGC-driven human digital twins. *IEEE Network, 39*(5), 238–246. https://doi.org/10.1109/MNET.2025.3526556

Osseni, M. A., Tossou, P., Laviolette, F., & Corbeil, J. (2022). MOT: A multi-omics transformer for multiclass classification tumour types predictions. *bioRxiv*.https://doi.org/10.11.1/2022.11.14.516459

Ouellet, F., & Charron, J.-B. (2013). Cold acclimation and freezing tolerance in plants. In *Encyclopedia of life sciences (eLS)*. John Wiley & Sons, Ltd. https://doi.org/10.1002/9780470015902.a0020093.pub2

Padshetty, S., & Umashetty, A. (2024). Agricultural innovation through deep learning: A hybrid CNN-Transformer architecture for crop disease classification. *Journal of Spatial Science*, 1–32. https://doi.org/10.1080/14498596.2024.2355225

Panda, S. K., Panayiotou, T., Ellinas, G., & Behera, S. (2025). A fair federated learning framework for collaborative network traffic prediction and resource allocation. *arXiv*. https://arxiv.org/abs/2502.06743

Park, G. W., & Bryson, K. (2021, August 1–4). LDEncoder: Reference deep learning-based feature detector for transfer learning in the field of epigenomics. *Proceedings of the*

12th ACM International Conference on Bioinformatics, Computational Biology, and Health Informatics (Article 58, p. 1). ACM. https://doi.org/10.1145/3459930.3469487

Perrone, A., & Martinelli, F. (2020). Plant stress biology in epigenomic era. *Plant Science, 294*, 110376. https://doi.org/10.1016/j.plantsci.2019.110376

Pineda, F. J. (1989). Recurrent backpropagation and the dynamical approach to adaptive neural computation. *Neural Computation, 1*(2), 161–172. https://doi.org/10.11.2/neco.1989.1.2.161

Podgorniak, T., Dhanasiri, A., Chen, X., Ren, X., Kuan, P.-F., & Fernandes, J. (2022). Early fish domestication affects methylation of key genes involved in the rapid onset of the farmed phenotype. *Epigenetics, 17*(10), 1281–1298. https://doi.org/10.1080/15592294.2021.2017554

Qin, Y., Maggio, A., Hawkins, D., Beaudry, L., Kim, A., Pan, D., Gong, T., Fu, Y., Yang, H., & Deng, Y. (2024). Whole-genome bisulfite sequencing data analysis learning module on google cloud platform. *Briefings in Bioinformatics, 25*(Supplement_1), bbae236. https://doi.org/10.1093/bib/bbae236

Qu, Z., Li, Y., Liu, B., Gupta, D., & Tiwari, P. (2023). DTQFL: A digital twin-assisted quantum federated learning algorithm for intelligent diagnosis in 5G mobile network. *IEEE Journal of Biomedical and Health Informatics*. https://doi.org/10.1109/JBHI.2023.3303401

Qu, Y., Yu, S., Gao, L., Sood, K., & Xiang, Y. (2024). Blockchained dual-asynchronous federated learning services for digital twin empowered edge-cloud continuum. *IEEE Transactions on Services Computing, 17*(3), 836–849. https://doi.org/10.1109/TSC.2024.3382952

Rahman, M. M., Keya, S. S., Bulle, M., Ahsan, S. M., Rahman, M. A., Roni, M. S., Al Noor, M. M., & Hasan, M. (2025). Past trauma, better future: How stress memory shapes plant adaptation to drought. *Functional Plant Biology, 52*(6), FP24355. https://doi.org/10.1071/FP24355

Rai, A. C., & Rai, K. K. (2020). Drought stress and its mitigation and management strategies in crop plants. In R. Roychowdhury, S. Choudhury, M. Hasanuzzaman, & S. Srivastava (Eds.), *Sustainable agriculture in the era of climate change* (pp. 143–162). Springer. https://doi.org/10.1007/978-3-030-45669-6_6

Ramu, S. P., Boopalan, P., Pham, Q.-V., Maddikunta, P. K. R., Huynh-The, T., Alazab, M., Nguyen, T. T., & Gadekallu, T. R. (2022). Federated learning enabled digital twins for smart cities: Concepts, recent advances, and future directions. *Sustainable Cities and Society, 79*, 103663. https://doi.org/10.1016/j.scs.2021.103663

Rao, X., Yang, S., Lü, S., & Yang, P. (2024). DNA methylation dynamics in response to drought stress in crops. *Plants, 13*(14), 1977. https://doi.org/10.3390/plants13141977

Ruprecht, N. A., Kennedy, J. D., Bansal, B., Singhal, S., Sens, D., Maggio, A., Doe, V., Hawkins, D., Campbel, R., O'Connell, K., Gill, J. S., Schaefer, K., & Singhal, S. K. (2024). Transcriptomics and epigenetic data integration learning module on google cloud. *Briefings in Bioinformatics, 25*(Supplement_1), bbae352. https://doi.org/10.1093/bib/bbae352

Satyakam, Zinta, G. Z., Singh, R. K., & Kumar, R. (2022). Cold adaptation strategies in plants—an emerging role of epigenetics and antifreeze proteins to engineer cold resilient plants. *Frontiers in Genetics, 13*, 909007. https://doi.org/10.3389/fgene.2022.909007

Seal, D. B., Das, V., Goswami, S., & De, R. K. (2020). Estimating gene expression from DNA methylation and copy number variation: A deep learning regression model for multi-omics integration. *Genomics, 112*(4), 2833–2841. https://doi.org/10.1016/j.ygeno.2020.03.021

Shaikh, A. A., Chachar, S., Chachar, M., Ahmed, N., Guan, C., & Zhang, P. (2022). Recent advances in DNA methylation and their potential breeding applications in plants. *Horticulturae, 8*(7), 562. https://doi.org/10.3390/horticulturae8070562

Shen, Y., Zhang, J., Liu, Y., & Zhao, Y. (2018). DNA methylation footprints during soybean domestication and improvement. *Genome Biology, 19*, 128. https://doi.org/10.1186/s13059-018-1516-z

Sheng, X., Zhou, Y., & Cui, X. (2024). Graph neural network based asynchronous federated learning for digital twin-driven distributed multi-agent dynamical systems. *Mathematics, 12*(16), 2469. https://doi.org/10.3390/math12162469

Shin, D., Lee, J., Lee, J., & Yoo, H. -J. (2017, February 5–9). 14.2 DNPU: An 8.1 TOPS/W reconfigurable CNN-RNN processor for general-purpose deep neural networks. *2017 IEEE International Solid-State Circuits Conference (ISSCC)* (pp. 240–241). IEEE. https://doi.org/10.1109/ISSCC.2017.7870350

Shittu, L., Ekundayo, F., & Olutimehin, D. (2024). Cancer diagnosis and prognosis using multi-omics data: A data science and machine learning approach. *World Journal of Advanced Research and Reviews, 24*(3), 2052–2069. https://doi.org/10.30574/wjarr.2024.24.3.3876

Singhal, R. K., Saha, D., Skalicky, M., Mishra, U. N., Chauhan, J., Behera, L. P., Lenka, D., Chand, S., Kumar, V., Dey, P., Indu, Pandey, S., Vachova, P., Gupta, A., Brestic, M., & El Sabagh, A. (2021). Crucial cell signaling compounds crosstalk and integrative multi-omics techniques for salinity stress tolerance in plants. *Frontiers in Plant Science, 12.* https://doi.org/10.3389/fpls.2021.67036

Sun, P., Hao, R., Fan, F., Wang, Y., & Zhu, F. (2025). Adaptation of high-altitude plants to plateau abiotic stresses: A case study of the Qinghai-Tibet Plateau. *International Journal of Molecular Sciences, 26*(5), 2292. https://doi.org/10.3390/ijms26052292

Thakur, K., Kumari, C., Zadokar, A., Sharma, P., & Sharma, R. (2023). Physiological and omics-based insights for underpinning the molecular regulation of secondary metabolite production in medicinal plants: UV stress resilience. *Plant Physiology and Biochemistry, 204*, 108060. https://doi.org/10.1016/j.plaphy.2023.108060

Uzun, Y., Wu, H., & Tan, K. (2021). Predictive modeling of single-cell DNA methylome data enhances integration with transcriptome data. *Genome Research, 31*(1), 101–109. https://doi.org/10.1101/gr.266536.120

Vogt, G. (2017). Facilitation of environmental adaptation and evolution by epigenetic phenotype variation: Insights from clonal, invasive, polyploid, and domesticated animals. *Environmental Epigenetics, 3*(1), dvx002. https://doi.org/10.1093/eep/dvx002

Voytetskiy, A., Herbert, A., & Poptsova, M. (2022, December 6–8). Graph neural networks for Z-DNA prediction in genomes. *2022 IEEE International Conference on Bioinformatics and Biomedicine (BIBM)* (pp. 3173–3178). IEEE. https://doi.org/10.1109/BIBM55620.2022.9995451

Wang, Y., Zhang, P., Guo, W., Liu, H., Li, X., Zhang, Q., Du, Z., Hu, G., Han, X., Pu, L., Tian, J., & Gu, X. (2021). A deep learning approach to automate whole-genome prediction of diverse epigenomic modifications in plants. *New Phytologist, 232*(2), 880–897. https://doi.org/10.1111/nph.17630

Wu, Y., & Xie, L. (2024, July 8). AI-driven multi-omics integration for multi-scale predictive modeling of causal genotype-environment-phenotype relationships. *arXiv*. https://arxiv.org/abs/2407.06405

Wu, C., Yang H, Li J, Geng F, Bai J, Liu C, Kao W. (2022, April 22–24). Prediction of DNA methylation site status based on fusion deep learning algorithm. *2022 5th International*

Conference on Advanced Electronic Materials, Computers and Software Engineering (AEMCSE) (pp. 180–183). IEEE. https://doi.org/10.1109/AEMCSE55572.2022.00044

Xue, Y., Yang, Y., Tian, H., Quan, H., Liu, S., Zhang, L., Yang, L., Zhu, H., Wu, H., & Gao, Y. Q. (2022). Computational characterization of domain-segregated 3D chromatin structure and segmented DNA methylation status in carcinogenesis. *Molecular Oncology*, *16*(3), 699–716. https://doi.org/10.1002/1878-0261.13127

Yao, M., Hu, J., Hu, T., Xu, Y., Zhou, Z., Tian, Y., Xu, B., & Li, G. (2024, May 7–11). Spike-driven transformer V2: Meta spiking neural network architecture inspiring the design of next-generation neuromorphic chips. *Proceedings of the International Conference on Learning Representations (ICLR 2024)*. https://proceedings.iclr.cc/paper_files/paper/2024/hash/e9882f7f7c44a10acc01132302bac9d8-Abstract-Conference.html

Yassi, M., Chatterjee, A., & Parry, M. (2023). Application of deep learning in cancer epigenetics through DNA methylation analysis. *Briefings in Bioinformatics*, *24*(6), bbad411. https://doi.org/10.1093/bib/bbad411

Zamanzadeh-Nasrabadi, S. M., Mohammadiapanah, F., Hosseini-Mazinani, M., & Sarikhan, S. (2023). Salinity stress endurance of the plants with the aid of bacterial genes. *Frontiers in Genetics*, *14*, 1049608. https://doi.org/10.3389/fgene.2023.1049608

Zhang, Z., Liu, Z., Zhou, Q., & Qu, Y. (2020, July 19–24). Cross message passing graph neural network. *Proceedings of the 2020 International Joint Conference on Neural Networks (IJCNN)* (pp. 1–7). IEEE. https://doi.org/10.1109/IJCNN48605.2020.9207357

Zhang, F., Wu, J., Sade, N., Wu, S., Egbaria, A., Fernie, A. R., Yan, J., Qin, F., Chen, W., Brotman, Y., & Dai, M. (2021a).). Genomic basis underlying the metabolome-mediated drought adaptation of maize. *Genome Biology*, *22*, 260. https://doi.org/10.1186/s13059-021-02481-1

Zhang, X., Li, C., Tie, D., Quan, J., Yue, M., & Liu, X. (2021b). Epigenetic memory and growth responses of the clonal plant *Glechoma longituba* to parental recurrent UV-B stress. *Functional Plant Biology*, *48*(8), 827–838. https://doi.org/10.1071/FP20303

Zhang, G., Song, C., Yin, M., Liu, L., Zhang, Y., Li, Y., Zhang, J., Guo, M., & Li, C. (2024a). TRAPT: A multi-stage fused deep learning framework for transcriptional regulators prediction via integrating large-scale epigenomic data. *bioRxiv*.https://doi.org/10.1101/2024.05.17.594242

Zhang, H., Cao, D., Chen, Z., Zhang, X., Chen, Y., Sessions, C., Cruchaga, C., Payne, P., Li, G., Province, M., & Li, F. (2024b). MosGraphGen: A novel tool to generate multi-omics signaling graphs to facilitate integrative and interpretable graph AI model development. *Bioinformatics Advances*, *4*(1), vbae151. https://doi.org/10.1093/bioadv/vbae151

Zhang, X., Chang, X., Li, M., Roy-Chowdhury, A., Chen, J., & Oym, S. (2024c). Selective attention: Enhancing transformer through principled context control. *arXiv*. https://arxiv.org/abs/2411.12892

Zhang, F., Zhu, M., Chen, Y., Wang, G., Yang, H., Lu, X., Li, Y., Chang, H.-M., Wu, Y., Ma, Y., Yuan, S., Zhu, W., Dong, X., Zhao, Y., Yu, Y., Wang, J., & Mu, L. (2025). Harnessing omics data for drug discovery and development in ovarian aging. *Human Reproduction Update*, *31*(3), 240–268. https://doi.org/10.1093/humupd/dmaf002

Zhao, Y., Gowda, M., Würschum, T., Longin, C. F. H., Korzun, V., Kollers, S., Schachschneider, R., Zeng, J., Fernando, R., Dubcovsky, J., & Reif, J. C. (2013). Dissecting the genetic architecture of frost tolerance in Central European winter wheat. *Journal of Experimental Botany, 64*(14), 4453–4460. https://doi.org/10.1093/jxb/ert259

Zhou, W., Fan, X., Li, L., Zheng, J., Liu, X., Jin, W., & Tian, L. (2025). Graph-based contrastive learning enables unified integration and niche transfer across single-cell and spatial multi-omics. *bioRxiv*. https://doi.org/10.1101/2025.02.19.638965

Zhu, L., Tang, L., Zhang, K., Nie, H., Gou, X., Kong, X., & Deng, W. (2025). Genetic and epigenetic adaptation mechanisms of sheep under multi-environmental stress environment. *International Journal of Molecular Sciences, 26*(7), 3261. https://doi.org/10.3390/ijms26073261

5

AI Models for Studying Plant Epigenetics and Epigenomics

Moahmed Kouighat[1], Abdelghani Bouchyoua[1,2], Anas Hamdani[1], and Yassine Mouniane[3]

[1] *Research Unit of Plant Breeding and Plant Genetic Resources Conservation, Regional Agricultural Research Center of Meknes, National Institute of Agricultural Research, Rabat, Morocco*
[2] *Laboratory of Natural Resources and Environment, Polydisciplinary Faculty of Taza, Sidi Mohamed Ben Abdallah University, Taza, Morocco*
[3] *Natural Resources and Sustainable Development Laboratory, Faculty of Sciences, Ibn Tofail University, Kenitra, Morocco*

5.1 Introduction

Epigenetics and epigenomics have emerged as fundamental areas of plant biology, providing insights into how plants regulate gene expression beyond the DNA sequence itself. Epigenetic mechanisms such as DNA methylation, histone modifications, chromatin remodeling, and non-coding RNAs orchestrate dynamic regulatory networks that shape plant development, environmental adaptation, and transgenerational inheritance (Bashir & Setyawati, 2025). Unlike purely genetic mutations, epigenetic marks are reversible and highly responsive to internal and external stimuli, offering plants a unique degree of plasticity (Lloyd & Lister, 2022).

Over the last two decades, next-generation sequencing (NGS) technologies have generated a vast body of epigenomic data, including genome-wide maps of DNA methylation (via whole-genome bisulfite sequencing [BS-seq]), histone modification landscapes (via chromatin immunoprecipitation followed by sequencing [ChIP-seq]), and chromatin accessibility profiles (via assay for transposase-accessible chromatin [ATAC-seq] and DNase I hypersensitive site sequencing [DNase-seq]) (Searle et al., 2023). These datasets provide unprecedented opportunities for understanding plant regulatory complexity. However, they also pose substantial analytical challenges: high dimensionality, noisy signals, interdependence between different layers of epigenetic regulation, and species-specific variations.

Artificial intelligence (AI), encompassing machine learning (ML) and deep learning (DL) methods, has emerged as a powerful ally in deciphering complex biological data. AI models are uniquely suited to detect nonlinear patterns, infer hidden structures, and make accurate predictions from massive datasets. In human epigenomics, AI has already contributed to methylation state prediction, enhancer identification, and chromatin conformation modeling (Alsayaydeh et al., 2025). In plant sciences, the adoption of AI for epigenomic analysis is rapidly expanding, with the potential to revolutionize breeding, stress resilience studies, and fundamental research on plant development (Gayathiri et al., 2025).

This chapter provides a comprehensive overview of the role of AI in plant epigenetics and epigenomics. It outlines fundamental principles of plant epigenetic regulation, introduces key AI techniques, explores specific applications in predicting and interpreting epigenomic features, and highlights challenges and future directions. The emphasis is placed on practical applications for crop improvement, environmental resilience, and systems biology, while also discussing the limitations and opportunities for next-generation AI models.

5.2 Plant Epigenetics and Epigenomics: An Overview

5.2.1 Major Epigenetic Mechanisms in Plants

Epigenetics of plants is the collection of molecular mechanisms that adjust the genome's gene expression by not changing the DNA sequence but by continuing to influence the activity of the genome and, in certain cases, its transgenerational inheritance (Figure 5.1).

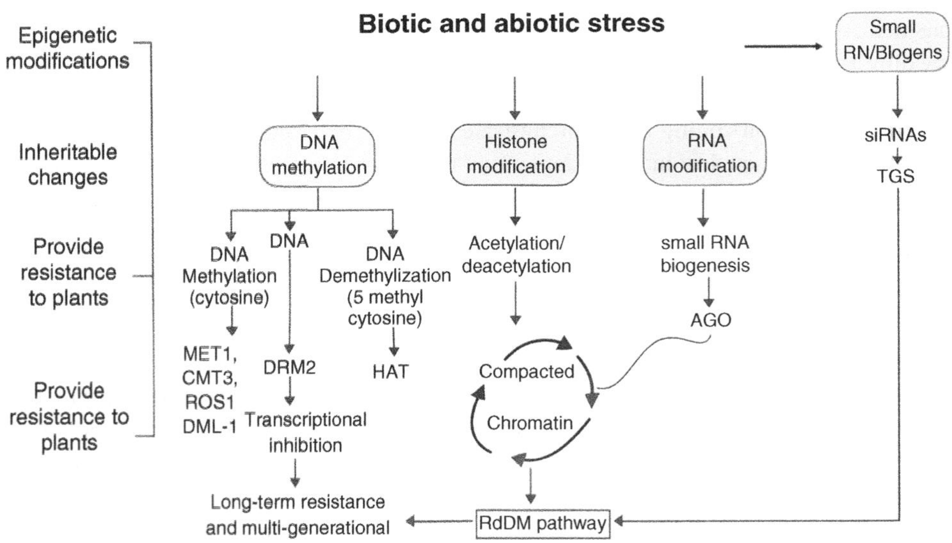

Figure 5.1 Major epigenetic mechanisms in plants under biotic and abiotic stresses. DNA methylation at cytosine residues, mediated by enzymes such as MET1, CMT3, DRM2, ROS1, and DML1, leads to transcriptional inhibition and gene silencing, thereby contributing to long-term and heritable stress resistance. Histone modifications, such as acetylation and deacetylation, regulate chromatin compaction and accessibility, dynamically influencing gene expression patterns. RNA modifications, particularly small RNA biogenesis, direct transcriptional gene silencing (TGS) through Argonaute (AGO) proteins and siRNAs, further integrating with the RdDM pathway. Together, these interconnected mechanisms provide plants with multigenerational epigenetic memory, enabling adaptive phenotypic plasticity and resilience against environmental fluctuations.

It is maintained by several complementing mechanisms, including DNA methylation, post-translational modifications of histones, activity of small regulatory RNAs, and chromatin movements. In plants, DNA methylation—especially in the cytosine–guanine (CG), cytosine–any nucleotide–guanine (CHG), and CHH (cytosine–any nucleotide–any nucleotide [H = A, T, or C]) contexts—acts as a stable epigenetic mark that can persist through both mitotic and meiotic cell divisions, preserving cellular memory and, in some cases, passing traits across generations (Gayathiri et al., 2025). This process is carefully regulated by DNA methyltransferases like MET1, CMT3, and DRM2, which are responsible for maintaining and establishing these methylation marks (Ashapkin et al., 2016). At the same time, histone modifications play a crucial role in shaping the three-dimensional structure of chromatin and controlling how accessible DNA is to transcription factors. Histone acetylation, typically linked to active transcription, along with specific methylation patterns on histone H3 lysines—such as H3K4me3 for activation and H3K27me3 or H3K9me2 for repression—help create dynamic chromatin landscapes that regulate gene flexibility (Bure et al., 2022). These marks do not work independently; instead, they interact with DNA methylation and small RNAs to either reinforce or loosen epigenetic states. Another key aspect of plant epigenetics is the role of small interfering RNAs (siRNAs) and long noncoding RNAs (lncRNAs). These molecules guide gene silencing through RNA interference (RNAi) and the RNA-directed DNA methylation (RdDM) pathway (Bhogireddy et al., 2021). Together, they help suppress transposable elements to maintain genomic stability, while also fine-tuning the regulation of genes involved in development and responses to stress. The flexible nature of the plant epigenome gives plants a remarkable ability to fine-tune their genetic programs in response to environmental changes such as nutrient levels, drought, temperature shifts, or pathogen attacks. This adaptability comes from reversible epigenetic modifications that provide plants with a form of adaptive memory and can shape the long-term evolution of traits. Therefore, plant epigenetics is not just an extra layer of regulation but a core element linking the genome, the environment, and evolutionary processes.

5.2.2 Major Epigenomic Mechanisms in Plants

Plant epigenomics is the comprehensive, genome-wide study of epigenetic marks and modifications that influence gene expression and the functional organization of chromatin. Unlike epigenetics, which focuses primarily on point regulatory mechanisms and their heritability, epigenomics takes a systems approach, aiming to map and interpret all epigenetic states in different tissues, developmental stages, or environmental conditions. This discipline relies on rapid advances in high-throughput sequencing technologies and bioinformatics approaches, making it possible to obtain comprehensive atlases of DNA modifications, histone modifications, chromatin accessibility, and small regulatory RNA profiles (Kouighat et al., 2025). DNA methylation is one of the most studied marks in epigenomics. BS-seq and its variants (whole genome bisulfite sequencing [WGBS] and reduced representation bisulfite sequencing [RRBS]) allow single-nucleotide resolution, revealing complex methylation patterns in CG, CHG, and CHH contexts specific to plants (Lu et al., 2023). In parallel, mapping of histone modifications by ChIP-seq has highlighted dynamic chromatin landscapes, such as H3K27me3 enrichment in developmental genes or H3K9me2 in transposable element-rich regions (Josipovic, 2022). These approaches are complemented by ATAC-seq, which allows the detection of accessible areas of chromatin and the identification of active regulatory regions (Grandi et al., 2022). Another key area of plant epigenomics concerns the transcriptomic profiles of non-coding RNAs, particularly small RNAs (siRNAs and miRNAs) involved in transcriptional and post-transcriptional regulation and lncRNAs that participate in structuring the chromatin landscape (Kouighat et al., 2025). The integration of these multiomic data is now possible thanks to advanced bioinformatics methods, allowing epigenetic modifications to be directly linked to gene expression variations and observed phenotypes. Plant epigenomics has highlighted the extraordinary plasticity of the plant genome in the face of biotic and abiotic stresses. Moreover, some of these modifications persist beyond the exposed generation, paving the way for transgenerational epigenomic inheritance that could be exploited for varietal improvement.

5.3 AI Models for Plant Epigenetics and Epigenomics

5.3.1 Why AI for Epigenomics?

AI plays a pivotal role in contemporary plant epigenetics and epigenomics research, as it provides the means to tackle the complexity and vast volume of data generated by NGS and high-throughput mapping technologies. Epigenetic profiles—including DNA methylation patterns, post-translational histone modifications, chromatin accessibility maps, and non-coding RNA signatures—generate multidimensional datasets that far exceed the capacity of traditional statistical analyses. AI, particularly ML and DL, offers approaches capable of uncovering intricate correlations, identifying hidden regulatory motifs, and constructing robust predictive models that link epigenetic marks to plant phenotypes (Dobránszki et al., 2025).

In plant epigenetics, regulatory systems are highly dynamic and influenced by both endogenous factors (such as development and cell differentiation) and exogenous stimuli (including abiotic and biotic stresses) (Ashapkin et al., 2020). AI algorithms enable the integration of these diverse layers of information, providing a systemic and functional perspective on epigenetic networks (Gayathiri et al., 2025). DL models applied to DNA methylation and promoter sequences have demonstrated remarkable accuracy in predicting cis-regulatory regions and anticipating gene expression shifts under stress conditions (Peleke et al., 2024). Similarly, convolutional neural network-based approaches have been employed to detect patterns of open chromatin and forecast their impact on transcriptional accessibility (Liu et al., 2018).

A further strength of AI lies in its ability to handle the heterogeneity and diversity of epigenomic data across plant species. Plant genomes are often large and enriched with transposable elements, making the characterization of their epigenetic landscapes particularly challenging (Galindo-González et al., 2018). Through transfer learning techniques, models trained on reference species such as *Arabidopsis thaliana* can be adapted to economically important crops, thereby reducing experimental costs and accelerating discovery. AI also enables the integration of epigenomic data with other omics layers—transcriptomics, metabolomics, and high-throughput phenotyping—fostering a more comprehensive understanding of plant adaptation and evolutionary mechanisms (Cembrowska-Lech et al., 2023).

Equally important are the practical applications of AI in plant improvement. The ability to predict epigenetic profiles associated with drought tolerance, pathogen resistance, or developmental plasticity opens new avenues for data-driven breeding strategies (Thingujam et al., 2025). Ultimately, AI may make it possible to design optimized plant genomes that account not only for genetic sequences but also for the epigenetic layers shaping gene expression and trait stability (Sun et al., 2024). In this way, AI transcends its role as a data analysis tool to become a strategic driver, transforming plant epigenetics and epigenomics into predictive disciplines capable of bridging molecular mechanisms with the urgent agricultural and environmental challenges of our time.

5.3.2 AI Models for Plant Epigenetics and Epigenomics

The application of AI to the study of plant epigenetics and epigenomics has expanded dramatically over the past decade, enabling the analysis of massive datasets derived from DNA methylation, histone modifications, chromatin accessibility, and the interplay between these regulatory layers (Figure 5.2). The range of AI models employed spans from classical ML approaches to more complex DL architectures, each tailored to address specific needs in interpretation or prediction (Table 5.1).

Different types of biological data are retrieved, including DNA methylation profiles, histone modifications, chromatin accessibility, non-coding RNA sequences (miRNA, siRNA, and lncRNA), and multiomic integration (transcriptomics, proteomics, and metabolomics). These heterogeneous data then undergo preliminary processing, including quality control, normalization, dimension reduction (principal component analysis [PCA], t-SNE, and uniform manifold approximation and projection [UMAP]), and extraction of specific features (motifs, chromatin states, and methylation sites). The next step involves various AI and ML models: regression models (linear, logistic, and least absolute shrinkage and selection operator [LASSO]) to associate phenotypic traits and epigenetic marks; random forest, support vector machines (SVM), and extreme gradient boosting (XGBoost) algorithms for classification, pattern detection, and multiomics prediction; as well as neural networks (convolutional neural networks [CNN], recurrent neural network [RNN], and graph neural networks [GNN]) and transformer/LLM models (bidirectional encoder representations from transformer [BERT] and generative pre-trained transformer [GPT]) for advanced sequence and regulatory network analysis. Finally, the generated results concern key applications such as the identification of stress-related epigenetic markers, epigenetic memory modeling, phenotypic plasticity prediction, gene regulatory network reconstruction, and multiomics integration for crop improvement.

5.4 Application of AI Models in Plant Epigenetics

Given the volume and complexity of epigenomic data, traditional statistical methods are increasingly being complemented—or even outperformed—by AI models (Figure 5.3). AI models, particularly ML and DL algorithms, provide powerful tools for analyzing these

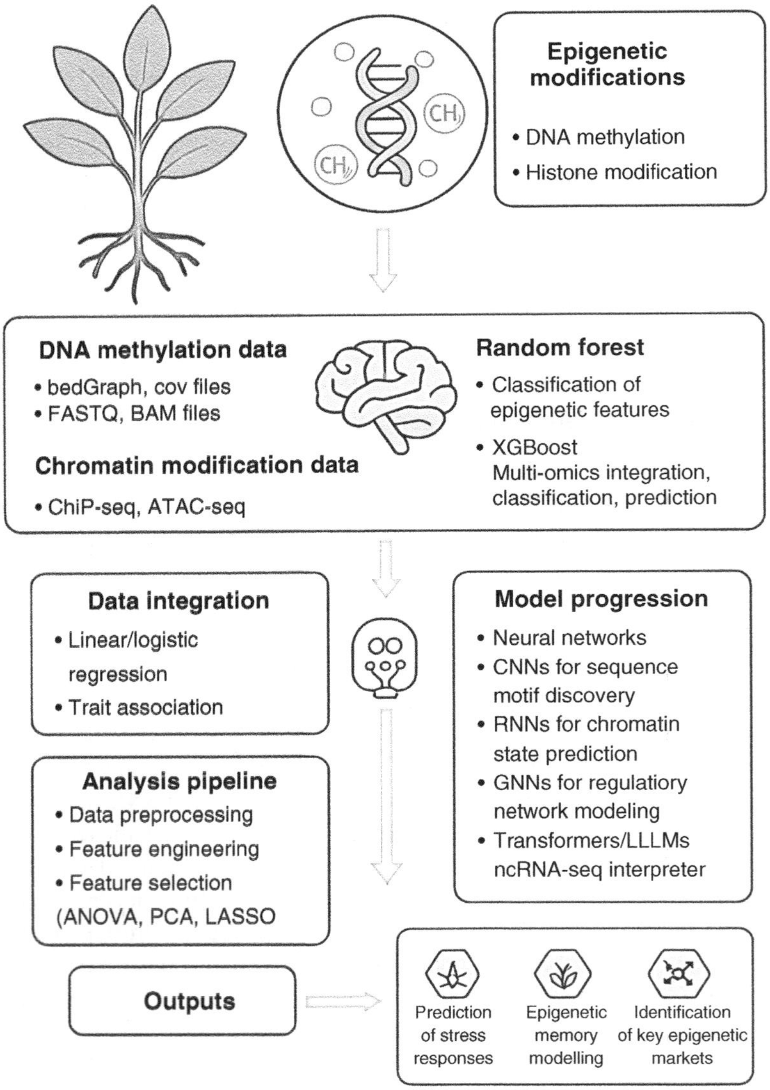

Figure 5.2 Integrated framework for the application of AI in the study of plant epigenetics and epigenomics.

datasets. Random forests (RF) and SVMs were among the earliest tools applied to plant epigenetics. These models proved particularly effective in classifying genomic regions based on their methylation status and in linking DNA methylation patterns to observable phenotypes (Eren et al., 2023). Similarly, penalized regression methods such as LASSO and elastic net have been widely used to reduce the dimensionality of large-scale epigenomic datasets and to identify the most relevant epigenetic variables associated with variation in traits of interest (Li & Sillanpää, 2012). As multiomics datasets have become increasingly available, DL methods have taken on a central role. CNNs are now indispensable for analyzing genomic sequences

Table 5.1 Comparison of AI models applied in plant epigenetics and epigenomics.

AI model	Type/approach	Applications	Strengths	Limitations	Reference
Linear and logistic regression	Supervised, statistical ML	Linking phenotypic traits to epigenetic marks (e.g., DNA methylation, stress phenotypes)	Simple, interpretable, requires fewer data	Poor performance on nonlinear, high-dimensional data	Briggs et al. (2010)
Random forest (RF)	Ensemble learning, decision trees	Classification of epigenetic marks, prediction of phenotypic traits	Robust, handles noisy data, provides variable importance	Less interpretable, sensitive to correlated features	Saavedra et al. (2023)
XGBoost (extreme gradient boosting)	Gradient boosting	Classification and multiomics integration, detection of complex epigenetic markers	High performance, fast, robust	Overfitting risk without strict validation, requires parameter tuning	Ma et al. (2020)
Support vector machines (SVM)	Supervised ML, kernel-based	Detecting DNA methylation, clustering epigenetic profiles	Effective on small datasets, good class separation	Less effective on very large datasets, requires tuning	Fernandez and Miranda-Saavedra (2012)
LASSO regression	Linear regression with penalization	Selecting significant epigenetic markers (e.g., methylation, histone marks)	Dimensionality reduction, interpretable	May discard correlated info, biased on small datasets	Greenwood et al. (2020)
CNN (convolutional neural networks)	Deep learning, pattern recognition	Detecting motifs in ATAC-seq and ChIP-seq data	Excellent motif recognition, captures local interactions	Requires large datasets, black-box model	Oh et al. (2020)
RNN (recurrent neural networks)	Deep learning, sequential data	Analyzing DNA/RNA sequences, predicting chromatin states and methylation	Captures temporal dependencies, works well with genomic sequences	Vanishing gradient problem, large data requirements	Trabelsi et al. (2019)
GNN (graph neural networks)	Deep learning on graphs	Modeling transcription factor interactions, chromatin architecture prediction	Models complex networks and gene-regulator interactions	Computationally intensive, low interpretability	Sun et al. (2023)
MLP (multilayer perceptron)	Feed-forward neural network	Classification, regression, modeling of epigenetic features	Flexible, adaptable	Overfitting risk, requires large datasets	Türkoğlu et al. (2023)
Transformers (BERT and GPT)	Deep learning, attention-based	ncRNA-seq analysis, interpretation of epigenetic regulation	Powerful for big sequential data, captures global context	Computationally expensive, requires massive datasets	Dobránszki et al. (2025)
DBN (deep belief networks)	Probabilistic generative DL	Detecting epigenetic markers, exploring marker–trait relationships	Strong exploratory capacity, unsupervised	Less commonly used today, difficult to train	Bu et al. (2017)
SOM (self-organizing maps)	Unsupervised clustering	Discovering new patterns in epigenomic data	Intuitive visualization, dimensionality reduction	Less precise, parameter-sensitive	Yang (2025)

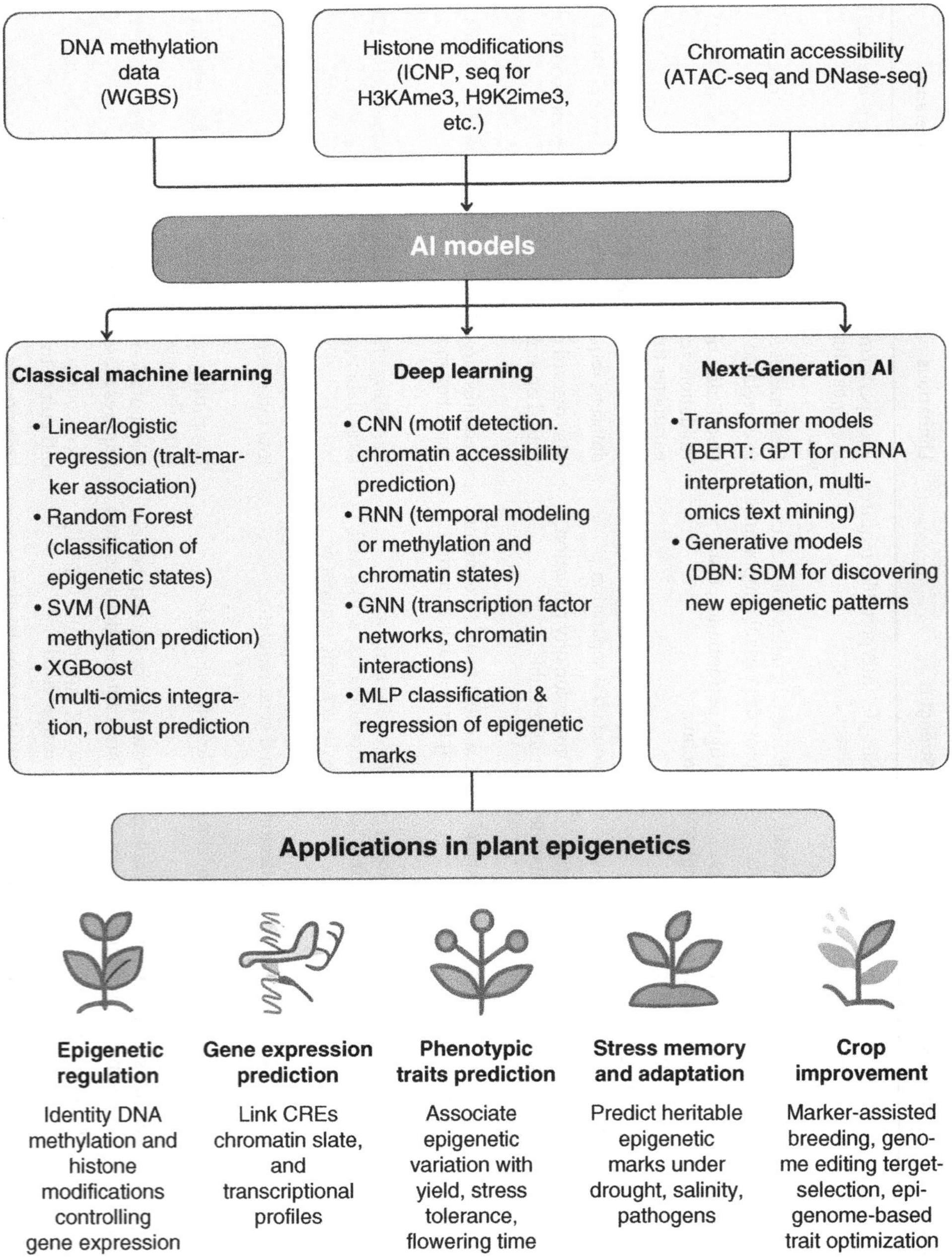

Figure 5.3 Application of AI in plants epigenetics and epigenomics.

enriched with epigenetic marks, as they can automatically detect complex regulatory motifs. CNN-based models have been successfully applied to predict de novo DNA methylation and to model chromatin states (Pimpalkar et al., 2025). RNNs, particularly long short-term memory (LSTM) models, have been used to capture sequential dependencies in epigenetic modifications, enabling the study of dynamic changes in marks related to development and stress responses (Kala et al., 2024). More recently, transformer-based models—originally developed for natural language processing—have been adapted to plant epigenomics. Their ability to handle long DNA sequences and integrate multiple layers of data (methylation, transcriptome, and chromatin accessibility) has markedly improved the accuracy of predicting phenotypic traits associated with epigenetic signatures (Consens et al., 2025).

In parallel, more specialized models are emerging; for example, graph neural networks (GNNs) represent a particularly innovative approach, allowing researchers to conceptualize the epigenome as a three-dimensional network. These models make it possible to explicitly account for spatial genome organization and directly link chromatin interactions to regulatory states and agronomic traits (Lazaros et al., 2024). Integrative approaches that combine AI with multiomics data (methylome, transcriptome, histone modifications, and metabolome) are increasingly making use of hybrid models—often based on architectures such as variational autoencoders (VAE) or deep canonical correlation analysis (DCCA)—to reveal global epigenetic signatures associated with adaptation and phenotypic plasticity in plants (Ali, 2023).

A critical frontier in this field is the interpretability of AI models. While DL offers unprecedented predictive accuracy, connecting predictions to biological mechanisms remains essential. Approaches such as eXplainable AI (XAI) and interpretability frameworks like SHapley Additive exPlanations (SHAP) are now being applied to bridge this gap (Vimbi et al., 2024). Several studies (Table 5.2) have used different AI models to identify and experimentally validate key epigenetic regions involved in plant development and environmental adaptation.

Table 5.2 Case studies of AI models applied to plant epigenomics and epigenetics.

AI model	Main application	Plant space	References
Random forests	Prediction of methylated versus unmethylated sites; association of epigenetic profiles with phenotypic traits	*Arabidopsis thaliana*; maize	Hu et al. (2015), Westhues et al. (2021)
Support vector machines (SVMs)	Identification of epigenetic signatures linked to stress and cell differentiation	Rice, wheat	Eren et al. (2023), Barel et al. (2014)
Convolutional neural networks (CNNs)	Motif detection in regulatory sequences and DNA methylation prediction	Rice, maize	Rahman et al. (2021), Wang et al. (2024)
Recurrent neural networks (RNNs/LSTMs)	Modeling temporal dynamics of epigenetic modifications	Wheat under abiotic stress	Le Roux et al. (2024)
Transformers (attention-based models)	Multiomics integration (DNA, methylation, histones, and chromatin) to predict expression and phenotypes	Multiplant datasets	Shafik et al. (2025)
Generative adversarial networks (GANs)	Simulation of epigenomic profiles and augmentation of rare datasets	Exploratory studies in *Arabidopsis*	Dobránszki et al. (2025)
Multi-branch deep learning (multimodal)	Fusion of epigenomic, transcriptomic, and phenotypic data for breeding applications	Maize and wheat in breeding programs	Gao et al. (2024)
Unsupervised clustering (k-means, t-SNE, and UMAP with AI)	Exploration of epigenetic landscapes and chromatin states grouping	*Arabidopsis*, tomato	Lalman et al. (2025)

5.5 Challenges and Limitations

The application of AI models to plant epigenetics and epigenomics offers tremendous potential, yet it remains constrained by several critical limitations and challenges. One of the most pressing issues is the availability and quality of data. Large-scale epigenomic profiles—such as DNA methylation, histone modifications, and chromatin states—are still relatively scarce, often limited to well-studied model species, and not easily transferable to agriculturally important crops (Mahajan et al., 2025). Moreover, these data are highly sensitive to environmental conditions, developmental stages, and tissue-specific variation, which complicates their integration into robust predictive models.

Another major challenge lies in the intrinsic complexity of AI models themselves. DL approaches, including CNNs and transformer architectures, provide remarkable predictive power, but they frequently operate as "black boxes," making biological interpretation of results difficult (Mowbray, 2025). This lack of interpretability limits researchers' ability to connect predictions with underlying molecular mechanisms, a crucial requirement in fields such as epigenetics, where mechanistic insight is indispensable. In addition, model generalizability remains problematic: algorithms trained on one plant species often lose predictive accuracy when applied to another due to differences in genome organization, chromatin landscapes, or species-specific epigenetic signatures (Lv et al., 2025).

Technical hurdles also persist in multiomics integration. Effectively combining genomic, transcriptomic, epigenomic, and phenotypic data into a coherent framework remains a formidable task, despite progress in hybrid models and generative networks (Lee, 2023). On top of this, computational demands pose practical barriers: processing large-scale datasets requires substantial hardware and software resources, often beyond the reach of many research laboratories. Looking ahead, one of the key challenges will be to design models that are not only more interpretable but also capable of handling heterogeneous data while delivering reliable predictions. Achieving this balance between predictive power, transparency, and biological relevance will be essential for fully leveraging the potential of AI in advancing plant improvement strategies.

5.6 Future Directions

Future directions in applying AI to plant epigenetics and epigenomics are likely to revolve around the integration of increasingly complex multiomics data and the development of models that are both accurate and biologically interpretable. A central avenue will be the use of single-cell epigenomic profiling combined with ML to capture cell-type-specific regulatory signatures, which are currently hidden in bulk data analyses (Raza, 2024). At the same time, advances in DL architectures, such as GNNs and attention-based models, could allow researchers to capture the spatial and temporal dynamics of chromatin interactions and histone modifications with far greater resolution. Another promising frontier lies in the integration of environmental and epigenomic datasets: developing predictive models that connect external stressors—such as drought, salinity, or pathogens—to specific epigenetic reprogramming events, thereby offering actionable insights for crop resilience and breeding programs.

Equally important will be the push toward XAI frameworks, capable of not only predicting epigenetic outcomes but also providing mechanistic clues that guide experimental validation (Chandra & Kumar, 2024). This interpretability will be essential for translating computational predictions into practical applications in agriculture. Furthermore, federated learning approaches may emerge as a solution to the challenges of data sharing across institutions and countries, enabling collaborative AI model training without compromising data privacy or ownership (Li et al., 2025). In parallel, the use of hybrid models that combine mechanistic plant biology with ML could bridge the gap between statistical accuracy and biological plausibility (Noordijk et al., 2024). Looking further ahead, quantum ML and advanced high-performance computing infrastructures may offer the computational power required to handle the immense complexity of plant epigenomic landscapes (Fernandes et al., 2024). Collectively, these directions point toward a future where AI not only accelerates discovery but also becomes a cornerstone of precision breeding strategies, ultimately supporting sustainable agriculture in the face of global climate change and food security challenges.

5.7 Conclusion

AI has become indispensable in modern plant epigenetics and epigenomics, offering unprecedented opportunities to predict, interpret, and manipulate complex regulatory landscapes. By integrating diverse omics layers, AI not only accelerates fundamental discoveries but also provides practical solutions for crop breeding and climate resilience. Despite challenges in data availability, interpretability, and computational demands, the synergy between AI and plant epigenomics is poised to drive transformative advances in agriculture and plant biology. Future developments in XAI, single-cell technologies, and generative modeling will further expand the boundaries of our understanding and enable epigenome-based innovations for sustainable food production.

References

Ali, H. (2023). Artificial intelligence in multi-omics data integration: Advancing precision medicine, biomarker discovery and genomic-driven disease interventions. *International Journal of Scientific Research Archive*, *8*(1), 1012–1030.

Alsayaydeh, D. A. J., Narayanamurthy, V., Futane, A. S., Alsayaydeh, J. A. J., & Najib, S. B. M. (2025). Advances of artificial intelligence-enabled epigenetics. *Health Technology*, *15*(4), 663–676.

Ashapkin, V. V., Kutueva, L. I., & Vanyushin, B. F. (2016). Plant DNA methyltransferase genes: Multiplicity, expression, methylation patterns. *Biochemistry (Moscow)*, *81*(2), 141–151.

Ashapkin, V. V., Kutueva, L. I., Aleksandrushkina, N. I., & Vanyushin, B. F. (2020). Epigenetic mechanisms of plant adaptation to biotic and abiotic stresses. *International Journal of Molecular Sciences*, *21*(20), 7457.

Barel, A. O., Paye, M., & Maibach, H. I. (2014). *Handbook of cosmetic science and technology.* CRC Press. Available from: https://books.google.com/books?hl=fr&lr=&id=FAYNAwAAQBAJ&oi=fnd&pg=PP1&dq=Barel,+A.+O.,+Paye,+M.,+%26+Maibach,+H.+I.+(2014).+Handbook+of+cosmetic+science+and+technology.+CRC+press.&ots=svvGpzk3eG&sig=kchZF8sS0wd9OLaKongukv98Daw

Bashir, H., & Setyawati, A. (2025). Decoding the mechanisms of plant epigenetic adaptations to stress. *Reviews in Agricultural Science, 13*(2), 1–19. https://doi.org/10.7831/ras.13.2_1

Bhogireddy, S., Mangrauthia, S. K., Kumar, R., Pandey, A. K., Singh, S., Jain, A., Budak, H., Varshney, R. K., & Kudapa, H. (2021). Regulatory non-coding RNAs: A new frontier in regulation of plant biology. *Functional & Integrative Genomics, 21*(3–4), 313–330. https://doi.org/10.1007/s10142-021-00787-8

Briggs, F. B. S., Ramsay, P. P., Madden, E., Norris, J. M., Holers, V. M., Mikuls, T. R., Sokka, T., Seldin, M. F., Gregersen, P. K., Criswell, L. A., & Barcellos, L. F. (2010). Supervised machine learning and logistic regression identifies novel epistatic risk factors with PTPN22 for rheumatoid arthritis. *Genes & Immunity, 11*(3), 199–208. https://doi.org/10.1038/gene.2009.110

Bu, H., Gan, Y., Wang, Y., Zhou, S., & Guan, J. (2017). A new method for enhancer prediction based on deep belief network. *BMC Bioinformatics, 18*(S12), 418. https://doi.org/10.1186/s12859-017-1828-0

Bure, I. V., Nemtsova, M. V., & Kuznetsova, E. B. (2022). Histone modifications and non-coding RNAs: Mutual epigenetic regulation and role in pathogenesis. *International Journal of Molecular Sciences, 23*(10), 5801. https://doi.org/10.3390/ijms23105801

Cembrowska-Lech, D., Krzemińska, A., Miller, T., Nowakowska, A., Adamski, C., Radaczyńska, M., Mikiciuk, G., & Mikiciuk, M. (2023). An integrated multi-omics and artificial intelligence framework for advance plant phenotyping in horticulture. *Biology, 12*(10), 1298. https://doi.org/10.3390/biology12101298

Chandra, O. R., & Kumar, V. (2024). *Explainable machine learning with epigenomic features for insights into regulatory and functional genomics* [PhD thesis]. IIIT-Delhi. Retrieved September 6 2025, from https://repository.iiitd.edu.in/jspui/handle/123456789/1710

Consens, M. E., Dufault, C., Wainberg, M., Forster, D., Karimzadeh, M., Goodarzi, H., Theis, F. J., Moses, A., & Wang, B. (2025). Transformers and genome language models. *Nature Machine Intelligence, 7*, 1–17. https://doi.org/10.1038/s42256-025-01007-9

Dobránszki, J., Vassileva, V., Agius, D. R., Moschou, P. N., Gallusci, P., Berger, M. M. J., Farkas, D., Basso, M. F., & Martinelli, F. (2025). Gaining insights into epigenetic memories through artificial intelligence and omics science in plants. *Journal of Integrative Plant Biology, 67*(9), 2320–2349. https://doi.org/10.1111/jipb.13953Digital Object Identifier (DOI)

Eren, B., Türkoğlu, A., Haliloğlu, K., Demirel, F., Nowosad, K., Özkan, G., Niedbała, G., Pour-Aboughadareh, A., Bujak, H., & Bocianowski, J. (2023). Investigation of the influence of polyamines on mature embryo culture and DNA methylation of wheat (*Triticum aestivum* L.) using the machine learning algorithm method. *Plants, 12*(18), 3261. https://doi.org/10.3390/plants12183261

Fernandes, L. P., Kharate, P., & Singh, B. (2024). The future of high performance computing in biomimetics and some challenges. In K. A. Ahmad, N. A. W. A. Hamid, M. Jawaid, T. Khan, & B. Singh (Eds.), *High performance computing in biomimetics* (pp. 287–303). Springer

Nature Singapore. (Series in BioEngineering). Available from: https://link.springer.com/10.1007/978-981-97-1017-1_15

Fernandez, M., & Miranda-Saavedra, D. (2012). Genome-wide enhancer prediction from epigenetic signatures using genetic algorithm-optimized support vector machines. *Nucleic Acids Research*, *40*(10), e77–e77. https://doi.org/10.1093/nar/gks149

Galindo-González, L., Sarmiento, F., & Quimbaya, M. A. (2018). Shaping plant adaptability, genome structure and gene expression through transposable element epigenetic control: Focus on methylation. *Agronomy*, *8*(9), 180. https://doi.org/10.3390/agronomy8090180

Gao, J., Xu, Y., Li, Y., Lu, F., & Wang, Z. (2024). Comprehensive exploration of multi-modal and multi-branch imaging markers for autism diagnosis and interpretation: Insights from an advanced deep learning model. *Cerebral Cortex*, *34*(2), bhad521. https://doi.org/10.1093/cercor/bhad521

Gayathiri, E., Viswanathan, D., Prakash, P., Pandiaraj, S., Kondapavuluri, B. K., Singh, C. D., Thiruvengadam, R., Balu, S., Anantharaman, R., Gaur, A., Govindasamy, R., & Thiruvengadam, M. (2025). A critical review on epigenetics and epigenomics in plant development and stress resilience. *Brazilian Journal Of Botany*, *48*(1), 51. https://doi.org/10.1007/s40415-025-01093-4

Grandi, F. C., Modi, H., Kampman, L., & Corces, M. R. (2022). Chromatin accessibility profiling by ATAC-seq. *Nature Protocols*, *17*(6), 1518–1552. https://doi.org/10.1038/s41596-022-00692-9

Greenwood CJ, Youssef GJ, Letcher P, Macdonald JA, Hagg LJ, Sanson A, Mcintosh J, Hutchinson DM, Toumbourou JW, Fuller-Tyszkiewicz M, Olsson CA A comparison of penalised regression methods for informing the selection of predictive markers. *PLoS One* 2020;15(11):e0242730. ttps://doi.org/https://doi.org/10.1371/journal.pone.0242730

Hu, Y., Morota, G., Rosa, G. J., & Gianola, D. (2015). Prediction of plant height in *arabidopsis thaliana* using DNA methylation data. *Genetics*, *201*(2), 779–793. https://doi.org/10.1534/genetics.115.177204

Josipovic, N. (2022). *Chromatin reorganization, signaling responses and non-coding RNA regulators in homeostasis and senescence*. Retrieved September 6 2025, from https://ediss.uni-goettingen.de/handle/11858/13997

Kala, S., Raj, H., Upadhyay, R. R., & Bishnu, P. S. (2024). Predicting stress levels using sequence modeling in deep learning. *AIP Conference Proceedings*, *3164*(1), 020004. https://doi.org/10.1063/5.0214575

Kouighat, M., Farhaoui, A., Assouguem, A., Taoussi, M., Belabess, Z., & Lahlali, R. (2025). Epigenetics for combating plant pathogens: Updated methods and current achievements. In J. T. Chen (Ed.), *CABI climate change series* (pp. 309–325). GB: CABI. https://doi.org/10.1079/9781800626102.0018

Kouighat, M., Chetto, O., & Nabloussi, A. (2025). Role of MicroRNA in the regulation of plant abiotic stress. In N. K. Sharma (Ed.), *Molecular markers and crop improvement: Principle and application* (pp. 41–51). Springer Nature Switzerland. https://doi.org/10.1007/978-3-031-87401-7_4

Lalman, C., Yang, Y., & Walker, J. L. (2025). Artificial intelligence in ocular transcriptomics: Applications of unsupervised and supervised learning. *Cells*, *14*(17), 1315. https://doi.org/10.3390/cells14171315

Lazaros, K., Koumadorakis, D. E., Vlamos, P., & Vrahatis, A. G. (2024). Graph neural network approaches for single-cell data: A recent overview. *Neural Computing and Applications, 36*(17), 9963–9987. https://doi.org/10.1007/s00521-024-09662-6

Le Roux, M. S., Kunert, K. J., Cullis, C. A., & Botha, A. M. (2024). Unlocking wheat drought tolerance: The synergy of omics data and computational intelligence. *Food and Energy Security, 13*(6), e70024. https://doi.org/10.1002/fes3.70024

Lee, M. (2023). Recent advances in generative adversarial networks for gene expression data: A comprehensive review. *Mathematics, 11*(14), 3055. https://doi.org/10.3390/math11143055

Li, Z., & Sillanpää, M. J. (2012). Overview of LASSO-related penalized regression methods for quantitative trait mapping and genomic selection. *Theoretical and Applied Genetics, 125*(3), 419–435. https://doi.org/10.1007/s00122-012-1892-9

Li, N., Lewin, A., Ning, S., Waito, M., Zeller, M. P., Tinmouth, A., & Shih, A. W. (2025). Privacy-preserving federated data access and federated learning: Improved data sharing and AI model development in transfusion medicine. *Transfusion, 65*(1), 22–28. https://doi.org/10.1111/trf.18077

Liu, Q., Xia, F., Yin, Q., & Jiang, R. (2018). Chromatin accessibility prediction via a hybrid deep convolutional neural network. *Bioinformatics, 34*(5), 732–738. https://doi.org/10.1093/bioinformatics/btx679

Lloyd, J. P., & Lister, R. (2022). Epigenome plasticity in plants. *Nature Reviews Genetics, 23*(1), 55–68. https://doi.org/10.1038/s41576-021-00407-y

Lu, R. J. H., Lin, P. Y., Yen, M. R., Wu, B. H., & Chen, P. Y. (2023). MethylC-analyzer: A comprehensive downstream pipeline for the analysis of genome-wide DNA methylation. *Botanical Studies, 64*(1), 1. https://doi.org/10.1186/s40529-022-00366-5

Lv, T., Han, Q., Li, Y., Liang, C., Ruan, Z., Chao, H., Chen, M., & Chen, D. (2025). Cross-species prediction of histone modifications in plants via deep learning. *bioRxiv*, pp. 2025–05. https://doi.org/10.1186/s13059-025-03929-4

Ma, B., Meng, F., Yan, G., Yan, H., Chai, B., & Song, F. (2020). Diagnostic classification of cancers using extreme gradient boosting algorithm and multi-omics data. *Computers in Biology and Medicine, 121*, 103761. https://doi.org/10.1016/j.compbiomed.2020.103761

Mahajan, A., Navarro, E. J., Poole, W., & Lopez, C. F. (2025). *Multiscale modeling primer: Focus on chromatin and epigenetics*. arXiv preprint arXiv:250403876. Retrieved September 6 2025, from https://arxiv.org/abs/2504.03876

Mowbray, T. (2025). A survey of deep learning architectures in modern machine learning systems: From CNNs to transformers. *Journal of Computer Technology and Software, 4*(8). https://doi.org/10.5281/zenodo.17035434

Noordijk, B., Garcia Gomez, M. L., ten Tusscher, K. H., de Ridder, D., van Dijk, A. D., & Smith, R. W. (2024). The rise of scientific machine learning: A perspective on combining mechanistic modelling with machine learning for systems biology. *Frontiers in Systems Biology, 4*, 1407994. https://doi.org/10.3389/fsysb.2024.1407994

Oh, D., Strattan, J. S., Hur, J. K., Bento, J., Urban, A. E., Song, G., & Cherry, J. M. (2020). CNN-Peaks: ChIP-Seq peak detection pipeline using convolutional neural networks that imitate human visual inspection. *Scientific Reports, 10*(1), 7933. https://doi.org/10.1038/s41598-020-64655-4

Peleke, F. F., Zumkeller, S. M., Gültas, M., Schmitt, A., & Szymański, J. (2024). Deep learning the cis-regulatory code for gene expression in selected model plants. *Nature Communications, 15*(1), 3488. https://doi.org/10.1038/s41467-024-47744-0

Pimpalkar, A., Gandhewar, N., Shelke, N., Patil, S., & Chhabria, S. (2025). An efficient deep convolutional neural networks model for genomic sequence classification. In S. Choudhary, S. Kumar, S. Gowroju, M. Gulhane, & R. Sri Lakshmi (Eds.), *Genomics at the nexus of AI, computer vision, and machine learning* (1st ed., pp. 345–375). Wiley. https://doi.org/10.1002/9781394268832.ch16

Rahman, C. R., Amin, R., Shatabda, S., & Toaha, M. S. I. (2021). A convolution based computational approach towards DNA N6-methyladenine site identification and motif extraction in rice genome. *Scientific Reports, 11*(1), 10357. https://doi.org/10.1038/s41598-021-89850-9

Raza, K. (2024). *Machine Learning in Single-Cell RNA-Seq Data Analysis; SpringerBriefs in Applied Sciences and Technology*. Springer Nature. ISBN 978-981-9767-02-1. https://doi.org/10.1007/978-981-97-6703-8

Saavedra, L. M., Pollastri, G., Riquelme, E., Sáez, M., & Martin, A. J. (2023). Random forest approach for the identification of relationships between epigenetic marks and its application to robust assignment of chromatin states. *bioRxiv*, pp. 2023–01. https://doi.org/10.1101/2023.01.12.523636

Searle, B., Müller, M., Carell, T., & Kellett, A. (2023). Third-generation sequencing of epigenetic DNA. *Angewandte Chemie, 135*(14), e202215704. https://doi.org/10.1002/ange.202215704

Shafik, W., Tufail, A., De Silva, L. C., & Apong, R. A. A. H. M. (2025). A lightweight deep learning model for multi-plant biotic stress classification and detection for sustainable agriculture. *Scientific Reports, 15*(1), 12195. https://doi.org/10.1038/s41598-025-90487-1

Sun, Y., Xu, X., Lin, L., Xu, K., Zheng, Y., Ren, C., Tao, H., Wang, X., Zhao, H., Tu, W., Bai, X., Wang, J., Huang, Q., Li, Y., Chen, H., Li, H., & Bo, X. (2023). A graph neural network-based interpretable framework reveals a novel DNA fragility–associated chromatin structural unit. *Genome Biology, 24*(1), 90. https://doi.org/10.1186/s13059-023-02916-x

Sun, L., Lai, M., Ghouri, F., Nawaz, M. A., Ali, F., Baloch, F. S., Nadeem, M. A., Aasim, M., & Shahid, M. Q. (2024). Modern plant breeding techniques in crop improvement and genetic diversity: From molecular markers and gene editing to artificial intelligence—a critical review. *Plants, 13*(19), 2676. https://doi.org/10.3390/plants13192676

Thingujam, D., Gouli, S., Cooray, S. P., Chandran, K. B., Givens, S. B., Gandhimeyyan, R. V., Tan, Z., Wang, Y., Patam, K., Greer, S. A., Acharya, R., Moseley, D. O., Osman, N., Zhang, X., Brooker, M. E., Tagert, M. L., Schafer, M. J., Jeong, C., Hoffseth, K. F., ... Mukhtar, M. S. (2025). Climate-resilient crops: Integrating AI, multi-omics, and advanced phenotyping to address global agricultural and societal challenges. *Plants, 14*(17), 2699. https://doi.org/10.3390/plants14172699

Trabelsi, A., Chaabane, M., & Ben-Hur, A. (2019). Comprehensive evaluation of deep learning architectures for prediction of DNA/RNA sequence binding specificities. *Bioinformatics, 35*(14), i269–i277. https://doi.org/10.1093/bioinformatics/btz339

Türkoğlu, A., Haliloğlu, K., Demirel, F., Aydin, M., Çiçek, S., Yiğider, E., Demirel, S., Piekutowska, M., Szulc, P., & Niedbała, G. (2023). Machine learning analysis of the impact of silver nitrate and silver nanoparticles on wheat (*Triticum aestivum* L.): Callus induction, plant regeneration, and DNA methylation. *Plants, 12*(24), 4151. https://doi.org/10.3390/plants12244151

Vimbi, V., Shaffi, N., & Mahmud, M. (2024). Interpreting artificial intelligence models: A systematic review on the application of LIME and SHAP in Alzheimer's disease detection. *Brain Informatics, 11*(1), 10. https://doi.org/10.1186/s40708-024-00222-1

Wang, Z., Peng, Y., Li, J., Li, J., Yuan, H., Yang, S., Ding, X., Xie, A., Zhang, J., Wang, S., & Li, K. (2024). Deepcba: A deep learning framework for gene expression prediction in maize based on DNA sequences and chromatin interactions. *Plant Communications, 5*(9), 100985. https://doi.org/10.1016/j.xplc.2024.100985

Westhues, C. C., Mahone, G. S., da Silva, S., Thorwarth, P., Schmidt, M., Richter, J. C., Simianer, H., & Beissinger, T. M. (2021). Prediction of maize phenotypic traits with genomic and environmental predictors using gradient boosting frameworks. *Frontiers in Plant Science, 12*, 699589. https://doi.org/10.3389/fpls.2021.699589

Yang, P. (2025). *Exploring self-organizing maps: Applications in omics data analysis and integration with a generative model* [PhD thesis]. Purdue University. Retrieved September 14 2025, from htt1?ps://search.1?proquest.com/o1?penview/a801a4cbbccdf6f1d87acaa379903ecd/1?1?pq-origsite=gscholar&cbl=18750&diss=y

6

AI-Based Approaches for Studying Plant Epigenetic Landscapes Under Abiotic Stress

Sangeeta Sarma[1], Arpita Talukdar[2], Abdul Jalil[2], Bhanu Priya Pegu[2], Nazneen Hussain[3], Abhik Gogoi[4], Dhanawantari L. Singha[2], and Manabendra Dutta Choudhury[2]

[1] Department of Botany, Jagiroad College, Jagiroad, Assam, India
[2] Department of Botany, Rabindranath Tagore University, Hojai, Assam, India
[3] Department of Biosciences, Assam Don Bosco University, Tapesia, Assam, India
[4] Department of Agricultural Biotechnology, Assam Agricultural University, Jorhat, Assam, India

6.1 Introduction

Plants are continually exposed to a wide array of environmental challenges. Their ability to respond to changing climatic conditions is crucial for survival, growth, and overall productivity. To navigate these dynamic environmental shifts, plants undergo a range of molecular-level modifications (Singroha et al., 2019). In the current scenario, climate change and extreme weather events have significantly affected agricultural output (Ma et al., 2025). Evidence suggests that stress exposure can trigger epigenetic modifications, which in turn regulate the expression of stress-responsive genes, thereby enhancing plant adaptability (Singroha et al., 2019). Key epigenetic mechanisms such as DNA methylation, histone modifications, and non-coding RNAs (ncRNAs) contribute to dynamic alterations in DNA, RNA, and histone proteins, influencing chromatin structure and conformation (Ma et al., 2025). These mechanisms collectively shape the epigenetic landscape, modulating gene expression without altering the underlying DNA sequence (Ma et al., 2025).

The field of epigenetics is rapidly expanding, generating vast datasets that present both formidable challenges and promising opportunities for future research. Artificial intelligence (AI) is materializing as a potent tool to traverse this huge data, allowing the extraction of meaningful information, even intricate gene–environment interactions (GxE) (Butera et al., 2023; Boye et al., 2024). AI models may encompass a variety of omic data layers such as genomics, transcriptomics, epigenomics, and proteomics (Dobránszki et al., 2025). Moreover, AI can integrate RNA and protein secondary or tertiary structures into these layers, which is difficult for traditional techniques (Sun et al., 2021). As a result, AI can have a big impact on epigenetics by adding new layers of information into omics data (Dobránszki et al., 2025).

This chapter will try to elucidate the implementation of AI in studying plant epigenetic landscapes under abiotic stress. It will also analyze the integration of omics data with AI technologies. However, AI techniques have several drawbacks. Understanding the strengths and limits of each AI model allows researchers to choose the best architecture. Hence, this chapter will also draw parallels between the challenges and future possibilities of the use of AI in the field of plant epigenetics, especially under abiotic stress.

6.1.1 Plant Epigenetic Mechanisms

Epigenetic mechanisms fundamentally constitute continuous alterations in the DNA, RNA, and histone protein levels in addition to chromatin structure and conformation (Ma et al., 2025). Furthermore, it also includes ncRNAs, notably long ncRNAs and short RNAs (like small interfering RNAs [siRNA] and microRNA [miRNA]), which control translation, chromatin modifications, and mRNA abundance (Chorostecki et al., 2023; Jha et al., 2020; Zhan & Meyers, 2023) (Figure 6.1). The epigenetic mechanisms are DNA methylation,

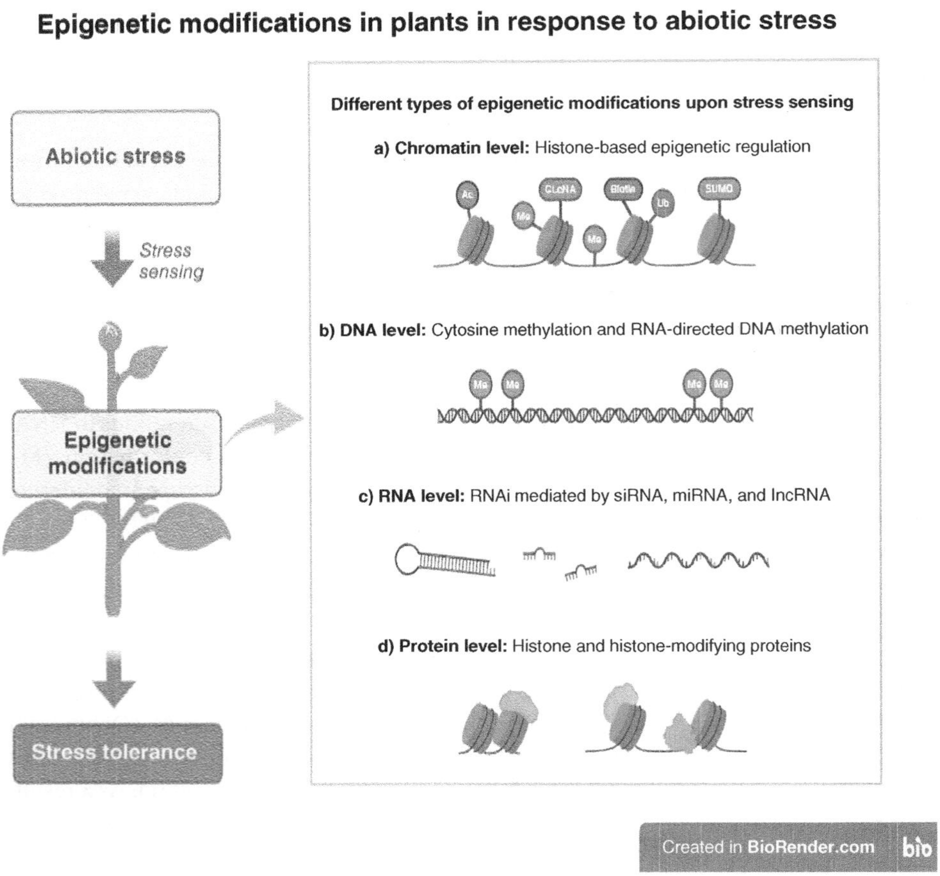

Figure 6.1 Epigenetic modification in plants under abiotic stress.

histone modifications, chromatin remodeling, RNA-mediated silencing, etc. In plants, the most prevalent kind of DNA methylation is 5-methylcytosine (5mC), which occurs in all cytosine sequence contexts: CG, CHG, and CHH (with H being A, C, or G) (Lister et al., 2008; Zhang et al., 2006). A variety of post-translational changes on histones, such as acetylation, methylation, ubiquitination, sumoylation, and phosphorylation, can alter chromatin activity, determine chromatin function, and regulate gene transcription (Millán-Zambrano et al., 2022). These alterations may have a direct effect on chromatin characteristics or necessitate the use of effector proteins that recognize specific histone modifications and mediate downstream activities (Yun et al., 2011). Histone variants contribute significantly to nucleosome diversity (Ma et al., 2025). Beyond their sequence divergence, these variants often exhibit distinct expression profiles and specific chromatin localization patterns (Probst, 2022). Small RNAs (sRNAs) are small ncRNA molecules that influence gene activity through both transcriptional (TGS) and post-transcriptional gene silencing (PTGS) (Zhan & Meyers, 2023). Plants utilize sRNAs, including miRNAs and siRNAs, to regulate gene silencing and orchestrate transcriptional reprogramming (Iqbal et al., 2021).

6.1.2 Epigenetic Regulation of Plant Responses to Abiotic Stress

Heat stress causes dramatic alterations in chromatin architecture. Heat stress leads to the decondensation of heterochromatin in *Arabidopsis*, disrupting its normally compact structure (Pecinka et al., 2010; Tittel-Elmer et al., 2010). This chromatin relaxation aligns with the global reorganization of three-dimensional (3D) chromatin architecture observed under heat stress conditions (Sun et al., 2020; Wang et al., 2023). Two plant-specific histone deacetylases, HD2B and HD2C, mitigate this decondensation by suppressing histone hyperacetylation and preventing DNA hypomethylation (Yang et al., 2023). In tomato, the heat stress transcription factor (HSFA1a) enhances the expression of heat-responsive genes by transiently promoting the formation of enhancer-promoter chromatin loops (Huang et al., 2023). Similarly, in rice, heat-induced changes in chromatin architecture are closely linked to alterations in chromatin accessibility (Liang et al., 2021). Its activation depends on the transcription factors HSFA1 and HSFA2 (Cavrak et al., 2014), although its overactivation and transposition are restrained by the RNA-directed DNA methylation (RdDM) pathway (Ito et al., 2011; Niu et al., 2022). Heat stress also results in reduced DNA methylation levels across multiple plant species, including *Arabidopsis*, soybean, and barley (Hossain et al., 2017; Korotko et al., 2021; Li et al., 2022), and is associated with histone modifications such as acetylation (Singroha et al., 2019). In contrast, cold stress influences epigenetic regulation through DNA methylation dynamics. Suppression of DNA methylation or deletion of the DNA methyltransferase DRM2 (domains rearranged methyltransferase 2) has been shown to enhance freezing tolerance in plants (Xie et al., 2019). Cold treatment also leads to a global reduction in DNA methylation in species such as winter rapeseed, sweet beetroot, and tartary buckwheat (Gutschker et al., 2022; Song et al., 2020; Zheng et al., 2022). These observations underscore the critical role of DNA methylation and chromatin remodeling in regulating gene expression during cold stress (Ma et al., 2023). RNA modification is also implicated in the plant cold stress response (Govindan et al., 2022).

In response to high salinity, *Arabidopsis* lineages exhibit approximately 45% more differentially methylated CG sites compared to unstressed controls, highlighting the epigenetic impact of salt stress (Jiang et al., 2014). High salinity settings also cause genome-wide DNA hypermethylation of TEs in the mangrove *Bruguiera gymnorhiza* (Miryeganeh et al., 2022). Salt causes distinct DNA methylation alterations in rice, wheat, and chickpea, implying a possible relationship between DNA methylation and salt stress resistance (Gupta & Garg, 2023; Wang et al., 2014, 2015). Indeed, DNA methylation regulates the expression of high-affinity potassium transporter (HKT) genes, which play a key role in salt resistance (Baek et al., 2011; Kumar et al., 2017; Wang et al., 2020). Histone acetyltransferases are known to activate salt stress resistance genes in *Arabidopsis* (Zheng et al., 2019), wheat, and soybean (Feng et al., 2022; Wu et al., 2011; Zheng et al., 2021). Histone deacetylases play a variety of roles in plant salt stress resistance. In *Arabidopsis*, the class II histone deacetylases HDA5/14/15/18 promote salt resistance, whereas the class I histone deacetylase HDA19 reduces salt tolerance by lowering histone acetylation and transcription at the ABA receptor genes PYL4, PYL5, and PYL6 (Mehdi et al., 2016; Ueda et al., 2017). Salt stress reduces the expression of the histone H1 variant H1.3/HIS1-3, which reduces H1.3-mediated suppression of SOS pathway genes SOS1, SOS2, and SOS3 in *Arabidopsis* (Wu et al., 2022). Another histone variation, H2A.Z, regulates gene expression in response to salt stress (Miao et al., 2024). Dynamic RNA changes also affect plant salt stress resistance (Ma et al., 2025).

6.1.3 Challenges in Handling and Integrating Large-Scale Epigenetic Datasets

Epigenetic profiles are tissue-specific, dynamic, and ongoing. The challenges now become developing data analysis techniques to enable the identification of coordinated epigenetic changes and the interpretation of their functional consequences in normal development and disease, as high-throughput sequencing and microarray technologies generate an increasing amount of epigenetic data (Cazaly et al., 2019). In the everyday study of omics data, integrating massive and heterogeneous data presents both a conceptual and practical barrier. Although freely accessible and adequately maintained data repositories provide a fair amount of genomics data availability (with the pertinent exception of clinical data), better (and innovative) annotation requirements and standards in data repositories are required to facilitate better integration and utilization of publicly available data. The use of existing knowledge and its effective storage, the development of statistical techniques to analyze heterogeneous datasets, and the development of data exploratory tools that combine both practical summary statistics and novel visualization tools are all part of the data exploitation component of data integration, which is arguably the one that needs the most attention (Gomez-Cabrero et al., 2014).

6.2 AI and Machine Learning in Epigenetics

Epigenetics encompasses a diverse array of mechanisms that regulate gene expression without altering the underlying nucleotide sequence. Core processes involved in epigenetic regulation include DNA methylation, post-translational histone modifications, chromatin remodeling, and the activity of ncRNAs. These molecular changes can significantly reshape gene expression profiles and are increasingly recognized as contributing factors in the development of various diseases and disorders (Tahir et al., 2024). Over the past few decades, high-throughput technologies have been instrumental in identifying and characterizing epigenetic modifications. However, the experimental procedures and biochemical analyses required for such investigations are often labor-intensive and time-consuming. To address these challenges, machine learning (ML) and AI have been integrated into epigenetic research to efficiently map molecular changes to phenotypic outcomes. ML algorithms enable computational systems to detect patterns and make predictions based on large datasets (Krittanawong et al., 2017). AI, leveraging ML capabilities, facilitates the analysis of complex epigenetic data, thereby deepening our understanding of how environmental cues and genetic variation influence plant traits such as stress tolerance and growth. By recognizing patterns in DNA methylation, histone modifications, and chromatin architecture, AI models can predict gene expression outcomes, accelerating the development of stress-resilient and high-yielding crop varieties (Farooq et al., 2024). The increasing availability of large-scale omics datasets has further enhanced the utility of ML in pattern recognition, positioning it as a powerful tool for data-intensive fields such as genomics and epigenetics. ML's ability to process and interpret complex data structures enables more refined insights and predictive modeling (Brasil et al., 2021).

Plants, despite lacking specialized organs, exhibit sophisticated capabilities for learning, communication, and decision-making through their complex cellular and biochemical networks. They adapt to environmental fluctuations by encoding and recalling recurring stimuli (Auge et al., 2023). The incorporation of advanced AI and ML technologies is revolutionizing biological research, particularly in forecasting plant responses to climate-induced stressors. Epigenetic mechanisms play a pivotal role in plant memory, enabling adaptive responses through chromatin remodeling and gene regulation. These dynamic processes enhance plant resilience and survival under changing environmental conditions (Dobránszki et al., 2025).

6.2.1 Types of AI Techniques

ML has emerged as a transformative AI technology, characterized by its focus on developing algorithms that facilitate the automation of analytical model development. These ML models are typically constructed from a combination of heuristic rules, procedural frameworks, and advanced transfer functions, all of which enable the identification of intricate data patterns and the prediction of behavioral trends (Sarker, 2021).

Supervised learning is a prominent technique in ML, particularly utilized for classification and predictive modeling within the field of epigenetics. This method involves training models on labeled datasets, enabling them to predict outcomes based on specified tasks. *Unsupervised learning*, a data-driven approach, focuses on uncovering patterns and structures in unlabeled datasets. The intricate and high-dimensional landscape of epigenetics, characterized by diverse components such as gene expression profiles, DNA methylation patterns, and chromatin accessibility, often necessitates the application of unsupervised ML methodologies for effectively identifying pertinent patterns and features (Arslan et al., 2021). Deep learning (DL) is a key AI technique that utilizes artificial neural networks (ANN). Its significance in computing stems from its capacity for hierarchical learning through multiple layers. A typical deep neural network (DNN) comprises several hidden layers situated between input and output layers, allowing it to model complex functions and discern intricate data patterns (Sarkar, 2021). *Recurrent neural networks* (RNNs) represent a widely used DL architecture distinguished by their cyclic connections, which allow them to incorporate memory mechanisms for processing sequential data such as genomic sequences. *Graph neural networks* (GNNs), on the other hand, excel in interpreting complex biological datasets, especially molecular interaction networks. These models are instrumental in uncovering relationships among transcription factors, key DNA methylation sites, enhancers, and their target genes. *Convolutional neural networks* (CNNs) are adept at recognizing spatial hierarchies and patterns within grid-like datasets, making them particularly effective for analyzing genomic data. *Generative adversarial networks* (GANs) consist of a generator and a discriminator. GANs offer a compelling analogy to epigenetic regulation. *Multi-layer perceptrons* (MLPs) are a type of feedforward neural network particularly well-suited for regression, pattern recognition, and classification tasks. Even though MLPs may lack the architectural complexity of models such as GANs, they are notably effective in capturing specific epigenetic features. *Transformer architectures*, on the other hand, are well-suited for processing large-scale datasets and capturing long-range dependencies, making them ideal for integrating multiomics data and generating data-driven biological insights. *Deep belief networks* (DBNs) are advanced probabilistic generative models that excel in unsupervised learning and feature extraction. These networks are particularly adept at uncovering latent relationships between epigenetic marks, such as DNA methylation and histone modifications, and gene expression, all while operating without the necessity for pre-labeled training data (Li et al., 2025). *Self-organizing maps* (SOMs) offer a powerful approach for uncovering complex patterns within high-dimensional biological data. They are especially useful for identifying clusters and revealing intricate relationships among variables related to DNA methylation, gene expression, and histone modifications. *Large language models* (LLMs), trained on domain-specific linguistic corpora, can be integrated with omics datasets to construct models capable of interpreting multilayered biological phenomena. These approaches show considerable potential for analyzing ncRNA sequencing (ncRNA-seq) data, with the capacity to capture nuanced aspects of epigenetic regulation and enhance our understanding of cellular diversity and function.

6.2.2 Data Preprocessing, Feature Extraction, Model Training, and Evaluation

AI and ML are increasingly being employed to decode complex plant epigenetic responses to abiotic stresses such as drought, salinity, heat, and cold. The initial phase of the analytical pipeline is data preprocessing, which entails cleaning and preparing high-throughput sequencing datasets. This begins with quality assessment of raw reads using tools such as FastQC and TrimGalore, followed by adapter trimming and removal of low-quality sequences via Cutadapt or Fastp (Dobránszki et al., 2025). Subsequent steps involve aligning the reads to reference genomes using specialized tools like Bismark for whole-genome bisulfite sequencing (WGBS), Bowtie2 for ChIP-seq, and STAR (spliced transcripts alignment to a reference) for RNA-seq. After alignment, methylation calling is typically conducted using the Bismark methylation extractor, with low-coverage and ambiguous sites filtered out to improve data integrity. For datasets with sparse coverage, imputation tools like METHimpute are employed to infer methylation states, thereby enhancing reliability (Shahryary et al., 2020). To ensure consistency across experimental conditions, normalization and batch effect correction are applied. Annotation tools are then used to associate methylation data with genomic features such as promoters, gene bodies, and transposable elements (TEs), enabling more targeted downstream analyses. Following preprocessing, the pipeline advances to feature extraction, where biologically relevant variables are derived from the curated datasets. These features may include average methylation levels across sequence contexts (CG, CHG, and CHH), identification of differentially methylated regions (DMRs), histone modification patterns, chromatin accessibility signals, and gene expression profiles. For example, in *Medicago ruthenica*, stress-induced hypomethylation of promoter and gene body regions was found to correlate with the upregulation of drought-responsive genes such as those in the ABA and proline biosynthesis pathways (Zi et al., 2024).

6.3 Application of AI in Studying Plant Epigenetic Responses to Abiotic Stress

The application of AI in biology holds immense potential, offering revolutionary ways to understand complex biological systems and accelerate research since it democratizes access to analytical tools and computational power (Koh et al., 2024). AI can significantly impact epigenetics by incorporating additional information layers into omics datasets (Dobranszki et al., 2025). Some of the key application areas of AI are discussed below.

6.3.1 Pattern Recognition and Classification

One of the important applications of AI in plant epigenetics is detecting and classifying various patterns in high-dimensional epigenomic data under abiotic stress, such as drought, salinity, and heat. Advanced AI and ML algorithms excel at recognizing patterns within complex datasets. A variety of ML and DL techniques have been applied to predict key crop characteristics. These include algorithms like random forest (RF), neural networks (NN), k-nearest neighbor (KNN), partial least squares (PLS), and support vector machines (SVM), as well as DL architectures such as residual network (ResNet), ResNetXt-101, AlexNet, DenseNet, GoogLeNet, VGG-16, and YOLOv5 (Farooq et al., 2024).

ML models are designed to identify patterns and generate predictive models. For instance, a widely used ML approach, supervised learning, can be employed to identify genes associated with drought tolerance. By training on known drought-tolerant genes and their expression profiles, the model can predict the relevance of other genes in conferring drought resistance (Singhal et al., 2025). Supervised learning is extensively used in epigenetics for classification and prediction tasks. In contrast, unsupervised learning techniques such as clustering and principal component analysis (PCA) are applied to decipher complex, high-dimensional epigenetic data, including gene expression, DNA methylation, and chromatin accessibility (Cembrowska-Lech et al., 2023; Dobranszki et al., 2025).

DL classifiers, which utilize NN, categorize data by learning hierarchical features from large datasets. In this approach, features are automatically extracted from RGB images,

Table 6.1 Machine learning and deep learning techniques for pattern recognition and classification in plant epigenetics under abiotic stress.

Algorithm/models	AI techniques	Task	Application/examples
Random forest (RF)	ML-supervised	Classification	Identifying important genes under drought conditions
Support vector machine (SVM)	ML-supervised	Classification and prediction	Drought and salt tolerance in arabidopsis and rice
k-nearest neighbor (KNN)	ML-supervised	Classification and prediction	Predicting and classification of gene function
Artificial neural network (ANN)	ML-supervised	Prediction and identification	Salt tolerance and drought tolerance in maize
Convolutional neural network (CNN)	DL-supervised	Image classification, pattern recognition	Stress detection in cucumber with 99.85% accuracy
Deep neural network	DL-supervised	High level pattern extraction, classification	Extract meaningful feature from RGB image
ResNet	DL-supervised	Image feature extraction	Prediction of crop traits
AlexNet	DL-supervised	Image classification and identification	Used in remote sensing to identify stress
YOLOv5	DL-supervised	Object detection	Detecting plant under stress condition via image analysis
Clustering	ML-unsupervised	Recognition and pattern detection	Methylation and chromatin accessibility, grouping gene expression
Bayesian algorithm	ML-supervised	Identification	Drought resistance in maize

enabling the model to identify meaningful patterns for classification. Common DL techniques include CNNs, DNNs, and various pretrained models (Mulugeta et al., 2024). CNNs, in particular, are well-suited for classification tasks due to their ability to perform automatic feature extraction. For example, studying CNNs in fruit crops such as stone fruit and cucumber showed that it helps in image classification and early visual stress detections with high accuracy (up to 99.85%), excelling in pattern recognition, and they are found to be computationally intensive (Mushtaq et al., 2025). Some other applications of CNNs include image classification, object detection, object tracking, pose estimation, text detection, and recognition. CNNs are also applicable in remote sensing for object detection and pattern recognition. The various algorithms used and their application are listed in Table 6.1.

6.3.2 Predictive Modeling

AI techniques such as ML and DL are used to develop predictive models that can stimulate crop responses to various stress conditions and can classify stress types, i.e., salt, drought, heat, and waterlogging, from morphological images and predict potential yield reduction

Table 6.2 AI-based predictive modeling approaches for plant epigenetic responses.

AI techniques	Learning type	Algorithm	Epigenetic relevance/application
Machine learning	Mostly supervised and some unsupervised	RF, SVM, XBoost; clustering	Yield prediction, stress classification, and remote sensing integration. Identifies associations between abiotic stress factor and epigenetic regulation
Deep learning	Primarily supervised; some unsupervised	CNN, RNN, ReNet, transformers; autoencoders, GNNs, PlantDeepMeth	Predicts DNA methylation, histone modifications, RNA modifications, for example SMEP model

(Mushtaq et al., 2025). ML models are capable of analyzing large datasets to identify patterns and relationships between environmental variables and crop productivity. ML models can be used with remote sensing data, such as satellite or drone-based images, to assess the spatial extent and severity of abiotic stress across large agricultural areas (Paul et al., 2024). Algorithms such as RF, extreme gradient boosting (XGBoost), and SVMs are widely implemented in plant trait prediction due to their flexibility and ability to model complex nonlinear relationships (Amin et al., 2025) and also include those involving epigenetic regulation. DL models such as CNNs and RNNs excel in processing high-dimensional data and capturing complex relationships, leading to more accurate predictions of plant stress responses (Ali et al., 2024). Recent studies have shown the use of some advanced, powerful tools of DL models, such as RNNs, ResNet, transformers, GNNs, and autoencoders in genomic prediction in plant breeding. These advanced genomic prediction methods allow plant breeders to make more informed decisions, speeding up the breeding process and enhancing the development of improved performance, resilience, and yield (Montesnos-Lopez et al., 2024) (Table 6.2).

6.3.3 Multiomics Data Integration

Multiomics approaches include genomics, epigenomics, metabolomics, transcriptomics, and proteomics; these approaches are indispensable for elucidating the genetic underpinning, environmental adaptability, and developmental processes of crops (Fan et al., 2025). The integration of multiomics data connects genotype to phenotype and provides a proper understanding of biological processes such as abiotic stress response. These data represent a robust methodology that merges insights from various high-throughput omics technologies and provides a more holistic view of how plants respond to abiotic stress (Roychowdhury et al., 2023) (Figure 6.2).

The integration of multiomics data with AI/ML is emerging as a powerful area of research, enabling the integration and interpretation of complex biological data to uncover meaningful patterns (Cembrowska-Lech et al., 2023). AI-assisted omics approaches combine advanced molecular technologies with AI to understand how plants respond to stresses, particularly in plant defense studies, rapidly identify crucial components in

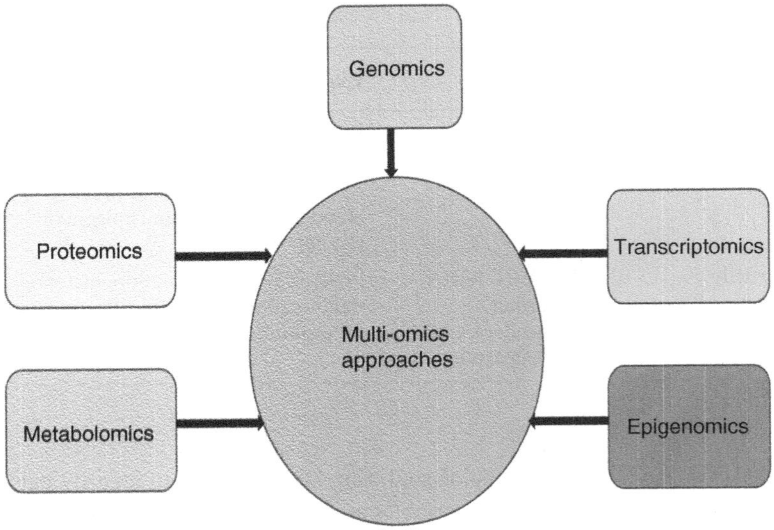

Figure 6.2 Multiomics data integration.

defense pathways, discover biomarkers, and reveal hidden patterns, enhancing plant defense processes (Murmu et al., 2024).

Bioinformatics plays a vital role by integrating with various omics approaches and provides a base for collecting information from plant abiotic stress, facilitating data mining and data production in support of different omics technologies (Yang et al., 2021). Platforms such as PlantGDB and Phytozome serve as bioinformatics resources for plant genomics research, providing access to genomic datasets related to abiotic stress responses (Sarfraz et al., 2025). The combination of epigenetic studies and genomic studies within an organism is a powerful approach and plays a crucial role in advancements of cellular processes and analysis of epigenetic modification, which is then applied to the enhancement of any phenotype characteristics of plants (Zhang et al., 2022).

According to Murmu et al. (2024), integrating remote sensing with AI-driven omics techniques strengthens stress monitoring by linking environmental inputs to molecular (genetic) responses and environmental stress. In a study by Argueuso et al. (2019), genomic, transcriptomic, and epigenetic data were integrated in *Arabidopsis thaliana* using AI and ML approaches. This integration uncovered complex interrelationships among omics layers and provided valuable insights into the mechanisms underlying plant stress responses. The incorporation of DNA methylation, ncRNAs, and chromatin modifications into interpretable AI frameworks holds significant promise for identifying adaptive epigenetic markers. These markers could play a pivotal role in advancing breeding programs and genetic engineering strategies aimed at developing climate-resilient crops (Dobranski et al., 2025). Additionally, Ngyun et al. (2019) introduced a multiview learning method called ManiNetCluster, which was applied to gene expression data in *Chlamydomonas reinhardtii*. Unlike earlier multiomics studies that relied primarily on basic statistical methods, ManiNetCluster facilitates the systematic integration of omics data through AI-driven analysis (Figure 6.3).

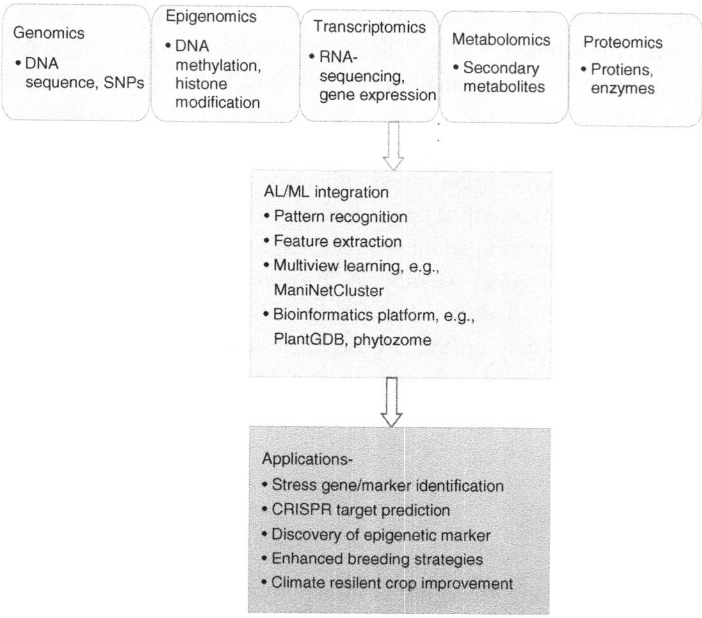

Figure 6.3 Integration of multiomics data using AI/ML for deciphering plant epigenetic response to abiotic stress.

6.4 Limitations and Drawbacks

The integration of AI into plant epigenetics has opened new avenues for understanding the complex regulatory mechanisms that govern plant responses to abiotic stresses. By leveraging the power of ML and DL, researchers are now capable of uncovering latent patterns within large-scale omics datasets ranging from DNA methylation and histone modifications to ncRNA dynamics and chromatin accessibility. These approaches enable not only the identification of stress-responsive epigenetic signatures but also the prediction of their functional outcomes on gene expression and phenotype. In doing so, AI facilitates a shift from traditional descriptive frameworks toward predictive and mechanistic models of plant–environment interactions (Guo et al., 2025). One of the key strengths of AI-based methods lies in their capacity to handle high-dimensional, multiomics data and uncover nonlinear relationships that may escape conventional statistical analyses. This has important implications for modeling the temporal and spatial dynamics of epigenetic modifications under stress conditions, as well as for identifying regulatory elements that contribute to phenotypic plasticity and stress memory. Additionally, the ability to generalize findings across species and genotypes offers promising opportunities for translational research and crop improvement, particularly when combined with emerging techniques such as transfer learning, attention-based models, and interpretable AI frameworks (Dobránszki et al., 2025). However, significant challenges are still there. Although AI and ML have diverse potential to apply in various ways in the area of plant epigenetics under abiotic stresses, their application has some limitations. The limitations may be the data scarcity, computational cost, and also the lack of wet lab validation. The data for the model crops such as rice, tomato, and Arabidopsis are available for epigenetic studies under abiotic stresses. However, the problem occurs with the nonmodel crops as comprehensive datasets for them are scarce restricting its validation of AI models. Further, the nonmodel plants lack well-annotated or incomplete reference sequences, resulting in difficulties with AI applications. Further, the available AI models were trained on small or species-specific datasets or specific stress, limiting the generalization across other species or stress conditions. Moreover, the ML models need careful handling of the parameters to avoid errors or misleading results. Epigenetic marks change temporally and spatially; capturing these dynamic patterns accurately is still challenging for static AI models. To make it more accurate, the AI-predicted epigenetic regulatory mechanism needs extensive molecular validation for its accuracy. Those molecular techniques may be the use of ChIP and bisulfite sequencing. Further large computational storage facility and power are required when DL-based multiomics shall be integrated.

6.5 Conclusion

In summary, AI-driven methodologies are emerging as indispensable assets in the field of plant epigenetics, providing scalable and data-centric strategies to unravel the intricate regulatory networks underlying abiotic stress responses. With ongoing advancements in computational tools and the growing availability of comprehensive, multilayered datasets, AI is set to significantly enhance the identification of epigenetic mechanisms, support functional validation efforts, and ultimately guide the breeding and development of crop varieties resilient to evolving environmental challenges.

References

Ali, T., Rehman, S. U., Ali, S., Mahmood, K., Obregon, S. A., Iglesias, R. C., Khurshaid, T., & Ashraf, I. (2024). Smart agriculture: Utilizing machine learning and deep learning for drought stress identification in crops. *Scientific Reports*, *14*(1), 30062. https://doi.org/10.1038/s41598-024-74127-8

Amin, A., Zaman, W., & Park, S. (2025). Harnessing multi-omics and predictive modelling for climate-resilient crop breeding: From genomes to fields. *Genes*, *16*(7), 809. https://doi.org/10.3390/genes16070809

Argueso, C. T., Assmann, S. M., Birnbaum, K. D., Chen, S., Dinneny, J. R., Doherty, C. J., Eveland, A. L., Friesner, J., Greenlee, V. R., Law, J. A., Marshall-Colón, A., Mason, G. A., O'Lexy, R., Peck, S. C., Schmitz, R. J., Song, L., Stern, D., Varagona, M. J., Walley, J. W., & Williams, C. M. (2019). Directions for research and training in plant omics: Big Questions and Big Data. *Plant Direct*, *3*(4), e00133. https://doi.org/10.1002/pld3.133

Arslan, E., Schulz, J., & Rai, K. (2021). Machine learning in epigenomics: Insights into cancer biology and medicine. *Biochimica Et Biophysica Acta. Reviews on Cancer*, *1876*(2), 188588. https://doi.org/10.1016/j.bbcan.2021.188588

Auge, G., Hankofer, V., Groth, M., Antoniou-Kourounioti, R., Ratikainen, I., & Lampei, C. (2023). Plant environmental memory: Implications, mechanisms and opportunities for plant scientists and beyond. *AoB Plants*, *15*(4), plad032. https://doi.org/10.1093/aobpla/plad032

Baek, D., Jiang, J., Chung, J. S., Wang, B., Chen, J., Xin, Z., & Shi, H. (2011). Regulated *AtHKT1* gene expression by a distal enhancer element and DNA methylation in the promoter plays an important role in salt tolerance. *Plant and Cell Physiology*, *52*(1), 149–161. https://doi.org/10.1093/pcp/pcq182

Boye, C., Nirmalan, S., Ranjbaran, A., & Luca, F. (2024). Genotype × environment interactions in gene regulation and complex traits. *Nature Genetics*, *56*(6), 1057–1068. https://doi.org/10.1038/s41588-024-01776-w

Brasil, S., Neves, C. J., Rijoff, T., Falcão, M., Valadão, G., Videira, P. A., & Dos Reis Ferreira, V. (2021). Artificial intelligence in epigenetic studies: Shedding light on rare diseases. *Frontiers in Molecular Biosciences*, *8*, 648012. https://doi.org/10.3389/fmolb.2021.648012

Butera, A., Smirnova, L., Ferrando-May, E., Hartung, T., Brunner, T., Leist, M., & Amelio, I. (2023). Deconvoluting gene and environment interactions to develop an 'epigenetic score meter' of disease. *EMBO Molecular Medicine*, *15*(9), e18208. https://doi.org/10.15252/emmm.202318208

Cavrak, V. V., Lettner, N., Jamge, S., Kosarewicz, A., Bayer, L. M., & Mittelsten Scheid, O. (2014). How a retrotransposon exploits the plant's heat stress response for its activation. *PLoS Genetics*, *10*(1), e1004115. https://doi.org/10.1371/journal.pgen.1004115

Cazaly, E., Saad, J., Wang, W., Heckman, C., Ollikainen, M., & Tang, J. (2019). Making sense of the epigenome using data integration approaches. *Frontiers in Pharmacology*, *10*, 126. https://doi.org/10.3389/fphar.2019.00126

Cembrowska-Lech, D., Krzemińska, A., Miller, T., Nowakowska, A., Adamski, C., Radaczyńska, M., Mikiciuk, G., & Mikiciuk, M. (2023). An integrated multi-omics and artificial intelligence framework for advance plant phenotyping in horticulture. *Biology*, *12*(10), 1298. https://doi.org/10.3390/biology12101298

Chorostecki, U., Bologna, N. G., & Ariel, F. (2023). The plant noncoding transcriptome: A versatile environmental sensor. *The EMBO Journal*, *42*(20), e114400. https://doi.org/10.15252/embj.2023114400

Dobránszki, J., Vassileva, V., Agius, D. R., Moschou, P. N., Gallusci, P., Berger, M. M. J., Farkas, D., Basso, M. F., & Martinelli, F. (2025). Gaining insights into epigenetic memories through artificial intelligence and omics science in plants. *Journal of Integrative Plant Biology*, *67*(9), 2320–2349. https://doi.org/10.1111/jipb.13953

Fan, B. L., Chen, L. H., Chen, L. L., & Guo, H. (2025). Integrative multi-omics approaches for identifying and characterizing biological elements in crop traits: Current progress and future prospects. *International Journal of Molecular Sciences*, *26*(4), 1466. https://doi.org/10.3390/ijms26041466

Farooq, M. A., Gao, S., Hassan, M. A., Huang, Z., Rasheed, A., Hearne, S., Prasanna, B., Li, X., & Li, H. (2024). Artificial intelligence in plant breeding. *Trends in Genetics*, *40*(10), 891–908. https://doi.org/10.1016/j.tig.2024.07.001

Feng, P., Sun, X., Liu, X., Li, Y., Sun, Q., Lu, H., Li, M., Ding, X., & Dong, Y. (2022). Epigenetic regulation of plant tolerance to salt stress by histone acetyltransferase *GsMYST1* from wild soybean. *Frontiers in Plant Science*, *13*. https://doi.org/10.3389/fpls.2022.860056

Gomez-Cabrero, D., Abugessaisa, I., Maier, D., Teschendorff, A., Merkenschlager, M., Gisel, A., ... Tegnér, J. (2014). Data integration in the era of omics: Current and future challenges. *BMC Systems Biology*, *8*(Suppl 2), I1. https://doi.org/10.1186/1752-0509-8-S2-I1

Govindan, G., Sharma, B., Li, Y. F., Armstrong, C. D., Merum, P., Rohila, J. S., Gregory, B. D., & Sunkar, R. (2022). mRNA N6-methyladenosine is critical for cold tolerance in arabidopsis. *The Plant Journal, 111*(4), 1052–1068. https://doi.org/10.1111/tpj.15872

Guo, Z., Fan, W., Cai, C., Zhang, K., Hou, X., Li, Y., & Cheng, F. (2025). Plantdeepmeth: A deep learning model for predicting DNA methylation states in plants. *Plants, 14*(11), 1724. https://doi.org/10.3390/plants14111724

Gupta, K., & Garg, R. (2023). Unravelling differential DNA methylation patterns in genotype dependent manner under salinity stress response in chickpea. *International Journal of Molecular Sciences, 24*(3), 1863. https://doi.org/10.3390/ijms24031863

Gutschker, S., Corral, J. M., Schmiedl, A., Ludewig, F., Koch, W., Fiedler-Wiechers, K., Czarnecki, O., Harms, K., Keller, I., Martins Rodrigues, C., Pommerrenig, B., Neuhaus, H. E., Zierer, W., Sonnewald, U., & Müdsam, C. (2022). Multi-omics data integration reveals link between epigenetic modifications and gene expression in sugar beet (Beta vulgaris subsp. vulgaris) in response to cold. *BMC Genomics, 23*(1), 144. https://doi.org/10.1186/s12864-022-08312-2

Hossain, M. S., Kawakatsu, T., Kim, K. D., Zhang, N., Nguyen, C. T., Khan, S. M., Batek, J. M., Joshi, T., Schmutz, J., Grimwood, J., Schmitz, R. J., Xu, D., Jackson, S. A., Ecker, J. R., & Stacey, G. (2017). Divergent cytosine DNA methylation patterns in single-cell, soybean root hairs. *New Phytologist, 214*(2), 808–819. https://doi.org/10.1111/nph.14421

Huang, Y., An, J., Sircar, S., Bergis, C., Lopes, C. D., He, X., da Costa, B., Tan, F. Q., Bazin, J., Antunez-Sanchez, J., Mammarella, M. F., Devani, R., Brik-Chaouche, R., Bendahmane, A., Frugier, F., Xia, C., Rothan, C., Probst, A. V., Mohamed, Z., ... Benhamed, M. (2023). HSFA1a modulates plant heat stress responses and alters the 3D chromatin organization of enhancer-promoter interactions. *Nature Communications, 14*(1), 469. https://doi.org/10.1038/s41467-023-36227-3

Iqbal, Z., Iqbal, M. S., Hashem, A., Abd Allah, E. F., & Ansari, M. I. (2021). Plant defense responses to biotic stress and its interplay with fluctuating dark/light conditions. *Frontiers in Plant Science, 12*, 631810. https://doi.org/10.3389/fpls.2021.631810

Ito, H., Gaubert, H., Bucher, E., Mirouze, M., Vaillant, I., & Paszkowski, J. (2011). An siRNA pathway prevents transgenerational retrotransposition in plants subjected to stress. *Nature, 472*(7341), 115–119. https://doi.org/10.1038/nature09861

Jha, U. C., Nayyar, H., Jha, R., Khurshid, M., Zhou, M., Mantri, N., & Siddique, K. H. M. (2020). Long non-coding RNAs: Emerging players regulating plant abiotic stress response and adaptation. *BMC Plant Biology, 20*(1), 466. https://doi.org/10.1186/s12870-020-02595-x

Jiang, C., Mithani, A., Belfield, E. J., Mott, R., Hurst, L. D., & Harberd, N. P. (2014). Environmentally responsive genome-wide accumulation of de novo arabidopsis thaliana mutations and epimutations. *Genome Research, 24*(11), 1821–1829. https://doi.org/10.1101/gr.177659.114

Koh, E., Sunil, R. S., Lam, H. Y. I., & Mutwil, M. (2024). Confronting the data deluge: How artificial intelligence can be used in the study of plant stress. *Computational and Structural Biotechnology Journal, 23*, 3454–3466. https://doi.org/10.1016/j.csbj.2024.09.010

Korotko, U., Chwiałkowska, K., Sańko-Sawczenko, I., & Kwasniewski, M. (2021). DNA demethylation in response to heat stress in arabidopsis thaliana. *International Journal of Molecular Sciences, 22*(4), 1555. https://doi.org/10.3390/ijms22041555

Krittanawong, C., Zhang, H., Wang, Z., Aydar, M., & Kitai, T. (2017). Artificial intelligence in precision cardiovascular medicine. *Journal of the American College of Cardiology, 69*(21), 2657–2664. https://doi.org/10.1016/j.jacc.2017.03.571

Kumar, S., Beena, A. S., Awana, M., & Singh, A. (2017). Salt-induced tissue-specific cytosine methylation downregulates expression of HKT genes in contrasting wheat (triticum aestivum L.) genotypes. *DNA and Cell Biology, 36*(4), 283–294. https://doi.org/10.1089/dna.2016.3505

Li, C., Xu, M., Cai, X., Han, Z., Si, J., & Chen, D. (2022). Jasmonate signaling pathway modulates plant defense, growth, and their trade-offs. *International Journal of Molecular Sciences, 23*(7), 3945. https://doi.org/10.3390/ijms23073945

Li, Z. P., Du, Z., Huang, D. S., & Teschendorff, A. E. (2025). Interpretable deep learning of single-cell and epigenetic data reveals novel molecular insights in aging. *Scientific Reports, 15*, 5048. https://doi.org/10.1038/s41598-025-89646-1

Liang, Z., Zhang, Q., Ji, C., Hu, G., Zhang, P., Wang, Y., Yang, L., & Gu, X. (2021). Reorganization of the 3D chromatin architecture of rice genomes during heat stress. *BMC Biology, 19*(1), 53. https://doi.org/10.1186/s12915-021-00996-4

Lister, R., O'Malley, R. C., Tonti-Filippini, J., Gregory, B. D., Berry, C. C., Millar, A. H., & Ecker, J. R. (2008). Highly integrated single-base resolution maps of the epigenome in *arabidopsis*. *Cell, 133*(3), 523–536. https://doi.org/10.1016/j.cell.2008.03.029

Ma, T., Wang, S., Sun, C., Tian, J., Guo, H., Cui, S., & Zhao, H. (2023). *Arabidopsis* LFR, a SWI/SNF complex component, interacts with ICE1 and activates ICE1 and CBF3 expression in cold acclimation. *Frontiers in Plant Science, 14*. https://doi.org/10.3389/fpls.2023.1097158

Ma, L., Xing, L., Li, Z., & Jiang, D. (2025). Epigenetic control of plant abiotic stress responses. *Journal of Genetics and Genomics, 52*(2), 129–144. https://doi.org/10.1016/j.jgg.2024.09.008

Mehdi, S., Derkacheva, M., Ramström, M., Kralemann, L., Bergquist, J., & Hennig, L. (2016). The WD40 domain protein MSI1 functions in a histone deacetylase complex to fine-tune abscisic acid signaling. *The Plant Cell, 28*(1), 42–54. https://doi.org/10.1105/tpc.15.00763

Miao, R., Zhang, Y., Liu, X., Yuan, Y., Zang, W., Li, Z., Yan, X., Pang, Q., & Zhang, A. (2024). Histone variant H2A.Z is required for plant salt response by regulating gene transcription. *Plant, Cell & Environment, 47*(7), 2691–2707. https://doi.org/10.1111/pce.14908

Millán-Zambrano, G., Burton, A., Bannister, A. J., & Schneider, R. (2022). Histone post-translational modifications—cause and consequence of genome function. *Nature Reviews Genetics, 23*(9), 563–580. https://doi.org/10.1038/s41576-022-00468-7

Miryeganeh, M., Marlétaz, F., Gavriouchkina, D., & Saze, H. (2022). De novo genome assembly and in natura epigenomics reveal salinity-induced DNA methylation in the mangrove tree Bruguiera gymnorhiza. *New Phytologist, 233*(5), 2094–2110. https://doi.org/10.1111/nph.17738

Montesinos-Lopez, O. A., Chavira-Flores, M., Kismiantini, C.-H. L., Saint Piere, C., Li, H., Fritsche-Neto, R., Al-Nowibet, K., Montesinos-López, A., & Crossa, J. (2024). A review of multimodal deep learning methods for genomic-enabled prediction in plant breeding. *Genetics, 228*(4), iyae161. https://doi.org/10.1093/genetics/iyae161

Mulugeta, A. K., Sharma, D. P., & Mesfin, A. H. (2024). Deep learning for medicinal plant species classification and recognition: A systematic review. *Frontiers in Plant Science, 14*, 1286088. https://doi.org/10.3389/fpls.2023.1286088

Murmu, S., Sinha, D., Chaurasia, H., Sharma, S., Das, R., Jha, G. K., & Archak, S. (2024). A review of artificial intelligence-assisted omics techniques in plant defense: Current trends and future directions. *Frontiers in Plant Science, 15*, 1292054. https://doi.org/10.3389/fpls.2024.1292054

Mushtaq, M. A., Ateeq, M., Ikram, M., Alam, S. M., Kaleem, M. M., Ashraf, M. A., ... Shireen, F. (2025). Securing fruit trees future: AI-driven early warning and predictive systems for abiotic stress in changing climate. *Plant Stress, 17*, 100953. https://doi.org/10.1016/j.stress.2025.100953.

Nguyen, N. D., Blaby, I. K., & Wang, D. (2019). Maninetcluster: A novel manifold learning approach to reveal the functional links between gene networks. *BMC Genomics, 20*(Suppl 12), 1003. https://doi.org/10.1186/s12864-019-6329-2

Niu, X., Chen, L., Kato, A., & Ito, H. (2022). Regulatory mechanism of a heat-activated retrotransposon by DDR complex in *arabidopsis thaliana*. *Frontiers in Plant Science, 13*: https://doi.org/10.3389/fpls.2022.1048957

Paul, N. C., Gaikwad, B. B., Nangare, D. D., & Reddy, K. S. (2024). Enhancing crop resilience: Harnessing machine learning models for abiotic stress management.

Pecinka, A., Dinh, H. Q., Baubec, T., Rosa, M., Lettner, N., & Mittelsten Scheid, O. (2010). Epigenetic regulation of repetitive elements is attenuated by prolonged heat stress in arabidopsis. *The Plant Cell, 22*(9), 3118–3129. https://doi.org/10.1105/tpc.110.078493

Probst, A. V. (2022). Deposition and eviction of histone variants define functional chromatin states in plants. *Current Opinion in Plant Biology, 69*, 102266. https://doi.org/10.1016/j.pbi.2022.102266

Roychowdhury, R., Das, S. P., Gupta, A., Parihar, P., Chandrasekhar, K., Sarker, U., Kumar, A., Ramrao, D. P., & Sudhakar, C. (2023). Multi-omics pipeline and omics-integration approach to decipher plant's abiotic stress tolerance responses. *Genes, 14*(6), 1281. https://doi.org/10.3390/genes14061281

Sarfraz, Z., Zarlashat, Y., Ambreen, A., Mujahid, M., Iqbal, M., Fatima, S., Iqbal, M., Iqbal, R., & Fiaz, S. (2025). Plant biochemistry in the era of omics: Integrated omics approaches to unravel the genetic basis of plant stress tolerance. *Plant Breeding*. https://doi.org/10.1111/pbr.13277

Sarker, I. H. (2021). Machine learning: Algorithms, real-world applications and research directions. *SN Computer Science, 2*(3), 160. https://doi.org/10.1007/s42979-021-00592-x

Shahryary, Y., Hazarika, R. R., & Johannes, F. (2020). Methylstar: A fast and robust pre-processing pipeline for bulk or single-cell whole-genome bisulfite sequencing data. *BMC Genomics, 21*(1), 479. https://doi.org/10.1186/s12864-020-06886-3

Singhal, R., Izquierdo, P., Ranaweera, T., Segura Abá, K., Brown, B. N. I., Lehti-Shiu, M. D., & Shiu, S. H. (2025). Using supervised machine-learning approaches to understand abiotic stress tolerance and design resilient crops. *Philosophical Transactions of the Royal Society of London. Series B, Biological Sciences, 380*(1927), 20240252. https://doi.org/10.1098/rstb.2024.0252

Singroha, G., & Sharma, P. (2019). Epigenetic Modifications in Plants under Abiotic Stress. Rosaria Meccariello (ed.). Pages 1–14. https://doi.org/10.5772/intechopen.84455

Song, Y., Jia, Z., Hou, Y., Ma, X., Li, L., Jin, X., & An, L. (2020). Roles of DNA methylation in cold priming in Tartary buckwheat. *Frontiers in Plant Science, 11*. https://doi.org/10.3389/fpls.2020.608540

Sun, L., Jing, Y., Liu, X., Li, Q., Xue, Z., Cheng, Z., Wang, D., He, H., & Qian, W. (2020). Heat stress-induced transposon activation correlates with 3d chromatin organization rearrangement in arabidopsis. *Nature Communications*, *11*(1), 1886. https://doi.org/10.1038/s41467-020-15809-5

Sun, L., Xu, K., Huang, W., Yang, Y. T., Li, P., Tang, L., Xiong, T., & Zhang, Q. C. (2021). Predicting dynamic cellular protein–RNA interactions by deep learning using in vivo RNA structures. *Cell Research*, *31*(5), 495–516. https://doi.org/10.1038/s41422-021-00476-y

Tahir, M., Norouzi, M., Khan, S. S., Davie, J. R., Yamanaka, S., & Ashraf, A. (2024). Artificial intelligence and deep learning algorithms for epigenetic sequence analysis: A review for epigeneticists and AI experts. *Computers in Biology and Medicine*, *183*, 109302. https://doi.org/10.1016/j.compbiomed.2024.109302

Tittel-Elmer, M., Bucher, E., Broger, L., Mathieu, O., Paszkowski, J., & Vaillant, I. (2010). Stress-induced activation of heterochromatic transcription. *PLoS Genetics*, *6*(10), e1001175. https://doi.org/10.1371/journal.pgen.10.1175

Ueda, M., Matsui, A., Tanaka, M., Nakamura, T., Abe, T., Sako, K., Sasaki, T., Kim, J. M., Ito, A., Nishino, N., Shimada, H., Yoshida, M., & Seki, M. (2017). The distinct roles of class I and II RPD3-like histone deacetylases in salinity stress response. *Plant Physiology*, *175*(4), 1760–1773. https://doi.org/10.1104/pp.17.01332.

Wang, M., Qin, L., Xie, C., Li, W., Yuan, J., Kong, L., Yu, W., Xia, G., & Liu, S. (2014). Induced and constitutive DNA methylation in a salinity-tolerant wheat introgression line. *Plant and Cell Physiology*, *55*(7), 1354–1365. https://doi.org/10.1093/pcp/pcu059

Wang, W., Huang, F., Qin, Q., Zhao, X., Li, Z., & Fu, B. (2015). Comparative analysis of DNA methylation changes in two rice genotypes under salt stress and subsequent recovery. *Biochemical and Biophysical Research Communications*, *465*(4), 790–796. https://doi.org/10.1016/j.bbrc.2015.08.089

Wang, J., Nan, N., Li, N., Liu, Y., Wang, T. J., Hwang, I., Liu, B., & Xu, Z. Y. (2020). A DNA methylation reader–chaperone regulator–transcription factor complex activates *OsHKT1;5* expression during salinity stress. *The Plant Cell*, *32*(11), 3535–3558. https://doi.org/10.1105/tpc.20.00301

Wang, N., Wang, Z., Tzourtzou, S., Wang, X., Bi, X., Leimeister, J., Xu, L., Sakamoto, T., Matsunaga, S., Schaller, A., Jiang, H., & Liu, C. (2023). The plant nuclear lamina disassembles to regulate genome folding in stress conditions. *Nature Plants*, *9*(7), 1081–1093. https://doi.org/10.1038/s41477-023-01457-2

Wu, T., Pi, E. X., Tsai, S. N., Lam, H. M., Sun, S. M., Kwan, Y. W., & Ngai, S. M. (2011). GmPHD5 acts as an important regulator for crosstalk between histone H3K4 di-methylation and H3K14 acetylation in response to salinity stress in soybean. *BMC Plant Biology*, *11*, 178. https://doi.org/10.1186/1471-2229-11-178

Wu, X., Xu, J., Meng, X., Fang, X., Xia, M., Zhang, J., Cao, S., & Fan, T. (2022). Linker histone variant HIS1-3 and WRKY1 oppositely regulate salt stress tolerance in arabidopsis. *Plant Physiology*, *189*(3), 1833–1847. https://doi.org/10.1093/plphys/kiac174.

Xie, H., Sun, Y., Cheng, B., Xue, S., Cheng, D., Liu, L., Meng, L., & Qiang, S. (2019). Variation in *ICE1* methylation primarily determines phenotypic variation in freezing tolerance in *arabidopsis thaliana*. *Plant and Cell Physiology*, *60*(1), 152–165. https://doi.org/10.1093/pcp/pcy197

Yang, Y., Saand, M. A., Huang, L., Abdelaal, W. B., Zhang, J., Wu, Y., Li, J., Sirohi, M. H., & Wang, F. (2021). Applications of multi-omics technologies for crop improvement. *Frontiers in Plant Science*, *12*, 563953. https://doi.org/10.3389/fpls.2021.563953

Yang, F., Sun, Y., du, X., Chu, Z., Zhong, X., & Chen, X. (2023). Plant-specific histone deacetylases associate with ARGONAUTE4 to promote heterochromatin stabilization and plant heat tolerance. *New Phytologist*, *238*(1), 252–269. https://doi.org/10.1111/nph.18729

Yun, M., Wu, J., Workman, J. L., & Li, B. (2011). Readers of histone modifications. *Cell Research*, *21*(4), 564–578. https://doi.org/10.1038/cr.2011.42

Zhan, J., & Meyers, B. C. (2023). Plant small RNAs: Their biogenesis, regulatory roles, and functions. *Annual Review of Plant Biology*, *74*, 21–51. https://doi.org/10.1146/annurev-arplant-070122-035226

Zhang, X., Yazaki, J., Sundaresan, A., Cokus, S., Chan, S. W., Chen, H., Henderson, I. R., Shinn, P., Pellegrini, M., Jacobsen, S. E., & Ecker, J. R. (2006). Genome-wide high-resolution mapping and functional analysis of DNA methylation in *arabidopsis*. *Cell*, *126*(6), 1189–1201. https://doi.org/10.1016/j.cell.2006.08.003

Zhang, R., Zhang, C., Yu, C., Dong, J., & Hu, J. (2022). Integration of multi-omics technologies for crop improvement: Status and prospects. *Frontiers in Bioinformatics*, *2*, 1027457. https://doi.org/10.3389/fbinf.2022.1027457

Zheng, M., Liu, X., Lin, J., Liu, X., Wang, Z., Xin, M., Yao, Y., Peng, H., Zhou, D. X., Ni, Z., Sun, Q., & Hu, Z. (2019). Histone acetyltransferase GCN5 contributes to cell wall integrity and salt stress tolerance by altering the expression of cellulose synthesis genes. *The Plant Journal*, *97*(3), 587–602. https://doi.org/10.1111/tpj.14144

Zheng, M., Lin, J., Liu, X., Chu, W., Li, J., Gao, Y., An, K., Song, W., Xin, M., Yao, Y., Peng, H., Ni, Z., Sun, Q., & Hu, Z. (2021). Histone acetyltransferase *TaHAG1* acts as a crucial regulator to strengthen salt tolerance of hexaploid wheat. *Plant Physiology*, *186*(4), 1951–1969. https://doi.org/10.1093/plphys/kiab187.

Zheng, G., Dong, X., Wei, J., Liu, Z., Aslam, A., Cui, J., Li, H., Wang, Y., Tian, H., & Cao, X. (2022). Integrated methylome and transcriptome analysis unravel the cold tolerance mechanism in winter rapeseed (Brassica napus L.). *BMC Plant Biology*, *22*(1), 414. https://doi.org/10.1186/s12870-022-03797-1

Zi, N., Ren, W., Guo, H., Yuan, F., Liu, Y., & Fry, E. (2024). DNA methylation participates in drought stress memory and response to drought in *medicago ruthenica*. *Genes*, *15*(10), 1286. https://doi.org/10.3390/genes15101286

7

AI-Based Whole-Genome Prediction for Diverse Plant Epigenetic Modulations

Manoj Mani[1], *Nahidha Parveen Mohammed Koya*[1], *Vivek Patel*[1], *Vivek Kumar Varshney*[1], *Antony Prabhu Jeyabal Philomenathan*[1], *Maria Jerline Babu*[2], *Akilandeswari Govindraj*[3], *and Vijaya Anand Arumugam*[1]

[1] *Department of Human Genetics and Molecular Biology, Bharathiar University, Coimbatore, Tamil Nadu, India*
[2] *Department of Biochemistry, Dr. R.A.N.M Arts and Science College, Erode, Tamil Nadu, India*
[3] *Department of Biochemistry, CMS College of Science & Commerce, Coimbatore, Tamil Nadu, India*

7.1 Introduction

Epigenetics is key to crop improvement and stress tolerance by enabling plants to respond adaptively to environmental stress impulses without altering their underlying DNA sequence. Plants fine-tune their gene expressions and form stress memories by mechanisms of DNA methylation, histone reprogramming, and RNA-directed avenues, thus adapting to diverse biotic and abiotic pressures (Abdulraheem et al., 2024). These epigenetic traits are often heritable, with the vertical transmission of stress tolerance through generations—an attribute that is becoming all the more pivotal for achieving climate change-proof crop productivity (Ahtisham & Obaid, 2024). Novel technologies in genome-wide profiling and genome-editing systems, such as CRISPR-Cas9, have revamped our potential in characterizing, discovering, and manipulating epigenetic traits. Integration of these molecular technologies with classical breeding methods opens unprecedented opportunities for developing stress-tolerant cultivars that could withstand environmental pressures of heat, drought, and salinity (Jin et al., 2024). However, concerns persist in characterizing the stability, reversibility, and inheritance of epigenetic marks and their complex relationships in the plant genome, thus calling for further studies in realizing the full potential of epigenetics in crop improvement (Saeed et al., 2022).

The development of whole-genome prediction (WGP), especially when supplemented with artificial intelligence (AI), is revolutionizing plant biology by accurately predicting phenotypic results from genotypic data (Figure 7.1). Deep learning (DL) models fortify this predictive potential by simulating complex genotype–phenotype relationships, enabling fast gene expression prediction and real-time phenotyping-prime drivers of improved sustainable agronomic systems (Kajrolkar, 2025). Machine learning (ML) models have shown remarkable competence in predicting functional genes, utilizing large-scale genomic datasets to discern structural features, guess gene functions, and forecast organismal

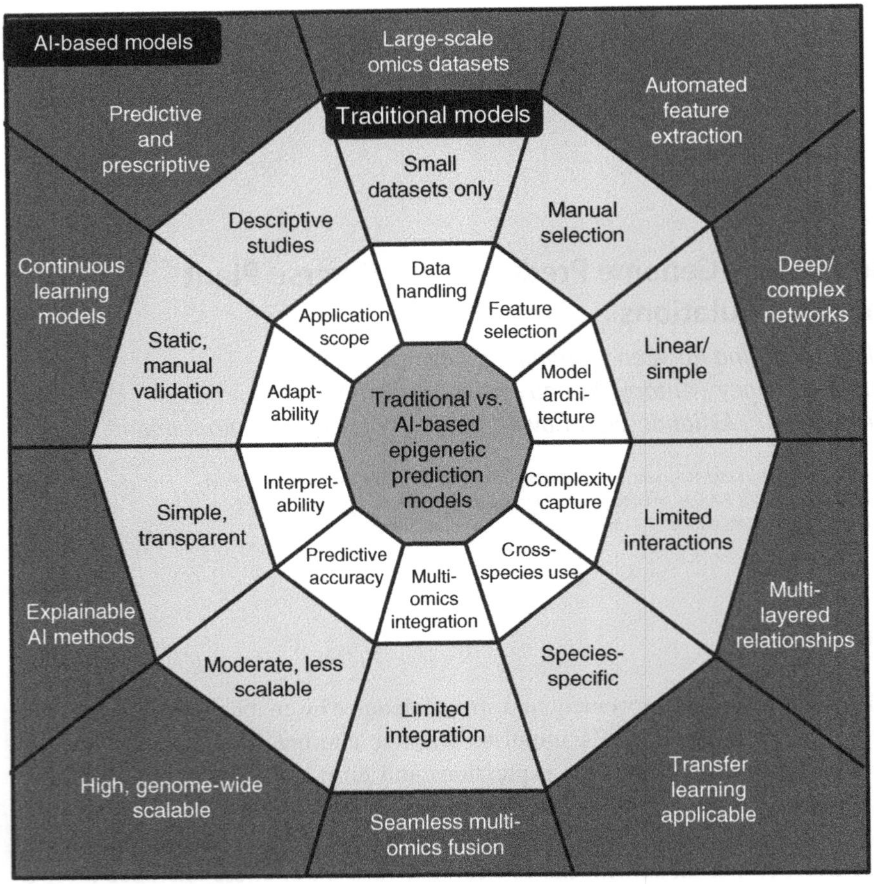

Figure 7.1 Traditional versus AI-based epigenetic prediction models.

phenotypes, thereby inscribing insightful knowledge into plant biology (Mahood et al., 2020). The PlantConnectome project illustrates this confluence by utilizing AI to dig into and process voluminous scientific literature, predicting gene functions and molecular interactions that increase our knowledge of plant systems (Fo et al., 2023). Additionally, the integration of crop growth models (CGMs) into WGP systems optimally manages genotype-by-environment-by-management (G×E×M) interactions, consequently boosting yield prediction precision and guiding breeding strategies in multispecific agroecosystems like the US corn belt (Messina et al., 2018).

The combination of AI with epigenomics in plant sciences is a transformative development for dissecting plant memory, stress adaptation, and phenotypic plasticity. ML and DL algorithms help decipher the multilayered mechanisms of epigenetic control that manage plant responses to environmental cues, thereby leading to climate resilience in crops (Dobránszki et al., 2025). State-of-the-art DL models were successfully used to forecast chromatin reorganization events under restrictive environmental scenarios, e.g., cold-mediated dormancy of apple bud bundles-showcasing AI applicability for detecting genetic and epigenetic regulators of adaptive traits (Saito et al., 2024). Additionally, the concurrency of AI with genomics and Internet of Things technologies is the foundation of precision agriculture through early stress perception, followed by adaptive management measures that maximize crop productivity outputs (Polegopu et al., 2025). The combination of multiomics data with AI models is creating an integrated strategy for high-resolution plant phenotyping, going beyond the realm of conventional observation-based methods and accelerating predictive breeding and precision horticulture (Cembrowska-Lech et al., 2023). Altogether, AI-based strategies are redefining plant breeding and phenotyping pipelines, permitting the production of high-yielding, resilient cultivars and supporting global food safety objectives (Gupta et al., 2024).

7.2 Plant Epigenetic Modifications

Epigenetic marks constitute a foundation circuit of plant gene regulation, enabling plants to fine-tune gene expression, adapt to environmental differences, and transmit adaptive traits from generation to generation while leaving their DNA sequence unchanged. These mechanisms comprise DNA methylation, histone modifications, and noncoding RNAs (ncRNAs), and they encompass key functioning processes that, in concert, outline plant development, stress tolerance, and phenotypic plasticity.

7.2.1 DNA Methylation and Predictive Epigenomics

DNA methylation is a stable but reversible epigenetic mark that controls gene expression, transposon silencing, and genome stability. In recent years, DNA methylation predictive models have shown great promise in dissecting complex traits, biological aging, and epigenomic control in plants. In barley, combining methylation data with genomic and transcript information has improved the predictive power of complex traits such as grain yield and N uptake, explaining more of the phenotypic variance than genomic models by themselves (Hansen et al., 2022). Analogously, in *Populus balsamifera*, DL models based on natural DNA methylation variation have been used to predict industrially important traits, discovering tissue-specific methylation signatures associated with wood property control (Champigny et al., 2020). In *Pinus taeda* (loblolly pine), epigenetic clocks based on CpG and CHG methylation accurately predicted biological age within 8% of the species' life length, highlighting the involvement of methylation in developmental timing and aging (Gardner et al., 2023). Additionally, the Smart Model for Epigenetics in Plants (SMEP) exhibited high predictive power for DNA and RNA methylation sites among species, aiding in functional genomics and genome engineering applications (Wang et al., 2021).

7.2.2 Histone Modifications and Chromatin Remodeling

Histone posttranslational modifications, as well as chromatin remodeling, confer dynamic events that control plant genome transcriptional regulation. Chromatin conformation as well as DNA access of transcriptional regulators is controlled by acetylation, methylation, phosphorylation, as well as ubiquitination. The activities assume particular functions in stress adaptation, with plants reprogramming histone modifications in accordance with environmental signals. During abiotic stress, such as cold as well as salinity, histone modifications activate transcriptional networks that ensure osmotic homeostasis as well as ion homeostasis (Hyun, 2022; Yung et al., 2021). Genome-wide mapping of rice revealed that 130 histone modification genes (OsHMs) participated in tissue specification as well as in responses to diverse abiotic stresses, including anther development (Huang et al., 2024). Upon pathogen ingress, histone modifications rewire defense gene expressions as an attempt to fortify immune responses (Kang et al., 2022). In addition, these activities assume important roles in seed germination as well as in seed dormancy, controlling developmental transformations in accordance with environmental signals (Ding et al., 2022). Involved elaborate crosstalk of histone modifications as well as signaling mechanisms, including hormonal regulation, sheds light on their complexity, as well as their potential as genome-editing-mediated crop improvement targets (Kang et al., 2022; Yung et al., 2021).

7.2.3 Noncoding RNAs and Epigenetic Regulation

NcRNAs, specifically, long noncoding RNAs (lncRNAs), are crucial regulators of plant epigenomes, fine-tuning gene expression by manipulating chromatin remodeling, histone-modifying activities, and DNA methylation (Lakhani et al., 2025; Sun, 2025). These molecules fine-tune abiotic stress-responsive transcriptional responses, such as drought, salinity, and heavy metals, by guiding chromatin-modifying complexes to specific genomic sites (Tremblay & Qüesta, 2023). LncRNAs also participate in the establishment of epigenetic memories, so that plants can “memorize” previous exposures to stress and respond more quickly, efficaciously when again exposed (Magar et al., 2024). Their roles in guiding transcriptional regulation—with RNA polymerase II processivity modulation and loop formation by chromosomes—increasingly improve transcriptomic accuracy and efficiency in stressful environments (Tremblay & Qüesta, 2023). Integrating lncRNA-driven information into crop breeding holds vast potential for breeding stress-resistant cultivars by directed manipulation of epigenetic networks (Magar et al., 2024). Although huge advances have been made in the roles of ncRNA-mediated regulation and interaction of these with plants, epigenomic machineries remain incomprehended in part; increased research aided by high-throughput sequencing and functional annotation technologies is necessary (Lakhani et al., 2025; Ramirez-Prado et al., 2017).

7.2.4 Transgenerational Epigenetic Inheritance

Epigenetic inheritance allows plants to pass stress responses and developmental adaptations acquired in life to subsequent generations without changing nucleotide sequences. The process of the stable passing of DNA methylation marks, histone modifications, and of ncRNA-mediated regulation gives a molecular basis for adaptive evolution in sessile organisms (Annacondia & Martinez, 2019; Ito, 2014). Environmental signals such as drought, temperature, and pathogen perception may trigger epimutations that pass in a heritable manner through meiosis and mitosis, encoding in progeny “stress memory” that allows plants to better withstand environmental variations (Ito, 2014; Spanou et al., 2025). These heritable marks give phenotypic plasticity, and plant resilience is enhanced in changing environments (Annacondia & Martinez, 2019). Elucidating mechanisms governing epigenetic memory stability as well as passing is key to developing new breeding strategies for creating crops with increased yield and climate resilience (Cao & Chen, 2024; Dobránszki et al., 2025). The use of AI and ML in epigenomic research is fast-tracking the identification as well as prediction of heritable epigenetic marks, generating new opportunities for breeding program optimization and increased sustainability in agriculture (Dobránszki et al., 2025).

7.3 Data Sources and Profiling Technologies

Recent advances in high-throughput sequencing and integrative omics have completely revolutionized plant epigenome analysis, enabling genome-wide mapping of DNA methylation, chromatin accessibility, histone modifications, and RNA-mediated regulation at unprecedented resolution. As a whole, these technologies, in combination, bring about overarching insights into the molecular mechanisms behind gene regulation, stress adaptation, and transgenerational plant inheritance.

7.3.1 Whole-Genome Bisulfite Sequencing

Whole-genome bisulfite sequencing (WGBS) is still the best method for single-base resolution analysis of DNA methylation, providing an unwavering view of plant genome landscapes of methylation. Via bisulfite-mediated conversion of unmethylated cytosine residues, WGBS allows for discrimination of methylated from unmethylated cytosine residues with base resolution, although further steps are necessitated to distinguish hydroxymethylation (Derbala et al., 2024). The process is key in deciphering heritable epigenetic marks of environmental interaction, such as stress or pathogen encounter (Gupta & Garg, 2020). The technique of WGBS has been successfully used in various plant species, such as large genome crops of maize and barley, and can further be improved via sequence capture for reduced region-focused analysis of the methylome (Li et al., 2017). Bioinformatics software such as MethylScore utilizes ML to process WGBS datasets, discerning differentially methylated regions (DMRs) and defining genotype–epigenotype relationships, as shown in the *Arabidopsis thaliana* 1,001 Genomes project (Hüther et al., 2022).

7.3.2 Chromatin Accessibility and Protein–DNA Interaction Profiling

Epigenomic methods like assay for transposase-accessible chromatin using sequencing (ATAC-seq) and chromatin immunoprecipitation sequencing (ChIP-seq) became indispensable for deciphering chromatin architecture and transcriptional regulation in plants. ATAC-seq offers genome-wide information on accessible chromatin regions (ACRs), marking regulatory elements associated with active transcription. In chickpea, ATAC-seq detected 11,555 ACRs enriched close to transcription start sites, colocalizing with activating histone marks including H3K27ac and H3K4me3 (Yang et al., 2025). Analogously, grapevine epigenome profiling indicated more than 16,000 open chromatin regions, encompassing distant enhancers, in reflection of the enhanced genomic complexity of long-lived species (Schwope et al., 2021). To complement, ChIP-seq allows for mapping of protein–DNA contacts and histone-modifying marks. In *Arabidopsis*, for instance, the INTACT protocol for Isolation of Nuclei TAgged in specific Cell Types allows for cell-type-specialized ChIP-seq profiling, delineating differential dynamics of chromatin across developmental fates (Wang & Deal, 2015). Integrative platforms such as ChIP-Atlas integrate ATAC-seq and ChIP-seq datasets, providing for cross-species comparative analysis of chromatin access, DNA epigenomes, and binding landscapes of proteins (Zou et al., 2022).

7.3.3 RNA-Based Epigenomic Profiling

RNA-seq and small RNA sequencing (sRNA-seq) present primary insights into RNA-based plant epigenetic regulation, particularly for RNA-directed DNA methylation (RdDM) and genome stability. Small noncoding RNAs, including microRNAs and siRNAs, regulate gene expression and transposon silence, as in rice, in which they govern stress responses and developmental transitions (Park et al., 2022). Additional global perspectives from mammalian systems also highlight the functions of RNA families such as PIWI-interacting RNAs (piRNAs) in transcriptional and posttranscriptional regulation of genes (Hashim et al., 2014; Slack, 2014). These small RNAs, in association with lncRNAs, guide chromatin-modifying complexes to newly synthesized RNA scaffolds, recruiting histone and DNA methyltransferases for the establishment of transcriptional silence (Holoch & Moazed, 2015). Functional divergence of RNA species emphasizes their hub functions in coordinating multilayered epigenetic control mechanisms beyond kingdoms.

7.3.4 Single-Cell Epigenomics

Single-cell epigenomic profiling is a groundbreaking method for deciphering cell-type-exclusive chromatin states and gene regulation networks in plants. Overcoming obstacles of stiff, immobile cell walls and cell totipotency, latest technologies, including combinatorial indexing and dual PCR barcoding, make it feasible to profile single-cell chromatin in the absence of specially crafted instruments (Ouyang et al., 2022). Such technologies facilitated delineating distinctive cis-regulatory element (CRE) landscapes and chromatin states in single-cell types, as shown in *Arabidopsis thaliana* (Mendieta et al., 2023). snCUT&Tag technology also facilitates high-resolution mapping of histone marks in single cells, interrogating heterogeneity of the chromosome and enhancer–promoter contacts commensurate with bulk-cell epigenomic data (Corbin & Lopez, 2020). A combination of single-cell epigenomic data with transcriptomics and space-based omics further refines insights into plant cell differentiation, stress responses, and cell-to-cell signaling (Thibivilliers & Libault, 2021).

7.3.5 Long-Read Sequencing and Direct Epigenetic Detection

The advent of long-read sequencing technologies, such as PacBio and Oxford Nanopore technologies, increased the capabilities for direct detection of epigenetic alterations sans chemical conversion. The technologies allow for real-time identification of methylation events such as 5-methylcytosine (5mC), 5-hydroxymethylcytosine (5hmC), and N6-methyladenine (m6A), as they move along uninterrupted DNA molecules and allow for integrative analysis of genetic as well as epigenetic architecture (Xie et al., 2021). PacBio's fibertools, utilizing convolutional neural networks, surpass 90% accuracy in m6A identification while combining multiomic data streams (Jha et al., 2024). The ultra-long reads from Nanopore sequencing find their greatest benefit in regions that consist of repetitive and transposon-dense regions, also known as the hotspots of methylation (Chera et al., 2024). Still, benchmarking of pipelines used for calling of methylation in a standardized manner is crucial for obtaining optimally correct platform choices in large-scale epigenomic studies and for maintaining the aspect of replicability (Liu et al., 2022).

7.3.6 Time Series and Multiomics Integration

Time series multiomics datasets offer dynamic perspectives on epigenetic regulation by environmental and developmental inputs. Software packages such as iDREM integrate temporal transcriptomic, proteomic, and epigenomic datasets to reconstruct gene-regulatory networks as well as master regulators of plant stress responses (Mishra et al., 2021). Spatiotemporal single-cell and gene-regulatory atlases in response to pathogen challenge in Arabidopsis thaliana offer transcription factor dynamics and activities of cis-regulatory elements governing plant immune responses (Nobori et al., 2023). Similarly, time course methylome analysis in tea plants determines seasonally dynamic DNA methylation that impacts secondary metabolite pathways, such as biosynthesis of flavonoids and theanine biosynthesis (Han et al., 2024). In grapevine, developmentally regulated transition from bud dormancy to budburst involves complex epigenetic and miRNA-mediated regulation of gene networks (Shinde et al., 2023). Integrating ML and AI into the multiomics platform further enhances predictive power, with epigenetic alteration associating with phenotypic adaptation and enhancing precision in predictive breeding strategy (Dobránszki et al., 2025).

7.4 AI and Machine Learning Frameworks for Epigenomic Prediction

ML and AI models have proven indispensable in dissecting plant epigenomes, enabling precise prediction of complex traits, regulatory factors, and stress responses. Classical ML models like support vector machines (SVM), random forest (RF), and XGBoost exhibited great predictive efficiency and interpretability in various plant genomic and epigenomic applications. As an example, SVM models were effectively used in predicting sequence-derived features for root-associated proteins, thus complementing experimental strategies including genome-wide association studies (GWAS) and transcriptome profiling (Kumar Meher et al., 2024). RF and XGBoost models also emerged as superior in plant phenotypic traits prediction, outperforming DL models in some applications due to their excellent feature selection properties—particularly for soybean crops complex trait prediction in crops (Gill et al., 2022). In addition to prediction, the models also provide mechanistic insights into plant memory and adaptation, explaining epigenetic mechanisms of stress resilience in plants. Their synergy with multiomics datasets also facilitates parental training strategy optimization for the production of crops resistant to multifactorial environmental stress, thus addressing climate change imperatives (Dobránszki et al., 2025). Additionally, RF demonstrated unwavering accuracy in coding versus ncRNA discrimination, enhancing knowledge of plant transcriptomic and epigenetic regulation (Majumdar et al., 2022).

7.4.1 Deep Learning Architectures for Epigenomic Inference

DL revolutionized plant epigenomics by providing sequence-based, high-continuity inference of regulatory modes and epigenetic alterations. The SMEP illustrates such advances, predicting six types of epigenomic variations—DNA and RNA methylation and histone modifications—with accuracy in various plant species such as rice and maize, thus accelerating functional genome studies and genome engineering (Wang et al., 2021). Analogously, GenoRetriever utilizes deep convolutional models for deciphering mechanisms of transcriptional initiation by recognizing key promoter motifs and variations in wild and domestic soybeans (Gao et al., 2025). DeepCRE further extends the spectrum by directly predicting gene expressions from regulatory DNA sequences, thus permitting assessment of variant activities on the promoter and advancing genotype-to-phenotype linkage mapping (Zumkeller et al., 2025). In *Populus balsamifera*, DL models, when trained with DNA methylation data, accounted for substantial phenotypic variance, bringing to light tissue-differential epigenetic signatures correlated with wood properties (Champigny et al., 2020). Complementary to these, PlantDeepSEA offers a web-based platform to analyze the regulatory effector potential of genomic variants on chromatin accessibilities, thus supporting comparative epigenomics in multiple species (Zhao et al., 2021).

7.4.2 Transformer Models and Context-Aware Genomic Learning

Transformer architectures such as Enformer and DNABERT recently reimagined genomic modeling by learning the contextual "language" of DNA. DNABERT, a pretrained bidirectional encoder representation, performs best in learning noncoding DNA syntax, reaching best-in-class accuracy in predicting plant genomes' diverse promoters and transcription factor binding sites (Ji et al., 2021). Extending from that, transformer models trained on maize and sorghum datasets were successfully used to predict tissue-specific gene expression patterns and phenotypic diversity-influencing regulatory regions, demonstrating scalability with or without retraining through different species (Tross et al., 2024). Deployment of optimal k-mer tokenization strategies, particularly overlap-based tokenization, also improved transformer performance by maintaining local sequence dependencies that are important for applications such as splice site prediction (Suzuki et al., 2025).

7.4.3 Graph Neural Networks for 3D Genome Modeling

Graph neural networks (GNNs) brought a spatial aspect to plant epigenomics by allowing modeling of chromatin topology from Hi-C interaction data. Using graph convolutional layers, the HiC-GNN architecture extracts three-dimensional genomic coordinates from

Hi-C contact matrices with better reconstruction accuracy and cross-resolution generality (Hovenga et al., 2023). Multigraph models' application to barley and rice genomes revealed classic chromatin configurations, including Rabl and Rosetta structures with an accent on GNNs' ability to find conserved plant genome organizational architecture at the spatial level (Makai et al., 2022). Coupling of polymer physics models with GNNs enables precise estimation of chromatin structures through ML integrated with physical constraints, an idea further extendable to plant genomes for mapping structure–function relationships (Schultz et al., 2025). The Principal Node Aggregation model further advances the principle by integrating into gene expression predictions with chromosomal distances and 3D genome features, providing a complete framework for spatially resolved epigenomic delineation (Yan et al., 2024).

7.4.4 Generative Adversarial Networks for Synthetic Epigenomic Simulation

Generative Adversarial Networks (GANs) present a new platform for plant epigenetic landscape simulation and manipulation. The Transcriptome State Perturbation Generator (TSPG), based on GANs, models simulations of RNA expression state transition, shedding light on perturbation-responsive genes and biologically meaningful transcriptomic transformations (Targonski et al., 2019). ExpressionGAN also shows the potential for the design of synthetic regulatory DNA sequences that could fine-tune gene expression outside of natural boundaries, signifying the potential of GANs for plants' precision regulation engineering (Zrimec et al., 2022). State-of-the-art reviews present the advancements of GANs in contributing to filling limited sets of genomic data, generating diversity, and improving model robustness—an important step closer to accurate epigenomic prediction in understudied plant species (Lee, 2023).

7.4.5 Federated Learning for Secure and Decentralized Epigenomic Prediction

Federated learning (FL) provides a transformative collaborative plant epigenomic modeling framework by enabling distributed model training in decentralized datasets with data privacy protection. Briefly proven in human genomics with complex traits like Crohn's disease, FL ensures the same predictive accuracy as centralized methods while data sovereignty is ensured (Raimondi et al., 2023). FL in crop systems facilitated smart crop management with functionalities like disease identification and yield estimation in spatiotemporally varying datasets, yielding region- or location-oriented knowledge (Dwarampudi & Yogi, 2024). Implementations of FL with higher-order models like Graph Attention Networks improved predictive precision and safety with crop health profiling further (Kusharki et al., 2026). The models, such as Fed-PPI, ensure statistically correct inferences from the distributed datasets (Luo et al., 2024), while the PRAISE study shows FL competing with centralized models in accuracy, with potential for mass-scale, privacy-enforcing plant epigenomic predictions (Arfat et al., 2023).

7.5 Multiomics Integration Using AI

The integration of genomics, epigenomics, and transcriptomics with AI is transforming plant biology and crop improvement by creating a system-level understanding of plant adaptation and stress resilience. AI and ML models make it feasible for managing and interpreting large datasets of multiomics information to accurately predict complex traits of stress tolerance as well as optimizing yield (Cembrowska-Lech et al., 2023; Thingujam et al., 2025). Integrative models of this type unveil genetic and epigenetic modules of significance to plant adaptation to abnormally harsh environmental conditions, making it possible to develop climate-smart and stress-resistant crops (Raza et al., 2024; Roychowdhury et al., 2023). At the same time, advanced phenomics platforms such as drone-based assessment and hyperspectral imaging supplement classical breeding pipelines through enabling rapid, quantitative assessment of plant traits in dynamic environmental scenarios (Thingujam et al., 2025). AI-based multiomics analysis also unveils significant details of the molecular basis of plant defense mechanisms in biotic as well as abiotic stress, making it possible to design predictive and preventive crop protection strategies (Murmu et al., 2024). An integration of genomics, transcriptomics, proteomics, metabolomics, and epigenomics thereby guarantees multidimensionality of stress-responsive paths, directly informing breeding efforts for the purpose of stabilizing yield as well as fortifying crops with respect to nutrition in the midst of climatic variations (Roychowdhury et al., 2023). As a complete system, such AI-driven multiomics integration promotes sustainability in agriculture by enhancing food safety as well as minimizing financial risk related to environmental variability (Thingujam et al., 2025).

7.5.1 Systems Biology Approaches for Phenotype–Epigenome Prediction

Systems biology offers a conceptual and computational framework for integrating multiomics data with plant phenotype by network-centric modeling and AI-driven reasoning. The platform of EpiNet demonstrates it through the deployment of machine learning–based network inference for reconstructing gene-regulatory networks that comprise epigenetic regulators and thereby capture prominent molecular actors involved in developmental processes such as shoot and root apical meristem differentiation (McCoy et al., 2021). At the same time, systems biology strategies strongly accentuated by Kumar et al. stress integrated genomics, transcriptomics, and metabolomics for comprehensively modeling plant systems, thereby enabling decipherment of molecular mechanisms governing main agronomic traits such as stress tolerance, nutrient use, and plant growth control (Kumar et al., 2015). Computational modeling of epigenetic memory also enabled further insights into feedback dynamics in GRNs, explaining DNA methylation and histone modifications in terms of transcriptional stability as well as adaptive plasticity (Feng et al., 2024). DL-led SMEP further demonstrates AI's potential in predicting multiple epigenomic modifications in a set of species with excellent accuracy, enabling comparative study of plant systems between species as well as directed genome editing (Wang et al., 2021). In sum, these systems biology- and AI-led strategies close the genotype-to-phenotype knowledge gap, enabling predictive modeling of complex plant traits.

7.5.2 Knowledge Graphs and Epigenetic Trait Modeling

Knowledge graph embeddings are a new frontier in integrating multiple "omes," allowing for the structured connection of genomic, epigenetic, and phenotypic data for predictive modeling. Integrating multisource information, such as DNA methylation, histone modifications, and gene expression, knowledge graphs can make inferences about relationships between epigenetic markers and phenotypic outputs. For example, research in *Populus balsamifera* showed that natural variations in DNA methylation make significant contributions to phenotypic variance in biomass and wood density, highlighting the predictive value of epigenetic data for modeling traits (Champigny et al., 2020). These results complement more extensive evidence for the involvement of epigenetic mechanisms, such as DNA methylation and histone modifications, in improving stress tolerance and agronomic performance (Bhuvaneswari et al., 2020). Integrating AI and ML with epigenetic datasets further enhances predictions of plant responses to environmental stress, with plant memory and adaptive mechanisms as main drivers of resilience being used as predictors (Dobránszki et al., 2025). New evidence of transgenerational epigenetic inheritance buttresses the potential of epigenomic data in breeding crop species with the ability to sustain productivity in unpredictable climatic conditions (Chen, 2025). In addition, mapping of epigenetic quantitative attribute loci (epiQTLs) in *Populus* gives us a powerful tool for dissecting genotype–epigenotype interaction, with insight into the possibility that interaction of genetic and epigenetic variants drives modulation of complex traits in long-lived perennials (Lu et al., 2020).

7.6 AI-Based Whole-Genome Prediction Models

AI model development for predicting plant DNA methylation landscapes has advanced steadily with DL models that integrate heterogeneous genomic features for enhanced accuracy in predictions (Figure 7.2). In *Populus balsamifera*, DL models appropriately utilized natural variations of DNA methylation to estimate plant phenotypes, capturing large variance in wood-related traits and demonstrating the potential of methylation-guided models for forestry applications (Champigny et al., 2020). On such a footing, MEDCNN models introduced a multidimensional feature encoding framework that accounts for positional, biological, and chemical properties in order to make DNA methylation site predictions with increased accuracy compared to default algorithms (Hu et al., 2023). Similarly, the SMEP is trained to give predictions of diverse epigenomic modifications such as DNA methylation in diverse species with high accuracy, signifying its potential in genome function as well as genome engineering applications (Wang et al., 2021). Furthermore, integrative ML methods revealed that CG and CHG methylation, as well as nucleosome occupancy, were important predictors of recombination site location in the process of meiosis, hinting at typical plant taxa-chromatin organization, with evolutionary conservation of recombination regulation mechanisms in plant genomes (Wang et al., 2022). The MHCpG model continues the development in the same course by integrating MeDIP-seq and histone-modifying data with sequence-based features, outperforming default models such as CpGenie as well as DeepCpG in predicting CpG methylation states (Fu et al., 2019).

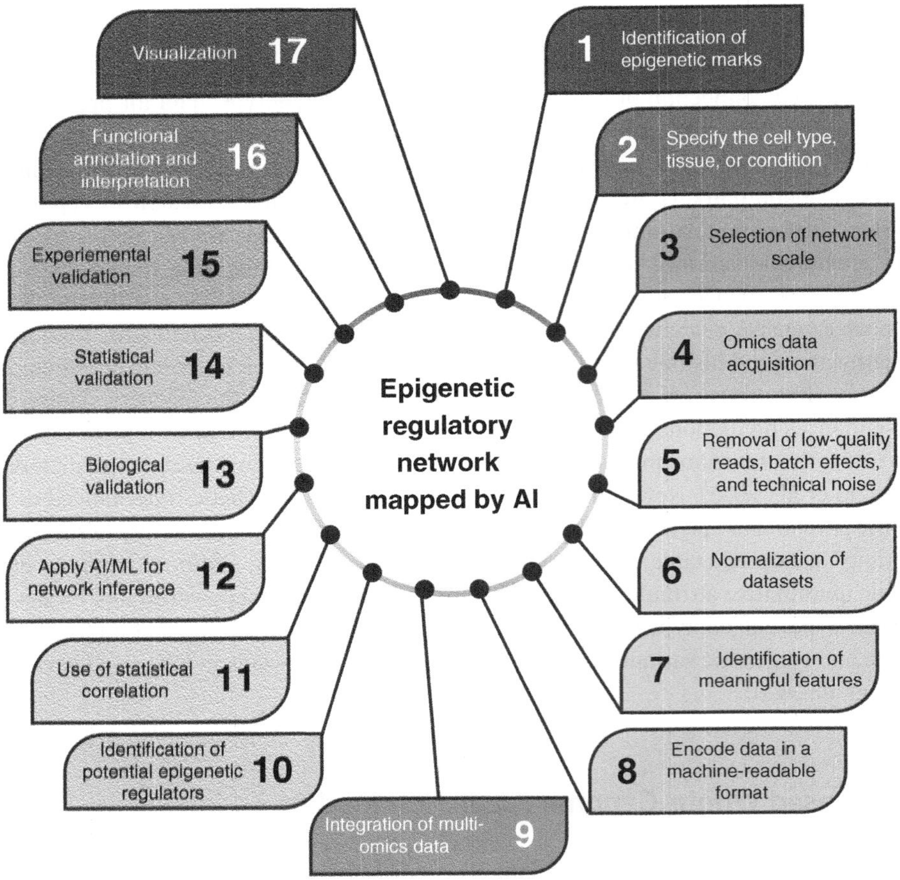

Figure 7.2 Epigenetic regulatory network mapped by AI.

Genome-wide histone modification prediction became a parallel area of interest, utilizing methods of ChIP-seq, in association with DL-based inferences. Genome-wide mapping of histone methylation (H3K9me2) and acetylation (H4K12ac) in *Panicum virgatum* (switchgrass) revealed widespread sites of modification in nongenic regions, illuminating their role in metabolic processes associated with biomass and biofuel production (Ayyappan et al., 2019). Corresponding investigations in *Sorghum bicolor* elucidated the regulatory connection between histone modifications of the type of H3K4me3 and H3K36me3 with coding and ncRNA expressions, thus bringing into perspective important aspects of chromatin-dependent transcriptional regulation (Zhou et al., 2021). The ChIP-seq protocol continues to form the basis of mapping such connections, inasmuch as it bridges protein–DNA cross-linkage and immunoprecipitation with deep-sequencing for creating genome-wide histone-binding maps (Ricci et al., 2019). Combining these experimental datasets with DL models such as SMEP facilitates precise histone modification pattern prediction in crop species such as rice and maize, expediting knowledge on systems-level epigenetic regulation (Wang et al., 2021).

AI-based modeling also transformed the prediction of ncRNA regulation, a key feature of epigenetic control, as well as posttranscriptional regulation of plants. Feature-engineering techniques hold no promise in describing the complexity of ncRNA sequence dependencies and secondary structures, thereby requiring deep neural models. The CIRNN model using convolutional neural networks (CNN) as well as independently recurrent neural networks (IndRNN) successfully predicts miRNA–lncRNA interactions by learning sequence and structural features in an independent manner, improving on previous ML models (Zhang et al., 2020). Complementary to it, the PlncRNA-HDeep model utilizes combined DL with dual encoding schemes to successfully predict plant lncRNAs with enhanced accuracy and sensitivity, improving on traditional software (Meng et al., 2021). Extending it further, the PmliGKKS model uses k-mer cross-encoding to represent miRNA–lncRNA pairs as two-dimensional encoded matrices, improving interaction predictions across species (Yong et al., 2023). Extending accuracy further, the PmliPEMG model also includes multilevel information augmentation as well as fuzzy decision strategies, achieving robust species cross-generalization (Kang et al., 2021). Finally, DL models trained with untranslated region (UTR) and promoter sequence data also began deciphering the cis-regulatory code of gene regulation, achieving precise species cross-predictions and stimulating regulatory genomics (Peleke et al., 2024).

The integration of epigenetic datasets in whole-genome predictive models facilitates improved understanding of the heritable regulatory processes for phenotypic variation and environmental adaptability. DNA methylation, histone modifying, and chromatin remodeling direct transcriptional activity and induce phenotypic plasticity in excess of that of genomic variation (Xue et al., 2025). Multilayer datasets can be utilized by sophisticated AI and ML models to predict plant responses in abiotic and biotic stress, elucidating plant systems of memory and permitting precision breeding of stress-tolerant crops (Dobránszki et al., 2025). Integrating chromatin accessibility and methylation profiling also facilitates improved interpretability of complex crop genomes, permitting design of genome-editing strategies based on CRISPR/Cas9 for increased yield and stress tolerance (Zhang et al., 2022). Although knowledge translation issues from model plants remain, new systems that combine modeling, in vitro regeneration systems, and data-driven feature extraction continue to advance the promise of predicting and manipulating epigenetically controlled adaptive traits (Kakoulidou et al., 2021).

7.7 Applications in Plant Science and Agriculture

AI-based prediction of stress-mediated epigenetic reprogramming in plants is an area of plant systems biology that is rapidly expanding. Plants, in contrast to animals, achieve perception, storage, and recall of environmental signals by means of complex cellular signaling and molecular memory networks. The adaptive processes of plants are centrally regulated by epigenetic mechanisms such as DNA methylation, histone modifications, and small RNA pathways. ML, in combination with AI (AI), allows for fusion and large-scale interrogation of epigenomic datasets, enabling the prediction of dynamic methylation cycles and stress-responsive regulation of plant memory formation and adaptation (Dobránszki et al., 2025). RdDM, an important regulator of such reprogramming, facilitates directed events of methylation contributing to conditionally heritable phenotypic plasticity and environmental stress responses (Kundariya et al., 2022). Experimental data from *Arabidopsis thaliana* show that stress-mediated alterations of methylation are stably inherited through the female germline, displaying conditionally heritable epigenetic mechanisms of memory that persist through generations of adaptation (Wibowo et al., 2016). Empirical data of these results indicate the promise of AI to predict and manipulate plant transgenerational memory states, informing cultivar development for stress resilience (Dobránszki et al., 2025; Kinoshita & Seki, 2014).

Epigenetic plasticity, in combination with AI-based prediction of traits, is transformative for crop breeding by synthesizing epigenomic control and phenotypic performance. Noncoding DNA epigenetic signatures that guide gene expression offer a mechanistic explanation for environmental adaptation and stress tolerance. AI and ML applications of these sorts, in turn, redefined crop trait prediction by untangling intricate relationships among germplasm architecture, environment, and epigenetic states (Dobránszki et al., 2025; Kakoulidou et al., 2021). AI-capable models turn predictive breeding into a reality by mining large-scale, multidimensional data for phenotypic outputs under changing climatic scenarios (Fu et al., 2025). In addition, AI-based quantitative trait loci mapping reinforces the identification of polygenic interactions that build complex traits, enhancing the accuracy and efficiency of marker-assisted decision-making (Kassem, 2025). Incorporating these instruments in breeding models such as Breeding 5.0 streamlines the design-to-deployment process from ideotype conceptualization through elite line development, aligning precision cultivation with sustainability mandates (Fu et al., 2025). Data sparsity, heterogeneous datasets, and limited model interpretability, nevertheless, represent the main obstacles necessitating methodological breakthroughs for more reliable cross-species traits by predicting data (Kassem, 2025).

Recent findings of stress-memory epialleles by AI systems further augmented the knowledge corpus of plant adaptive responses. Plant stress memory—fairly well delineated in *Arabidopsis*—comprises reversible or heritable epigenetic changes, primarily mediated by DNA methylation and histone reorganization, which fine-tune gene expressions in repeated stress exposures (Lagiotis et al., 2023; Wibowo et al., 2016). ML systems such as RF and XGBoost demonstrated strong predictive capability in distinguishing stress responses' epigenetic biomarkers, thereby forming computational bases for resilience mechanisms modeling. Superior models such as eDICE, based on DL for making individual-specific imputations of epigenomic profiles, give rise to tissue- and condition-dependent predictions of epigenetic differences such that plant breeding and stress adaptation science gain increased accuracy (Hawkins-Hooker et al., 2023; Schiele et al., 2024).

Epigenome modeling is an effective strategy for crop yield prediction, stress tolerance optimization, and disease resistance improvement. The epigenetic regulatory layers, including DNA methylation and histone modification landscapes, fine-tune gene networks governing plant defense responses and developmental fates (Zhi & Chang, 2021). Predictable modeling of these interactions by AI allows the delineation of molecular circuits mediating defense activation and growth, a key factor in the reduction of trade-offs between disease resistance and yield potential. Recent literature has revealed the epigenetic regulation of immune receptor gene expression and calcium sensor-mediated defense suppression, some of these regulation mechanisms becoming potential targets for breeding programs (Wang & Dong, 2021). Precision epigenome editing technologies, including the targeted DNA methylation of promoters, were shown to be capable of repressing pathogen-driven gene activation and thus enhancing disease resistance while preserving growth vigor (Veley et al., 2022). Their use, however, needs to take into account the ecological and agronomic background, as resistant cultivars often exhibit decreased yield in the absence of stress (Vyska et al., 2016). Finally, the AI-driven interrogation of epigenetic viable variations with heritability—combined with predictive modeling—can lead to the development of climate-adaptive, high-yielding, and disease-responsive crop types, with important applications in climate-driven food insecurity worldwide (Ahtisham & Obaid, 2024).

7.8 AI-Guided Epigenome Editing and Synthetic Biology

The combination of CRISPR/dCas9-based epigenome editing with ML or computer vision, known as AI-guided CRISPR/dCas9, enhances plant gene engineering accuracy and predictability. CRISPR/dCas9, using a dead Cas9 that is not active in terms of catalysis, allows site-directed modulation of gene expression and epigenetic modification—DNA methylation and histone acetylation, for instance—without creating double-strand breaks or permanent DNA lesions in the genome (Moradpour & Abdulah, 2020). As a programmable and reversible system, it provides an influential platform for fine-tuning plant transcriptional networks. It was demonstrated as being used for reprogramming plant epigenomes for enhancing disease defense and stress tolerance, thereby enhancing resilience to abiotic and biotic stress (Suruthi et al., 2025).

The integration of AI with CRISPR-based systems further enhanced their effectiveness and precision. ML systems also enabled optimization of guide RNA (gRNA) design by predicting binding affinity, secondary structure stability, and off-target interaction, thereby increasing genome and epigenome editing specificity (Fal & Carles, 2024). Furthermore, AI-guided CRISPR multiplex approaches make it feasible for simultaneous modulation of a genomic locus of more than one, thus enabling complex agronomic traits such as optimization of yield and multi-stress tolerance to be engineered. Emerging literature also cites AI-aided predictions of landscapes of chromatin accessibility for guiding the locus choice for targeting, optimizing the spatiotemporal accuracy of epigenetic regulation control. Main issues persist in spite of these advances, particularly in terms of developing compact, combinatorial systems of editing capable of achieving live imaging of the chromatin with temporal activation with high accuracy. Of particular note, empirical literature established that CRISPR/Cas9-mediated manipulations do not create an unintentional epigenetic footprint, thereby underscoring their biosafety and credibility for large-scale plant breeding (Lee et al., 2020).

The development of synthetic epigenetic circuits brought a new dimension to plant synthetic biology, providing programmable regulation of gene expression through engineered transcriptional logic. These circuits, based on recombinase or synthetic promoter-based systems, drive stable, inheritable, and site-specific transcriptional states that allow for site- and time-precise reprogramming of development and physiology processes (Lloyd et al., 2022). Synthetic circuits, for example, in *Arabidopsis thaliana*, were used to restructure the architecture of the root system for optimizing nutrient uptake and improving the efficiency of water acquisition (Brophy et al., 2022). By implementing Boolean combinations of multiple endogenous promoters, as in the case of AND/OR gates, it is possible to obtain spatiotemporally specific activation of the gene in specific tissues. Predictive follow-up frameworks that make use of computational modeling and AI-driven optimization, in turn, design and test synthetic circuits governing multistate phenotypes in the form of responses to external signals with high accuracy of in silico predictions versus in vivo results (Kong et al., 2025). Meanwhile, synthetic-directed evolution (SDE) methods are being used for creating new genetic diversity in underrepresented plant systems, thus boosting adaptive potential and accelerating the emergence of resilient genotypes adapted to various agro-ecological settings (Kababji et al., 2024).

AI-chromatin modifier optimization is an emerging technique of precision breeding within the larger Breeding 5.0 panorama. It couples AI, big data analytical technologies, and new-era genome-editing technologies to create more predictable and efficient plant improvement pipelines. Through simulation of complex germplasm architecture and genotype–environment interaction, AI enables the unfolding of optimal parental sets and inheritance units of traits (Fang, 2024; Fu et al., 2025). Some of the advanced models of genomic prediction, such as SoyDNGP, demonstrate the power of DL in selecting optimal parents and predicting traits, thus streamlining manipulations of chromatin state for optimal breeding outputs. Integrating AI with speed breeding and genomics-assisted breeding strategies offers a transformative platform for creating high-yield, stress-tolerant cultivars, addressing the inadequacies of traditional systems of selection-based breeding (Bhat et al., 2023). In addition, generative models of AI, such as DNA diffusion, permit synthetic de novo design of synthetic regulatory components, fine-tuning of gene and chromatin accessibility for precision regulation of transcriptional regimes in crops (DaSilva et al., 2024).

7.9 Explainable and Interpretable AI

Explainable ML methods, such as SHAP (Shapley Additive exPlanations) as well as LIME (Local Interpretable Model-Agnostic Explanations), are potent tools for explaining intricate ML models, particularly in high-dimensional biological datasets, such as in plant epigenomic prediction. These systems aim to change the native obscurity of DL models into clear, interpretable systems that achieve scientific credibility as well as user trust. SHAP as well as LIME present local (instance-based) as well as global (model-based) interpretability such that researchers can understand feature contribution as well as predictive reasoning on heterogeneous datasets (Aldughayfiq et al., 2023; Carmichael & Scheirer, 2023). Although these methods have had widespread application, they pose some constraints, such as sensitivity to model design as well as data distribution. Feature collinearity, as well as nonlinear interaction of genomic as well as environmental factors, can distort feature importance scores, bringing about issues for accurate interpretation in plant epigenomic applications. Because of the intricate interaction among plants' genotype, methylation pattern, as well as environmental stress, the correct calibration of XAI instruments still remains for the development of biologically relevant insights (Salih et al., 2025).

The integration of XAI models with DL networks has demonstrated much potential in facilitating biological interpretability. In biomedical images, for instance, the integration of spatial attention mechanisms with SHAP and LIME enhanced both retinoblastoma classification accuracy and interpretability (Aldughayfiq et al., 2023), with transferable application to plant epigenomic prediction. Hybrid models of such integrations could uncover locus-specific functions in epigenetic variations, with interpretable models for plant stress responses, transcriptional regulation, and chromatin remodeling.

Within plant epigenetics, DL architecture-based AI models have become irreplaceable for dissecting complex regulatory relationships. Hybrid and CNNs are also utilized for predicting and deciphering genome-wide epigenetic signatures involving DNA and RNA methylation and histone posttranslational modifications (Kajrolkar, 2025; Küçükrecep et al., 2024). The models integrate multiomics data-genotypic, transcriptomic, and phenotypic data, in order to make gene expression dynamics predictions with unparalleled accuracy. The SMEP, for example, achieved over 95% accuracy for certain epigenetic marks in rice, illustrating the promise of AI-based models for identifying the underlying crop adaptability as well as yield regulation mechanisms (Wang et al., 2021). DL paradigms can additionally learn positional binding preference of transcription factors and make direct predictions of expressions from nucleotide sequences, thereby allowing identification of the regulation motifs as well as enhancer–promoter interaction even in polyploid crops (Rockenbach et al., 2025). It is applicable for a better understanding of the epigenetic regulation of gene expression as well as for the rational design of improved resilience crops.

Connecting computational predictions with molecular validation is a seminal development in plant systems biology. Integrative approaches now bridge AI-driven inference with empirical validation, closing the distance between predictive modeling and experimental biology. DL models, such as those explained by Kajrolkar, allow for real-time gene expression and outcome of phenotype prediction, accelerating experimental pipelines in optimizing crop traits (Kajrolkar, 2025). The PlantConnectome project illustrates such compatibility by utilizing AI for large-scale scientific literature mining, unveiling functional connections among metabolites, proteins, and genes with exceptional accuracy (Fo et al., 2023). Likewise, higher-order models, such as AlphaFold and DeepFRI, have furthered functional annotation of nonmodel plant genes, substantiated by molecular docking and biochemical assays (Stephan et al., 2021). Bai et al. also illustrated ML integrated with multiomics data for predicting specialized metabolite biosynthesis genes, with successful cross-species validation and enhanced predictive robustness (Bai et al., 2024). To complement these computational breakthroughs, ML-based quantitative trait loci mapping further optimizes the identification of genetic regions for key agronomic traits, with feature selection and integration of multiomics data important for precise trait dissection (Kassem, 2025).

7.10 Digital Twins and In Silico Plant Epigenomics

AI-based digital twins (DTs) for dynamic plant epigenomes represent a cutting-edge paradigm in modeling, predicting, and optimizing plant responses to environmental stimuli (Figure 7.3). The integration of AI and ML with multiomics datasets—encompassing

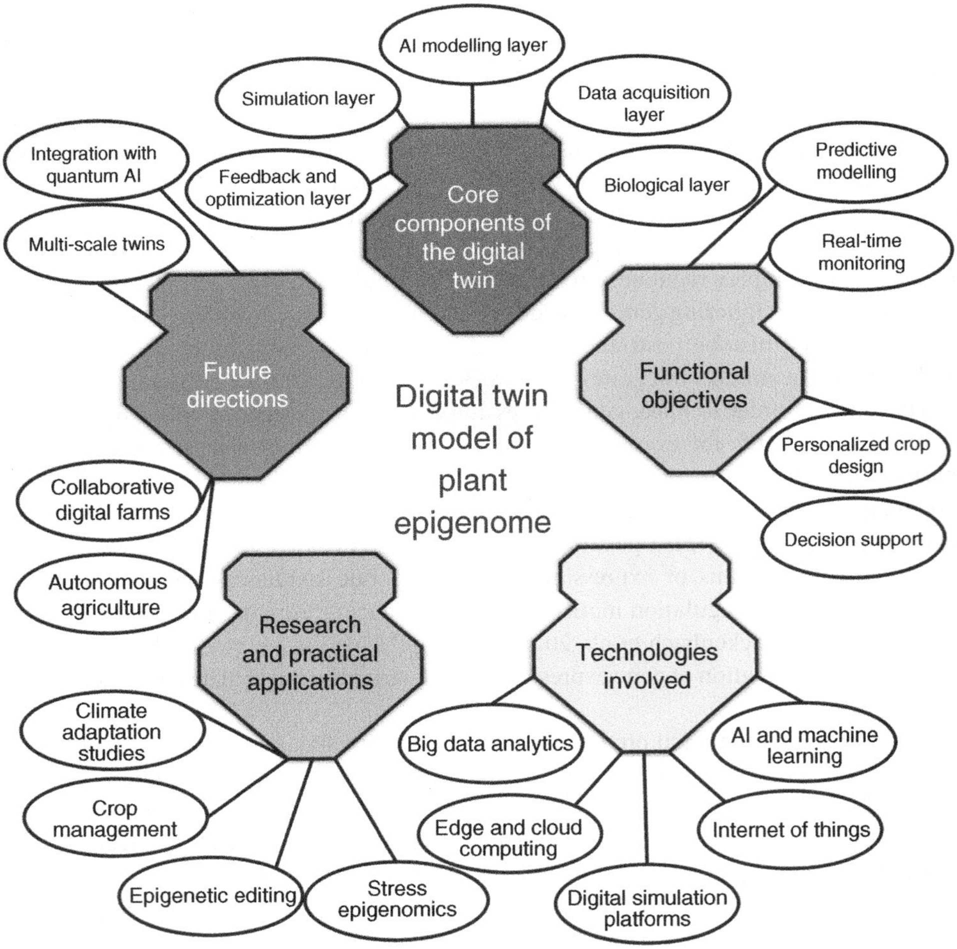

Figure 7.3 Digital twin model of plant epigenome.

genomics, transcriptomics, and proteomics—provides a robust analytical framework to decipher the complex molecular mechanisms underlying plant phenotypes and epigenetic memory (Cembrowska-Lech et al., 2023). In plant science, DTs are conceptualized as intelligent, multiagent systems capable of simulating growth, development, and adaptive responses by integrating physiological, morphological, and environmental data. These systems employ ontological modeling and emergent intelligence to construct adaptive, self-organizing simulations that replicate real-world plant dynamics under diverse environmental conditions (Skobelev et al., 2023).

The application of AI in these virtual twin systems facilitates real-time dynamic modeling of plant epigenomes for the predictive interrogation of molecular and phenotypic adaptations to abiotic stress conditions such as drought, salinity, and temperature differences. Through sequential feedback loops and iterative learning, such systems revise their predictive outputs, creating a transformative system of inquiry into plant memory, stress adaptation, and resilience. This modeling system, in turn, becomes significant not only in terms of faster theoretical understanding of epigenetic regulation but also in terms of unlocking practical crop development of stress tolerance and climate resilience through predictive breeding and precision horticulture interventions (Dobránszki et al., 2025).

Epigenetic in silico modeling of environmental stress responses supplements DT strategies by using computational algorithms for simulating and dissecting gene regulation mechanisms that are determined by epigenetic alterations. For instance, research on *Arabidopsis thaliana* subjected to concurrent heat and drought stress demonstrates that ML models utilizing putative CREs as predictive features may reliably forecast transcriptional outcomes. DL enhances these predictions further by integrating datasets for chromatin accessibility as well as histone modifications, thus enhancing model accuracy as well as interpretability. The integrative strategy highlights the potential of in silico systems in unraveling intricate regulatory codes in addition to discerning important transcription factors that participate in plant stress responses, including factors that were not known to participate in environmental stress adaptation (Azodi et al., 2020).

Plant adaptive responses feature epigenetic mechanisms such as histone alterations and DNA methylation in key roles, with effects on stress-memory formation as well as reprogramming of transcription. Consistent with the Epigenetic Disease Adaptational Model (EDAM), an organismic failure to achieve appropriate epigenetic adaptation would introduce physiological impairment or disease predisposition (Krasteva, 2024). Here, computational models in particular have come to be used increasingly to analyze modes of transgenerational inheritance, by means of which parental stress imposes stress-driven, epigenetically heritable transformations in progeny. Consistent with studies, the methylation of particular sites, for example, Sfmbt2, could present a molecular doorway for stress adaptations that become heritable, thus introducing a mechanism whereby environmental exposure could intersect with epigenetic inheritance (Lei et al., 2018). Beyond plants, the identical mechanisms were revealed in nervous systems, such that chronic stress conveys lasting epigenetic reprogramming of neural circuitry, contributing to neuropsychiatric disease. These results, with gene dysregulatory functions, show the universality of epigenetic plasticity and also the roles of epigenomic modeling in diverse biological systems (Chakravarty et al., 2014).

The fusion of AI-driven epigenomics with precision farming is a distinguishing step in advanced farming practice. The integration of AI, multiomics data, and IoT-based sensor systems facilitates the elucidation of genetic and epigenetic mechanisms of stress resilience, generating useful information for sustainable crop production (Polegopu et al., 2025; Thingujam et al., 2025). The systems capitalize on real-time data acquisition-integrating environmental monitoring, spectral imaging, and molecular characterizations direct adaptive interventions such as optimal irrigation, nutrient supply, and site-specific breeding strategies. Through these technologies, AI-driven DTs extend their application from theoretical modeling to practical precision farming, boosting resource efficiency and improving food production sustainability. Furthermore, the fusion of autonomous decision-support systems with predictive modeling facilitates the deployment of smart farming technologies, including automated phenotyping, yield estimation, and resource optimization, while optimizing ecological balance, as well as food safety in the midst of climate stress globally (Vetrivel & Arun, 2025).

7.11 Quantum Computing and AI for Epigenetic Predictions

Quantum Machine Learning (QML) is a new frontier in computational biology with transformative power for large-scale and high-dimensional epigenomic data analysis in plants. Its combination with plant epigenomics is a natural extension of established AI-based methodologies, including the SMEP, in DL architecture for predicting epigenomic alterations in crops such as rice and maize (Wang et al., 2021). In contrast with traditional DL systems, QML exploits the laws of quantum superposition and entanglement for exponential efficiency in processing huge datasets. More importantly, qz random quantum measurements and fast data encoding achieved quadratic time-speedups in computations, validating QML as a computationally superior methodology for integrating complex multiomic datasets (Haug et al., 2021).

This computational power is particularly relevant considering the complexity of plant epigenomic regulation, including various molecular mechanisms like DNA methylation, histone posttranslational modifications, and chromatin remodeling of which operate in concert to modulate gene expression and environmental adaptability (Dobránszki et al., 2025; Xu et al., 2010). Due to effective encoding of high-dimensional biological data, QML can capture low-dimensional, nonlinear relationships between epigenetic states and phenotypic traits that would be lost with traditional algorithms. Additionally, combining QML with genomic selection methodologies has exhibited enhanced predictive ability in discovering disease resistance-linked genetic markers, providing a potential tool for further developing precision breeding methods (Albert et al., 2023).

7.11.1 Hybrid Quantum-Classical AI Models in Epigenomic Prediction

Hybrid quantum–classical models appear as a strong intermediate solution, bridging the representational power of DL with the computational benefits of quantum processing. These models take advantage of the complementary benefits of both camps: parallelized data exploration of quantum computing and classical AI's well-matured optimization methods (Farshi, 2023). In that hybrid system, quantum attention mechanisms and quantum-supplemented neural networks were implemented in biomedical classification issues, with their results outperforming classical equivalents in accuracy, generalization, and computational cost (Roosan et al., 2025). Their successful application in cancer epigenomics and molecular diagnostics emphasizes their potential use in plant epigenetic prediction systems.

Applications of hybrid quantum–classical AI range from drug discovery and protein–ligand binding prediction, where quantum-inspired DL-such as GNNs and CNNs, exhibited enhanced data scalability and reduced parameter requirements (Domingo et al., 2024; Dong et al., 2023). The same principles can also be transferred to the description of plant system epigenetic interactions, where the complexity of multilayered alteration patterns of chromatin topology calls for computational entities that can sustain hierarchical dependencies. As pointed out by Tahir et al., the integration of quantum computing in classical AI models could transcend current drawbacks of classical ML models, such as high computational resource requirements and instability in large-scale prediction, and thereby make way for more stable and accurate epigenetic modeling (Tahir et al., 2024).

7.11.2 Quantum-Enhanced Predictive Epigenomics in Agriculture

Quantum-enhanced predictive epigenomics is a new model of agricultural biotechnology that combines quantum computing, AI, and systems biology to address the twin issues of climate change and sustainable food production. Quantum computational methods, when used, can make such systems not only simulate complex agro-ecological interactions but also extrapolate crop behavior with varying stress regimes and refine management strategies for enhanced productivity (Bansod et al., 2024). Interfaced quantum AI enables large-scale simulations of crop systems that process huge environmental and molecular datasets and forecast crop yield, soil nutrient dynamics, and pest emergence patterns with hitherto unprecedented accuracy.

Recent breakthroughs illustrate that potential quantum-enhanced algorithms were used to maximize soil microbiome relationships, achieving a 32% increase in nutrient efficiency and a 19% crop yield boost through fast data analysis and adaptive feedback cycles (Mukherjee, 2025). Likewise, the Quantum-AI Genomic Frontier Platform brings quantum-inspired processing to multiomics datasets to boost vegetable crop resilience, achieving up to 15% stress tolerance increases through predictive modeling of plant stress response networks (Kaushik, 2025). These breakthroughs mark a transformative move toward intelligent, quantum-led agriculture, where predictive epigenomics guides breeding, management, and resource allocation choices in real time. Nevertheless, while there is a problem of technological hurdles, such as restricted access to quantum hardware and data standardization hurdles, the combination of quantum computing with AI-led epigenomics presents a hitherto unprecedented chance for the attainment of sustainable, climate-resilient farming. This combination allows for not only crop improvement with precision but also for furthering global initiatives for food security and environmental preservation (Bansod et al., 2024).

7.12 Challenges and Limitations

7.12.1 Data Sparsity, Noise, and Analytical Complexity

One of the significant hurdles in plant epigenomic studies is the natural sparsity and noise of single-cell datasets, thereby complicating downstream analysis and biological insight. Single-cell DNA methylation assays, while achieving unprecedented resolution into intercellular heterogeneity, also present highly sparse data matrices with appreciable missing values. Data sparsity masks true biological variability and imposes technical bias. Classical methods for discovering variably methylated regions (VMRs) often depend on predefined genome intervals or sliding window strategies that do not discern fine-scale, cell type-specific variability. The vmrseq algorithm rectifies these issues by utilizing a statistically robust VMR-detection framework that improves sensitivity and specificity while permitting correlation analyses, dependent on contextual information, between methylation and gene expression (Shen & Korthauer, 2024).

Equivalently, single-cell chromatin accessibility assays also do not offer extensive coverage, such that new computational strategies must recover lost information in order to reconstruct absent data. Some methods that encompass sparse functional data analysis, including in particular sparse functional principal component analysis (S-FPCA), successfully fill in missing chromatin accessibility by information borrowing from related experimental conditions or cell types (Madrigal et al., 2018). Furthermore, variance in depths of sequencing as well as inconstant signal-to-noise ratios from experiment to experiment further obscures biological signals. S3norm is an ideal normalization method that, through applying a monotonic nonlinear transform simultaneously and compensating for signal intensity as well as depth of sequencing, optimizes true detection of biological variations in gene expression (Xiang et al., 2020).

7.12.2 Computational and Biological Challenges in Genome-Wide Prediction

The computational environment for genome-wide epigenetic prediction in plants is multilayered, with technical constraints meeting the biological complexity of plant genomes. Predictions of specific epigenomic marks in DNA, RNA methylation, and histone modifications remain the greatest challenge, as these marks often change in a dynamic, contextual way, dependent on environmental stimulus. The SMEP, in deploying DL models, is a particular example of being successful in predicting a variety of different epigenetic modifications with considerable accuracy (Wang et al., 2021). The complexity of the plant epigenomes, in concert with restricted access to large datasets that are well annotated, still constrains model generalizability.

Epigenetic memory and regulation of gene networks present even more computational problems. Mathematical modeling of the feedback loops and nonlinear dynamics of memory formation necessitates higher-order systems biology models that combine theory with empirical observations (Feng et al., 2024). Furthermore, k-mer–based GWAS are a potential tool for finding epigenetically significant variants beyond single-nucleotide variants. While useful, such models engender data dimensionality, computational, and spurious association problems (Karikari et al., 2023). While AI and ML achieved grand potential for predicting plant stress responses based on epigenetic signatures, the practical application of such technologies into large breeding programs is restricted and needs further translational maturation (Dobránszki et al., 2025).

7.12.3 Cross-Species Transferability and Model Generalization

Cross-species generalization of AI-driven epigenome models is a significant limitation in plant bioinformatics. More recent works on DNA language models attempted to break through such limitations by utilizing evolutionary conservation and common genomic features among species. PlantCaduceus, a pretrained DNA LM, successfully transferred learned representations of *Arabidopsis* genomes to heterologous genomes of Zea mays, a species that separated around 160 million years ago, and exhibited better performance in predicting splice donor sites and translation initiation sites compared with classical models. Similarly, FloraBERT utilizes cross-species encoding of genomes for gene expression prediction in nonmodel plants, efficiently capturing long-range dependency as well as taxonomic similarities. In much the same fashion, PlantCAD2, pretrained with 65 genomes of angiosperms, exhibits improved ability to describe evolutionary conservation as well as functional annotation among broadly diverse taxa (Zhai et al., 2025). These works, in combination, highlight deep contextual embeddings' as well as transfer learning's role in enhancing the scalability of models as well as functional inference among plant species. The big potential of such models resides in epigenetic adaptation and environmental responses, whereby machine-driven approaches might shed light on the plant resilience at the molecular level. Integrating such cross-species models maximizes the predictive ability of the epigenetic and transcriptomic networks for more efficacious crop adaptation strategies in the changing climatic circumstances (Dobránszki et al., 2025).

7.12.4 Ethical, Regulatory, and Biosafety Constraints

The integration of AI into plant epigenetics and breeding is coupled with a series of ethics, regulations, and biosafety concerns that must be confronted with appropriate governance. Ethical issues imply considering data protection, algorithmic bias, and model interpretability from AI factors that happen to be paramount for the sake of maintaining transparency, public trust, and equitable access in biotechnology among crops (Dara & Azarpira, 2025). The crossing of AI with genome-editing technologies such as CRISPR-Cas systems in horticultural crops heightens issues of regulation, more so with unintentional off-target effects, ecological gene flow, and biosafety in the long run (Munawar et al., 2024).

The intricacy of epigenetic inheritance also makes it complex for regulators to oversee, as these results of epigenomic reprogramming that happen through epigenetic inheritance may occur unpredictably in generations that follow (Bhuvaneswari et al., 2020). It is for these reasons that more agreement exists around implementing integrated biosecurity measures in AI-based platforms to curb misuse and ensure that technologies are used responsibly (Wang et al., 2025). Defining common standards for the use of ethical AI, data management, and risk management will help in balancing innovation and sustainability with acceptability in society.

7.13 Future Perspectives

The integration of AI with CRISPR genome editing in plants is a groundbreaking development in plant biotechnology, combining computational intelligence with genome precision engineering. CRISPR/Cas systems have redrafted plant genome editing by introducing site-directed alterations that are critical for the development of desired agronomic traits, such as higher yield, stress tolerance, and disease resistance (Ahmar et al., 2024; Pandita & Pandita, 2023). Extending beyond genetic modification, CRISPR-guided epigenome editing makes it possible for directed control of gene expression to modulate DNA methylation and histone modifications, significant plant defense and adaptation mechanisms (Suruthi et al., 2025). AI is crucial in optimizing these procedures by optimizing guide RNA (gRNA) design, predicting off-target activities, and optimizing editing efficiency. DL models, trained on large sets of genomic data, can precisely forecast CRISPR-mediated alterations' functional outcome and thus ensure better reliability and specificity of epigenetic interventions (Kim et al., 2025; Li et al., 2025). Furthermore, AI-assisted identification of new CRISPR variants and optimization algorithms accelerate functional genomics studies and ensure the creation of sustainable, stress-resistant crop variants (Ahmar et al., 2024). While it holds huge potential, combining AI with CRISPR also generates new concerns about ethics, such as algorithmic explicability, data bias, and biosecurity threats, which need to be addressed by rigorous regulation and interdisciplinary collaboration (Li et al., 2025).

Advances in real-time predictive epigenomics will revolutionize field-deployable monitoring and disease diagnosis systems in agriculture. The convergence of AI, portable genomic sequencing technologies, and data sciences has enabled on-site assessments of near-real-time epigenetic and genomic changes. The Oxford Nanopore Technologies' MinilON has paved the way for fast field-based DNA and RNA sequencing with the promise of direct access to plant health reactions and stress reaction monitoring in real time (Udegbe et al., 2023). Integration with field-programmable gate array (FPGA) technologies makes the processing even faster, allowing for advanced bioinformatic computations under limited-resource scenarios (Ramachandra et al., 2021). Selective sequencing tactics such as the "Read Until" approach make the best of efficiency, abolishing redundant read discards while sequencing and thereby reducing computational workload but increasing data throughput. Hardware–software code-design architectures, such as HARU, are designed to have minimal latency and power consumption when performing genomic analysis across real-time monitoring (Shih et al., 2023). At the same time, secure architectures, such as the Secure Bloom-Filter Analysis and Compression (SB-FAC) architecture, address data privacy issues so that sensitive genome data, as well as epigenetic data, are kept secure when data are transmitted and processed (Wyschogrod et al., 2023). The mutual complementarity between technologies brings to reality the possibility of precision epigenomics in agriculture, where devices based in the field will be able to monitor environmental responses, anticipate stress in crops, and guide timely interventions.

Global epigenetic knowledge graph development for crops holds the key to the emerging data-driven, systems-level approach to crop biology. DNA methylations, histone tail modifications, and ncRNA-regulated control are important players in plant growth, stress, and phenotypic flexibility (Kakoulidou et al., 2021; Xue et al., 2025). Environment-activated heritable but reversible epigenetic signals are essential regulatory networks that control gene expression in the absence of DNA change, leaving a window for a further dimension of targeted improvement in crops (Kakoulidou et al., 2021; Ostrowska-Mazurek et al., 2020). Integration of data from genome-regulated regions in genome-based data to the breeding pipeline—epigenome-led crop improvement—enables finding functional chromatin zones and preferable epialleles that bring a better set of important traits like yield, stress tolerance, and flowering behavior (Zhang et al., 2022). Integration of this information, coupled with genome-editing tools like CRISPR/Cas9, gives predictive targeting of breeding approach to selected epigenetic loci. Although translation from model plant paradigms remains a challenge due to the diversity of crops, the push toward predictive modeling, single-cell epigenomics, and tissue-specific regenerating systems is enhancing this transition process (Kakoulidou et al., 2021; Xue et al., 2025). Furthermore, unraveling the role of epigenetic control of developmental phases like leaf senescence also gives directions toward enhancing the longevity and postharvest quality of crops, thus supporting the idea of "epibreeding" toward sustainable crops (Ostrowska-Mazurek et al., 2020).

In tandem, machine-based personalized plant breeding for climatic resilience—visualized as Breeding 5.0—is reimagining modern agriculture by combining computational intelligence with improved breeding methods. This approach applies AI and big data analytics to study intricate phenotypic and genotypic interactions, allowing for the fast discovery of the best combinations of genes tailored to particular climatic conditions (Fang, 2024). By utilizing AI-aided trait prediction and multimodal data fusion, breeders are able to speed up selection cycles and hasten the development of crops with climatic resilience capable of withstanding drought, salinity, and extreme temperatures. The urgency for this innovation is given by the estimated global population approaching nearly 10 billion by 2050, requiring an almost doubling of crop yields per unit area by the same year (Beans, 2020). Innovations like speed breeding, when combined with AI-assisted decisions, allow for fast-paced crop generation and selection for desirable characteristics (Rai, 2022). AI-based frameworks like Breeding 5.0 utilize explainable AI and expert systems to decipher the data hidden in the germplasm, conduct predictive modeling of traits, and design the best parental crosses (Fu et al., 2025). This convergence of AI, multiomics data, and precision breeding is a landmark moment toward the goal of global food security through sustainable, adaptive, and data-based agriculture.

7.14 Conclusion

AI-driven epigenomic prediction has been revolutionizing plant systems biology and sustainable agriculture. Integrating DL, multiomic information, and genome-editing technologies, AI allows for accurate modeling of DNA methylation and histone modification mechanisms and ncRNA regulation of phenotypic plasticity and adaptability to stress. Combined AI and CRISPR/dCas9 joint potential yields targeted editing of the epigenome, allowing unprecedented control over gene regulation and phenotypic improvement. Real-time predictive epigenomics and deployable genomic systems connect laboratory exploration and translational crop management, delivering prescriptive information for adaptive behavior to dynamically fluctuating environmental conditions. Development of global epigenetic knowledge graphs, AI-driven personalized breeding programs, and DT descriptions accurately increases prediction accuracy and decision-making when designing crops. QML further accelerates this front door by remedying data dimensionality and scaling computations to enable cross-species prediction modeling. With the intersection point between AI and plant biotechnology gaining prominence, the establishment of ethical, transparent, and secure frameworks will become necessary to unlock its complete potential to boost food security and climate-resilient agriculture.

Abbreviations

ACRs	Accessible Chromatin Regions
AI	Artificial Intelligence
ATAC-seq	Assay for Transposase-Accessible Chromatin Using Sequencing
CGMs	Crop Growth Models
ChIP-seq	Chromatin Immunoprecipitation Sequencing
CNN	Convolutional Neural Networks
CRE	Cis-Regulatory Element

DL	Deep Learning
DMRs	Differentially Methylated Regions
DTs	Digital Twins
EDAM	Epigenetic Disease Adaptational Model
epiQTLs	Epigenetic Quantitative Attribute Loci
FL	Federated Learning
FPGA	Field-Programmable Gate Array
G×E×M	Genotype-by-Environment-by-Management
GANs	Generative Adversarial Networks
GNNs	Graph Neural Networks
gRNA	Guide RNA
GWAS	Genome-Wide Association Studies
IndRNN	Independently Recurrent Neural Networks
LIME	Local Interpretable Model-Agnostic Explanations
lncRNAs	Long Noncoding RNAs
ML	Machine Learning
ncRNAs	Noncoding RNAs
piRNAs	PIWI-Interacting RNAs
QML	Quantum Machine Learning
RdDM	RNA-Directed DNA Methylation
RF	Random Forest
SB-FAC	Secure Bloom-filter Analysis and Compression
SDE	Synthetic-Directed Evolution
SHAP	Shapley Additive Explanations
SMEP	Smart Model for Epigenetics in Plants
sRNA-seq	Small RNA Sequencing
SVM	Support Vector Machines
TSPG	Transcriptome State Perturbation Generator
UTR	Untranslated Region
VMRs	Variably Methylated Regions
WGBS	Whole-genome Bisulfite Sequencing
WGP	Whole-genome Prediction

Acknowledgement

All the authors are thankful to their respective Universities and Institutions for their valuable supports.

Data Availability

The authors confirm that the data supporting the findings of this study are available within the book/chapter.

References

Abdulraheem, M. I., Xiong, Y., Moshood, A. Y., Cadenas-Pliego, G., Zhang, H., & Hu, J. (2024). Mechanisms of plant epigenetic regulation in response to plant stress: Recent discoveries and implications. *Plants, 13*(2), 163. https://doi.org/10.3390/plants13020163

Ahmar, S., Usman, B., Hensel, G., Jung, K. H., & Gruszka, D. (2024). CRISPR enables sustainable cereal production for a greener future. *Trends in Plant Science, 29*(2), 179–195. https://doi.org/10.1016/j.tplants.2023.10.016

Ahtisham, M., & Obaid, Z. (2024). Harnessing epigenetic mechanisms for crop resilience: A comprehensive review of plant responses to biotic and abiotic stresses. *Plant Biology, 1*, 100007. https://doi.org/10.70389/PJPB.100007

Albert, E. T. A., Bille, N. H., Martin, B. J., & Leonard, N. M. E. (2023). Integrating genetic markers and adiabatic quantum to improve disease resistance-based marker assisted plant selection. *Journal of Scientific Agriculture, 7*, 63–76. https://doi.org/10.25081/jsa.2023.v7.8556

Aldughayfiq, B., Ashfaq, F., Jhanjhi, N. Z., & Humayun, M. (2023). Explainable AI for retinoblastoma diagnosis: Interpreting deep learning models with LIME and SHAP. *Diagnostics, 13*(11), 1932. https://doi.org/10.3390/diagnostics13111932

Annacondia, M. L., & Martinez, G. (2019). Plant models of transgenerational epigenetic inheritance. In T. O. Tollefsbol (Ed.), *Transgenerational epigenetics* (pp. 263–282). Academic Press.

Arfat, Y., Mittone, G., Colonnelli, I., D'Ascenzo, F., Esposito, R., & Aldinucci, M. (2023, March 1–3). Pooling critical datasets with federated learning. *2023 31st Euromicro International Conference on Parallel, Distributed and Network-Based Processing (PDP),* Naples, Italy (pp. 329–337). IEEE.

Ayyappan, V., Sripathi, V.R., Kalavacharla, V., Saha, M.C., Thimmapuram, J., Bhide, K.P. and Fiedler, E., 2019. Genome-wide identification of histone methylation (H3K9me2) and acetylation (H4K12ac) marks in two ecotypes of switchgrass (*Panicum virgatum* L.). *BMC Genomics*, 20(1), p.667. https://doi.org/10.1186/s12864-019-6038-x

Azodi, C.B., Lloyd, J.P. and Shiu, S.H., 2020. The *cis*-regulatory codes of response to combined heat and drought stress in *arabidopsis thaliana*. *NAR Genomics and Bioinformatics*, 2(3), p.lqaa049. https://doi.org/10.1093/nargab/lqaa049

Bai, W., Li, C., Li, W., Wang, H., Han, X., Wang, P. and Wang, L., 2024. Machine learning assists prediction of genes responsible for plant specialized metabolite biosynthesis by integrating multi-omics data. *BMC Genomics*, 25(1), p.418. https://doi.org/10.1186/s12864-024-10258-6

Bansod, P. J., Usharani, R., Oliver, A. S., Prasad, S. S., Sharma, D. M., & Myilsamy, S. (2024). Quantum computing-powered agricultural transformation: Optimizing performance in farming. In S. Sharma, A. Prakash, & V. Sugumaran (Eds.), *Technological advancements in data processing for next generation intelligent systems* (pp. 169–195). IGI Global Scientific Publishing.

Beans, C. (2020). Crop researchers harness artificial intelligence to breed crops for the changing climate. *Proceedings of the National Academy of Sciences, 117*(44), 27066–27069. https://doi.org/10.1073/pnas.2018732117

Bhat, J. A., Feng, X., Mir, Z. A., Raina, A., & Siddique, K. H. (2023). Recent advances in artificial intelligence, mechanistic models, and speed breeding offer exciting opportunities for precise and accelerated genomics-assisted breeding. *Physiologia Plantarum*, *175*(4), e13969. https://doi.org/10.1111/ppl.13969

Bhuvaneswari, R., Saravanan, K. R., Vennila, S., & Suganthi, S. (2020). The role of epigenetics in plant breeding understanding heritable changes beyond DNA sequence. *Plant Science Archives*, *75*(1), 265–290. https://doi.org/10.51470/PSA.2020.5.1.22

Brophy, J. A., Magallon, K. J., Duan, L., Zhong, V., Ramachandran, P., Kniazev, K., & Dinneny, J. R. (2022). Synthetic genetic circuits as a means of reprogramming plant roots. *Science*, *377*(6607), 747–751. https://doi.org/10.1126/science.abo4326

Cao, S., & Chen, Z. J. (2024). Transgenerational epigenetic inheritance during plant evolution and breeding. *Trends in Plant Science*, *29*(11), 1203–1223. https://doi.org/10.1016/j.tplants.2024.04.007

Carmichael, Z., & Scheirer, W. J. (2023). How well do feature-additive explainers explain feature-additive predictors?*arXiv preprint arXiv:2310.18496*. https://doi.org/10.48550/arXiv.2310.18496

Cembrowska-Lech, D., Krzemińska, A., Miller, T., Nowakowska, A., Adamski, C., Radaczyńska, M., Mikiciuk, G., & Mikiciuk, M. (2023). An integrated multi-omics and artificial intelligence framework for advance plant phenotyping in horticulture. *Biology*, *12*(10), 1298. https://doi.org/10.3390/biology12101298

Chakravarty, S., Pathak, S. S., Maitra, S., Khandelwal, N., Chandra, K. B., & Kumar, A. (2014). Epigenetic regulatory mechanisms in stress-induced behavior. *International Review of Neurobiology*, *115*, 117–154. https://doi.org/10.1016/B978-0-12-801311-3.00004-4

Champigny, M. J., Unda, F., Skyba, O., Soolanayakanahally, R. Y., Mansfield, S. D., & Campbell, M. M. (2020). Learning from methylomes: Epigenomic correlates of Populus balsamifera traits based on deep learning models of natural DNA methylation. *Plant Biotechnology Journal*, *18*(6), 1361–1375. https://doi.org/10.1111/pbi.13299

Chen, Z. J. (2025). Empowering plant epigenetics to breed resilience of crops: From nucleolar dominance to transgenerational epigenetic inheritance. *The Plant Genome*, *18*(2), e70064. https://doi.org/10.1002/tpg2.70064

Chera, A., Stancu-Cretu, M., Zabet, N. R., & Bucur, O. (2024). Shedding light on DNA methylation and its clinical implications: The impact of long-read-based nanopore technology. *Epigenetics & Chromatin*, *17*(1), 39. https://doi.org/10.1186/s13072-024-00558-2

Corbin, K. R., & Lopez, C. M. R. (2020). Library preparation for whole genome bisulfite sequencing of plant genomes. In M. Budak, M. Yildiz, & A. M. Carmona-Ribeiro (Eds.), *DNA methylation mechanism*. IntechOpen. https://doi.org/10.5772/intechopen.90716

Dara, M., & Azarpira, N. (2025). Ethical considerations emerge from artificial intelligence (AI) in biotechnology. *Avicenna Journal Of Medical Biotechnology*, *17*(1), 80–81. https://doi.org/10.18502/ajmb.v17i1.17680

DaSilva, L. F., Senan, S., Patel, Z. M., Reddy, A. J., Gabbita, S., Nussbaum, Z., Córdova, C. M. V., Wenteler, A., Weber, N., Tunjic, T. M., & Khan, T. A. (2024). DNA-diffusion: Leveraging generative models for controlling chromatin accessibility and gene expression via synthetic regulatory elements. *biorxiv*. https://doi.org/10.1101/2024.02.01.578352

Derbala, D., Garnier, A., Bonnet, E., Deleuze, J. F., & Tost, J. (2024). Whole-genome bisulfite sequencing protocol for the analysis of genome-wide DNA methylation and hydroxymethylation patterns at single-nucleotide resolution. In *Epigenome editing: Methods and protocols* (pp. 353–382). Springer US.

Ding, X., Jia, X., Xiang, Y., & Jiang, W. (2022). Histone modification and cshromatin remodelling during the seed life cycle. *Frontiers in Plant Science, 13*, 865361. https://doi.org/10.3389/fpls.2022.865361

Dobránszki, J., Vassileva, V., Agius, D. R., Moschou, P. N., Gallusci, P., Berger, M. M., Farkas, D., Basso, M. F., & Martinelli, F. (2025). Gaining insights into epigenetic memories through artificial intelligence and omics science in plants. *Journal of Integrative Plant Biology*, *67*(9), 2320–2349. https://doi.org/10.1111/jipb.13953

Domingo, L., Chehimi, M., Banerjee, S., Yuxun, S.H., Konakanchi, S., Ogunfowora, L., Roy, S., Selvarajan, S., Djukic, M., & Johnson, C. (2024, September 15–20). A hybrid quantum-classical fusion neural network to improve protein-ligand binding affinity predictions for drug discovery. *2024 IEEE International Conference on Quantum Computing and Engineering (QCE),* Montreal, QC, Canada (Vol. 2, pp. 126–131). IEEE.

Dong, L., Li, Y., Liu, D., Ji, Y., Hu, B., Shi, S., Zhang, F., Hu, J., Qian, K., Jin, X. and Wang, B., 2023. Prediction of protein-ligand binding affinity by a hybrid quantum-classical deep learning algorithm. *Advanced Quantum Technologies*, 6(9), p.2300107. https://doi.org/10.1002/qute.202300107

Dwarampudi, A., & Yogi, M. K. (2024). Application of federated learning for smart agriculture system. *International Journal of Information Technology & Computer Engineering, 4*(3), 2455–5290.

Fal, K., & Carles, C. C. (2024). dCas-Based tools to visualize chromatin or modify epigenetic marks at specific plant genomic loci. In C. Baroux & C. Tatout (Eds.), *Methods for plant nucleus and chromatin studies: Methods and protocols* (pp. 305–332). Springer US.

Fang, J. (2024). Breeding 5.0: AI-driven revolution in designed plant breeding. *Molecular Plant Breeding, 15*(1), 27–33. https://doi.org/10.5376/mpb.2024.15.0004

Farshi, E. (2023). Hybrid quantum-classical approach: Quantum-inspired deep learning using classical simulation. https://doi.org/10.21203.rs.3.rs.3021644/v1

Feng, W., Gao, P., & Wang, X. (2024). AI breeder: Genomic predictions for crop breeding. *New Crops, 1*, 100010. https://doi.org/10.1016/j.ncrops.2023.12.005

Fo, K., Chuah, Y. S., Foo, H., Davey, E. E., Fullwood, M., Thibault, G., & Mutwi, M. (2023). PlantConnectome: Knowledge networks encompassing> 100,000 plant article abstracts. *bioRxiv*, pp. 2023–07. https://doi.org/10.1101/2023.07.11.548541

Fu, L., Peng, Q., & Chai, L. (2019). Predicting DNA methylation states with a hybrid information based deep-learning model. *IEEE/ACM Transactions on Computational Biology and Bioinformatics, 17*(5), 1721–1728. https://doi.org/10.1109/TCBB.2019.2909237

Fu, J., Zheng, S., Fan, L., Zheng, X., & Qian, Q. (2025). Breeding 5.0: Artificial intelligence (AI)-decoded germplasm for accelerated crop innovation. *Journal of Integrative Plant Biology*, 1–13. https://doi.org/10.1111/jipb.70008

Gao, P., Lian, L., Feng, W., Ma, Y., Lin, J., Qin, L., Hao, S., Zhao, H., Liu, X., Yuan, J., Lin, Z., Li, X., Guan, Y., & Wang, X. (2025). Deciphering the sequence basis and application of transcriptional initiation regulation in plant genomes through deep learning. *Genome Biology, 26*(1), 293. https://doi.org/10.1186/s13059-025-03782-5

Gardner, S. T., Bertucci, E. M., Sutton, R., Horcher, A., Aubrey, D., & Parrott, B. B. (2023). Development of DNA methylation-based epigenetic age predictors in loblolly pine (*Pinus taeda*). *Molecular Ecology Resources, 23*(1), 131–144. https://doi.org/10.1111/1755-0998.13698

Gill, M., Anderson, R., Hu, H., Bennamoun, M., Petereit, J., Valliyodan, B., Nguyen, H. T., Batley, J., Bayer, P. E., & Edwards, D. (2022). Machine learning models outperform deep learning models, provide interpretation and facilitate feature selection for soybean trait prediction. *BMC Plant Biology*, *22*(1), 180. https://doi.org/10.1186/s12870-022-03559-z

Gupta, K., & Garg, R. (2020). Method for bisulfite sequencing data analysis for whole-genome level DNA methylation detection in legumes. In M. Jain & R. Garg (Eds.), *Legume genomics: Methods and protocols* (pp. 127–145). Springer US.

Gupta, D. K., Pagani, A., Zamboni, P., & Singh, A. K. (2024). AI-powered revolution in plant sciences: Advancements, applications, and challenges for sustainable agriculture and food security. *Exploration of Foods and Foodomics*, *2*(5), 443–459. https://doi.org/10.37349/eff.2024.00045

Han, M., Lin, S., Zhu, B., Tong, W., Xia, E., Wang, Y., Yang, T., Zhang, S., Wan, X., Liu, J., Niu, Q., Zhu, J., Bao, S., & Zhang, Z. (2024). Dynamic DNA methylation regulates season-dependent secondary metabolism in the new shoots of Tea plants. *Journal of Agricultural and Food Chemistry*, *72*(8), 3984–3997. https://doi.org/10.1021/acs.jafc.3c08568

Hansen, P. B., Ruud, A. K., de Los Campos, G., Malinowska, M., Nagy, I., Svane, S. F., Thorup-Kristensen, K., Jensen, J. D., Krusell, L., & Asp, T. (2022). Integration of DNA methylation and transcriptome data improves complex trait prediction in *hordeum vulgare*. *Plants*, *11*(17), 2190. https://doi.org/10.3390/plants11172190

Hashim, A., Rizzo, F., Marchese, G., Ravo, M., Tarallo, R., Nassa, G., Giurato, G., Santamaria, G., Cordella, A., Cantarella, C., & Weisz, A. (2014). RNA sequencing identifies specific PIWI-interacting small non-coding RNA expression patterns in breast cancer. *Oncotarget*, *5*(20), 9901–9910. https://doi.org/10.18632/oncotarget.2476

Haug, T., Dumke, R., Kwek, L. C., Miniatura, C., & Amico, L. (2021). Machine-learning engineering of quantum currents. *Physical Review Research*, *3*(1), 013034. https://doi.org/10.1103/PhysRevResearch.3.013034

Hawkins-Hooker, A., Visonà, G., Narendra, T., Rojas-Carulla, M., Schölkopf, B., & Schweikert, G. (2023). Getting personal with epigenetics: Towards individual-specific epigenomic imputation with machine learning. *Nature Communications*, *14*(1), 4750. https://doi.org/10.1038/s41467-023-40211-2

Holoch, D., & Moazed, D. (2015). RNA-mediated epigenetic regulation of gene expression. *Nature Reviews Genetics*, *16*(2), 71–84. https://doi.org/10.1038/nrg3863

Hovenga, V., Kalita, J., & Oluwadare, O. (2023). Hic-gnn: A generalizable model for 3d chromosome reconstruction using graph convolutional neural networks. *Computational and Structural Biotechnology Journal*, *21*, 812–836. https://doi.org/10.1016/j.csbj.2022.12.051

Hu, W., Guan, L., & Li, M. (2023). Prediction of DNA methylation based on multi-dimensional feature encoding and double convolutional fully connected convolutional neural network. *PLoS Computational Biology*, *19*(8), e1011370. https://doi.org/10.1371/journal.pcbi.10.1370

Huang, Y., Liu, J., Cheng, L., Xu, D., Liu, S., Hu, H., Ling, Y., Yang, R., & Zhang, Y. (2024). Genome-wide analysis of the histone modification gene (HM) family and expression investigation during anther development in rice (*Oryza sativa* L.). *Plants*, *13*(17), 2496. https://doi.org/10.3390/plants13172496

Hüther, P., Hagmann, J., Nunn, A., Kakoulidou, I., Pisupati, R., Langenberger, D., Weigel, D., Johannes, F., Schultheiss, S. J., & Becker, C. (2022). MethylScore, a pipeline for accurate and context-aware identification of differentially methylated regions from population-scale plant whole-genome bisulfite sequencing data. *Quantitative Plant Biology*, *3*, e19. https://doi.org/10.1017/qpb.2022.14

Hyun, T. K. (2022). Plant histone modifications in response to cold stress. *Botanica Serbica, 46*(1), 1–6. https://doi.org/10.2298/BOTSERB2201001H

Ito, H. (2014). Plant models of transgenerational epigenetic inheritance. In T. Tollefsbol (Ed.), *Transgenerational epigenetics* (pp. 147–161). Academic Press.

Jha, A., Bohaczuk, S. C., Mao, Y., Ranchalis, J., Mallory, B. J., Min, A. T., Hamm, M. O., Swanson, E., Dubocanin, D., Finkbeiner, C., Li, T., Whittington, D., Noble, W. S., Stergachis, A. B., & Vollger, M. R. (2024). DNA-m6A calling and integrated long-read epigenetic and genetic analysis with fibertools. *Genome Research, 34*(11), 1976–1986. https://www.genome.org/cgi/doi/10.1101/gr.279095.124

Ji, Y., Zhou, Z., Liu, H., & Davuluri, R. V. (2021). DNABERT: Pre-trained bidirectional encoder representations from transformers model for DNA-language in genome. *Bioinformatics, 37*(15), 2112–2120. https://doi.org/10.1093/bioinformatics/btab083

Jin, Q., Chachar, M., Ali, A., Chachar, Z., Zhang, P., Riaz, A., Ahmed, N., & Chachar, S. (2024). Epigenetic regulation for heat stress adaptation in plants: New horizons for crop improvement under climate change. *Agronomy, 14*(9), 2105. https://doi.org/10.3390/agronomy14092105

Kababji, A. M., Butt, H., & Mahfouz, M. (2024). Synthetic directed evolution for targeted engineering of plant traits. *Frontiers in Plant Science, 15*, 1449579. https://doi.org/10.3389/fpls.2024.1449579

Kajrolkar, A. (2025). Deep learning applications in plant biology: Bridging genotype and phenotype. *Plant Biology, 3*, 100011. https://doi.org/10.70389/PJPB.100011

Kakoulidou, I., Avramidou, E. V., Baránek, M., Brunel-Muguet, S., Farrona, S., Johannes, F., Kaiserli, E., Lieberman-Lazarovich, M., Martinelli, F., Mladenov, V., & Testillano, P. S. (2021). Epigenetics for crop improvement in times of global change. *Biology, 10*(8), 766. https://doi.org/10.3390/biology10080766

Kang, Q., Meng, J., Shi, W., & Luan, Y. (2021). Ensemble deep learning based on multi-level information enhancement and greedy fuzzy decision for plant miRNA-lncRNA interaction prediction. *Interdisciplinary Sciences: Computational LIfe Sciences, 13*(4), 603–614. https://doi.org/10.1007/s12539-021-00434-7

Kang, H., Fan, T., Wu, J., Zhu, Y., & Shen, W. H. (2022). Histone modification and chromatin remodelling in plant response to pathogens. *Frontiers in Plant Science, 13*, 986940. https://doi.org/10.3389/fpls.2022.986940

Karikari, B., Lemay, M. A., & Belzile, F. (2023). K-mer-based genome-wide association studies in plants: Advances, challenges, and perspectives. *Genes, 14*(7), 1439. https://doi.org/10.3390/genes14071439

Kassem, M. A. (2025). Harnessing artificial intelligence and machine learning for identifying quantitative trait loci (QTL) associated with seed quality traits in crops. *Plants, 14*(11), 1727. https://doi.org/10.3390/plants14111727

Kaushik, P. (2025). The quantum-AI genomic frontier platform: Revolutionizing vegetable crop resilience in a changing climate. https://doi.org/10.20944/preprints202503.1073.v1

Kim, M. G., Go, M. J., Kang, S. H., Jeong, S. H., & Lim, K. (2025). Revolutionizing CRISPR technology with artificial intelligence. *Experimental & Molecular Medicine, 57*, 1–13. https://doi.org/10.1038/s12276-025-01462-9

Kinoshita, T., & Seki, M. (2014). Epigenetic memory for stress response and adaptation in plants. *Plant and Cell Physiology, 55*(11), 1859–1863. https://doi.org/10.1093/pcp/pcu125

Kong, C., Yang, Y., Qi, T., & Zhang, S. (2025). Predictive genetic circuit design for phenotype reprogramming in plants. *Nature Communications*, *16*(1), 715. https://doi.org/10.1038/s41467-025-56042-2

Krasteva, M. E. (2024). Environmental stress, epigenetic modifications, adaptation, and disease: A fine interplay. In K. Ikehara & M. Seifi (Eds.), *Beyond the blueprint-decoding the elegance of gene expression: Decoding the elegance of gene expression* (p. 27).

Küçükrecep, A., Yıldız, Ş., Tekdal, D., & Lucas, S. (2024). Deep learning for genomics and epi-genomics in horticulture crops improvement. In K. A. Abd-Elsalam, A. Ahmad, & B. Zhang (Eds.), *CRISPRized horticulture crops* (pp. 217–232). Academic Press.

Kumar Meher, P., Hati, S., Sahu, T. K., Pradhan, U., Gupta, A., & Rath, S. N. (2024). SVM-root: Identification of root-associated proteins in plants by employing the support vector machine with sequence-derived features. *Current Bioinformatics*, *19*(1), 91–102. https://doi.org/10.2174/1574893618666230417104543

Kumar, A., Pathak, R. K., Gupta, S. M., Gaur, V. S., & Pandey, D. (2015). Systems biology for smart crops and agricultural innovation: Filling the gaps between genotype and phenotype for complex traits linked with robust agricultural productivity and sustainability. *OMICS: A Journal of Integrative Biology*, *19*(10), 581–601. https://doi.org/10.1089/omi.2015.0106

Kundariya, H., Sanchez, R., Yang, X., Hafner, A., & Mackenzie, S. A. (2022). Methylome decoding of RdDM-mediated reprogramming effects in the arabidopsis MSH1 system. *Genome Biology*, *23*(1), 167. https://doi.org/10.1186/s13059-022-02731-w

Kusharki, M. B., Muhammad-Bello, B. L., Misra, S., Liman, M. M., & Blamah, N. (2026). Federated graph attention neural networks (FeGAN) for privacy-preserving crop disease detection. In S. Misra, P. Kvalvik, K. M. Kjølerbakken, & P.-A. Jørgensen (Eds.), *Empowering digital sovereignty: Balancing privacy and global connectivity* (pp. 151–192). IGI Global Scientific Publishing.

Lagiotis, G., Madesis, P., & Stavridou, E. (2023). Echoes of a stressful past: Abiotic stress memory in crop plants towards enhanced adaptation. *Agriculture*, *13*(11), 2090. https://doi.org/10.3390/agriculture13112090

Lakhani, K. G., Hamid, R., & Marviya, G. V. (2025). Perspective chapter: Long non-coding RNAs in plants-unlocking hidden regulators for stress tolerance and crop resilience. In Y. Tutar, L. Tutar, & K. Ikehara (Eds.), *Long non-coding RNAs-function, mechanisms, and applications*. IntechOpen.

Lee, M. (2023). Recent advances in generative adversarial networks for gene expression data: A comprehensive review. *Mathematics*, *11*(14), 3055. https://doi.org/10.3390/math11143055

Lee, J. H., Mazarei, M., Pfotenhauer, A. C., Dorrough, A. B., Poindexter, M. R., Hewezi, T., Lenaghan, S. C., Graham, D. E., & Stewart, C. N., Jr. (2020). Epigenetic footprints of CRISPR/Cas9-mediated genome editing in plants. *Frontiers in Plant Science*, *10*, 1720. https://doi.org/10.3389/fpls.2019.01720

Lei, J., Nie, Q., & Chen, D. B. (2018). A single-cell epigenetic model for paternal psychological stress-induced transgenerational reprogramming in offspring. *Biology of Reproduction*, *98*(6), 846–855. https://doi.org/10.1093/biolre/ioy050

Li, Q., Hermanson, P. J., & Springer, N. M. (2017). Detection of DNA methylation by whole-genome bisulfite sequencing. In L. M. Lagrimini (Ed.), *Maize: Methods and protocols* (pp. 185–196). Springer New York.

Li, Z., Khan, W. U., Bai, G., Dong, C., Wang, J., Zhang, Y., Wang, C., Zhang, H., Wang, W., Luo, M., & Chen, F. (2025). From code to life: The AI-driven revolution in genome editing. *Advanced Science, 12*, e17029. https://doi.org/10.1002/advs.202417029

Liu, Y., Liu, Z., Jiang, T., Zang, T., & Wang, Y. (2022, December 6–8). Comparison of the nanopore and PacBio sequencing technologies for DNA 5-methylcytosine detection. *2022 IEEE International Conference on Bioinformatics and Biomedicine (BIBM)* (pp. 220–225), Las Vegas, NV, USA. IEEE.

Lloyd, J. P., Ly, F., Gong, P., Pflüger, J., Swain, T., Pflüger, C., Khan, M. A., Kidd, B., & Lister, R. (2022). Synthetic memory circuits for programmable cell reconfiguration in plants. *bioRxiv*, pp. 2022-02. https://doi.org/10.1101/2022.02.11.480167

Lu, W., Xiao, L., Quan, M., Wang, Q., El-Kassaby, Y. A., Du, Q., & Zhang, D. (2020). Linkage-linkage disequilibrium dissection of the epigenetic quantitative trait loci (epiQTLs) underlying growth and wood properties in Populus. *New Phytologist, 225*(3), 1218–1233. https://doi.org/10.1111/nph.16220

Luo, P., Deng, X., Wen, Z., Sun, T., & Li, D. (2024). Federated prediction-powered inference from decentralized data. *arXiv preprint arXiv:2409.01730*. https://doi.org/10.48550/arXiv.2409.01730

Madrigal, P., Dai, X., & Hadjipantelis, P. Z. (2018). Sparse functional data analysis accounts for missing information in single-cell epigenomics. *bioRxiv*, p. 504365. https://doi.org/10.1101/504365

Magar, N. D., Shah, P., Barbadikar, K. M., Bosamia, T. C., Madhav, M. S., Mangrauthia, S. K., Pandey, M. K., Sharma, S., Shanker, A. K., Neeraja, C. N., & Sundaram, R. M. (2024). Long non-coding RNA-mediated epigenetic response for abiotic stress tolerance in plants. *Plant Physiology and Biochemistry, 206*, 108165. https://doi.org/10.1016/j.plaphy.2023.10.165

Mahood, E. H., Kruse, L. H., & Moghe, G. D. (2020). Machine learning: A powerful tool for gene function prediction in plants. *Applications in Plant Sciences, 8*(7), e11376. https://doi.org/10.1002/aps3.11376

Majumdar, P. G., Rao, A. R., Kairi, A., Meher, P. K., & Sahu, S. (2022). Identification of efficient learning classifiers for discrimination of coding and non-coding RNAs in plant species. *Indian Journal Of Genetics And Plant Breeding, 82*(3), 280–288. https://doi.org/10.31742/ISGPB.82.3.2

Makai, D., Cseh, A., Sepsi, A., & Makai, S. (2022). A multigraph-based representation of Hi-C data. *Genes, 13*(12), 2189. https://doi.org/10.3390/genes13122189

McCoy, R. M., Julian, R., Kumar, S. R., Ranjan, R., Varala, K., & Li, Y. (2021). A systems biology approach to identify essential epigenetic regulators for specific biological processes in plants. *Plants, 10*(2), 364. https://doi.org/10.3390/plants10020364

Mendieta, J. P., Sangra, A., Yan, H., Minow, M. A., & Schmitz, R. J. (2023). Exploring plant cis-regulatory elements at single-cell resolution: Overcoming biological and computational challenges to advance plant research. *The Plant Journal, 115*(6), 1486–1499. https://doi.org/10.1111/tpj.16351

Meng, J., Kang, Q., Chang, Z., & Luan, Y. (2021). Plncrna-Hdeep: Plant long noncoding RNA prediction using hybrid deep learning based on two encoding styles. *BMC Bioinformatics, 22*(Suppl 3), 242. https://doi.org/10.1186/s12859-020-03870-2

Messina, C. D., Technow, F., Tang, T., Totir, R., Gho, C., & Cooper, M. (2018). Leveraging biological insight and environmental variation to improve phenotypic prediction: Integrating crop growth models (CGM) with whole genome prediction (WGP). *European Journal of Agronomy, 100*, 151–162. https://doi.org/10.1016/j.eja.2018.01.007

Mishra, B., Kumar, N., Liu, J., & Pajerowska-Mukhtar, K. M. (2021). Dynamic regulatory event mining by iDREM in large-scale multi-omics datasets during biotic and abiotic stress in plants. In S. Mukhtar (Ed.), *Modelling transcriptional regulation: Methods and protocols* (pp. 191–202). Springer US.

Moradpour, M., & Abdulah, S. N. A. (2020). CRISPR/dC as9 platforms in plants: Strategies and applications beyond genome editing. *Plant Biotechnology Journal, 18*(1), 32–44. https://doi.org/10.1111/pbi.13232

Mukherjee, A. (2025, May 17–18). A novel quantum-enhanced algorithm for dynamic soil microbiome optimization to boost agriculture. *EPJ Web of Conferences,* Amravati, Maharashtra, India (Vol. 328, p. 01056). EDP Sciences.

Munawar, N., Faheem, M., Niamat, A., Munir, A., Khan, S. H., Zahoor, M. K., Aslam, A., & Ahmad, A. (2024). Regulatory, ethical, social, and biosafety concerns in genome-edited horticultural crops. In K. A. Abd-Elsalam, A. Ahmad, & B. Zhang (Eds.), *CRISPRized horticulture crops* (pp. 421–438). Academic Press.

Murmu, S., Sinha, D., Chaurasia, H., Sharma, S., Das, R., Jha, G. K., & Archak, S. (2024). A review of artificial intelligence-assisted omics techniques in plant defense: Current trends and future directions. *Frontiers in Plant Science, 15*, 1292054. https://doi.org/10.3389/fpls.2024.1292054

Nobori, T., Monell, A., Lee, T. A., Zhou, J., Nery, J., & Ecker, J. R. (2023). Time-resolved single-cell and spatial gene regulatory atlas of plants under pathogen attack. *bioRxiv*, pp. 2023-04. https://doi.org/10.1101/2023.04.10.536170

Ostrowska-Mazurek, A., Kasprzak, P., Kubala, S., Zaborowska, M., & Sobieszczuk-Nowicka, E. (2020). Epigenetic landmarks of leaf senescence and crop improvement. *International Journal of Molecular Sciences, 21*(14), 5125. https://doi.org/10.3390/ijms21145125

Ouyang, W., Luan, S., Xiang, X., Guo, M., Zhang, Y., Li, G., & Li, X. (2022). Profiling plant histone modification at single-cell resolution using snCUT&tag. *Plant Biotechnology Journal, 20*(3), 420–422. https://doi.org/10.1111/pbi.13768

Pandita, D., & Pandita, A. (Eds.). (2023). *CRISPR/Cas-mediated genome editing in plants.* CRC Press.

Park, S. Y., Cho, J., & Jeong, D. H. (2022). Small regulatory RNAs in rice epigenetic regulation. *Biochemical Society Transactions, 50*(3), 1215–1225. https://doi.org/10.1042/BST20210336

Peleke, F. F., Zumkeller, S. M., Gültas, M., Schmitt, A., & Szymański, J. (2024). Deep learning the cis-regulatory code for gene expression in selected model plants. *Nature Communications, 15*(1), 3488. https://doi.org/10.1038/s41467-024-47744-0

Polegopu, R., Vanapalli, S. S., Vanapalli, S. V., Diyya, N. P., Vandila, M., Valluri, D., Peddinti, A., Saladi, S., Pyla, M., & Gelli, P. (2025). Integrating genomics & AI for precision crop monitoring and adaptive stress management. *Informatyka, Automatyka, Pomiary w Gospodarce i Ochronie Środowiska, 15*(2), 18–26. https://doi.org/10.35784/iapgos.7484

Rai, K. K. (2022). Integrating speed breeding with artificial intelligence for developing climate-smart crops. *Molecular Biology Reports, 49*(12), 11385–11402. https://doi.org/10.1007/s11033-022-07769-4

Raimondi, D., Chizari, H., Verplaetse, N., Loescher, B. S., Franke, A., & Moreau, Y. (2023). Genome interpretation in a federated learning context allows the multi-center exome-based risk prediction of Crohn's disease patients. *Scientific Reports, 13*(1), 19449. https://doi.org/10.1038/s41598-023-46887-2

Ramachandra, C. N., Nag, A., Balasubramonion, R., Kalsi, G., Pillai, K., & Subramoney, S. (2021, May 9–12). ONT-X: An FPGA approach to real-time portable genomic analysis. *2021 IEEE 29th Annual International Symposium on Field-Programmable Custom Computing Machines (FCCM),* Orlando, FL, USA (pp. 268–269). IEEE.

Ramirez-Prado, J. S., Ariel, F., Benhamed, M., & Crespi, M. (2017). Plant epigenetics: Non-coding RNAs as emerging regulators. In N. Rajewsky, S. Jurga, & J. Barciszewski (Eds.), *Plant epigenetics* (pp. 129–147). Springer International Publishing.

Raza, A., Bashir, S., Khare, T., Karikari, B., Copeland, R. G., Jamla, M., Abbas, S., Charagh, S., Nayak, S. N., Djalovic, I., & Rivero, R. M. (2024). Temperature-smart plants: A new horizon with omics-driven plant breeding. *Physiologia Plantarum, 176*(1), e14188. https://doi.org/10.1111/ppl.14188

Ricci, W. A., Levin, L., & Zhang, X. (2019). Genome-Wide profiling of histone modifications with ChIP-Seq. In L. M. Vaschetto (Ed.), *Cereal genomics: Methods and protocols* (pp. 101–117). Springer US.

Rockenbach, K. C., Zanini, S. F., Morris, R. J., Wells, R. J., & Golicz, A. A. (2025). A deep learning model recapitulates position specific effects of plant regulatory sequences and suggests genes under complex regulation. *bioRxiv*, pp. 2025-08. https://doi.org/10.1101/2025.08.30.673246

Roosan, D., Khan, R., Ashakin, M. R., Khou, T., Nirzhor, S., & Haider, M. R. (2025). Quantum variational transformer model for enhanced cancer classification. *arXiv preprint arXiv:2506.21641*. https://doi.org/10.48550/arXiv.2506.21641

Roychowdhury, R., Das, S. P., Gupta, A., Parihar, P., Chandrasekhar, K., Sarker, U., Kumar, A., Ramrao, D. P., & Sudhakar, C. (2023). Multi-omics pipeline and omics-integration approach to decipher plant's abiotic stress tolerance responses. *Genes, 14*(6), 1281. https://doi.org/10.3390/genes14061281

Saeed, F., Chaudhry, U. K., Bakhsh, A., Raza, A., Saeed, Y., Bohra, A., & Varshney, R. K. (2022). Moving beyond DNA sequence to improve plant stress responses. *Frontiers in Genetics, 13*, 874648. https://doi.org/10.3389/fgene.2022.874648

Saito, T., Wang, S., Ohkawa, K., Ohara, H., & Kondo, S. (2024). Deep learning with a small dataset predicts chromatin remodelling contribution to winter dormancy of apple axillary buds. *Tree Physiology, 44*(7), tpae072. https://doi.org/10.1093/treephys/tpae072

Salih, A. M., Raisi-Estabragh, Z., Galazzo, I. B., Radeva, P., Petersen, S. E., Lekadir, K., & Menegaz, G. (2025). A perspective on explainable artificial intelligence methods: SHAP and LIME. *Advanced Intelligent Systems, 7*(1), 2400304. https://doi.org/10.1002/aisy.202400304

Schiele, M., Salvador, O. C., Köttgen, A., Baune, B., Deckert, J., & Domschke, K. (2024). 'Therapyepigenetics': Epigenetic predictors and signatures of treatment response in stress-related disorders. *Biological Psychiatry, 95*(10), S41. https://doi.org/10.1016/j.biopsych.2024.02.103

Schultz, E. R., Kyhl, S., Willett, R., & de Pablo, J. J. (2025). Chromatin structures from integrated AI and polymer physics model. *PLoS Computational Biology, 21*(4), e1012912. https://doi.org/10.1371/journal.pcbi.1012912

Schwope, R., Magris, G., Miculan, M., Paparelli, E., Celii, M., Tocci, A., Marroni, F., Fornasiero, A., De Paoli, E., & Morgante, M. (2021). Open chromatin in grapevine marks candidate CREs and with other chromatin features correlates with gene expression. *The Plant Journal, 107*(6), 1631–1647. https://doi.org/10.1111/tpj.15404

Shen, N., & Korthauer, K. (2024). vmrseq: Probabilistic modelling of single-cell methylation heterogeneity. *Genome Biology, 25*(1), 321. https://doi.org/10.1186/s13059-024-03457-7

Shih, P. J., Saadat, H., Parameswaran, S., & Gamaarachchi, H. (2023). Efficient real-time selective genome sequencing on resource-constrained devices. *GigaScience, 12*, giad046. https://doi.org/10.1093/gigascience/giad046

Shinde, H., Raihan, T., Anand, L., Perry, S. E., Geneve, R. L., & Rodriguez Lopez, C. M. (2023). Multi-omics insights into grapevine ecodormancy to budburst transition: Interplay of gene expression, miRNA regulation, and DNA methylation. *bioRxiv*, pp. 2023-10. https://doi.org/10.1101/2023.10.21.563414

Skobelev, P., Tabachinskiy, A., Simonova, E., Strizhakov, A., Ermakov, V., Goryanin, O., & Zhuravel, Y. (2023). Developing digital twin of plant based on emergent intelligence concept, ontology and multi-agent technology. In T. Borangiu, D. Trentesaux, P. Leitão, L. Berrah, & J.-F. Jimenez (Eds.), *International workshop on service orientation in holonic and multi-agent manufacturing* (pp. 430–441). Springer Nature Switzerland.

Slack, F. J. (2014). In this issue of epigenetics: Special focus on non-coding RNAs in epigenetic regulation. *Epigenetics, 9*(1), 1–2. https://doi.org/10.4161/epi.27579

Spanou, V. M., Andriopoulou, T. P., Giamarellos-Bourboulis, E. J., & Netea, M. G. (2025). Improving the odds of survival: Transgenerational effects of infections. *EMBO Molecular Medicine, 17*(4), 609–624. https://doi.org/10.1038/s44321-025-00192-9

Stephan, G., Dugdale, B., Deo, P., Harding, R., Dale, J., & Visendi, P. (2021). Bridging functional annotation gaps in non-model plant genes with AlphaFold, DeepFRI and small molecule docking. *bioRxiv*, pp. 2021-12. https://doi.org/10.1101/2021.12.22.473925

Sun, M. (2025). Advances in plant epigenetic regulation of abiotic stress response. *Theoretical and Natural Science, 90*, 81–87. https://doi.org/10.54254/2753-8818/2025.GU20455

Suruthi, G. S., Umera Begam, A. K., Arun, M., Robert, A. A., Sampathkumar, P., Gowtham, P., Arthi, B., & Anand, A. V. (2025). Reprogramming epigenomes by CRISPR-based technology against plant diseases. In J.-T. Chen (Ed.), *Stress-resilient crops: Coordinated omics-CRISPR-nanotechnology strategies* (pp. 325–341). CABI.

Suzuki, S., Horie, K., Amagasa, T., & Fukuda, N. (2025). Genomic language models with k-mer tokenization strategies for plant genome annotation and regulatory element strength prediction. *Plant Molecular Biology, 115*(4), 100. https://doi.org/10.1007/s11103-025-01604-7

Tahir, M., Norouzi, M., Khan, S. S., Davie, J. R., Yamanaka, S., & Ashraf, A. (2024). Artificial intelligence and deep learning algorithms for epigenetic sequence analysis: A review for epigeneticists and AI experts. *Computers in Biology and Medicine, 183*, 109302. https://doi.org/10.1016/j.compbiomed.2024.109302

Targonski, C., Shealy, B. T., Smith, M. C., & Feltus, F. A. (2019). Cellular state transformations using generative adversarial networks. *arXiv preprint arXiv:1907.00118.*

Thibivilliers, S., & Libault, M. (2021). Enhancing our understanding of plant cell-to-cell interactions using single-cell omics. *Frontiers in Plant Science, 12*, 696811. https://doi.org/10.3389/fpls.2021.696811

Thingujam, D., Gouli, S., Cooray, S. P., Chandran, K. B., Givens, S. B., Gandhimeyyan, R. V., Tan, Z., Wang, Y., Patam, K., Greer, S. A., & Acharya, R. (2025). Climate-resilient crops: Integrating AI, multi-omics, and advanced phenotyping to address global agricultural and societal challenges. *Plants*, *14*(17), 2699. https://doi.org/10.3390/plants14172699

Tremblay, B. J., & Qüesta, J. I. (2023). Mechanisms of epigenetic regulation of transcription by lncRNAs in plants. *IUBMB Life*, *75*(5), 427–439. https://doi.org/10.1002/iub.2681

Tross, M. C., Duggan, G., Shrestha, N., & Schnable, J. C. (2024). Models trained to predict differential expression across plant organs identify distal and proximal regulatory regions. *bioRxiv*, pp. 2024-06. https://doi.org/10.1101/2024.06.04.597477

Udegbe, F. C., Nwankwo, E. I., Igwama, G. T., & Olaboye, J. A. (2023). Real-time data integration in diagnostic devices for predictive modelling of infectious disease outbreaks. *Computer Science & IT Research Journal*, *4*(3), 525–545. https://doi.org/10.51594/csitrj.v4i3.1502

Veley, K. M., Elliott, K., Jensen, G., Zhong, Z., Feng, S., Yoder, M., Gilbert, K. B., Berry, J. C., Ghoshal, B., Bartolomé, J. G., & Norton, J. (2022). Improving disease resistance in plants by editing the epigenome. bioRxiv, pp. 2022-01. https://doi.org/10.1101/2022.01.04.474801

Vetrivel, S. C., & Arun, V. P. (2025). Smart farming and precision agriculture using AI technologies. In S. Mallik, S. K. Mathivanan, S. K. B. Sangeetha, & B. O. Soufiene (Eds.), *Real-world applications of AI innovation* (pp. 85–106). IGI Global Scientific Publishing.

Vyska, M., Cunniffe, N., & Gilligan, C. (2016). Trade-off between disease resistance and crop yield: A landscape-scale mathematical modelling perspective. *Journal of the Royal Society Interface*, *13*(123), 20160451. https://doi.org/10.1098/rsif.2016.0451

Wang, D., & Deal, R. B. (2015). Epigenome profiling of specific plant cell types using a streamlined INTACT protocol and ChIP-seq. In J. M. Alonso & A. N. Stepanova (Eds.), *Plant functional genomics: Methods and protocols* (pp. 3–25). Springer New York.

Wang, Y., & Dong, S. (2021). A new roadmap for the breeding of disease-resistant and high-yield crops. *Stress Biology*, *1*(1), 21. https://doi.org/10.1007/s44154-021-00023-0

Wang, Y., Zhang, P., Guo, W., Liu, H., Li, X., Zhang, Q., Du, Z., Hu, G., Han, X., Pu, L., & Tian, J. (2021). A deep learning approach to automate whole-genome prediction of diverse epigenomic modifications in plants. *New Phytologist*, *232*(2), 880–897. https://doi.org/10.1111/nph.17630

Wang, M., Shilo, S., Zhou, A., Zelkowski, M., Olson, M. A., Azuri, I., Shoshani-Hechel, N., Melamed-Bessudo, C., Marand, A.P., Jiang, J., & Schnable, J. C. (2022). Machine learning reveals conserved chromatin patterns determining meiotic recombination sites in plants. *bioRxiv*, pp. 2022-07. https://doi.org/10.1101/2022.07.11.499557

Wang, M., Zhang, Z., Bedi, A. S., Velasquez, A., Guerra, S., Lin-Gibson, S., Cong, L., Qu, Y., Chakraborty, S., Blewett, M., Ma, J., Xing, E., & Church, G. (2025). A call for built-in biosecurity safeguards for generative AI tools. *Nature Biotechnology*, *43*(6), 845–847. https://doi.org/10.1038/s41587-025-02650-8

Wibowo, A., Becker, C., Marconi, G., Durr, J., Price, J., Hagmann, J., Papareddy, R., Putra, H., Kageyama, J., Becker, J., & Weigel, D. (2016). Hyperosmotic stress memory in arabidopsis is mediated by distinct epigenetically labile sites in the genome and is restricted in the male germline by DNA glycosylase activity. *eLife*, *5*, e13546. https://doi.org/10.7554/eLife.13546

Wyschogrod, D., Murphy, S.T., Beal, J., & Taggart, A. (2023, December 5–8). Securing fieldable bioinformatics. *2023 IEEE International Conference on Bioinformatics and Biomedicine (BIBM),* Istanbul, Turkey (pp. 3390–3397). IEEE.

Xiang, G., Keller, C. A., Giardine, B., An, L., Li, Q., Zhang, Y., & Hardison, R. C. (2020). S3norm: Simultaneous normalization of sequencing depth and signal-to-noise ratio in epigenomic data. *Nucleic Acids Research, 48*(8), e43–e43. https://doi.org/10.1093/nar/gkaa105

Xie, S., Leung, A. W. S., Zheng, Z., Zhang, D., Xiao, C., Luo, R., Luo, M., & Zhang, S. (2021). Applications and potentials of nanopore sequencing in the (epi) genome and (epi) transcriptome era. *The Innovation, 2*(4), 100153. https://doi.org/10.1016/j.xinn.2021.10.153

Xu, X., Hoang, S., Mayo, M. W., & Bekiranov, S. (2010). Application of machine learning methods to histone methylation ChIP-Seq data reveals H4R3me2 globally represses gene expression. *BMC Bioinformatics, 11*(1), 396. https://doi.org/10.1186/1471-2105-11-396

Xue, Y., Cao, X., Chen, X., Deng, X., Deng, X. W., Ding, Y., Dong, A., Duan, C. G., Fang, X., Gong, L., & Gong, Z. (2025). Epigenetics in the modern era of crop improvements. *Science China Life Sciences, 68*, 1570–1609. https://doi.org/10.1007/s11427-024-2784-3

Yan, F., Jiang, L., Chen, D., Ceccarelli, M., & Guo, Y. (2024). Reinventing gene expression connectivity through regulatory and spatial structural empowerment via principal node aggregation graph neural network. *Nucleic Acids Research, 52*(13), e60–e60. https://doi.org/10.1093/nar/gkae514

Yang, L., Liang, Q., Zhang, L., Lv, X., Tian, S., Xiong, Y., Zhang, Z., Huang, M., Feng, C., Yang, H., & Hu, Y. (2025). Genome-wide characterization of the accessible chromatin regions in chickpea. https://doi.org/10.20944/preprints202502.0571.v1

Yong, S., Li, W., Tong, L., Sun, D., & Yang, X. (2023). De novo miRNA-lncRNA interactions prediction via K-square encoding and mixed network. In Y. Yu, B. P. Nguyen, & J. Sang (Eds.), *Advances in biomedical and bioinformatics engineering* (pp. 496–504). IOS Press.

Yung, W. S., Li, M. W., Sze, C. C., Wang, Q., & Lam, H. M. (2021). Histone modifications and chromatin remodelling in plants in response to salt stress. *Physiologia Plantarum, 173*(4), 1495–1513. https://doi.org/10.1111/ppl.13467

Zhai, J., Gokaslan, A., Hsu, S. K., Chen, S. P., Liu, Z. Y., Marroquin, E., Czech, E., Cannon, B., Berthel, A., Romay, C., & Pennell, M. (2025). PlantCAD2: A long-context DNA language model for cross-species functional annotation in angiosperms. *bioRxiv*, pp. 2025-08. https://doi.org/10.1101/2025.08.27.672609

Zhang, P., Meng, J., Luan, Y., & Liu, C. (2020). Plant miRNA-lncRNA interaction prediction with the ensemble of CNN and IndRNN. *Interdisciplinary Sciences: Computational LIfe Sciences, 12*(1), 82–89. https://doi.org/10.1007/s12539-019-00351-w

Zhang, Y., Andrews, H., Eglitis-Sexton, J., Godwin, I., Tanurdžić, M., & Crisp, P. A. (2022). Epigenome guided crop improvement: Current progress and future opportunities. *Emerging Topics in Life Sciences, 6*(2), 141–151. https://doi.org/10.1042/ETLS20210258

Zhao, H., Tu, Z., Liu, Y., Zong, Z., Li, J., Liu, H., Xiong, F., Zhan, J., Hu, X., & Xie, W. (2021). PlantDeepSEA, a deep learning-based web service to predict the regulatory effects of genomic variants in plants. *Nucleic Acids Research, 49*(W1), W523–W529. https://doi.org/10.1093/nar/gkab383

Zhi, P., & Chang, C. (2021). Exploiting epigenetic variations for crop disease resistance improvement. *Frontiers in Plant Science*, *12*, 692328. https://doi.org/10.3389/fpls.2021.692328

Zhou, C., Zhou, H., Ma, X., Yang, H., Wang, P., Wang, G., Zheng, L., Zhang, Y., & Liu, X. (2021). Genome-wide identification and characterization of main histone modifications in sorghum decipher regulatory mechanisms involved by mRNA and long noncoding RNA genes. *Journal of Agricultural and Food Chemistry*, *69*(7), 2337–2347. https://doi.org/10.1021/acs.jafc.0c07035

Zou, Z., Ohta, T., Miura, F., & Oki, S. (2022). ChIP-Atlas 2021 update: A data-mining suite for exploring epigenomic landscapes by fully integrating ChIP-seq, ATAC-seq and Bisulfite-seq data. *Nucleic Acids Research*, *50*(W1), W175–W182. https://doi.org/10.1093/nar/gkac199

Zrimec, J., Fu, X., Muhammad, A. S., Skrekas, C., Jauniskis, V., Speicher, N. K., Börlin, C. S., Verendel, V., Chehreghani, M. H., Dubhashi, D., & Siewers, V. (2022). Controlling gene expression with deep generative design of regulatory DNA. *Nature Communications*, *13*(1), 5099. https://doi.org/10.1038/s41467-022-32818-8

Zumkeller, S., Peleke, F., Psaroudakis, D., Schmitz, G., & Szymański, J. (2025). Accessible deep learning for plant biology: Predicting gene expression from regulatory DNA with deepCRE. https://doi.org/10.21203.rs.3.rs-6196739/v1

8

Intelligent Priming System for Combating Biotic Stress

Fatima-Ezzahra Soussani[1,2], *Fatima-Zahra Akensous*[1], *Abdelhamid Aouabe*[1], *Naira Sbbar*[1], *Rachid Lahlali*[3], *and Abdelilah Meddich*[1,4]

[1] *Centre d'Agrobiotechnologie et Bioingénierie, Unité de Recherche labellisée CNRST (Centre AgroBiotech-URL-CNRST-05), Université Cadi Ayyad, Marrakech, Morocco*
[2] *Faculty of Sciences, Laboratory of Botany, Biotechnology and Plant Protection, University Ibn Toufail, Kenitra, Morocco*
[3] *Department of Plant Protection, Phytopathology Unit, Ecole Nationale d'Agriculture de Meknès, Meknès, Morocco*
[4] *African Sustainable Agriculture Research Institute (ASARI), Mohammed VI Polytechnic University (UM6P), Laâyoune, Morocco*

8.1 Introduction

Plant biotic stressors, such as fungi, bacteria, viruses, nematodes, and insect pests, are becoming a bigger problem for agriculture around the world. These biotic factors are responsible for an estimated 20–40% reduction in global crop yields each year, threatening food security and economic stability (Newton et al., 2011; Pandey et al., 2017). Conventional disease managemental methods, including chemical pesticides and genetic resistance breeding, have yielded only incomplete and frequently transient solutions (Khan et al., 2020). The swift advancement of pathogen virulence, alongside the detrimental environmental effects of agrochemicals, highlights the critical necessity for novel and sustainable defense strategies (Islam et al., 2024). In recent years, plant scientists have turned their attention to plant immune priming—a natural adaptive mechanism through which plants develop an enhanced state of defense preparedness following exposure to certain stimuli (Mauch-Mani et al., 2017). Primed plants display accelerated and amplified immune responses when re-exposed to pathogens or abiotic stressors, while avoiding the growth penalties associated with constitutive defense activation (Gurr et al., 2022; Wagh et al., 2025). This phenomenon involves intricate signal perception systems, phytohormonal regulation (notably salicylic acid, jasmonic acid, ethylene, and abscisic acid, and reactive oxygen species (ROS)-mediated signaling, which collectively coordinate immune readiness at both local and systemic levels (Ghosh & Roychoudhury, 2024; Lukan & Coll, 2022). Moreover, priming is increasingly recognized as an epigenetically regulated process where deoxyribonucleic acid (DNA) methylation, histone modifications, and transcriptional reprogramming contribute to long-lasting "stress memory" that can even be transmitted across generations (Liu et al., 2022**a**). The concept of an intelligent priming system (IPS) has recently emerged as a multidisciplinary approach integrating molecular biology, nanotechnology, omics platforms, biosensors, and artificial intelligence (AI). The IPS framework seeks to emulate natural priming but with smart, data-driven precision—enabling plants to anticipate, detect, and respond to biotic threats in real time (Banerjee & Arora, 2025; Sobti et al., 2022). Therefore, this chapter aims to explore the concept, mechanisms, and applications of the IPS as an advanced strategy for combating biotic stress in crops. It seeks to bridge fundamental plant immunity with emerging technologies—highlighting how IPS can serve as a cornerstone for next-generation, sustainable, and intelligent crop protection systems in the context of climate change and digital agriculture.

8.2 Mechanisms Underlying Priming-Based Immunity

Priming-based immunity in plants relies on a complex interplay of early perception, signal transduction, and coordinated activation of hormonal and redox networks that enable faster and stronger defense responses upon subsequent pathogen attack. These mechanisms collectively enhance the plant's capacity to recognize, remember, and respond efficiently to biotic stress, forming the molecular foundation of induced systemic resistance (ISR) and systemic acquired resistance (SAR) (Mauch-Mani, Brigitte et al., 2017; Pappas, Maria L. et al., 2020).

8.2.1 Signal Perception and Transduction

The priming process begins with the perception of external stimuli—microbial signals and elicitors or abiotic cues—by pattern recognition receptors (PRRs), located on the plasma membrane. These receptors detect microbe- or damage-associated molecular patterns (MAMPs/DAMPs) such as flagellin (flg22), chitin, or oligogalacturonides, initiating pattern-triggered immunity (PTI) (Saijo et al., 2018). Signal perception leads to rapid calcium ion (Ca^{2+}) influx, activation of mitogen-activated protein kinase (MAPK) cascades, and phosphorylation of transcription factors like WRKYs and Nonexpressor of Pathogenesis-Related Genes 1 (NPR1) that regulate downstream defense genes (Ali et al., 2024b; Sultana et al., 2025). During priming, plants exhibit sensitized signaling circuits, meaning that MAPKs and Ca-dependent protein kinases (CDPKs) are reactivated more rapidly and at greater amplitude during secondary challenges (Atif et al., 2019). This "sensory memory" is often mediated by posttranslational modifications and changes in chromatin state that lower the activation threshold of key immune components (Bahdar et al., 2025). Additionally, systemic signal propagation via mobile molecules such as pipecolic acid, azelaic acid, and glycerol-3-phosphate enables long-distance communication, coordinating local and systemic defense priming (Priya Reddy & Oelmüller, 2024).

8.2.2 Role of Phytohormones

Plant defense priming involves intricate hormonal cross-talk that fine-tunes immune outcomes based on the type of attacker. SA plays a central role in SAR against biotrophic pathogens, enhancing the accumulation of pathogenesis-related (PR) proteins and reinforcing cell walls (Roychowdhury et al., 2024). In contrast, JA and ET signaling pathways are primarily activated in response to necrotrophic pathogens and insect herbivores, regulating the synthesis of secondary metabolites such as alkaloids and phenolics (Upadhyay et al., 2024). Importantly, primed plants display enhanced hormonal sensitivity, allowing faster activation of SA- or JA/ET-dependent pathways upon challenge (Gamir et al., 2014). For instance, β-aminobutyric acid (BABA) primes plants for stronger SA-mediated defense gene expression, whereas beneficial microbes trigger JA/ET-mediated ISR (Vidhyasekaran, 2020). ABA, traditionally known for its role in biotic stress tolerance, also interacts with these pathways: low levels of ABA can promote priming, whereas excessive ABA may antagonize SA-mediated responses (Dar et al., 2017). This dynamic hormonal network ensures that priming responses are context dependent, maintaining a balance between growth and defense.

8.2.3 ROS Generation and Redox Signaling

ROS, such as hydrogen peroxide (H_2O_2) and superoxide radicals, are among the earliest defense signals following pathogen perception. The oxidative burst generated by nicotinamide adenine dinucleotide phosphate (NADPH) oxidases respiratory burst oxidase homologs (RBOHs) and cell wall peroxidases serves as both a direct antimicrobial mechanism and a secondary messenger in priming (Nadarajah, 2023). Primed plants exhibit enhanced ROS production upon pathogen exposure, leading to faster activation of defense gene transcription, callose deposition, and localized cell death that restricts pathogen spread (Hönig et al., 2023). Moreover, ROS interacts with redox-regulated transcription factors and signaling proteins, modulating their activity through oxidative modifications. The ascorbate-glutathione cycle and redox-sensitive proteins such as NPR1 act as key sensors of the cellular redox state, linking oxidative signaling to transcriptional reprogramming (Locato et al., 2018). Priming also enhances the antioxidant capacity of cells, ensuring ROS act as controlled signals rather than damaging agents. This fine-tuned redox homeostasis underlies the durability and efficiency of primed immunity, bridging local responses and systemic defense readiness (Domínguez-Andrées et al., 2022).

8.3 Epigenetic and Molecular Basis of Priming Memory

Plant priming is not a transient phenomenon—it often establishes a molecular memory that enables plants to "remember" previous stress exposure and respond more efficiently to subsequent challenges (Hilker & Schmülling, 2019). This memory is underpinned by epigenetic modifications, transcriptional reprogramming, and systemic signaling networks that together create a long-lasting defensive state (Agostini et al., 2018; Cooper & Ton, 2022).

8.3.1 DNA Methylation

DNA methylation, particularly in cytosine residues in CG, CHG, and CHH contexts, plays a critical role in regulating the transcription of stress-responsive genes during priming

(Talarico et al., 2024). Upon exposure to pathogens or elicitors, localized hypomethylation in promoter regions of defense genes such as *PR1* or *WRKY* can enhance their responsiveness to subsequent attacks (Mierziak & Wojtasik, 2024). Conversely, hypermethylation of negative regulators may silence susceptibility genes, contributing to durable resistance (Torres-Garcia et al., 2020). DNA-methylation changes are maintained through the action of enzymes like MET1, DRM2, and ROS1, allowing for intergenerational transmission of priming effects, as observed in *Arabidopsis* and *Oryza sativa* (Lanciano & Mirouze, 2017; Shaikh et al., 2022). Such transgenerational memory indicates that priming can serve as an adaptive mechanism for evolutionary resilience.

8.3.2 Histone Modification

Histone posttranslational modifications, such as acetylation, methylation, phosphorylation, and ubiquitination—are pivotal regulators of chromatin dynamics and transcriptional memory in primed plants. These epigenetic marks reshape nucleosome configuration, modulating DNA accessibility and allowing immune-related genes to remain in a "poised" state ready for rapid activation upon subsequent stress exposure (Ramirez-Prado et al., 2018). Histone acetylation, notably at H3K9ac and H4K16ac, enhances chromatin relaxation at promoters of defense genes such as *PR1*, *WRKY70*, and *NPR1*, promoting their swift transcription during pathogen attack (Kang et al., 2022). Acetyltransferases like GCN5 and HAC1 have been shown to catalyze these marks during SAR, while histone deacetylases (HDACs) such as HDA6 and HDA19, transiently repress them under basal conditions (Zhang et al., 2024b).

In parallel, histone methylation functions as a stable epigenetic signature sustaining immune memory. The activating mark H3K4me3, catalyzed by COMPASS-like methyltransferases such as ATX1 and SDG2, accumulates at defense loci following infection and persists beyond recovery, providing a heritable record of prior stress (Rai & Rai, 2024). Conversely, repressive marks like H3K27me3, controlled by the Polycomb Repressive Complex 2 (PRC2), are selectively removed during priming to unlock defense gene expression (Lämke & Bäurle, 2017; Li et al., 2023). Histone demethylases, including JMJ27 and JMJ14, modulate the turnover of these methylation states, balancing activation and repression across immune cycles. Collectively, these histone modifications act as molecular bookmarks, maintaining transcriptional readiness and enabling a faster, stronger response upon secondary challenge. Emerging evidence also suggests that such chromatin modifications can propagate systemically and even across generations, underpinning transgenerational immune memory that enhances resilience in progeny (Korolenko & Skinner, 2024; Legoff et al., 2019; Sen & Barnes, 2021).

8.3.3 Transcriptional Reprogramming

Epigenetic changes interface closely with transcriptional reprogramming, collectively remodeling the plant transcriptome to establish and maintain a "primed" physiological state. During defense priming, plants accumulate dormant messenger ribonucleic acid (mRNAs), chromatin-bound RNA polymerase II, and inactive signaling proteins that can be rapidly activated upon pathogen challenge, ensuring a faster transcriptional response with minimal energy consumption (Bashir & Setyawati, 2025; Jaskiewicz et al., 2010). This reprogramming is orchestrated by a complex network of transcription factors (TFs), including members of the WRKY, MYB, NAC, and TGA families, which function as master regulators of immune signaling. For instance, WRKY70 integrates SA and JA cross-talk, while TGA1/TGA2 interacts with NPR1 to activate *PR* gene transcription during SAR (Thilakarathne et al., 2025; Tsuda & Somssich, 2015). The Nonexpressor of Pathogenesis-Related Genes 1 (NPR1) protein acts as a central transcriptional coactivator, modulating redox-dependent nuclear translocation and interaction with TGA TFs to fine-tune defense gene expression (Kovacs et al., 2015; Liu et al., 2020). Upon priming, NPR1-mediated transcriptional networks remain sensitized, allowing rapid induction of downstream immune genes upon reinfection. Moreover, chromatin remodeling complexes, such as SWI/SNF and histone chaperones (e.g., NAP1 and FACT), contribute to transcriptional memory by maintaining open chromatin conformations at key defense loci (Ramirez-Prado et al., 2018). In parallel, small RNAs (sRNAs) including microRNAs (miRNAs) and long noncoding RNAs (lncRNAs), serve as posttranscriptional regulators that reinforce priming memory. For example, miR393 and miR398 modulate hormone signaling and ROS homeostasis, while lncRNAs such as *ELENA1* and *COOLAIR* interact with transcriptional machinery and DNA methylation enzymes to sustain stress-responsive gene expression (Liu et al., 2020; Sintaha, 2025). These RNA-mediated layers act in concert with histone and DNA modifications, forming a multilayered regulatory network that stabilizes transcriptional memory and ensures adaptive immunity across cellular and generational boundaries (Hilker & Schmülling, 2019).

8.3.4 Systemic Signal Transmission

Primed immunity extends beyond locally infected tissues through systemic signal transmission, leading to the establishment of whole-plant resistance known as SAR or ISR (Nadarajah, 2017; Nie et al., 2017). This long-distance communication relies on a diverse network of mobile signaling molecules, including methyl salicylate (MeSA), azelaic acid (AzA), dehydroabietinal (DA), glycerol-3-phosphate (G3P), and pipecolic acid (Pip) and its potent derivative N-hydroxypipecolic acid (NHP). These metabolites act synergistically to amplify systemic defense gene expression and reinforce immune memory across distant tissues (Gao et al., 2021; Kim & Lim, 2023). Upon local infection, ALD1 (AGD2-Like Defense Response Protein 1) and FMO1 (Flavin-dependent Monooxygenase 1) catalyze the biosynthesis of Pip and NHP, respectively, which serve as pivotal SAR regulators. These compounds move through the phloem vasculature, activating defense-related transcription factors such as TGA1/TGA2 and WRKYs in distal leaves, where they stimulate the accumulation of SA and PR proteins (Hartmann & Six, 2022). In addition to chemical mediators, systemic signaling is reinforced by hydraulic, electrical, and ROS waves that propagate through the plant vasculature at high speed, coupling local recognition events to systemic priming (Fichman & Mittler, 2020). Calcium ion (Ca^{2+}) oscillations and MAPK cascades act as intracellular transducers of these systemic cues, integrating biochemical and biophysical signals to maintain synchronized activation of immune genes throughout the plant body (Hilleary & Gilroy, 2018). Importantly, priming enhances the efficiency and fidelity of systemic signaling, lowering the threshold for activation and shortening response time upon subsequent pathogen encounters. This integrated "immunological network" forms the physiological foundation of the IPS—a concept linking molecular memory with spatial communication and adaptive coordination across the whole plant.

8.4 Priming Agents and Triggers

8.4.1 Biotic: PGPR, Mycorrhizae, and *Trichoderma* spp.

Beneficial microbes, particularly plant growth-promoting rhizobacteria (PGPR), arbuscular mycorrhizal fungi (AMF), and *Trichoderma* species, are among the most effective biological primers for enhancing plant immunity (Krzyzaniak et al., 2021). PGPR, such as *Pseudomonas fluorescens*, *Bacillus subtilis*, and *Rhizobium* spp., activate ISR through JA and ET signaling pathways, leading to enhanced expression of defense-related genes and secondary metabolites that deter pathogens (Meena et al., 2020). Similarly, AMF establish a symbiotic interface that enhances nutrient and water uptake while simultaneously inducing mycorrhiza-induced resistance (MIR). This involves activation of priming-related transcription factors and modulation of ROS homeostasis, preparing the host for rapid defense against soil-borne pathogens, such as *Fusarium oxysporum* and *Verticillium dahliae* (Copeland et al., 2025; Llorens et al., 2020).

Trichoderma harzianum and related species contribute through the secretion of elicitor proteins (e.g., Sm1, Epl1) and volatile compounds like 6-pentyl-α-pyrone, which influence hormonal cross-talk between JA, ET, and auxin pathways (Alfiky & Weisskopf, 2021). They also promote root system architecture remodeling and epigenetic adjustments, reinforcing long-term defense memory. Collectively, these beneficial microbes establish a dynamic rhizosphere-based priming network, forming the biological foundation of the IPS. Through synergistic interactions with native microbiota and potential integration with biosensor technologies, these microbial systems offer a sustainable and adaptive platform for smart plant immunization.

8.4.2 Abiotic: Temperature Shifts, UV Light, and Chemical Inducers (e.g., BABA and SA)

Abiotic factors like temperature fluctuations, ultraviolet (UV) radiation, osmotic stress, and drought can act as mild preconditioning factors, activating plant defense signaling pathways that result in cross-protection against future stress events (Hossain et al., 2018). These stresses induce transient increases in ROS and Ca signaling, which serve as secondary messengers to activate MAPK cascades and transcriptional regulators like WRKYs and NPR1, central to priming-associated immunity (Ravi et al., 2023). Such physiological "stress memories" enhance the plant's preparedness for subsequent biotic or abiotic challenges without compromising growth. Beyond natural cues, chemical elicitors including BABA, SA, jasmonates, and benzothiadiazole (BTH) have been widely explored for their ability to induce systemic resistance in diverse crops (Tripathi et al., 2019). Among these, BABA functions as a versatile priming agent by modulating SA-, JA-, and ABA-dependent pathways, strengthening both basal and systemic defense responses while maintaining metabolic efficiency

(Baccelli & Mauch-Mani, 2016). Integrating these abiotic and chemical priming elicitors into the IPS offers a promising avenue for precision agriculture (Zhang et al., 2024a).

8.4.3 Nanomaterials and Engineered Inducers

Recent advances in nanotechnology have paved the way for a new generation of engineered priming inducers that enhance delivery precision, signal amplification, and stress resilience in plants. Metal- and carbon-based nanomaterials—including silver, zinc oxide, graphene oxide, and carbon nanotubes—have demonstrated the ability to activate ROS-mediated signaling and MAPK cascades, leading to the upregulation of defense-related genes such as *PR1*, *WRKY70*, and *NPR1* (Singh et al., 2025). These nanomaterials can induce mild oxidative bursts that mimic natural elicitation without causing toxicity, effectively "training" plants for faster defense activation upon pathogen or abiotic challenge (Zhou et al., 2021).

In parallel, polymeric and biodegradable nanoparticles—such as chitosan, PLGA, and starch-based carriers—are being engineered to deliver phytohormones (SA, JA, and ABA), elicitors (BABA and BTH), or microbial metabolites in controlled-release or stimuli-responsive systems (Pal et al., 2025). This ensures localized and sustained defense activation directly at infection or stress sites while reducing chemical loss and environmental impact (Kumaraswamy et al., 2018). Importantly, chitosan-based nanocomposites exhibit dual functions as both elicitors and carriers, enhancing immune priming by reinforcing cell wall defenses and modulating redox balance (Riaz et al., 2025). When coupled with the Internet of Things (IoT)-enabled biosensors, machine learning algorithms, and AI-driven feedback systems, these nanoplatforms can form the technological backbone of a smart IPS. Such systems enable real-time monitoring of plant physiological status, dynamic adjustment of elicitor dosage, and precise spatiotemporal deployment of priming agents (Liu et al., 2022**a**). This synergistic convergence of nanotechnology, biotechnology, and digital agriculture marks a transformative step toward sustainable, automated plant immunity with minimal environmental footprint.

8.5 Integration with Omics and Smart Technologies

8.5.1 Transcriptomics, Proteomics, and Metabolomics in IPS Monitoring

IPS in plants, a mechanistic process that occurs in plant cells, allowing them to *save a memory* of a previous stressful factor, when associated with generated omics data, omics like transcriptomics, proteomics, and metabolomics, and AI, can help *priming* plants to strengthen tolerance versus biotic stressors (Kotkar & Giri, 2020). This approach can track premolecular signs and project plant stress responses, leading to adjusting plant stress responses and selecting specific crop genotypes with the potential for resiliency to biotic stress (González Guzmán et al., 2022). Such approaches are progressively being used in precision agriculture, transitioning toward premanaging agricultural systems, especially under the ongoing versus projected climate crisis trends (Liu et al., 2022**a**). Developments in omics, particularly transcriptomics, proteomics, and metabolomics, have transformed plant science, allowing for an accurate modulation of plant responses to stress signals and biotic stress (Aruna Kumara & Thiruchchelvan, 2021). In this sense, transcriptomics enables the complete profiling of gene expression, where pre-transcriptional modifications occur in crops like maize and rice, subject to biotic stress (Nazari et al., 2023; Singh & Mukhopadhyay, 2021). Gene expression undergoes spatiotemporal regulations, implicating crucial transcription factors like WRKY, MYB, and dehydration responsive element binding (DREB), instrumenting defense networks (Erpen et al., 2017). Following transcriptomics, proteomics profile posttranscriptional regulating events, which reveal alterations in protein posttranscriptional adjustments, proteins that play a key role in responding to the environing stress (Vo et al., 2021). For instance, proteomics investigation revealed the role of heat shock proteins and proteins linked to pathogenesis *throughout* priming versus *Phytophthora sojae* in soybean (Liu et al., 2021). This represents an important observation, since transcript rates do not align with the abundance of functional proteins all the time, given translational modulation and shifts in degradation of proteins (Muleya et al., 2022; Vo et al., 2021). Metabolomics provides insights into the involved priming-engendered cellular metabolites (e.g., plant hormones, flavonoids) that monitor plant SAR and tolerance to biotic stress (Gao et al., 2020).

8.5.2 Biosensors and Precision Agriculture Platforms in IPS Modulating

Owing to the climate crisis's current versus projected trends, IPS approaches, strengthened by omics and precision agriculture, can provide a sustainable solution to foster crop productivity despite biotic stress (Varadharajan et al., 2025). By adjusting priming approaches to particular genotypes and stress levels, IPS approaches can guarantee vigorous and resistant crops to biotic stressors, which represents a major shift to smart managing agriculture (Kotkar & Giri, 2020). Plant biotechnologies and smart farming progress have been exploring new heights owing to the incorporation of IPS, biosensors, and precision agriculture frameworks to improve crop resistance to biotic stress agents, such as pests and viruses (Kashyap & Kumar, 2021; Liu et al., 2022**b**). IPS enables the modulation of crop immune responses through signal molecules (e.g., elicitors), strengthening the crop's immune system through an energy economy-based stress tolerance (Jatana et al., 2024). Exemplifying this, the deployment of BABA and chitosan nanoparticles showcased potential in prompting SAR in crops, such as tomato and *Arabidopsis*, while promoting crop growth and productivity (Mukarram et al., 2023). Additionally, biosensors provide real-time, *in situ* uncovering of biotic stress markers (e.g., ROS), in addition to effectors and plant hormones (Kotkar & Giri, 2020). Wearable biosensors applied to plant leaves can track volatile organic compounds (VOCs) produced amidst *early pathogen attack*, assisting with a swift detection of biotic agents' preinvasion signs and focal action (Gan et al., 2023). Precision agriculture's technological features, such as AI-powered, remote sensing, IoT systems, and machine learning, further strengthen the efficiency of IPS approaches by allowing in-site and time-efficient priming and biocontrol versus biotic stress (Gómez-Lama Cabanás & Mercado-Blanco, 2025). Algorithms, powered by machine learning, can handle hyperspectral images to prevent the aggravation of symptoms in crops attacked by biotic agents like those observed in potato versus *Phytophthora infestans* under field conditions (Appeltans et al., 2022). Furthermore, digital twin systems are being progressively deployed to prompt crop responses to varied biotic stressors, which represents a promising avenue in preventive agriculture and timely priming adjustment (Chala et al., 2024; Zhang et al., 2025). The association of these technological features with IPS can enable the optimization of crop resistance to biotic stress, especially under an exacerbating climate crisis.

8.5.3 AI and Machine Learning for Predictive IPS Modeling

The incorporation of omics with advanced technologies like bioinformatics and machine learning allows real-time control and model projection of plant biotic stress responses, hence backing IPS approaches (Nazari et al., 2023). This system can be efficient, especially under controlled environmental settings (e.g., vertical farming). Associating AI and machine learning with omics outputs can establish priming signs versus biotic stress (Gou et al., 2024). By deploying large datasets that encompass genomic data, environmental fluctuations, soil profile, and pathogenic agent dynamics, besides real-time data registered by sensors, algorithms powered by machine learning can detect pathways and project the early signs of biotic stressors with high accuracy and notable precision (Al-Otaibi et al., 2024; Ali et al., 2024a). In this sense, the obtained data enable the activation of projecting IPS, which prompts plants to stimulate their immune system and trigger stress-related responses before stress occurs, thus encountering potential yield and water losses (Hazarika et al., 2024; Yang et al., 2022). In the sphere of projective IPS, crops are *reprogrammed* to detect particular stress signs *via* elicitors and environmental indicators (Cooper & Ton, 2022; Hazarika et al., 2024). AI models can ensure the optimization of IPS frameworks and protocols by modifying them depending on the crop status in question (e.g., species, variety, and growth stage) as well as the environing conditions (Al-Otaibi et al., 2024; Yeasmin et al., 2024). Exemplifying this, convolutional neural networks (CNNs) as well as recurrent neural networks (RNNs) can be deployed for hyperspectral image analysis and time-based data, which can allow the precognition of substantial morphophysiological and biochemical alterations pointing to biotic stress (Bandeppa & Vg, 2025; Zheng et al., 2025). Moreover, supervising machine learning features (e.g., *clustering*) can assist with disease classification and deciphering intricate interactions that span host plants and pathogen agents (Qasrawi et al., 2021). Regarding agronomy, IPS directed toward combating biotic stress in crops can lead to strengthening crop resistance to biotic stressors and contribute to sustainability (González Guzmán et al., 2022; Jatana et al., 2024). Notably, these innovative IPS approaches are gaining momentum, advancing with the advent of cloud computing and developed interfaces that are progressing agricultural systems (Sarkar et al., 2024). Climate crisis is exacerbating biotic stress in terms of frequency and intensity, which highlights the cruciality of adopting IPS, strengthened by AI and machine learning, in the sphere of precision agriculture, promoting a resilient and highly productive agriculture globally (Ben Ayed & Hanana, 2021).

8.6 Applications in Major Crops

8.6.1 Case Studies in Tomato, Rice, Wheat, and Arabidopsis

IPS versus biotic stress was explored in several crops, such as tomato, rice, wheat, and Arabidopsis (Table 8.1). As a result, the intelligently primed crops subject to biotic stressors can efficiently exert defensive responses that can still be present either within stressed crops

Table 8.1 Example studies of IPS in dealing with biotic stress in tomato, rice, wheat, and Arabidopsis.

Crop/plant species	Pathogenic agent	Disease/symptoms	Priming agent	Notable results	Reference
Tomato	*Botrytis cinerea*	Gray mold	BABA	Methylation events possibly resulting in transregulation events and regulation of stress-defense genes	Catoni et al. (2022)
Tomato	*Clavibacter michiganensis* ssp. *michiganensis*	Bacterial canker	BABA	Increased NPR1, PR1, and PR5 genes' expression	Özkurt et al. (2025)
Tomato	*Botrytis cinerea*	Gray mold	BABA	Decreased pathogen infection and increased metabolomic defense	Wilkinson et al. (2018)
Rice	*Cnaphalocrocis medinalis*	Leaf folding, whitening, and drying	Si	Reduced survival of leaf folder larvae, enhanced defense-related enzymes, and declined malondialdehyde levels	Han et al. (2016)
Rice	*Meloidogyne graminicola*	Root knot	Si	Decreased nematodes and delayed their development, important transcriptional events of ethylene pathway-linked genes, namely OsERF1, OsEIN2, and OsACS1	Zhan et al. (2018)
Wheat	*Puccinia striiformis* f. sp. *tritici*	Stripe rust	PGPR	Notable differentiated reprogramming of primary and secondary metabolites and upregulation of phenols	Mashabela et al. (2023)
Wheat	*Puccinia striiformis* f. sp. *tritici*	Stripe rust	Melatonin	Elicitation of wheat resistance versus stripe rust and upregulation of the melatonin biosynthesis-related gene TaASMT3	Jiang et al. (2025)
Arabidopsis Thaliana	*Pseudomonas syringae* pv *tomato*, *Hyaloperonospora parasitica*	Speck disease, downy mildew	BABA	Triggered protection against the pathogenic agents	Ton et al. (2005)
Arabidopsis Thaliana	*Pseudomonas syringae* pv *tomato*, *Hyaloperonospora parasitica*	Speck disease, downy mildew	BABA	Primed state transferred to progeny (stress memory)	Slaughter et al. (2012)

BABA, β-Aminobutyric Acid; NPR1, Nonexpressor of Pathogenesis-Related-1; OsACS1, Oryza sativa Acetoacetyl-CoA Synthetase 1; OsEIN2, Oryza sativa Ethylene Insensitive 2; OsERF1, Oryza sativa Ethylene Response Factor 1; PGPR, Plant Growth-Promoting Rhizobacteria; PR1, Pathogenesis-Related protein-1; PR5, Pathogenesis-Related protein-5; Si, Silicon; TaASMT3, Melatonin biosynthesis-related gene.

or throughout generations (e.g., epigenetic, transcriptional events, and stress memory) (Kotkar & Giri, 2020). In tomato, IPS studies were conducted in dealing with bacterial, fungal, and oomycete pathogen agents. For instance, tomato seedlings treated with BABA had decreased cytosine methylation, particularly within promoters and nearby transposable elements. Certain differentially methylated regions (DMRs) and stress-related genes showed upregulation in response to *Botrytis cinerea*. Tomato seedlings that were BABA-primed manifested pretranscriptional responses, highlighting the role of IPS in earlier biotic stress detection (Catoni et al., 2022). In another study, tomato seed priming decreased the severity of bacterial canker disease, caused by *Clavibacter michiganensis* ssp. *michiganensis*, with the upregulation of immunity genes like nonexpressor of pathogenesis-related (NPR) and PR in BABA-primed tomato plants (Özkurt et al., 2025). BABA-primed tomato seedlings against *Botrytis cinerea* had notable accumulation of metabolites (e.g., soluble sugars and secondary metabolites) in fruits, underscoring improved disease resistance post-harvest (Wilkinson et al., 2018). In rice, the priming effect of exogenous silicon (Si)-primed rice plants to strengthen Jasmonate (Ja)-mediated stress defense against rice leaf folder (*Cnaphalocrocis medinalis*) through the increase defense-stress genes expression, enhanced defense enzymes, and decreased rice leaf folder larvae survival, highlighting the efficiency of IPS in the crop dealing with pathogenic insects (Han et al., 2016). Si-primed rice roots intrigued defensive responses versus *Meloidogyne graminicola*, the nematode agent of root knot (Zhan et al., 2018). ET genes implicated in the ET pathway enhanced the production of stress-responsive compounds, such as phenols, and led to important callose deposition. In wheat, priming against biotic stress represents a promising approach. Priming agents like microbes (e.g., PGPR) and elicitors are being examined for their increasing resistance against pathogenic agents, implicating signaling plant hormones, defense-related genes' expression, and modification of metabolites (Mashabela et al., 2023). Wheat plants are being primed to strengthen defensive responses against fungal pathogenic agents (e.g., *Puccinia striiformis* f. sp. *tritici*), especially implicating breeding for resistance traits (Jiang et al., 2025). In *Arabidopsis*, studies on primed plants led to insightful information on stress memory and the transgenerational impact. BABA-primed *Arabidopsis* plants or exposed to avirulent *Pseudomonas syringae* could manifest resistance to the virulent bacterium *Pseudomonas syringae* pv *tomato* and oomycete *Hyaloperonospora arabidopsidis* in the *Arabidopsis* ascendents, highlighting swift defense gene activation (e.g., SA signaling pathway) and stress memory transfer to progeny (Slaughter et al., 2012; Ton et al., 2005).

8.6.2 IPS Deployment Under Field and Greenhouse Conditions

IPS represent potential approaches in improving crop resistance to biotic stress, escalated by pathogenic microbial and viral agents, as well as herbivorous insects and pests under greenhouse and field conditions, owing to progressed technological features (Gou et al. 2024) (Table 8.2). The particularity of these approaches resides in the integration of omics, AI-powered technological features, and biosensors, which strengthen decision-making related to crop biotic stress, toward an efficient precision agriculture (Murmu et al., 2024). Developments in omics have showcased priming-linked alterations in defense-related genes' expression (e.g., PR and NPR genes) in several crops like vegetables, cereals, and *Arabidopsis*, as well as primary and secondary metabolites' concentration, indicating an enhanced

Table 8.2 Examples of AI and machine learning technological features and their notable roles in combating biotic stress under greenhouse/field.

Artificial Intelligence/ Machine learning technology	Notable role	Setting	Reference
Convolutional neural networks (CNNs)	Imagery detection, imagery classification, biotic stress symptoms' identification	Field	Malvade et al. (2022)
Deep convolutional neural networks (DCNNs)	Massive datasets interpretation, accurate disease diagnosis pertaining to biotic stress	Field	Swaminathan and Vairavasundaram (2024)
Recurrent neural networks (RNNs)	Sequential data analysis pertaining to biotic stress	Field	Lee et al. (2020)
Random forests (RF)	Heterogeneous data analysis pertaining to biotic stress	Greenhouse	Khan et al. (2022)
Support vector machines (SVMs)	Disease detection and classification pertaining to biotic stress	Greenhouse	Wang et al. (2020)
Generative adversarial networks (GANs)	Image recognition and classification pertaining to biotic stress	Greenhouse	Tan et al. (2025)

CNNs, Convolutional Neural Networks; DCNNs, Deep Convolutional Neural Networks; GANs, Generative Adversarial Networks; RF, Random Forests; RNNs, Recurrent Neural Networks; SVMs, Support Vector Machines.

immune system of the primed crops versus a range of biotic stressors (Jatana et al., 2024; Yadav & Mathur, 2025). Under greenhouse conditions, IPS has been showing remarkable success, especially when associated with real-time detecting of pathogenic agents (e.g., biosensors, (hyper)spectral imagery), in localizing and opportune activating induced immune/resistance pathways, optimizing energetic and metabolomic processes within stressed crops (Kotkar & Giri, 2020). Furthermore, exogenous use of priming agents like BABA, microbes, and elicitors can be further improved by deploying algorithms powered by machine learning to project crop responses versus biotic stress conditions (Sanchez-Lucas et al., 2025). Under field conditions, challenging conditions governed by the diversity of climate, soils, and pathogens can be tackled using biosensors and IoT-powered platforms, enabling adaptive IPS frameworks (Kashyap & Kumar, 2021). Besides, seed priming, for instance, is progressing to guarantee efficiently monitored priming elements (Jatana et al., 2024; Yadav & Mathur, 2025). Remarkably, field studies encompassing cereals and solanaceous crops, for instance, have showcased increased resistance to pathogenic microorganisms (e.g., *Fusarium* and *Phytophthora*) as well as insects and pests, thereby decreasing crop yield losses and advancing the agronomic promising potential (Kashyap & Kumar, 2021).

8.7 Transgenerational Priming and Long-Term Immunity

Plant priming is considered one of the most effective strategies for protecting crops against various biotic and abiotic stresses. Beyond providing an immediate defense response, it endows plants with an immune memory, enabling them to react more quickly and robustly to subsequent attacks. A study conducted by Ross (1961) on induced resistance in tobacco revealed that the protective state triggered by a primary attack by a pathogen could last for several weeks and even extend to newly formed tissues, suggesting that initiation signals are transmitted through successive cell divisions, thereby enabling the long-term maintenance of induced resistance. Moreover, recent findings have expanded this concept to include transgenerational priming and immune memory mechanisms. This means that the first contact triggers a sort of "memorization" that serves to increase tolerance to subsequent stressful occurrences. For instance, BABA-induced resistance has been associated with chromatin reprogramming and the establishment of an "open chromatin" configuration at defense gene promoters, a modification that keeps these genes in a heightened state of readiness for upcoming stressful situations (Jaskiewicz et al., 2010). Another study conducted as part of a multigenerational experiment showed that wild radish plants exhibit defense responses to herbivory both within and between generations. Damage caused by herbivores triggered DNA methylation in both parent plants and their offspring. Over generations, herbivory experienced by mother plants caused a strong and direct induction of physical defenses in their respective offspring, with effects that lasted from the seedling stage through to the reproductive stage (Sobral et al., 2021). The mechanisms by which such memories are built and kept are still under debate and study, although epigenetics plays a key role (Mirouze & Paszkowski, 2011). Epigenetic modifications are an important mechanism in the regulation of gene expression during development and in reaction to environmental stimuli, particularly those resulting from biotic and abiotic stress (Iwasaki & Paszkowski, 2014; Zhang & Hsieh, 2013). Luna et al. (2012) also showed that priming can be inherited by the next generation when parental Arabidopsis plants are subjected to repeated inoculations with virulent *Pseudomonas syringae*. In this study, not only was the priming state inherited by the next generation, but increased disease resistance was also observed in the grandchildren of the initially infected plants, in a generation that had not been directly exposed to stress. These results suggest that maternal effects alone cannot explain this increased resistance, highlighting that the phenomenon is truly regulated by epigenetic mechanisms (Luna et al., 2012). Overall, these studies clearly reveal that priming for improved resistance to stress continues into subsequent generations and suggest that epigenetic mechanisms, including DNA methylation, chromatin remodeling, and small RNA signaling, play a key role in regulating transgenerational immunity in plants.

8.8 Advantages, Limitations, and Risk Assessment

Transgenerational priming is a sophisticated plant defense strategy in which stress experiences in one generation can induce heritable physiological or epigenetic changes that enhance the resistance of future progeny (Cooper & Ton, 2022). Unlike conventional methods that rely heavily on genetic selection or agrochemical inputs, this approach mobilizes endogenous mechanisms, such as DNA methylation, histone modifications, and small RNA signaling, to prepare offspring for anticipated biotic or abiotic challenges (Iwasaki & Paszkowski, 2014; Mirouze & Paszkowski, 2011). Its integration into smart priming systems offers several compelling advantages. Once established, priming states can persist beyond a single life cycle, providing lasting protection against a wide range of stresses without the need for repeated external treatments (Tabassum et al., 2017). By reducing dependence on pesticides and stress-mitigating chemicals, transgenerational priming lowers production costs and reduces the environmental footprint of agriculture. Furthermore, its reversible nature allows plants to adapt quickly to fluctuating environmental conditions, an important feature in the context of climate change. This new approach can be adopted alongside conventional breeding programs, complementing genetic resistance without permanently altering the genome, and thus offering a multipurpose and cost-effective tool for sustainable crop management (Filippou et al., 2013; Luna et al., 2012).

Despite these advantages, there are some limitations that make it hard for large scale applications. The persistence and intensity of primed traits can decrease after one or two generations if there are no more stressors, which makes it hard to know the stability of the pattern (Liu et al., 2022**a**). The effectiveness of priming varies considerably across plant genotypes and environmental contexts, complicating its predictability in diverse agroecosystems (Bilichak & Kovalchuk, 2016). The underlying molecular mechanisms remain poorly understood, and epigenetic marks may be unstable or reversed, particularly under stress-free conditions (Lucibelli et al., 2022; Wang et al., 2021). Detecting and quantifying the effects of transgenerational priming is also technically challenging, hindering the standardization of testing and large-scale implementation. Unwanted epigenetic changes can affect nonstress-related traits, influencing yield or marketability, while altered defenses could disrupt plant-microbe or plant-insect interactions, harming beneficial symbionts or pollinators. Inconsistent expression without triggering conditions can reduce reliability and farmer confidence.

Transgenerational priming offers significant advantages over conventional strategies. Breeding programs take years to establish resistance traits, and chemical control depends on repeated and costly applications (Alemu et al., 2024). Priming can be induced rapidly, offers broad-spectrum defense without genetic modification or chemical residues, and remains reversible: Plants can maintain enhanced defenses for several generations and then revert to their initial state under favorable conditions, thus avoiding permanent compromises (Ganie et al., 2024; Yadav et al., 2024). This adaptability makes it a valuable component of intelligent priming systems that integrate multiple levels of defense. Regarding environmental and regulatory considerations, reduced pesticide use is likely to improve soil and water quality and preserve beneficial microbial communities (Nandillon et al., 2024; Pieterse et al., 2014). However, potential ecological changes, such as alterations in pest populations or species interactions, must be carefully monitored. As regulatory frameworks for epigenetically modified crops are still underdeveloped, it is essential to establish clear standards for testing, performance evaluation, and labeling in order to maintain public confidence. Although generally less controversial than genetic modifications, potential heritable modifications underscore the importance of transparent communication, thorough risk assessment, and ongoing ecological monitoring.

8.9 Future Directions and Prospects

In the present day, IPSs are at the center of changes in agriculture, particularly in view of climate-smart agriculture and digital farming (Prajapati & Pandya, 2025). SPI integration should be used to effectively merge weather services and farmer needs (Mabhaudhi et al., 2025). This aims to integrate growers, transport companies, service institutions, and consumers to provide a complete vertical system of weather services for the agricultural sector. Their integration into these dynamics opens up new avenues for crop management strategies, adapted to today's environmental and technological challenges (Taha et al., 2025). Indeed, in the context of expanding climatic hazards, SPI is a fundamental lever for anticipating phytosanitary risks, based on agroclimatic models and meteorological information (Mangat & Kaundal, 2025). This predictive capacity not only helps reduce crop vulnerability but also consolidates the resilience of agricultural production systems.

AI represents a major lever for transforming and supporting the development of climate-smart agriculture, paving the way for a sustainable agricultural future (Kumari et al., 2025). AI enables farmers and the various players in the sector to make more informed decisions, optimize the use of resources, and limit the negative impacts of climate change on food production (Indu et al., 2024). AI technologies facilitate the collection and interpretation of massive data, such as weather conditions, soil quality, crop development stages, or even the presence of pests (Ali et al., 2024). Their processing provides strategic information and real-time recommendations, enabling farmers to adjust their practices and adopt more sustainable choices (Akintuyi, 2024). A central application of AI in this context is precision agriculture. Tools such as remote sensing and satellite imagery make it possible to monitor crop health with great precision and detect areas requiring targeted interventions (Sishodia et al., 2020). These approaches enable rational management of irrigation, fertilization, and pest control, while reducing resource consumption and the environmental footprint. These technologies, through increased accuracy and timeliness of intervention, enable the optimization of resources and reduction of environmental footprint (Padhiary et al., 2025). The introduction of these innovations will require greater interaction between researchers, farmers, technology companies, and political decision-makers. Only a concerted and inclusive effort will guarantee the adoption of these large-scale systems, which will respond to the economic, social, and environmental realities of agricultural regions. IPS are therefore presented as a strategic element in achieving a more resilient, sustainable, and competitive agriculture to meet the challenges of the future (McCown, 2002).

8.10 Conclusion

The development of the IPS represents a transformative advance in plant protection, uniting the principles of biological priming with modern digital and nanotechnological innovations. As detailed throughout this chapter, IPS builds upon the natural capacity of plants to develop an enhanced state of defense preparedness—mediated by complex signaling cascades, phytohormonal cross-talk, and redox regulation. These molecular events are reinforced by epigenetic and transcriptional memory, allowing plants to retain "immunological experience" and respond rapidly to recurrent pathogen challenges. By leveraging beneficial microbes (such as PGPR, AMF, and *Trichoderma* spp.), chemical elicitors, and engineered nanomaterials, IPS enables a multifaceted and sustainable approach to plant immunization. Beyond biological foundations, its integration with omics platforms, biosensor technologies, and AI-driven modeling paves the way for predictive, adaptive, and self-regulating crop defense systems. In the context of climate-smart and digital agriculture, IPS offers a paradigm shift from reactive pest management toward proactive, precision-based resilience. It minimizes the reliance on chemical pesticides, promotes ecological balance, and supports global food security under increasing environmental pressures. Future research should focus on refining IPS frameworks through field validation, multiomics integration, and scalable data analytics to fully realize its potential as the cornerstone of next-generation sustainable agriculture.

Acknowledgments

This book chapter was financially supported by the Phytopathology Unit within the Department of Plant Protection and Environment at the Ecole Nationale d'Agriculture de Meknès, Morocco.

References

Agostini, R. B., Postigo, A., Rius, S. P., Rech, G. E., Campos-Bermudez, V. A., & Vargas, W. A. (2018). Long-lasting primed state in maize plants: Salicylic acid and steroid signaling pathways as key players in the early activation of immune responses in silks. *Molecular Plant-Microbe Interactions*, *32*(1), 95–106. https://doi.org/10.1094/MPMI-07-18-0208-R

Akintuyi, O. B. (2024). Adaptive AI in precision agriculture: A review: Investigating the use of self-learning algorithms in optimizing farm operations based on real-time data. *Research Journal of Multidisciplinary Studies*, *7*(2), 16–30. https://doi.org/10.53022/oarjms.2024.7.2.0023

Alemu, A., Åstrand, J., Montesinos-López, O. A., Isidro y Sánchez, J., Fernández-Gónzalez, J., Tadesse, W., Vetukuri, R. R., Carlsson, A. S., Ceplitis, A., Crossa, J., Ortiz, R., & Chawade, A. (2024). Genomic selection in plant breeding: Key factors shaping two decades of progress. *Molecular Plant*, *17*(4), 552–578. https://doi.org/10.1016/J.MOLP.2024.03.007

Alfiky, A., & Weisskopf, L. (2021). Deciphering *Trichoderma*–plant–pathogen interactions for better development of biocontrol applications. *Journal of Fungi*, *7*, 61. https://doi.org/10.3390/JOF7010061

Ali, F., Rehman, A., Hameed, A., Sarfraz, S., Rajput, N. A., & Atiq, M. (2024). Climate change impact on plant pathogen emergence: Artificial intelligence (AI) approach. In K. A. Abd-Elsalam & S. M. Abdel-Momen (Eds.), *Plant quarantine challenges under climate change anxiety* (pp. 281–303). Springer Nature Switzerland.

Ali, S., Tyagi, A., & Mir, Z. A. (2024b). Plant immunity: At the crossroads of pathogen perception and defense response. *Plants*, *13*, 1434. https://doi.org/10.3390/PLANTS13111434

Al-Otaibi, S., Khan, R., Ali, J., & Ahmed, A. (2024). Artificial intelligence and internet of things-enabled decision support system for the prediction of bacterial stalk root disease in maize crop. *Computational Intelligence*, *40*. https://doi.org/10.1111/coin.12632

Appeltans, S., Pieters, J. G., & Mouazen, A. M. (2022). Potential of laboratory hyperspectral data for in-field detection of *Phytophthora infestans* on potato. *Precision Agriculture*, *23*, 876–893. https://doi.org/10.1007/s11119-021-09865-0

Aruna Kumara, U. M., & Thiruchchelvan, N. (2021). Omics technology for plant stress management. In R. Soni, D. C. Suyal, P. Bhargava, & R. Goel (Eds.), *Microbiological activity for soil and plant health management* (pp. 375–412). Springer Singapore.

Atif, R. M., Shahid, L., Waqas, M., Ali, B., Rashid, M. A., Azeem, F., Nawaz, M. A., Wani, S. H., & Chung, G. (2019). Insights on calcium-dependent protein kinases (CPKs) signaling for abiotic stress tolerance in plants. *International Journal of Molecular Sciences, 20*(21), 5298. https://doi.org/10.3390/IJMS20215298

Baccelli, I., & Mauch-Mani, B. (2016). Beta-aminobutyric acid priming of plant defense: The role of ABA and other hormones. *Plant Molecular Biology, 91*, 703–711. https://doi.org/10.1007/S11103-015-0406-Y/METRICS

Bahdar, Z. I., Abu-El-Rub, E., Almazari, R., Alzu'bi, A., & Al-Zoubi, R. M. (2025). The molecular mechanism of nitric oxide in memory consolidation and its role in the pathogenesis of memory-related disorders. *Neurogenetics, 26*, 1–12. https://doi.org/10.1007/S10048-025-00803-0/FIGURES/3

Bandeppa, L., & Vg, B. T. (2025, June 13–14). Early detection and classification of biotic and abiotic stress in tomato leaves using AI-based approaches-review. *2025 International Conference on Computing Technologies (ICOCT)*, Bengaluru, India (pp. 1–6). IEEE.

Banerjee, A., & Arora, D. (2025). Nanobiomaterials at the nexus. *Stem Cell Ther, 159–226*. https://doi.org/10.1002/9781394313785.CH4

Bashir, H., & Setyawati, A. (2025). Decoding the mechanisms of plant epigenetic adaptations to stress. *Review of Agricultural Science, 13*, 1–19. https://doi.org/10.7831/RAS.13.2_1

Ben Ayed, R., & Hanana, M. (2021). Artificial intelligence to improve the food and agriculture sector. *Journal of Food Quality, 2021*, 1–7. https://doi.org/10.1155/2021/5584754

Bilichak, A., & Kovalchuk, I. (2016). Transgenerational response to stress in plants and its application for breeding. *Journal of Experimental Botany, 67*(7), 2081–2092. https://doi.org/10.1093/JXB/ERW066

Catoni, M., Alvarez-Venegas, R., Worrall, D., Holroyd, G., Barraza, A., Luna, E., Ton, J., & Roberts, M. R. (2022). Long-lasting defence priming by β-aminobutyric acid in tomato is marked by genome-wide changes in DNA methylation. *Frontiers in Plant Science, 13*, 1–13. https://doi.org/10.3389/fpls.2022.836326

Chala, D., Kusch, E., Weiland, C., Andrew, C., Grieb, J., Rossi, T., Martinovic, T., & Endresen, D. (2024). Prototype biodiversity digital twin: Crop wild relatives genetic resources for food security. *Research Ideas & Outcomes, 10*, 1–14. https://doi.org/10.3897/rio.10.e125192

Cooper, A., & Ton, J. (2022). Immune priming in plants: From the onset to transgenerational maintenance. *Essays in Biochemistry, 66*, 635–646. https://doi.org/10.1042/EBC20210082

Copeland, C., Schulze-Lefert, P., & Ma, K.-W. (2025). Potential and challenges for application of microbiomes in agriculture. *The Plant Cell* https://doi.org/10.1093/PLCELL/KOAF185, *37*.

Dar, N. A., Amin, I., Wani, W., Wani, S. A., Shikari, A. B., Wani, S. H., & Masoodi, K. Z. (2017). Abscisic acid: A key regulator of abiotic stress tolerance in plants. *Plant Gene, 1*(11), 106–111. https://doi.org/10.1016/J.PLGENE.2017.07.003

Domínguez-Andrées, J., Dos Santos, J. C., Bekkering, S., Mulder, W. J., van der Meer, J. W., Riksen, N. P., Joosten, L. A., & Netea, M. G. (2022). Trained immunity: Adaptation within innate immune mechanisms, *103*, 313–346. https://doi.org/10.1152/PHYSREV.00031.2021

Erpen, L., Devi, H. S., Grosser, J. W., & Dutt, M. (2017). Potential use of the DREB/ERF, MYB, NAC and WRKY transcription factors to improve abiotic and biotic stress in transgenic plants. *Plant Cell, Tissue and Organ Culture, 32*, 1–25. https://doi.org/10.1007/S11240-017-1320-6

Fichman, Y., & Mittler, R. (2020). Rapid systemic signaling during abiotic and biotic stresses: Is the ROS wave master of all trades? *The Plant Journal, 102*, 887–896. https://doi.org/10.1111/TPJ.14685

Filippou, P., Tanou, G., Molassiotis, A., & Fotopoulos, V. (2013). Plant acclimation to environmental stress using priming agents. In N. Tuteja & S. S. Gill (Eds.), *Plant acclimation to environmental stress* (pp. 1–27). https://doi.org/10.1007/978-1-4614-5001-6_1/FIGURES/00011

Gamir, J., Sánchez-Bel, P., & Flors, V. (2014). Molecular and physiological stages of priming: How plants prepare for environmental challenges. *Plant Cell Reports, 33*, 1935–1949. https://doi.org/10.1007/S00299-014-1665-9/METRICS

Gan, Z., Zhou, Q., Zheng, C., & Wang, J. (2023). Challenges and applications of volatile organic compounds monitoring technology in plant disease diagnosis. *Biosensors & Bioelectronics, 237*, 115540. https://doi.org/10.1016/j.bios.2023.115540

Ganie, S. A., McMulkin, N., & Devoto, A. (2024). The role of priming and memory in rice environmental stress adaptation: Current knowledge and perspectives. *Plant, Cell & Environment, 47*(5), 1895–1915. https://doi.org/10.1111/PCE.14855

Gao, H., Zhou, Q., Yang, L., Zhang, K., Ma, Y., & Xu, Z. Q. (2020). Metabolomics analysis identifies metabolites associated with systemic acquired resistance in arabidopsis. *PeerJ, 8*, e10047. https://doi.org/10.7717/peerj.10047

Gao, H., Guo, M., Song, J., Ma, Y., & Xu, Z. (2021). Signals in systemic acquired resistance of plants against microbial pathogens. *Molecular Biology Reports, 48*, 3747–3759. https://doi.org/10.1007/S11033-021-06344-7/METRICS

Ghosh, P., & Roychoudhury, A. (2024). Molecular basis of salicylic acid–phytohormone crosstalk in regulating stress tolerance in plants. *Brazilian Journal of Botany, 47*, 735–750. https://doi.org/10.1007/S40415-024-00983-3/METRICS

Gómez-Lama Cabanás, C., & Mercado-Blanco, J. (2025). Groundbreaking technologies and the biocontrol of fungal vascular plant pathogens. *Journal of Fungi, 11*, 77. https://doi.org/10.3390/JOF11010077

González Guzmán, M., Cellini, F., Fotopoulos, V., Balestrini, R., & Arbona, V. (2022). New approaches to improve crop tolerance to biotic and abiotic stresses. *Physiologia Plantarum, 174*, 1–18. https://doi.org/10.1111/ppl.13547

Gou, C., Zafar, S., Fatima, Hasnain, Z., Aslam, N., Iqbal, N., Abbas, S., Li, H., Li, J., Chen, B., & Ragauskas, A. J. (2024). Machine and deep learning: Artificial intelligence application in biotic and abiotic stress management in plants. *Frontiers in Bioscience, 29*. https://doi.org/10.31083/j.fbl2901020

Gurr, S. J., Trigg, S. A., Vadopalas, B., Roberts, S. B., & Putnam, H. M. (2022). Acclimatory gene expression of primed clams enhances robustness to elevated pCO2. *Molecular Ecology, 31*, 5005–5023. https://doi.org/10.1111/MEC.16644

Han, Y., Li, P., Gong, S., Yang, L., Wen, L., & Hou, M. (2016). Defense responses in rice induced by silicon amendment against infestation by the leaf folder *Cnaphalocrocis medinalis*. *PLoS One, 11*, 1–14. https://doi.org/10.1371/journal.pone.0153918

Hartmann, M., & Six, J. (2022). Soil structure and microbiome functions in agroecosystems. *Nature Reviews Earth and Environment, 41*(4), 4–18. https://doi.org/10.1038/s43017-022-00366-w

Hazarika, P., Singh, H. R., Das, D. K., & Das, S. (2024). Priming of plant's immune system: The future sustainable approach for tea improvement. *Discovery Plants, 1*, 1–25. https://doi.org/10.1007/s44372-024-00035-w

Hilker, M., & Schmülling, T. (2019). Stress priming, memory, and signalling in plants. *Plant, Cell & Environment, 42*, 753–761. https://doi.org/10.1111/PCE.13526

Hilleary, R., & Gilroy, S. (2018). Systemic signaling in response to wounding and pathogens. *Current Opinion in Plant Biology, 43*, 57–62. https://doi.org/10.1016/J.PBI.2017.12.009

Hönig, M., Roeber, V. M., Schmülling, T., & Cortleven, A. (2023). Chemical priming of plant defense responses to pathogen attacks. *Frontiers in Plant Science, 14*, 1146577. https://doi.org/10.3389/FPLS.2023.1146577/FULL

Hossain, M. A., Li, Z. G., Hoque, T. S., Burritt, D. J., Fujita, M., & Munné-Bosch, S. (2018). Heat or cold priming-induced cross-tolerance to abiotic stresses in plants: Key regulators and possible mechanisms. *Protoplasma, 255*, 399–412. https://doi.org/10.1007/S00709-017-1150-8/METRICS

Indu, P. V., Nanjundan, P., & Thomas, L. (2024). AI for optimization of farming resources and their management. In S. Potluri, S. Satpathy, S. S. Basa, & A. Zuorro (Eds.), *AI in agriculture for sustainable and economic management* (pp. 42–52). CRC Press.

Islam, T., Danishuddin, Tamanna, N. T., Matin, M. N., Barai, H. R., & Haque, M. A. (2024). Resistance mechanisms of plant pathogenic fungi to fungicide, environmental impacts of fungicides, and sustainable solutions. *Plants, 13*, 2737. https://doi.org/10.3390/PLANTS13192737

Iwasaki, M., & Paszkowski, J. (2014). Identification of genes preventing transgenerational transmission of stress-induced epigenetic states. *Proceedings of the National Academy of Sciences of the United States of America, 111*(23), 8547–8552. https://doi.org/10.1073/PNAS.1402275111

Jaskiewicz, M., Conrath, U., & Peterhälnsel, C. (2010). Chromatin modification acts as a memory for systemic acquired resistance in the plant stress response. *EMBO Reports, 12*, 50–55. https://doi.org/10.1038/EMBOR.2010.186

Jatana, B. S., Grover, S., Ram, H., & Baath, G. S. (2024). Seed priming: Molecular and physiological mechanisms underlying biotic and abiotic stress tolerance. *Agronomy, 14*, 2901. https://doi.org/10.3390/agronomy14122901

Jiang, L., Yuan, Z., Yan, W., Tang, P., Yuan, P., Zheng, P., Chu, J., Xin, P., Cheng, S., Kang, Z., & Liu, J. (2025). Transcriptomic and metabolomic analyses unveil *TaASMT3*-mediated wheat resistance against stripe rust by promoting melatonin biosynthesis. *The Plant Journal, 122*. https://doi.org/10.1111/tpj.70182

Kang, H., Fan, T., Wu, J., Zhu, Y., & Shen, W. H. (2022). Histone modification and chromatin remodeling in plant response to pathogens. *Frontiers in Plant Science, 13*, 986940. https://doi.org/10.3389/FPLS.2022.986940/FULL

Kashyap, B., & Kumar, R. (2021). Sensing methodologies in agriculture for monitoring biotic stress in plants due to pathogens and pests. *Inventions, 6*, 29. https://doi.org/10.3390/inventions6020029

Khan, A. H., Hassan, M., & Khan, M. N. (2020). Conventional plant breeding program for disease resistance. In I. U. Haq & S. Ijaz (Eds.), *Plant disease management strategies for sustainable agriculture through traditional and modern approaches* (pp. 27–51). Springer. https://doi.org/10.1007/978-3-030-35955-3_3

Khan, A. R., Wicaksono, W. A., Ott, N. J., Poret-Peterson, A. T., & Browne, G. T. (2022). Random forest analysis reveals taxa predictive of Prunus replant disease in peach root microbiomes. *PLoS One, 17*, e0275587. https://doi.org/10.1371/journal.pone.0275587

Kim, T. J., & Lim, G. H. (2023). Salicylic acid and mobile regulators of systemic immunity in plants: Transport and metabolism. *Plants, 12*, 1013. https://doi.org/10.3390/PLANTS12051013

Korolenko, A., & Skinner, M. K. (2024). Generational stability of epigenetic transgenerational inheritance facilitates adaptation and evolution. *Epigenetics, 19*. https://doi.org/10.1080/15592294.2024.2380929

Kotkar, H., & Giri, A. (2020). Plant epigenetics and the 'intelligent' priming system to combat biotic stress. In S. K. Sharma (Ed.), *Epigenetics of the immune system* (pp. 25–38). Elsevier, Elsevier.

Kovacs, I., Durner, J., & Lindermayr, C. (2015). Crosstalk between nitric oxide and glutathione is required for NONEXPRESSOR OF PATHOGENESIS-RELATED GENES 1 (NPR1)-dependent defense signaling in *arabidopsis thaliana*. *New Phytologist, 208*, 860–872. https://doi.org/10.1111/NPH.13502

Krzyzaniak, Y., Magnin-Robert, M., Randoux, B., Fontaine, J., & Lounès-Hadj Sahraoui, A. (2021). Combined use of beneficial bacteria and arbuscular mycorrhizal fungi for the biocontrol of plant cryptogamic diseases: Evidence, methodology, and limits. *Symbiotic Soil Microorganisms: Biology and Applications*, 429–468. https://doi.org/10.1007/978-3-030-51916-2_24

Kumaraswamy, R. V., Kumari, S., Choudhary, R. C., Pal, A., Raliya, R., Biswas, P., & Saharan, V. (2018). Engineered chitosan based nanomaterials: Bioactivities, mechanisms and perspectives in plant protection and growth. *International Journal of Biological Macromolecules, 113*, 494–506. https://doi.org/10.1016/J.IJBIOMAC.2018.02.130

Kumari, K., Mirzakhani Nafchi, A., Mirzaee, S., & Abdalla, A. (2025). AI-driven future farming: Achieving climate-smart and sustainable agriculture. *AgriEngineering, 7*(3), 89. https://doi.org/10.3390/agriengineering7030089

Lämke, J., & Bäurle, I. (2017). Epigenetic and chromatin-based mechanisms in environmental stress adaptation and stress memory in plants. *Genome Biology, 18*, 1–11. https://doi.org/10.1186/S13059-017-1263-6/FIGURES/2

Lanciano, S., & Mirouze, M. (2017). DNA methylation in rice and relevance for breeding. *Epigenomes, 1*, 10. https://doi.org/10.3390/EPIGENOMES1020010

Lee, S. H., Goëau, H., Bonnet, P., & Joly, A. (2020). Attention-based recurrent neural network for plant disease classification. *Frontiers in Plant Science, 11*, 1–8. https://doi.org/10.3389/fpls.2020.601250

Legoff, L., D'Cruz, S. C., Tevosian, S., Primig, M., & Smagulova, F. (2019). Transgenerational inheritance of environmentally induced epigenetic alterations during mammalian development. *Cells, 8*(12), 1559. https://doi.org/10.3390/CELLS8121559

Li, J., Cai, B., Chang, S., Yang, Y., Zi, S., & Liu, T. (2023). Mechanisms associated with the synergistic induction of resistance to tobacco black shank in tobacco by arbuscular mycorrhizal fungi and β-aminobutyric acid. *Frontiers in Plant Science, 14*, 1195932. https://doi.org/10.3389/FPLS.2023.1195932/BIBTEX

Liu, Y., Sun, T., Sun, Y., Zhang, Y., Radojičić, A., Ding, Y., Tian, H., Huang, X., Lan, J., Chen, S., Orduna, A. R., Zhang, K., Jetter, R., Li, X., & Zhang, Y. (2020). Diverse roles of the salicylic acid receptors NPR1 and NPR3/NPR4 in plant immunity. *Plant Cell, 32*, 4002–4016. https://doi.org/10.1105/TPC.20.00499

Liu, T., Wang, H., Liu, Z., Pang, Z., Zhang, C., Zhao, M., Ning, B., Song, B., Liu, S., He, Z., & Wei, W. (2021). The 26S proteasome regulatory subunit GmPSMD promotes resistance to *Phytophthora sojae* in soybean. *Frontiers in Plant Science, 12*, 1–14. https://doi.org/10.3389/fpls.2021.513388

Liu, H., Able, A. J., & Able, J. A. (2022a). Priming crops for the future: Rewiring stress memory. *Trends in Plant Science, 27*, 699–716. https://doi.org/10.1016/j.tplants.2021.11.015

Liu, X., Quan, W., & Bartels, D. (2022b). Stress memory responses and seed priming correlate with drought tolerance in plants: An overview. *Planta, 255*(2), 1–14. https://doi.org/10.1007/S00425-022-03828-Z/TABLES/2

Llorens, E., González-Hernández, A. I., Scalschi, L., Fernández-Crespo, E., Camañes, G., Vicedo, B., & García-Agustín, P. (2020). Priming mediated stress and cross-stress tolerance in plants: Concepts and opportunities. *Priming-Mediated Stress Cross-Stress Toler Crop Plants*, 1–20. https://doi.org/10.1016/B978-0-12-817892-8.00001-5

Locato, V., Cimini, S., & De Gara, L. (2018). ROS and redox balance as multifaceted players of cross-tolerance: Epigenetic and retrograde control of gene expression. *Journal of Experimental Botany, 69*, 3373–3391. https://doi.org/10.1093/JXB/ERY168

Lucibelli, F., Valoroso, M. C., & Aceto, S. (2022). Plant DNA methylation: An epigenetic mark in development, environmental interactions, and evolution. *International Journal of Molecular Sciences, 23*(15), 8299. https://doi.org/10.3390/IJMS23158299

Lukan, T., & Coll, A. (2022). Intertwined roles of reactive oxygen species and salicylic acid signaling are crucial for the plant response to biotic stress. *International Journal of Molecular Sciences, 23*, 5568. https://doi.org/10.3390/IJMS23105568

Luna, E., Bruce, T. J. A., Roberts, M. R., Flors, V., & Ton, J. (2012). Next-generation systemic acquired resistance. *Plant Physiology, 158*(2), 844–853. https://doi.org/10.1104/PP.111.187468

Mabhaudhi, T., Dirwai, T. L., Taguta, C., Senzanje, A., Abera, W., Govid, A., . . . Chimonyo, V. G. P. (2025). Linking weather and climate information services (WCIS) to Climate-Smart Agriculture (CSA) practices. *Climate Services, 37*, 100529. https://doi.org/10.1016/j.cliser.2024.100529

Malvade, N. N., Yakkundimath, R., Saunshi, G., Elemmi, M. C., & Baraki, P. (2022). A comparative analysis of paddy crop biotic stress classification using pre-trained deep neural networks. *Artificial Intelligence in Agriculture, 6*, 167–175. https://doi.org/10.1016/j.aiia.2022.09.001

Mangat, G. K., & Kaundal, M. (2025). The agrometeorological edge: Transforming agriculture in a changing climate. *Journal of Agrometeorology, 27*(2), 258–266. https://doi.org/10.54386/jam.v27i2.2615

Mashabela, M. D., Tugizimana, F., Steenkamp, P. A., Piater, L. A., Dubery, I. A., Terefe, T., & Mhlongo, M. I. (2023). Metabolomic evaluation of PGPR defence priming in wheat (*Triticum aestivum* L.) cultivars infected with *Puccinia striiformis f. sp. tritici* (stripe rust). *Frontiers in Plant Science, 14*, 1–11. https://doi.org/10.3389/fpls.2023.1103413

Mauch-Mani, B., Baccelli, I., Luna, E., & Flors, V. (2017). Defense priming: An adaptive part of induced resistance. *Annual Review of Plant Biology*, *68*, 485–512. https://doi.org/10.1146/ANNUREV-ARPLANT-042916-041132/CITE/REFWORKS

McCown, R. L. (2002). Changing systems for supporting farmers' decisions: Problems, paradigms, and prospects. *Agricultural Systems*, *74*(1), 179–220. https://doi.org/10.1016/S0308-521X(02)00026-4

Meena, M., Swapnil, P., Divyanshu, K., Kumar, S., Harish, Tripathi, Y. N., Zehra, A., Marwal, A., & Upadhyay, R. S. (2020). PGPR-mediated induction of systemic resistance and physiochemical alterations in plants against the pathogens: Current perspectives. *Journal of Basic Microbiology*, *60*, 828–861. https://doi.org/10.1002/JOBM.202000370

Mierziak, J., & Wojtasik, W. (2024). Epigenetic weapons of plants against fungal pathogens. *BMC Plant Biology*, *24*, 1–23. https://doi.org/10.1186/S12870-024-04829-8/FIGURES/6

Mirouze, M., & Paszkowski, J. (2011). Epigenetic contribution to stress adaptation in plants. *Current Opinion in Plant Biology*, *14*(3), 267–274. https://doi.org/10.1016/J.PBI.2011.03.004

Mukarram, M., Ali, J., Dadkhah-Aghdash, H., Kurjak, D., Kačík, F., & Ďurkovič, J. (2023). Chitosan-induced biotic stress tolerance and crosstalk with phytohormones, antioxidants, and other signalling molecules. *Frontiers in Plant Science*, *14*, 1217822. https://doi.org/10.3389/FPLS.2023.1217822/BIBTEX

Muleya, V., Lois, L. M., Chahtane, H., Thomas, L., Chiapello, M., & Marondedze, C. (2022). (De)activation (Ir)reversibly or degradation: Dynamics of post-translational protein modifications in plants. *Life*, *12*, 324. https://doi.org/10.3390/life12020324

Murmu, S., Sinha, D., Chaurasia, H., Sharma, S., das, R., Jha, G. K., & Archak, S. (2024). A review of artificial intelligence-assisted omics techniques in plant defense: Current trends and future directions. *Frontiers in Plant Science*, *15*, 1292054. https://doi.org/10.3389/FPLS.2024.1292054/FULL

Nadarajah, K. K. (2017). Induction of systemic resistance for disease suppression. In S. N. A. Abdullah, H. Chai-Ling, & C. Wagstaff (Eds.), *Crop improv sustain through leading-edge technology* (pp. 335–357). https://doi.org/10.1007/978-3-319-65079-1_15

Nadarajah, K. K. (2023). Defensive strategies of ROS in plant–pathogen interactions. *Plant-Pathogen Interactions*, 163–183. https://doi.org/10.1007/978-981-99-4890-1_6

Nandillon, R., Guinet, M., & Munier-Jolain, N. (2024). Crop management strategy redesign enables a reduction in reliance on pesticides: A diachronic approach based on a diversity of French commercial farms. *Agriculture, Ecosystems & Environment*, *366*, 108949. https://doi.org/10.1016/J.AGEE.2024.108949

Nazari, L., Aslan, M. F., Sabanci, K., & Ropelewska, E. (2023). Integrated transcriptomic meta-analysis and comparative artificial intelligence models in maize under biotic stress. *Scientific Reports*, *13*, 15899. https://doi.org/10.1038/s41598-023-42984-4

Newton, A. C., Johnson, S. N., & Gregory, P. J. (2011). Implications of climate change for diseases, crop yields and food security. *Euphytica*, *179*, 3–18. https://doi.org/10.1007/S10681-011-0359-4/METRICS

Nie, P., Li, X., Wang, S., Guo, J., Zhao, H., & Niu, D. (2017). Induced systemic resistance against *Botrytis cinerea* by *Bacillus cereus* AR156 through a JA/ET- and NPR1-dependent signaling pathway and activates PAMP-triggered immunity in arabidopsis. *Frontiers in Plant Science*, *8*, 247629. https://doi.org/10.3389/FPLS.2017.00238/BIBTEX

Özkurt, N., Bektas, H., & Bektas, Y. (2025). Enhancing tomato (*Solanum lycopersicum* L.) resistance against bacterial canker disease (*Clavibacter michiganensis* ssp. *michiganensis*) via seed priming with β-aminobutyric acid (BABA). *Horticulturae, 11*, 1–16. https://doi.org/10.3390/horticulturae11060587

Padhiary, M., Hoque, A., Prasad, G., Kumar, K., & Sahu, B. (2025). Precision agriculture and AI-driven resource optimization for sustainable land and resource management. In J. A. Ruiz-Vanoye & O. Díaz-Parra (Eds.), *Smart water technology for sustainable management in modern cities* (pp. 197–232). IGI Global Scientific Publishing.

Pal, P., Prakash, O., Parveen, A., Singh, A. K., Gupta, R., Sarangi, P. K., Sahoo, U. K., Rathore, S. S., & Singh, R. K. (2025). Nanoparticle-driven plant signaling for advancing stress resilience and agricultural productivity—a review. *Journal of Nanoparticle Research, 27*, 1–28. https://doi.org/10.1007/S11051-025-06449-1/METRICS

Pandey, P., Irulappan, V., Bagavathiannan, M. V., & Senthil-Kumar, M. (2017). Impact of combined abiotic and biotic stresses on plant growth and avenues for crop improvement by exploiting physio-morphological traits. *Frontiers in Plant Science, 8*, 237767. https://doi.org/10.3389/FPLS.2017.00537/BIBTEX

Pieterse, C. M. J., Zamioudis, C., Berendsen, R. L., Weller, D. M., Van Wees, S. C. M., & Bakker, P. A. H. M. (2014). Induced systemic resistance by beneficial microbes. *Annual Review of Phytopathology, 52*, 347–375. https://doi.org/10.1146/ANNUREV-PHYTO-082712-102340/CITE/REFWORKS

Prajapati, G. V., & Pandya, P. A. (Eds.). (2025). *Trailblazing trends in sustainable climate-resilient precision agriculture*. Allied Publishers.

Priya Reddy, Y. N., & Oelmüller, R. (2024). Lipid peroxidation and stress-induced signalling molecules in systemic resistance mediated by azelaic acid/*AZELAIC ACID INDUCED1*: Signal initiation and propagation. *Physiology and Molecular Biology of Plants, 30*, 305–316. https://doi.org/10.1007/S12298-024-01420-1/FIGURES/2

Qasrawi, R., Amro, M., Zaghal, R., Sawafteh, M., & Polo, S.V. (2021, November 17–18). Machine learning techniques for tomato plant diseases clustering, prediction and classification. *2021 International Conference on Promising Electronic Technologies (ICPET)*, Deir El-Balah, State of Palestine (pp. 40–45). IEEE

Rai, M. N., & Rai, R. (2024). H3K4 methylation and demethylation in fungal pathogens: The epigenetic toolbox for survival and adaptation in the host. *Pathogens, 13*, 1080. https://doi.org/10.3390/PATHOGENS13121080

Ramirez-Prado, J. S., Piquerez, S. J. M., Bendahmane, A., Hirt, H., Raynaud, C., & Benhamed, M. (2018). Modify the histone to win the battle: Chromatin dynamics in plant–pathogen interactions. *Frontiers in Plant Science, 9*, 339070. https://doi.org/10.3389/FPLS.2018.00355/FULL

Ravi, B., Foyer, C. H., & Pandey, G. K. (2023). The integration of reactive oxygen species (ROS) and calcium signalling in abiotic stress responses. *Plant, Cell & Environment, 46*, 1985–2006. https://doi.org/10.1111/PCE.14596

Riaz, M., Yasmeen, E., Saleem, B., Hameed, M. K., Saeed Almheiri, M. T., Saeed Al Mir, R. O., Alameri, G., Khamis Alghafri, J. S., & Gururani, M. A. (2025). Evolution of agricultural biotechnology is the paradigm shift in crop resilience and development: A review. *Frontiers in Plant Science, 16*, 1585826. https://doi.org/10.3389/FPLS.2025.1585826/FULL

Ross, A. F. (1961). Systemic acquired resistance induced by localized virus infections in plants. *Virology, 14*(3), 340–358. https://doi.org/10.1016/0042-6822(61)90319-1

Roychowdhury, R., Mishra, S., Anand, G., Dalal, D., Gupta, R., Kumar, A., & Gupta, R. (2024). Decoding the molecular mechanism underlying salicylic acid (SA)-mediated plant immunity: An integrated overview from its biosynthesis to the mode of action. *Physiologia Plantarum, 176*, e14399. https://doi.org/10.1111/PPL.14399

Saijo, Y., Loo, E. P., & Yasuda, S. (2018). Pattern recognition receptors and signaling in plant–microbe interactions. *The Plant Journal, 93*, 592–613. https://doi.org/10.1111/TPJ.13808

Sanchez-Lucas, R., Bosanquet, J. L., Henderson, J., Catoni, M., Pastor, V., & Luna, E. (2025). Elicitor specific mechanisms of defence priming in oak seedlings against powdery mildew. *Plant, Cell & Environment, 48*, 4455–4474. https://doi.org/10.1111/pce.15419

Sarkar, S., Ganapathysubramanian, B., Singh, A., Fotouhi, F., Kar, S., Nagasubramanian, K., Chowdhary, G., das, S., Kantor, G., Krishnamurthy, A., Merchant, N., & Singh, A. K. (2024). Cyber-agricultural systems for crop breeding and sustainable production. *Trends in Plant Science, 29*, 130–149. https://doi.org/10.1016/j.tplants.2023.08.001

Sen, R., & Barnes, C. (2021). Do transgenerational epigenetic inheritance and immune system development share common epigenetic processes? *Journal of Developmental Biology, 9*, 20. https://doi.org/10.3390/JDB9020020

Shaikh, A. A., Chachar, S., Chachar, M., Ahmed, N., Guan, C., & Zhang, P. (2022). Recent advances in DNA methylation and their potential breeding applications in plants. *Horticulturae, 8*, 562. https://doi.org/10.3390/HORTICULTURAE8070562

Singh, P., & Mukhopadhyay, K. (2021). Comprehensive molecular dissection of TIFY transcription factors reveal their dynamic responses to biotic and abiotic stress in wheat (*Triticum aestivum L.*). *Scientific Reports, 11*, 9739. https://doi.org/10.1038/s41598-021-87722-w

Singh, K. M., Jha, A. B., Dubey, R. S., & Sharma, P. (2025). Nanoparticle-mediated mitigation of salt stress-induced oxidative damage in plants: Insights into signaling, gene expression, and antioxidant mechanisms. *Environmental Science: Nano, 12*, 2983–3017. https://doi.org/10.1039/D5EN00174A

Sintaha, M. (2025). Molecular mechanisms of plant stress memory: Roles of non-coding RNAs and alternative splicing. *Plants, 14*, 2021. https://doi.org/10.3390/PLANTS14132021

Sishodia, R. P., Ray, R. L., & Singh, S. K. (2020). Applications of remote sensing in precision agriculture: A review. *Remote Sensing, 12*(19), 3136. https://doi.org/10.3390/rs12193136

Slaughter, A., Daniel, X., Flors, V., Luna, E., Hohn, B., & Mauch-Mani, B. (2012). Descendants of primed arabidopsis plants exhibit resistance to biotic stress. *Plant Physiology, 158*, 835–843. https://doi.org/10.1104/pp.111.191593

Sobral, M., Sampedro, L., Neylan, I., Siemens, D., & Dirzo, R. (2021). Phenotypic plasticity in plant defense across life stages: Inducibility, transgenerational induction, and transgenerational priming in wild radish. *Proceedings of the National Academy of Sciences of the United States of America, 118*(33). https://doi.org/10.1073/PNAS.2005865118

Sobti, R. C., Ali, A., Dolma, P., Kadwalia, A., Dolma, T., Rai, J., & Chauhan, A. (2022). Emerging techniques in biological sciences. *Advanced Animal Experimental Model: Understanding Life Phenomena*, 3–18. https://doi.org/10.1016/B978-0-323-90583-1.00013-1

Sultana, R., Imam, Z., Kumar, R. R., Banu, V. S., Nahakpam, S., Bharti, R., Bharadwaj, C., Singh, A. K., Pasala, R. K., Singh, D. R., & Siddiqui, M. W. (2025). Signaling and defence mechanism of jasmonic and salicylic acid response in pulse crops: Role of WRKY transcription factors in stress response. *Journal of Plant Growth Regulation*, *44*, 5–21. https://doi.org/10.1007/S00344-023-11203-9/METRICS

Swaminathan, B., & Vairavasundaram, S. (2024). D2CNN: Double-staged deep CNN for stress identification and classification in cropping system. *Agricultural Systems*, *216*, 103886. https://doi.org/10.1016/j.agsy.2024.103886

Tabassum, T., Farooq, M., Ahmad, R., Zohaib, A., & Wahid, A. (2017). Seed priming and transgenerational drought memory improves tolerance against salt stress in bread wheat. *Plant Physiology and Biochemistry*, *118*, 362–369. https://doi.org/10.1016/J.PLAPHY.2017.07.007

Taha, M. F., Mao, H., Zhang, Z., Elmasry, G., Awad, M. A., Abdalla, A., . . . Elsherbiny, O. (2025). Emerging technologies for precision crop management towards agriculture 5.0: A comprehensive overview. *Agriculture*, *15*(6), **582**. https://doi.org/10.3390/agriculture15060582

Talarico, E., Zambelli, A., Araniti, F., Greco, E., Chiappetta, A., & Bruno, L. (2024). Unravelling the epigenetic code: DNA methylation in plants and its role in stress response. *Epigenomes*, *8*, 30. https://doi.org/10.3390/EPIGENOMES8030030

Tan, F., Cang, H., Gao, X., Wu, N., di, R., Zhang, Y., Gao, P., Lv, X., & Zhang, C. (2025). Early detection of cotton Verticillium wilt based on generative adversarial networks and hyperspectral imaging technology. *Industrial Crops and Products*, *231*, 121167. https://doi.org/10.1016/j.indcrop.2025.121167

Thilakarathne, A. S., Liu, F., & Zou, Z. (2025). Plant signaling hormones and transcription factors: Key regulators of plant responses to growth, development, and stress. *Plants*, *14*, 1070. https://doi.org/10.3390/PLANTS14071070

Ton, J., Jakab, G., Toquin, V., Flors, V., Iavicoli, A., Maeder, M. N., Métraux, J. P., & Mauch-Mani, B. (2005). Dissecting the β-aminobutyric acid–induced priming phenomenon in arabidopsis. *The Plant Cell*, *17*, 987–999. https://doi.org/10.1105/tpc.104.029728

Torres-Garcia, S., Yaseen, I., Shukla, M., Audergon, P. N. C. B., White, S. A., Pidoux, A. L., & Allshire, R. C. (2020). Epigenetic gene silencing by heterochromatin primes fungal resistance. *Nature*, *585*, 453–458. https://doi.org/10.1038/S41586-020-2706-X;TECHMETA

Tripathi, A. D., Mishra, R., Maurya, K. K., Singh, R. B., & Wilson, D. W. (2019). Estimates for world population and global food availability for global health. In R. B. Singh, R. R. Watson, & T. Takahashi (Eds.), *The role of functional food security in global health* (pp. 3–24). **Academic Press**. https://doi.org/10.1016/B978-0-12-813148-0.00001-3

Tsuda, K., & Somssich, I. E. (2015). Transcriptional networks in plant immunity. *The New Phytologist*, *206*, 932–947. https://doi.org/10.1111/NPH.13286

Upadhyay, V. R., Ramesh, V., Kumar, H., Somagond, Y. M., Priyadarsini, S., Kuniyal, A., Prakash, V., & Sahoo, A. (2024). Phenomics in livestock research: Bottlenecks and promises of digital phenotyping and other quantification techniques on a global scale. *OMICS: A Journal of Integrative Biology*, *28*(8), 380–393. https://doi.org/10.1089/OMI.2024.0109

Varadharajan, V., Rajendran, R., Muthuramalingam, P., Runthala, A., Madhesh, V., Swaminathan, G., Murugan, P., Srinivasan, H., Park, Y., Shin, H., & Ramesh, M. (2025).

Multi-omics approaches against abiotic and biotic stress—a review. *Plants, 14*, 865. https://doi.org/10.3390/PLANTS14060865

Vidhyasekaran, P. (2020). *Plant innate immunity signals and signaling systems.* **Springer Dordrecht**. https://doi.org/10.1007/978-94-024-1940-5

Vo, K. T., Rahman, M. M., Rahman, M. M., Trinh, K. T., Kim, S. T., & Jeon, J. S. (2021). Proteomics and metabolomics studies on the biotic stress responses of rice: An update. *Rice, 14*, 30. https://doi.org/10.1186/s12284-021-00461-4

Wagh, S. G., Patil, A. M., Patil, G. B., Mankar, S. P., Rastogi, K., & Nishiguchi, M. (2025). Small RNA and epigenetic control of plant immunity. *DNA* https://doi.org/10.3390/DNA 5040047, *5*.

Wang, Y., Du, X., Ma, G., Liu, Y., Wang, B., & Mao, H. (2020). Classification methods for airborne disease spores from greenhouse crops based on multifeature fusion. *Applied Sciences, 10*, 7850. https://doi.org/10.3390/app10217850

Wang, M. R., Bi, W., Shukla, M. R., Ren, L., Hamborg, Z., Blystad, D. R., Saxena, P. K., & Wang, Q. C. (2021). Epigenetic and genetic integrity, metabolic stability, and field performance of cryopreserved plants. *Plants, 10*(9), 1889. https://doi.org/10.3390/PLANTS10091889.

Wilkinson, S. W., Pastor, V., Paplauskas, S., Pétriacq, P., & Luna, E. (2018). Long-lasting β-aminobutyric acid-induced resistance protects tomato fruit against *Botrytis cinerea. Plant Pathology, 67*, 30–41. https://doi.org/10.1111/ppa.12725

Yadav, G., & Mathur, J. (2025). Seed priming with fruit and vegetable waste-based nanoparticles: A sustainable green approach. *Journal of Soil Science and Plant Nutrition, 25*, 303–321. https://doi.org/10.1007/s42729-024-02134-9

Yadav, N., Bora, S., Devi, B., Upadhyay, C., & Singh, P. (2024). Nanoparticle-mediated defense priming: A review of strategies for enhancing plant resilience against biotic and abiotic stresses. *Plant Physiology and Biochemistry, 213*, 108796. https://doi.org/10.1016/J.PLAPHY.2024.108796

Yang, Z., Zhi, P., & Chang, C. (2022). Priming seeds for the future: Plant immune memory and application in crop protection. *Frontiers in Plant Science, 13*, 1–11. https://doi.org/10.3389/fpls.2022.961840

Yeasmin, S., Dipto, A. R., Zakir, A. B., Shovan, S. D., Suvo, M. A. H., Bhuiyan, M. A., Amin, M. N., Rashid, T. U., Islam, S., & Habib, A. (2024). Nanopriming and AI for sustainable agriculture: Boosting seed germination and seedling growth with engineered nanomaterials, and smart monitoring through deep learning. *ACS Applied Nano Materials, 7*, 8703–8715. https://doi.org/10.1021/acsanm.4c00109

Zhan, L.-P., Peng, D.-L., Wang, X.-L., Kong, L. A., Peng, H., Liu, S. M., Liu, Y., & Huang, W. K. (2018). Priming effect of root-applied silicon on the enhancement of induced resistance to the root-knot nematode *Meloidogyne graminicola* in rice. *BMC Plant Biology, 18*, 1–12. https://doi.org/10.1186/s12870-018-1266-9

Zhang, C., & Hsieh, T.-F. (2013). Heritable epigenetic variation and its potential applications for crop improvement. *Plant Breeding and Biotechnology, 1*(4), 307–319. https://doi.org/10.9787/PBB.2013.1.4.307

Zhang, X., Ibrahim, Z., Khaskheli, M. B., Raza, H., Zhou, F., & Shamsi, I. H. (2024a). Integrative approaches to abiotic stress management in crops: Combining bioinformatics educational tools and artificial intelligence applications. *Sustainability, 16*, 7651. https://doi.org/10.3390/su16177651

Zhang, X., Zhou, Y., Liu, Y., Li, B., Tian, S., & Zhang, Z. (2024b). Research progress on the mechanism and function of histone acetylation regulating the interaction between pathogenic fungi and plant hosts. *Journal of Fungi, 10*, 522. https://doi.org/10.3390/JOF10080522

Zhang, R., Zhu, H., Chang, Q., & Mao, Q. (2025). A comprehensive review of digital twins technology in agriculture. *Agriculture, 15*, 903. https://doi.org/10.3390/agriculture15090903

Zheng, Y., Sarigul, E., & Agyepong, K. (2025). Automated assessment of field-grown crop biotic stresses using AI approaches. In M. S. Kim, B.-K. Cho, & F. Vasefi (Eds.), *Sensing for agriculture and food quality and safety XVII* (p. 32). SPIE.

Zhou, P., Adeel, M., Shakoor, N., Guo, M., Hao, Y., Azeem, I., Lei, M., Liu, M., & Rui, Y. (2021). Application of nanoparticles alleviates heavy metals stress and promotes plant growth: An overview. *Nanomaterials (Basel), 11*, 1–18. https://doi.org/10.3390/nano11010026

9

AI Technology for Studying Plant Noncoding RNAs

Amarjeet Singh Bhogal, Rinee Doley, Shibani Ritusmita Borah, Niharika Saharia, Olympica Das, Rekha Sharma, Satwik Subhankar, Shyamalin Rajmedhi, Dikshita Hazarika, and Debojit Sarma

Department of Plant Breeding and Genetics, Assam Agricultural University, Jorhat, Assam, India

9.1 Introduction

In the dynamic field of plant biology, the discovery and investigation of noncoding RNAs (ncRNAs) have challenged the long-held belief that proteins control the majority of cellular processes. These adaptable RNA molecules, which were previously considered as genomic "noise" or "junk DNA", have quickly become key factors in the coordination of plant gene regulation, adaptation, and stress resilience (Yadav et al., 2024). Today, as the rapidly expanding artificial intelligence (AI) provides unparalleled analytical depth to biological research, understanding the hidden world of plant ncRNAs holds the possibility of unlocking new routes for agricultural innovation and sustainability.

The classical central dogma of molecular biology delineates the transfer of genetic information from DNA to RNA to proteins, establishing RNA as a conduit between genetic storage and functional implementation. While this theory has long stressed messenger RNA (mRNA) as the essential intermediary, subsequent studies demonstrated that RNA roles extend much beyond just encoding proteins (Cao et al., 2024). Extensive experimental data later confirmed Britten and Davidson's early theories, which showed that different RNA species directly regulate gene expression (Britten & Davidson, 1969; Prasad et al., 2021). Advances in transcriptomics have since shown that more than 90% of the eukaryotic genome is transcribed into RNA, with only a tiny fraction actually translating into proteins. The great majority are ncRNAs, which were originally rejected as transcriptional noise but are now recognized as key regulators of transcription, posttranscriptional processing, translation, and epigenetic modification (Zhou et al., 2012).

The advancement in the field of machine learning and AI has opened up fascinating new possibilities for overcoming the challenges in plant ncRNA research. This has been especially revolutionary in plant biology, where ncRNA research lags behind that of animals due to higher transcriptome complexity and lower sequence conservation. Genome-wide RNA sequencing and bioinformatic pipelines have supplemented (and in many cases, surpassed) conventional experimental techniques, including Northern blotting, qRT-PCR, and in situ hybridization (Chao et al., 2022; Song et al., 2009). However, there are substantial challenges in the identification and characterization of ncRNAs because of their sequence characteristics, complex biogenesis methods, and species-specific expression patterns.

The development of machine learning and artificial intelligence has opened up fascinating new possibilities for overcoming the challenges in plant ncRNA research. Earlier computational tools, such as CPC and CPC2, showed limited effectiveness for plants because their models were based on animal-derived training datasets using support vector machines. In contrast, the development of plant-specific tools like PLncPRO, RNAplonc, PreLnc, and CNIT—which incorporate both sequence and structural features and are trained on plant data—has markedly enhanced prediction accuracy (Cao et al., 2020; Guo et al., 2019; Negri et al., 2019; Singh et al., 2017). Besides, AI-based pipelines like PINC, LncFinder-plant, and Plant-LncPipe are simplifying ncRNA classification, functional annotation, and integration with multiomics data (Han et al., 2019; Tian et al., 2024; Zhang et al., 2022).

Recognizing the pivotal role of ncRNAs in the regulation of gene expression, this review compiles existing knowledge about their classification, structure, and functional relevance in plants. It also looks at their many biological significances and how computational methods have changed over time, from basic bioinformatics tools to more powerful AI approaches. The comparison of AI-based prediction systems, new deep learning frameworks, and their potential uses in biotechnology and plant breeding is also reviewed. These developments present fresh chances to expand our knowledge of plant ncRNAs and realize their potential for long-term crop improvement by fusing molecular biology and computational techniques.

9.2 Overview of Plant Noncoding RNAs

9.2.1 Classification of ncRNA

The central dogma of molecular biology stated protein as the functional molecule in gene regulation. It also established mRNA as a pivotal component and a temporary intermediate in gene expression. Our genome is composed of coding and noncoding regions, and for a long time, it was believed that the majority of important genes lie in the coding regions and the rest portion was considered as junk. High-throughput sequencing and advancement in transcriptomics, proteomics, and bioinformatics have revolutionized our understanding of the RNA world, demonstrating that a large portion of the genome transcribes into RNA without coding for any protein. These transcripts are referred to as non-protein coding or ncRNA (Djebali et al., 2012). Recent studies also proved that the majority of genomes of complex organisms are initially transcribed into ncRNAs that are later spliced or processed into smaller products. These RNAs are now recognized as important regulators of gene expression, chromatin remodeling, genomic stability, and cellular signaling (Mattick & Makunin, 2006). These ncRNAs include siRNA, miRNA, lncRNA, etc.

Plants consist of a wide variety of plant ncRNAs that have unique functions. They are expressed under diverse conditions and responses. Due to their multiple functions and manifold diversity, it is often difficult to classify them into distinct classes. On the basis of nucleotide size, biogenesis, function, and processing, ncRNA can be classified as small ncRNA and long ncRNA (Saha et al., 2025). The small ncRNAs are 20–25 nt long and are derived from single-stranded or double-stranded RNA; these include microRNA and small interfering RNA (siRNA). Long ncRNAs are 100–200 nt long and perform distinct functions as compared to miRNA and siRNA. The structure, biogenesis, function, and gene regulation of these ncRNAs have been discussed below.

9.2.1.1 MicroRNA

MicroRNAs are short, 20–22 nt long, endogenous ncRNAs that are highly conserved across several species. These RNAs are responsible for negative gene regulation through mRNA cleavage or mRNA translation suppression. Multiple target mRNAs can be regulated by miRNAs that are encoded by Mir genes conserved in the DNA (Vaistij et al., 2010). MiRNA biogenesis involves several stages to produce mature miRNAs. MiRNA genes are initially transcribed into long primary miRNA (pri-mRNA) by PolIII. Pri-miRNA is cleaved into a precursor (pre-miRNA) by Dicer-like 1 (DCL1) enzyme in plants and Drosha in animals (Yadav et al., 2024). In plants, miRNA is cleaved into miRNA: miRNA duplex in the nucleus before transporting into the cytoplasm. These precursors are transported from the nucleus to the cytoplasm by HASTY. In the cytoplasm, DICER unwinds the duplex to form mature miRNA, which is then integrated into the RNA-induced silencing complex (RISC) (Bernstein et al., 2001; Lee et al., 2012). This RISC then binds to the target gene and inhibits gene expression. These small RNAs control several plant processes, including germination, leaf morphology, shoot development, flowering, fruiting, as well as stress response. Currently, around half of the known miRNA targets are identified as transcription factors (Bartel, 2004; Kurihara & Watanabe, 2004; Zhang et al., 2006). Among these, at least 12 miRNA families target mRNA which encodes transcriptional factors that control growth in plants.

9.2.1.2 Si-RNAs

SiRNAs are a class of 20–25 nt long, double-stranded RNA molecules. These are derived from exogenous double-stranded DNA (virus, transposons, signal transduction, immune response, etc.) or are produced by processing long, completely complementary, double-stranded RNA precursors by the action of DICER enzymes. The siRNA duplex is then loaded into RISC. One strand acts as the guide strand in directing RISC to the complementary mRNA sequence, which is further cleaved. This results in mRNA degradation by gene silencing at a posttranscriptional level. Plant siRNAs have been shown to direct sequence-dependent endolytic cleavage of mRNA to regulate gene processes (Tomari & Zamore, 2005). Besides their role in plant growth and metabolism, siRNAs have a significant role in stress response and function in defense against foreign particles. SiRNAs utilize perfect complementarity for cleavage and usually target one gene. Based on the mode of origin, siRNAs can be endogenous siRNAs (from repeat-rich genomic regions, transposable elements, and retro-elements) and exogenous siRNAs (viral RNA and transgenes).

9.2.1.3 Long Noncoding RNA

LncRNAs are the class of ncRNAs that are more than 200 nt in length and function either as primary or spliced transcripts. These mRNAs are primarily transcribed by RNA polymerase II similar to mRNA; however, many lncRNAs avoid the standard mRNA processing. In eukaryotes, RNA polymerases (Pol I, II, and III) are used to transcribe their nuclear genome into coding and noncoding transcripts, including the synthesis of lncRNA (Yadav et al., 2024). Along with this, Pol IV and Pol V are used to generate lncRNAs for transposable element recognition. Most of them are similarly multiexonic, 5′-capped, and 3′polyadenylated and have transcriptional activity comparable to mRNA. Despite being found and expressed in a wide variety of taxa, lncRNAs are not well conserved while in evolution (Ma et al., 2022). This uncertainty about functionality and low expression often makes them look like transcriptional noise. Substantial evidence suggests that they have a key role in various biological processes including splicing, transcription, translation, protein localization, cell cycle regulation, and apoptosis. LncRNAs are usually long, which enables them to form secondary structures and bind to DNA, RNA, and protein. They interact with almost-perfect complementarity to bind to targets and facilitate alternate splicing, control gene splicing, and modulate translation efficiency of mRNAs (Ma et al., 2022). LncRNAs are identified by certain sequence motifs and function via sequence complementarity to DNA or RNA. They can function as competing endogenous RNAs (ceRNAs) by regulating miRNA activity and modulating miRNA-mediated repression of target genes.

Four regulatory mechanisms of lncRNAs have been identified: signal, decoy, guide, and scaffold. Signal lncRNAs regulate gene expression by binding to transcription factors and by participating in signal pathways. Decoy lncRNAs use RNA-binding proteins, including transcription factors and chromosomal alterations, to indirectly control gene expression (Shen et al., 2022a). Guide lncRNAs direct chromatin-modifying enzymes to target genes or guide ribonucleoprotein complexes to particular sites. Scaffold lncRNAs enable simultaneous and spatial transcriptional activation or repression by combining many proteins to form ribonucleoprotein complexes. Apart from these four processes, lncRNAs can also function as building blocks for the production of short ncRNAs (siRNA and miRNA) in a variety of plants.

9.2.2 Functional Roles and Mechanisms in Plant Growth, Development, Stress Responses, and Adaptation

Plants, being sessile in nature, have developed an intricate gene regulation mechanism that synchronizes biological functions and adjusts to stressful conditions. This complex mechanism is imperative for growth, development, and stress-responsive signaling, contributing to plant sustainability. This complex mechanism is mediated by gene coding as well as noncoding elements. Developments in proteomics, bioinformatics, RNA sequencing, and sequencing technologies have highlighted the critical function of ncRNA in gene regulation. These ncRNAs serve as important regulators in this process by regulating gene expression at transcriptional, posttranscriptional, and epigenetic levels (Yadav et al., 2024).

9.2.2.1 Role in Plant Growth and Development

Root Development

In plants, several classes of ncRNAs play a vital role in controlling root formation. They are involved in tissue differentiation; organization of cell layers; and formation of meristem, cork, cambium, and other secondary tissues. TAS3-derived siRNAs alter root patterning and architecture by targeting auxin-response factors (ARFs) and leafbladeness1 (Marin et al., 2010). MiRNAs such as miR165/166, mir156, and miR165 regulate HD-ZIP III transcription factors, SPL10 and SHR to mediate vascular patterning and root meristem differentiation. Furthermore, CDR1 (Crown Root Defect 1) controls the growth of crown roots in rice through the CRD1-miR156-SPL pathway (Zhu et al., 2019). LncRNA Enod40 generates biological functions by generating peptides during root nodule formation. ASCO-lncRNA generated by Enod40 transcription in *Arabidopsis thaliana* competes with mRNA to bind AS regulators to regulate root development.

Shoot Development

Shoot development is an important process that determines a plant's growth and metabolism. During the vegetative stage, plants undergo rapid growth characterized by leaf expansion and stem elongation, processes that are intricately regulated by a diverse array of ncRNAs. Several siRNAs such as TAS3-siRNA (ARF3) and TAS3-siRNA (leafbladeness1) are responsible for leaf development in *A. thaliana* (Chitwood et al., 2009; Schwab et al., 2009). MiR164 plays a crucial role in shaping leaf morphology by targeting cup-shaped cotyledon (CUC) transcription factors, which are instrumental in establishing clear boundaries between emerging leaf primordia, thereby ensuring well-defined leaf margins and overall organ architecture (Zhang et al., 2006). Similarly, miR319 modulates the expression of TCP transcription factors that influence not only the size and curvature of leaves but also the proliferation of cells within the developing tissues, contributing to balanced and coordinated vegetative growth. Twisted leaf and age-related lncRNAs regulate gene expression throughout many stages of leaf development in rice (Liu et al., 2018).

Phase Transition and Flowering

MicroRNAs play a crucial role in plant growth, flowering, and circadian rhythm of plants. During the transition from the juvenile stage to the reproductive stage, several changes occur within the plant such as increase in girth, change in leaf morphology, flowering, and fruiting; these changes are triggered by ncRNAs. During the vegetative stage, miRNAs shape leaf morphology, maintain the juvenile state, and regulate the transition to flowering and fruit grain setting. High levels of miR156 during early vegetative development help maintain the juvenile stage by repressing the squamosa promoter protein-like (SPL) genes, ensuring sufficient biomass reproduction development. They interact with phytohormones like auxins, gibberellins, and abscisic acid (ABA) in a circadian-dependent manner, fine-tuning processes like root growth, flowering, and leaf movement in response to the time of day. LncRNAs regulate flowering in Arabidopsis by upregulating COOLAIR, an antisense lncRNA during vernalization by mediating histone modification in the FLC gene (Csorba et al., 2014, Shafiq et al., 2016). LncRNA LDMAR controls male sterility in rice by RNA-dependent DNA-methylation (RdDm) (Ding et al., 2012). MiRNA172 influences the formation of reproductive organs and coordinates processes leading to effective pollination and fruit or grain development. MiRNA also plays a key role in regulating the circadian rhythm of plants.

9.2.2.2 Role in Stress Response and Adaptation

Drought Stress

NcRNAs are essential for adaptation to drought stress. Under water-deficient conditions, siRNAs generated by dehydration help to maintain cellular homeostasis. MiRNAs play a key role in controlling how plants react to abiotic stress, which has enormous potential for agricultural enhancement. MiRNAs such as miR398 and miR156 regulate plant's stress response under water-deficit conditions. These miRNAs modify growth and trigger defense mechanisms by integrating signals from hormonal changes brought by drought. During seed germination in Arabidopsis, miR159 builds up and mediates the cleavage of MYB33 and MYB101 in response to ABA and dehydration (Reyes & Chua, 2007). In tomato, sly-miR159 enhances drought tolerance by accumulation of osmoprotective substances such as putrescine and proline (López-Galiano et al., 2019).

Salinity Stress

Plant growth and development are severely affected by salinity. NcRNAs play a vital role in plants' salinity tolerance by triggering genes for maintaining ionic balance, early flowering, fruiting, and increasing tiller number. SiRNAs such as OsSIDp301 employ stress-induced regulatory pathways to confer salinity tolerance (Ge et al., 2022). In peanuts, salt stress triggers the expression of miR160 and its ARF target genes, whereas in apples, miR156 both increases and decreases salt resistance (Ma et al., 2013). Different plants may respond differently to overexpression of the same miRNA. Overexpression of the miR393-resistant version of mTIr1 increases salt tolerance, whereas overexpression of Osa-miR393 decreases salt tolerance in rice and Arabidopsis (Gao et al., 2011; Xia et al., 2012). MiRNAs such as osa-miR12477 modulate lysine oxoglutarate reductase (LAO) to improve salt tolerance (Parmar et al., 2020). *Medicago truncatula* and several other crops include a number of salt-responsive lncRNAs that control gene expression and ion balance in saline conditions (Wang et al., 2015).

Temperature Stress

Plants are subjected to various abiotic stresses, and temperature stress is one of the most detrimental ones. Plants regulate their metabolism and physiology in response to such stress, which is highly mediated by ncRNAs. Several ncRNAs trigger plants' stress response to changing temperature. NcRNAs control flowering and vernalization under cold stress and target genes that produce heat shock proteins under heat stress. During heat stress, phased siRNAs including heat-induced TAS1 TARGET1 function as important regulators (Li et al., 2014). Thermotolerance is mediated by heat-responsive miRNAs, such as miR164, which target NAC transcription factors. Resistance to cold stress in rice is increased by the overexpression of Osa-miR319b, which upregulates genes such as dehydration-responsive element binding protein/C-repeat binding factor (DREB1/CBF), DREB2A, and Trehalose-6-phosphate phosphate 1/2 (TPP1/2) (Wang et al., 2014). Cold stress is also regulated by lncRNAs such as CIL1 (cold-induced lncRNA). In cotton, lncRNA XH123 has also been identified to contribute to cold stress (Cao et al., 2021). The Sucrose synthase4 gene is expressed under heat stress by the target lncRNA lnc-173 (Di et al., 2014).

Biotic Stress

Plant defense against disease and insects is greatly aided by ncRNAs, which control immune responses and guarantee a timely and efficient defense response. Genes involved in pathogen identification, signaling, and activation of plant defense systems, such as hypersensitive response, are all influenced by their expression. They also control the expression of genes involved in the production of antimicrobial substances through pattern-triggered immunity (PTI) and effector-triggered immunity (ETI). NcRNAs control the plant's pattern recognition receptors (PRRs).

For defense against viral infections, ncRNAs use gene silencing and encode viral suppressors against infection. They regulate host immunity response by targeting foreign nucleic acid and influencing gene expression by antiviral signaling. Such immunity is triggered against several plant viral diseases, including cauliflower mosaic virus (CMV), tobacco stunt virus (TSV), sweet potato chlorotic stunt virus (SPSCV), and tomato bushy stunt virus (TBSV). Small interfering RNAs are generated from viral DNA, which utilizes RNA interference for direct cleavage of foreign DNA. MicroRNAs bind to the viral genome and inhibit replication by degradation and translation inhibition of viral DNA. They also modulate interferon signaling to regulate immune-responsive genes. LncRNAs indirectly regulate stress response by expressing critical genes for defense or for inferring their function.

For other diseases, miRNAs target host immune genes, affect host-parasite interactions, and regulate inflammatory cytokines. In Arabidopsis, miR172b increases the transcription of the FLS2 gene by targeting TOE1 and TOE2 mRNAs, which inhibit FLKAGELLIN-sensitive 2 (FLS2) promoter (Zou et al., 2018). MiR166b regulates the synthesis of ethylene and jasmonate in tomatoes by targeting the SIHDZ34 and SIHDZ45 genes. Activation of these genes makes plants resistant to *Phytophthora infestans* (Hong et al., 2022).

LncRNAs act as scaffolds or decoys to upregulate immune signaling against viral infections. They are also involved in altering macrophage response and activation of IFN pathways. In blast disease of rice, ncRNAs such as lncRNAs and sRNAs control the expression of the R gene. RT-qPCR validated 161 lncRNAs that were upregulated and downregulated under disease incidence (Wang et al., 2020). These RNAs also regulate hormone signaling to confer disease resistance. In wheat, resistance to *Verticillium dahlia* is conferred by suppression of F-box auxin receptor expression, and transport inhibitor response (TIR1) through the inhibition of auxin signaling mediated by upregulation of miR393 (Shi et al., 2022). In *Gossypium hirsutum*, a transacting lncRNA (lnc_015092) targets the enzyme lipoxygenase (LOX3) (Ghlncl OX3), which is essential for the production of jasmonic acid.

9.2.3 Challenges in Studying Plant ncRNA

Study of plant ncRNA offers significant challenges owing to its unique characteristics and biological complexity. Owing to its large number and diverse functions, expression of ncRNA is very tissue specific, stage specific, and stress specific and therefore needs sophisticated functional genomic tools and enhanced computational tools for profiling, which is often time-consuming and technically demanding. They are also context specific and functionally redundant as loss of one ncRNA is concealed by another, obscuring phenotypic effects. Their expression and function often depend on stress conditions, developmental stages, and metabolism, and often, one target is regulated by multiple ncRNAs, which makes them difficult to distinguish. Some ncRNAs are highly unstable in expression, which makes isolation and sequencing difficult. In contrast to many conserved microRNAs, many lncRNAs do not exhibit species-to-species evolutionary sequence conservation, making it challenging to find orthologues or predict their activities across species. Lack of sequence conservation makes it difficult to study by predicting sequence homology and structure prediction tools. Identification algorithms depend on eliminating RNAs with mRNA-like and snRNA-like features and classifying remaining transcripts as lncRNA. Along with this, short, protein-coding transcripts and noncoding transcripts with large ORFs make it challenging to identify an organism's complete lncRNA diversity; inadequate ncRNA annotations and the absence of a reliable bioinformatics process for precise target mapping and classification obscure identification, as some lack open reading frames. Their study and identification demands advanced sequencing tools that are often challenging.

9.3 Traditional Approaches for ncRNA Analysis in Plants

9.3.1 Laboratory-Based Detection: RNA-seq, qRT-PCR, Northern Blot, In Situ Hybridization

9.3.1.1 Northern Blotting for RNA Detection

In order to identify ncRNAs in plants, Northern blotting begins with the isolation of total RNA from tissues using the phenol extraction method or kits and then DNase treatment to eliminate DNA. After being denatured by heating it with formaldehyde and formamide, the RNA is loaded onto an agarose gel for size separation (for longer ncRNAs) or a denaturing polyacrylamide gel for small RNAs. Following electrophoresis, RNA is moved to a positively charged membrane, such as nylon, and fixed by baking or UV crosslinking. After prehybridizing the membrane, a labeled probe unique to the target ncRNA is used to hybridize it overnight. The size and abundance of the target ncRNA are revealed by exposing the membrane to X-ray film (a radioactive probe) or a luminescent substrate (a nonradioactive probe) after thorough washes to eliminate nonspecific binding.

Zhao et al. used small RNA Northern blot analysis to validate both novel and conserved miRNAs in maize under nitrogen deficiency, confirming and expanding on results from deep sequencing techniques. Nitrogen deficiency was found to differentially regulate a number of conserved miRNAs, including MIR169c/r, MIR395s, MIR528a/b, and MIR827, as well as several newly discovered miRNAs, including miRC1, miRC3, miRC5, miRC7, miRC171o/p, miRC11, miRC12, miRC18, miRC169s, miRC19, and miRC37. Their approach using hydroponically grown seedlings, 32P-labeled probes, and careful loading controls solidified Northern blotting's role as a standard tool for miRNA studies in crop nutrient response research.

9.3.1.2 RT-PCR and qRT-PCR for Plant ncRNA Expression Validation

RT-PCR and qRT-PCR for validating plant ncRNA expression involve several critical steps, including RNA extraction, cDNA synthesis, and quantitative PCR setup. This process is essential for accurately measuring the expression levels of lncRNAs in various plant species without requiring complete genomic data.

Total RNA is extracted from young panicles or leaves, as high-quality RNA is crucial for genuine qRT-PCR results. Quality assessment of RNA can be performed using spectral measurements and electrophoresis to confirm integrity before proceeding to cDNA synthesis. Following RNA extraction, reverse transcription is performed to synthesize complementary DNA (cDNA). This step is vital as it converts RNA into a stable form that can be amplified during PCR (Sah et al., 2014). The qRT-PCR process utilizes SYBR green dye for quantification, which binds to double-stranded DNA and emits fluorescence proportional to the amount of PCR product generated (Dhamija & Menon, 2021).

Tribhuvan et al. demonstrated how RT-qPCR can effectively validate plant ncRNA expression by studying salt stress responses in Indian mustard (*Brassica juncea*). The researchers extracted RNA from stressed and control plants and then used RT-qPCR to confirm the expression of five lncRNAs and their target genes. Their results showed that four out of five lncRNAs were significantly upregulated (ranging from 5.2- to 9.4-fold increases) under salt stress conditions, which matched well with their sequencing data. This approach allowed them to verify that these lncRNAs genuinely respond to salt stress and play important roles in the plant's stress tolerance mechanisms through lncRNA-miRNA-mRNA regulatory networks.

9.3.1.3 In Situ Hybridization

The process of in situ hybridization (ISH) to detect ncRNA genes involves fixing cells or tissues to preserve RNA and structure, followed by hybridization with a labeled probe complementary to the target ncRNA. The probe, which is frequently labeled with radioactive labels, fluorophores, or digoxigenin (DIG), binds exclusively to the ncRNA within the fixed tissue. After incubation, excess probe is washed away by strict washing procedures to reduce background. Antibodies conjugated to enzymes or fluorescent dyes are used to detect the bound probe, which results in a visible signal at the location of the RNA. This method makes it possible to visualize the spatial distribution and abundance of particular ncRNAs within cells or tissues.

Mancini et al. (2020), for instance, provided comprehensive procedures for identifying lncRNAs in human skin tissue, which aided in determining the expression locations and potential functions of these RNAs. Trotter et al. (2021) used fluorescence in situ hybridization (FISH) in another study to identify particular lncRNAs within individual cells and reveal information about their functions. To determine precisely where and how much of the target ncRNA is present in the cells, all these methods involve fixing cells or tissues to preserve RNA and structure, using labeled probes that have been specially made to hybridize to the ncRNA, and then observing the signals under a microscope.

9.3.2 Computational Biology Before AI: Sequence Alignment, Comparative Genomics, and Motif Analysis

9.3.2.1 Cloning and Sanger Sequencing of ncRNA Genes

Specific primers are designed for lncRNAs. The extraction of RNA is done for the sample, followed by conversion to cDNA. PCR amplification is done to amplify the target sequences. After getting the PCR products, purification is done using a gel purification kit and cloned into cloning vectors. Finally, the clones are sent for Sanger sequencing to confirm the actual sequences. Comparison is done between Sanger results with the original predictions to check if high lncRNAs showed any high-sequence identity.

Tian et al. (2020) used PCR amplification, cloning into pEASY-T1 vectors, and Sanger sequencing to validate predicted lncRNAs in peanuts (*Arachis hypogaea*). By matching the Sanger-sequenced cloned products with their corresponding genomic sequences, they were able to successfully validate six of the seven predicted lncRNAs (lncRNA1–7). For five lncRNAs, they obtained sequence identities of 99.33–99.86%, confirming the accuracy of their lncRNA predictions through direct sequencing validation.

9.3.2.2 Comparative Genomics

Crucial steps are required for using comparative genomics to find ncRNA genes. Assuming that ncRNA genes (conserved sequences) are likely to have functional significance, researchers first identify these conserved genomic regions across several related species. After that, they find conserved noncoding regions by performing sequence alignments. Possible ncRNA candidates within these conserved regions are then predicted using bioinformatics tools and algorithms, such as covariance models or machine learning classifiers, based on characteristics like secondary structure, sequence motifs, and evolutionary conservation. To confirm their identity and function, these candidates undergo additional experimental validation using techniques like PCR, cloning, and sequencing.

Using intragenome comparisons, Chen et al. (2011) conducted a genome-wide analysis in rice to find duplicate ncRNA gene families. They predicted almost 4,000 ncRNA families by clustering repetitive sequences and using an SVM model. Many of these families were specific to rice and displayed sequence conservation footprints connected to recent tandem duplications. Their research demonstrated how, without requiring cross-species comparisons, comparative genomics within a single genome can identify species-specific ncRNA gene expansions and provide insight into ncRNA evolution.

9.3.2.3 Motif Analysis

Motif analysis for detecting ncRNA genes is a way of finding short, common sequence patterns in known ncRNAs. Researchers use computer programs to look for conserved short sequences or "motifs" that occur frequently in sequences that are known or predicted to be ncRNAs. These motifs often indicate important functional parts of the RNA. Once the motifs are found or identified, researchers look for them in new sequences to determine whether they could also be ncRNAs. After that, the structure and conservation of the putative ncRNAs are examined in more detail. This technique aids in the discovery of ncRNAs by using common sequence patterns that suggest their function in the cell.

The study done by Pierouli et al. (2025) used motif analysis for ncRNA genes. They looked into the possible functional roles of conserved motifs in genome regulation by using the MEME algorithm to find them in lncRNA sequences. The study also examined the interactions between known miRNAs and identified motifs, indicating that lncRNA motifs may act as sponges in regulatory networks by acting as miRNA binding sites. Functional insights into lncRNA-mediated gene regulation were made possible by this motif-based methodology.

9.4 Rise of AI in Plant ncRNA Research

The incorporation of machine learning (ML) and deep learning (DL) in the biological data analysis of plants represents a transformative and integrated approach to understanding and addressing key biological questions like the functioning of cells or differential expression of different genes. These technologies are of utmost importance due to their ability to handle complex and large datasets typical in plant biology research, enabled by advances in high-throughput omics technologies. The use of ML in plant systems biology is significant owing to its performance in analyzing big data effectively and more efficiently (Yan & Wang, 2022).

One of the key contributions of ML in plant biology is in plant-pathogen/microbiome interactions. Network-based analyses, when combined with ML, offer significant predictive power and help in comprehending the dynamic nature of plant-microbe interactions and their effects. These techniques can draw out specific important features from the vast biological data, which provides new insights into the molecular mechanisms underlying plant immunity and adaptability (Mishra et al., 2019). Furthermore, deep learning, a part of machine learning, has shown incredible outputs in domains like bioimage analysis, which is essential for extracting information from biological images. DL techniques improve processes such as preprocessing, segmentation, and classification of biological images, thereby enabling deeper insights into cellular structures, gene expressions, and processes in plant cells (Jan et al., 2024). In agricultural contexts, DL techniques have been applied effectively for image analysis in tomato plants, showcasing their superiority over traditional image processing methods and providing promising results in various tasks such as disease identification, categorization, and classification (Saranya et al., 2021).

ML techniques, including unsupervised and semisupervised learning, are especially crucial in scenarios where labeled data is scarce. These models enable the extraction of relevant motifs from large volumes of unlabeled databases, which is often a challenge in biological datasets. These techniques can improve phenotyping and other plant systems biology domains where thorough data labeling is impractical (Yan & Wang, 2022). The integration of DL techniques, especially convolutional neural networks (CNNs) and recurrent neural networks (RNNs), has significantly improved the classification and prediction of plant health and disease outcomes. CNNs are proficient in image analysis, thereby facilitating accurate disease identification even in complex backgrounds, while RNNs are instrumental in analyzing time series data, such as in differential gene expression profiles (Barhim et al., 2021). DL has proven its effectiveness in high-throughput phenotyping as well as in the integration of omics data and yielding new understanding into plant response to biotic and abiotic stresses and enhancing the effectiveness of precision agricultural practices. A novel explainable gradient-based convolutional neural network (EG-CNN) model has been designed that combines omics data and hyperspectral images to predict plant diseases, which is found to have achieved a test set accuracy of 95.5% when applied on datasets from plants affected by common diseases like powdery mildew and rust (Shen et al., 2022b). This approach highlights the capability of DL to combine multiple data types for improved predictive accuracy.

Overall, the incorporation of ML and DL in plant biology is revolutionizing the way researchers analyze and interpret complex data. These technologies facilitate more comprehensive analyses and predictive modeling, ultimately advancing our understanding of plant biology and enhancing agricultural productivity and sustainability. Given the rapid evolution of these technologies, they are set to play an increasingly pivotal role in future plant research endeavors.

9.5 AI-Based Tools and Pipelines for Plant ncRNA Analysis

NcRNAs, especially lncRNAs, have been recognized as crucial regulators of plant growth, development, environmental adaptation, and nutrient balance (Wierzbicki et al., 2021; Yajnik et al., 2024). Accurate identification of plant ncRNAs is essential for deepening our understanding of their biological functions. In recent years, advanced AI tools and computational pipelines have been developed to facilitate this task. Some prominent plant ncRNA analysis platforms are PINC, PlantLncBoost, LncFinder-plant, CPAT-plant, LncADeep, RNAplonc, DeepPlnc, and Plant-LncPipe. Below is an overview of these leading software and computational workflows tailored for plant ncRNA research, focusing on the tools mentioned (Tian et al., 2024; Zhang et al., 2022).

9.5.1 PINC

Plant ncRNA Identification by Automated Machine Learning or PINC is an important breakthrough in plant ncRNA research. PINC employs the AutoGluon framework, a sophisticated automated machine learning (AutoML) technique that allows for robust, adaptive modeling without the need for laborious manual optimization. A comprehensive set of 91 sequence-derived features is extracted by PINC, which then isolates the top 10 most discriminative features using statistically rigorous feature selection that combines variance thresholding and F-test. In sensitivity and precision, the tool significantly outperformed established tools such as CPC2, CPAT, CNIT, and CPPred with classification accuracy ranging from 92.74% to 96.42% after rigorous evaluation across nine different plant species (Zhang et al., 2022). In crops like *Zea mays* and *Sorghum bicolor*, for instance, PINC maintained sensitivity above 99.7% and F1-scores exceeding 95% demonstrating its superiority in minimizing false negatives; this high sensitivity is crucial for discovery pipelines, where missing ncRNAs could preclude functional studies (Zhang et al., 2022). PINC achieves high performance via multiple models stacking by combining predictions from different ML algorithms with strong cross-validation to minimize overfitting. It is especially suitable for genome annotation projects in less-studied plant species owing to its high sensitivity, which successfully reduces the omission of true ncRNAs, which is crucial for comprehensive annotation in both model and non-model plants. The tool employs a streamlined three-step workflow that eliminates the need for manual model tuning or expert feature selection.

PINC workflow:

I) **Feature Extraction**: Automatically derive 91 features from RNA sequence data, including k-mer frequencies (1–3-mer), coding sequence-related features (score, cdsStart, cdsStop, cdsSize, and cdsPercent), sequence length, and GC content.
II) **Feature Selection**: Use a combination of F-test statistical analysis and variance threshold filtering to select the 10 most informative features, such as GC content, score, cdsStop, cdsSize, and specific nucleotide and dinucleotide frequencies (T, C, GT, GC, ACG, and TAT).
III) **Model Construction and Selection**: Train and stack multiple ML models (random forest, XGBoost, and deep neural networks) using the AutoGluon system. Model

stacking reduces variance and improves prediction power, creating robust classifiers with minimal manual intervention.

Identification of potential coding and ncRNAs is a crucial step in this process. RNA sequences' linguistic features can be used to train a supervised ML model. Since this method necessitates carefully selected training data, these models are either trained on species that have been extensively researched, such as humans, mice, and fruit flies, or they are created using sequence data from organisms in every area of life to produce a general model (Schmal et al., 2022). PINC offers a user-friendly online web interface (http://www.pncrna.com/) that accepts FASTA files for rapid ncRNA prediction. It supports cross-species application and requires only sequence data, making it flexible for research across a broad spectrum of plant species, including cereals, legumes, dicots, and wild relatives.

Application:

1) Identify and annotate novel ncRNAs in newly sequenced plant genomes.
2) Enhance transcriptome assemblies through precise ncRNA classification.
3) Investigate regulatory networks and gene expression patterns in crops subjected to stress, nutrient variation, or developmental changes. (Zhang et al., 2022)

9.5.2 PlantLncBoost

PlantLncBoost employs the CatBoost algorithm, a highly efficient gradient boosting technique that handles diverse feature types and mitigates overfitting (Khalid et al., 2023; Li & Liang, 2022). It simplifies feature engineering by reducing the requisite input features to only three principal components, including novel mathematical descriptors like Fourier spectrum-based features, which capture intrinsic sequence patterns conserved across plant genomes (Tian et al., 2025). Among the numerous mathematical descriptors used for DNA/RNA sequences, Fourier transform features capture periodic patterns and frequency-domain characteristics (Messaoudi et al., 2014), whereas entropy-based measures such as Shannon and Tsallis entropy assess sequence complexity and randomness, as utilized in the deep learning-based *ab initio* lncRNA identification and functional annotation tool, LncADeep (Yang et al., 2018). In benchmarking across 20 plant species, PlantLncBoost consistently achieved accuracy rates surpassing 96%, setting a new standard for cross-species prediction. PlantLncBoost thus facilitates rapid, cross-species lncRNA identification without species-specific model retraining or calibration.

Methodology:

I) **Feature Extraction**: PlantLncBoost reduces complex RNA sequence information into three highly informative features. These include Fourier transform-derived numerical representations of sequences, which encapsulate periodicity and nucleotide arrangement patterns fundamental to plant lncRNAs.
II) **CatBoost Model Training**: The tool trains a CatBoost classifier using balanced datasets of plant lncRNAs and coding RNAs. CatBoost's handling of categorical features and robustness to overfitting enables precise discrimination between lncRNA and mRNA sequences even in complex plant transcriptomes.

III) **Cross-Species Validation**: Extensive validation on 20 diverse plant species shows that PlantLncBoost's simplified feature set retains powerful predictive capacity without loss of accuracy or sensitivity. This broad applicability makes it ideal for researchers studying newly sequenced or less-characterized plant genomes (Tian et al., 2025).

Applications:

a) Genome-wide annotation of plant lncRNAs.
b) Transcriptomic studies in crops and wild plant species.
c) Supporting functional genomics and regulatory network analyses by accurate lncRNA identification.
d) Assisting in the discovery of lncRNAs involved in plant development, stress responses, and nutrient metabolism.

9.5.3 LncFinder-Plant and CPATplant

Originally engineered for animal or generic lncRNA identification, tools such as LncFinder and CPAT have recently been retrained on comprehensive plant datasets. LncFinder-plant leverages a support vector machine trained on plant features, including sequence composition, structure, and physicochemical properties, while CPATplant uses logistic regression focused on open reading frame (ORF)-related sequence characteristics. Retraining with diverse, high-quality plant data greatly improves these classifiers' sensitivity, specificity, and overall accuracy. By integrating these diverse feature types, LncFinder-plant achieves accurate separation of lncRNAs from mRNAs in plants, addressing the challenge of overlapping sequence characteristics that sometimes confound generic predictors.

In benchmarking, LncFinder-plant achieved the best trade-off of sensitivity and precision, with CPAT-plant consistently ranking second (Han et al., 2019). These retrained models are especially valuable for their ability to capture plant-specific sequence signatures and to minimize false positives in complex transcriptomes, distinguishing them from their animal-trained predecessors (Tian et al., 2024; Wang et al., 2013).

9.5.4 LncADeep, RNAplonc, and DeepPlnc

LncADeep employs deep neural networks to identify lncRNAs, incorporating both primary sequence and secondary structure features. The tool also predicts lncRNA-protein interactions and functional associations, outperforming many standard classifiers in both identification and biological inference (Wang et al., 2013; Yang et al., 2018). LncADeep supports transcriptome-wide analyses, reducing false positives and facilitating downstream functional genomic studies.

RNAplonc applies an ML classifier trained solely on plant lncRNA and mRNA data, tuned for high accuracy (up to 92%). It is particularly noted for outperforming animal-based lncRNA predictors when applied to diverse plant species (Chen et al., 2018; Negri et al., 2019).

DeepPlnc combines dense convolutional networks (DenseNet) with CNNs to efficiently extract hierarchical features from RNA sequences and their predicted secondary structures

(Wang et al., 2021). This hybrid design allows the model to learn rich, multilevel representations of input data. DeepPlnc incorporates secondary structure data, such as base pairing and folding patterns, enhancing the accuracy of lncRNA classification. The deep networks automatically learn discriminative features without manual feature engineering, facilitating adaptability to complex, noisy datasets typical in plant transcriptomics. In recent benchmarking, DeepPlnc achieved up to 94.2% accuracy and F1-scores superior to most existing software, making it especially suitable for large, noisy transcriptomes (Pandey & Sharma, 2025b).

9.5.5 Plant-LncPipe

Plant-LncPipe is a comprehensive computational pipeline that seamlessly integrates all stages of plant lncRNA analysis from RNA-seq data preprocessing through transcript assembly, lncRNA prediction, and classification (by genomic context) to investigation of lncRNA origins (including transposable element intersections). Critically, plant-LncPipe leverages an ensemble approach, integrating top-performing predictors like LncFinder-plant and CPAT-plant to maximize predictive power and stringency. The pipeline also supports flexible classification of lncRNAs (e.g., intronic, intergenic, and antisense) and facilitates downstream analysis, making it a robust, reproducible platform for next-generation ncRNA discovery in plants (Tian et al., 2024).

Workflow:

1) **Transcript Assembly and Filtering**: Plant-LncPipe performs transcript reconstruction to generate candidate transcripts, filtering out short or low-quality sequences.
2) **Coding Potential Assessment**: It incorporates retrained versions of popular classifiers such as LncFinder-plant and CPAT-plant, leveraging cross-species validated models for robust coding potential prediction.
3) **Ensemble Prediction**: To increase classification confidence, Plant-LncPipe integrates predictions from multiple tools, including LncFinder-plant and CPAT-plant, using consensus or voting mechanisms.
4) **LncRNA Classification**: The pipeline categorizes predicted lncRNAs based on their genomic context into classes like intergenic, intronic, antisense, and sense-overlapping lncRNAs.
5) **Functional Annotation Support**: Plant-LncPipe supports downstream analyses such as identification of transposable element intersections and potential lncRNA origin analysis, enabling insights into lncRNA evolution and regulation.
6) **Reporting**: Comprehensive summary statistics and visualization modules provide users with detailed reports on prediction accuracy, transcript distribution, and classification results.

Independent benchmarking studies repeatedly demonstrate that plant-specific retraining and ensemble prediction strategies outclass generic or animal-trained methods in lncRNA identification. For example, in benchmarks covering 20 plant species, the ensemble of LncFinder-plant and CPAT-plant within PlantLncPipe delivered the highest accuracy, precision, and F1-scores, confirming the value of both specialization and combinatorial

Table 9.1 Comparison of major plant ncRNA Tools/pipelines.

Tool/pipeline	Core method	Main features	Plant-specific	Key benchmark highlights
PINC	AutoML	91 features, AutoGluon, robust ML	Yes	Up to 96% accuracy; best in 9 species Zhang et al. (2022)
PlantLncBoost	CatBoost ML	Three features, Fourier, cross-species	Yes	>96% accuracy in 20 species Tian et al. (2025)
LncFinder-plant	ML retrained	Multiple sequence-derived features	Yes	Highest sensitivity/AUC among 20 tools Han et al. (2019)
CPAT-plant	Logistic regression	Four sequence features	Yes	2nd best in accuracy/sensitivity Wang et al. (2013)
LncADeep	Deep learning	Sequence, structure, functional info	Adapted	Outperforms for ID and protein prediction Yang et al. (2018)
RNAplonc	ML classifier	16 features specially selected	Yes	92% accuracy, high reliability Negri et al. (2019)
DeepPlnc	DenseNet-CNN	Sequence and secondary structure	Yes	94.2% accuracy, best F1/MCC Pandey and Sharma (2025a)
Plant-LncPipe	Ensemble pipeline	Integrates above tools, end-to-end	Yes	Reliable ensemble across species Tian et al. (2024)

approaches. Tools like PINC, DeepPlnc, and PlantLncBoost further reinforce the importance of AI-powered, data-driven methodologies (Tian et al., 2025; Wang et al., 2013; Zhang et al., 2022) (Table 9.1).

These AI-based software and pipelines have made plant ncRNA analysis more accurate, versatile, and accessible. By leveraging ML, DL, and ensemble approaches, researchers can more effectively discover and study the diverse landscape of ncRNAs that drive plant development, adaptation, and evolution (Schmal et al., 2022; Tian et al., 2024, 2025; Zhang et al., 2022).

9.6 Applications of AI in ncRNA Research

AI has become one of the most significant changes in the field of plant ncRNA research, where the identification, classification, functional annotation, network reconstruction, and multiomics integration of ncRNAs have been taken to the next level. ML-based approaches have transformed the identification, classification, functional annotation and reconstruction of systems-level structures of plant ncRNAs. With the combination of sophisticated ML algorithms, DL models, and big data systems, it is now possible to explore the biology of ncRNAs, many fold more successfully than ever before, with great accuracy and scale, in plant science. The example of JustRNA and ncPlantDB is representative of the combination of AI and omics and is used to promote the emergence of new knowledge about the resilience of plant stress and development, as well as disease resistance.

9.6.1 AI for Identification and Classification of Novel Plant ncRNAs

This monumental change in the field of plant molecular biology and biotechnology is the use of AI to identify and classify novel plant ncRNAs. Precise mapping of ncRNAs (such as lncRNAs, miRNAs, and other regulatory species) is the basis of understanding how gene regulation, physiological reactions, and stress resilience and disease resistance in plants work (Cao et al., 2024; Fonouni-Farde et al., 2021; Li & Liang, 2022; Liu et al., 2025; Zhang et al., 2022). The growing accessibility of high-throughput sequencing data has triggered the construction of highly advanced AI approaches that can handle large-scale data effectively and with greater accuracy and precision, thus handling the complexity of plant transcriptomes (Liu et al., 2025; Mattick et al., 2023).

Traditionally, sequence conservation and comparative genomics played a major role in the discovery and characterization of plant ncRNAs. Although these methods provided some background knowledge, they were not very efficient in documenting the enormous functional and evolutionary diversity of ncRNAs in various plant species, especially those that are underrepresented in model systems (Dominguez-Rosas et al., 2023; Lucero et al., 2021; Yu et al., 2019). Wet-lab validation is also necessary to verify functionality but typically becomes a bottleneck due to its laboriousness and resource-intensive nature and some non-model or agronomically relevant plants that are lagging in genomic resources (Fonouni-Farde et al., 2021; Lucero et al., 2021). Such difficulties have led to the implementation of scalable AI-based computational methods that can accurately perform processing of complex transcriptomic data to provide candidate ncRNAs with accurate classification and prioritization to continue testing (Liu et al., 2025; Mattick et al., 2023; Tian et al., 2025).

9.6.1.1 Machine Learning Tools

Automated ML (AutoML) platforms like PINC or similar systems use feature extraction based on comprehensive and robust feature selection on large multispecies datasets to optimize plant ncRNA discovery (Liu et al., 2025; Zhang et al., 2022). PINC has shown a high level of classification of above 96%, which is better than that in the traditional tools such as CPC2, CPAT, and CNIT that have less dynamic sets of features and are not always accurate when applied to a wide range of genetic backgrounds (Li & Lang, 2022; Liu et al., 2025; Zhang et al., 2022). PINC applies sequence composition metrics, k-mer frequencies, secondary structure propensities, and evolutionary conservation parameters, as well as ensemble learning algorithms and cross-validation to prevent overfitting and achieve maximum robustness and transferability across taxa (Fonouni-Farde et al., 2021; Liu et al., 2025; Zhang et al., 2022).

ML algorithms can effectively analyze transcriptomic data to identify coding and non-coding components and identify distinctive sequence and structural patterns that are specific to plant ncRNAs (Cao et al., 2024; Dominguez-Rosas et al., 2023; Yu et al., 2019). Approaches that use support vector machines, random forests, and gradient boosting algorithms maximize classification boundaries, enhancing the precision of identifying phylogenetically diverse plant genomes and transcriptomes (Liu et al., 2025; Lucero et al., 2021; Tian et al., 2025).

9.6.1.2 Deep Learning Architectures

The ability of DL to learn hierarchical representations of raw sequence data without having to design its features manually has made it increasingly prominent in ncRNA annotation (Fonouni-Farde et al., 2021; Tian et al., 2025; Zhang et al., 2022). CNNs can identify localized sequence motifs and structural features by using abstraction-capable layered filters, whereas RNNs, especially LSTM and gated recurrent units, are effective at capturing long-range interactions and sequence order essential to the process of classifying lncRNAs and miRNAs (Fonouni-Farde et al., 2021; Liu et al., 2025; Lucero et al., 2021; Mattick et al., 2023). Addition of attention systems and transfer learning also improves the model in terms of adaptability and generalization to model and orphan plant species (Liu et al., 2025; Lucero et al., 2021; Mattick et al., 2023).

Deep neural networks are regularly able to reach higher than 95% classification, high sensitivity, and high specificity scores, outperforming other traditional ML and rule-based pipelines (Liu et al., 2025; Mattick et al., 2023; Yu et al., 2019). These enhancements significantly cut down on false positives and negatives, simplifying downstream functional validations and experimental design.

9.6.1.3 Plant-Specific Databases and High-Throughput Annotation

Significant progress in the infrastructure of ncRNA research has been made by the creation of AI-augmented and plant-specific databases, including ncPlantDB (over 350,000 annotated ncRNAs across 43 species) and JustRNA (3,500,000 annotated ncRNAs across 6 species). ncPlantDB and JustRNA are designed to synthesize heterogeneous data layers (genomic sequences, expression profiles, co-expression networks, epigenetic data, and experimentally validated interactions). These platforms are equipped with advanced AI models, which automate the process of filtering and clustering as well as functional annotation of ncRNAs and their interaction modules, accelerating biological relevance (Lucero et al., 2021; Mattick et al., 2023).

Correspondingly, JustRNA offers whole-organism expression reports and functional protein interaction networks of over a million lncRNAs in plants of about 80 species, transcriptomically expressed in thousands of RNA-seq datasets in you know who knows how many tissues across various developmental stages and environmental conditions (Cao et al., 2024; Fonouni-Farde et al., 2021; Yu et al., 2019). The regulatory interactions between ncRNAs and stress response pathways, growth regulation, and development have also been discovered in functional network reconstruction and genome-wide association studies.

9.6.2 Functional and Regulatory Role Prediction

The key to unlocking the functions of ncRNAs and their roles in abiotic and biotic stress response, regulation of plant developmental processes, and plant immunological reactions is in deciphering their inherent biologic and regulatory roles (Gill et al., 2024; Jin et al., 2024; Yadav et al., 2024). In recent years, newer developments in AI have boosted the ability to foresee ncRNA evolution and have empowered researchers to address challenges of sparse experimental annotations and many-to-many regulatory networks. To execute this approach, a major strategy would be to use sequence-based DNNs that predict the functional classes of ncRNAs using solely their primary nucleotide sequences and thus eliminate secondary structure predictions that are not only computationally expensive but also often inaccurate due to typically on-target roles (Orro & Trombetti, 2023). Long short-term memory (LSTM) networks are equally effective; they can store sequence patterns over time, of course considering context, and reach more than 97% prediction accuracy in various ncRNA classes. Such accuracy compared to a high level of fidelity of functional annotation promotes the discovery of novel regulatory ncRNAs involved in key physiological pathways (Creux et al., 2024; Orro & Trombetti, 2023). It has spurred increasingly the use of natural language processing (NLP) and large language models (LLMs) to extract and aggregate functional knowledge in the ever-growing corpus of scientific data about ncRNAs. By utilizing automated pipelines of text mining and relation extraction methods, AI systems come up with comprehensive findings on the association of certain ncRNAs and phenotypical traits as well as stress reactions (Correia et al., 2024). An example is distant supervision relation extraction methods, which have effectively generated very large ncRNA-phenotype relational corpora that are more reliable than drastically reducing the bottlenecks of manual curation and facilitate systematic summarization of discrete functional annotations (Bhasuran et al., 2024; Correia et al., 2024). Moreover, when training large language and code models (LLCMs,) the tasks of relational extraction do not require intense fine-tuning and present promising opportunities to mine knowledge at scale and only some errors (Green et al., 2025).

Another important application of AI in this field is elucidation of ncRNA-phenotype relationships applicable to stress resistance and resistance to disorders. Researchers mine multisource biological data, such as human microRNA disease repositories and plant lncRNA disease annotation datasets by having AI systematically propagate associations between synonymous ncRNA identifiers and complex phenotype ontologies by embedding models like SentenceTransformers with similarity searching libraries like FAISS (Correia et al., 2024). With the help of this integrative framework network, it is possible to infer the regulatory structures of ncRNAs and molecular pathways, which control the adaptive mechanisms in plants, and to create the means to approach the goal of plant improvement (Zhang et al., 2024a; Zhao et al., 2025).

9.6.3 Network Reconstruction and Discovery of Regulatory Modules

The analysis of sophisticated, systems-level regulation of plants by ncRNAs necessitates the explicit reconstruction of gene regulatory networks (GRNs) and the discovery of regulatory modules under cellular conditions. Graph-based AI models have now gained critical importance in the modeling of multiplex interactions of ncRNAs, miRNAs, proteins, or even drugs through compound modeling. More sophisticated models such as graph convolutional networks, DeepWalk, and graph neural networks (GNNs) can unite topological and sequence features and generate strong models with high accuracy measures, such as high area under the receiver operating characteristic curves (AUCs), to predict molecular interactions and regulatory effects (Shen et al., 2021; Zhang et al., 2024b; Zhuo et al., 2022).

GRN reconstruction further expands explainable AI methods that provide mechanistic understandings of interactions that have been predicted. As an example, scGeneRAI is a machine that uses hubs of relevance propagation on single-cell RNA sequencing data to deterministically estimate regulatory relationships between genes at the level of individual cell HG. The given methodology can detect strong and more cell-specific networks along with appreciating heterogeneity in regulatory networks across cell populations to reveal biologically significant subnetworks with high fidelity (Keyl et al., 2023; Yuan & Duren, 2025). The provided explainability plays an essential role in the validation of AI-based models as well as the generation of hypotheses on the functions and regulation of genes.

Experimental multiomics data integration is crucial to increase the performance and biological relevance of GRN reconstructions. Pioneering machines such as JustRNA use the comprehensive wealth of information in the form of RNA-sequences databank where over a million annotated associates exist across thousands of samples, in a variety of tissues, at different developmental phases, and in diverse environmental setups to provoke dynamic and context-sensitive network inference. With their mutual combination, these datasets usually work together to recap regulatory landscapes with unprecedented resolution using co-expression profiles, genomic association-based information, and chromatin immunoprecipitation sequencing (ChIP-seq) data on transcription factor binding (Agu, 2025; Kajrolkar, 2025; Van den Broeck et al., 2020).

AI-driven network analysis not only explains the hierarchical regulatory concertation of ncRNAs, but it also reveals cryptic or indirect interactions of ncRNAs affecting phenotypic traits of vital signals and adaptation. The insights drive the study of intricate plant phenotypes under stress and development stimuli, providing a rational objective in the direction of the breeding and genetic engineering for enhancing crops that can withstand problems and achieve sales (Bhogireddy et al., 2021; Roychowdhury et al., 2023).

9.6.4 Integration of AI with Omics and Big Data Platforms

AI integration, in combination with multiomics and big data environments, brings a paradigm shift in the field of plant science because it multiplies dramatically the ability to analyze, interpret, and visualize the larger and more intricate datasets produced by high-throughput technologies of today. This combination of AI with multiomics data supports precision agriculture and molecular breeding projects, which allow researchers and breeders to learn new information about the genetic and epigenetic architecture of complex traits in crops.

Multiomics data integration defines the process of integrating various molecular layers of transcriptomics, proteomics, epigenomics, and metabolomics into integrated analytical systems. These heterogeneous datasets can be digested by advanced AI algorithms, specifically DL and ensemble algorithms, and can help find latent phenomena and demonstrate complex gene-interaction networks and regulatory factors that play roles in the actual phenotypic variation and environmental responsiveness (Kajrolkar, 2025; Roychowdhury et al., 2023; Yuan & Duren, 2025). At the heart of catalyzing studies in this field stands specialized plant ncRNA databases, which are essentially a huge vault of transcriptomic and functional annotations. An example of such a system is JustRNA, which offers details on nearly a million lncRNAs that are distributed among some 80 different plant species, giving detailed information on expression profiles, co-expression networks, and regulatory interaction maps. It has inbuilt pipelines that help predict regulatory activity, miRNA interactions, and transcription factor interactions to expedite the generation of functional hypotheses and network modeling (Liu et al., 2025; Tseng et al., 2023). Likewise, ncPlantDB also provides substantial data of ncRNAs, regulatory peptides, and more than four million interactions, including cell type-specific interactions and co-translational regulatory networks. It helps to compare closely related genomics and evolutionary studies that help to present research efforts with ncRNA-mediated regulation in a wide range of plant taxa, and this database is a valuable resource to decode it (Liu et al., 2025). In addition to these repositories, complementary AI-based tools, like PlantLncBoost and ANNInter, allow scalable prediction of these ncRNA interactions. Consensus models and supervised learning methods are used on these platforms to make predictions on a range of plant species and data to enable cross-species comparative analyses and functional annotation tasks (Tian et al., 2025; Vivek et al., 2025). These integrative platforms, collectively, are curating and centralizing not only huge amounts of multiomics data but also forward-looking predictive modeling, biological insights, and visualizing data centers referred to as AI-enabled facilities, which enable the advancement of plant systems biology and translational breeding activities (Liu et al., 2025; Roychowdhury et al., 2023).

Although AI has been crucial in research related to ncRNA, some major challenges remain, which must be overcome to achieve all benefits of AI in the field. Quality and comprehensiveness of data annotation is one of the larger issues. High-fidelity datasets are at the core of AI models to train and achieve a high level of prediction accuracy. However, in non-model orphan crops (there are exceptions), thorough ground truth annotations of ncRNA functions and molecular interactions are highly limited due to the straying of models, which complicates their overall scalability and performance (Adjeroh et al., 2024; Tian et al., 2025). The situation is made worse by different annotation criteria, lack of benchmarking, and even variability across databases, including but not limited to CANTATAdb and GreeNC, several of which have helped to generate inconsistent or conflicting ncRNA data (Adjeroh et al., 2024). Another issue of great concern is the interpretability of the model. Due to the increasing complexity of AI systems, including DL models with millions of parameters, making biological plausibility, reproducibility, and transparency is becoming a critical requirement that must be met before these systems can be adopted by experimental biologists. Explainable AI architectures that furnish mechanistic cognizance about choices within models are crucial in addressing deployability and inspiring belief in the computational discoveries (Keyl et al., 2023; Mattick et al., 2023). Combining diverse omics datasets in heterogeneous genotypes, developmental stages, and tissues and in diverse environmental settings is a challenging endeavor. Data format and quality variability and presentation of experimental protocols with disparities create challenges in harmonizing data to correctly integrate multiomics and to create downstream AI models (Kajrolkar, 2025; Roychowdhury et al., 2023). To overcome these complexities, it is important to develop dynamic context-sensitive models to accurately model ncRNA regulatory functions in various conditions during diverse plant physiological states. The availability of resources is also dependent on continuous updates of databases, user-friendly interfaces, and compatibility with other quintessential genomic resources. Biological databases like JustRNA or ncPlantDB need to be continually curated and technologically improved to ensure that large collections of data provide new knowledge to ensure universal availability and effective application by the community of plant scientists (Liu et al., 2025; Tseng et al., 2023). In the future, AI-inspired studies of the ncRNA biome in plants will have access to novel expanded multimodal frameworks, which can analyze genomic, transcriptomic, epigenetic, and phenotypic data concurrently. It will be possible to advance federated ML techniques that will allow model training on global distributed data, considering privacy and sovereignty (Yang et al., 2025). An analytical system based on real-time phenotyping and high throughput will speed up the transfer of computational forecasts to actionable crop improvement tactics on the path toward sustainable agriculture and food security in diverse environmental conditions.

9.7 Major Obstacles and Constraints in AI-Driven ncRNA Research in Flora

AI has become a revolutionary instrument for the identification, categorization, and functional annotation of plant ncRNAs. Nonetheless, despite significant advancements, the discipline continues to encounter considerable challenges that hinder its practical implementation in crop science. The main problems are that there are not many high-quality, well-annotated datasets; it is hard to move predictive models from one species to another; and there is a gap between what computers say will happen and what experiments show will happen. In addition to these, AI models are hard to understand, multi-omics datasets are hard to combine, access to computational resources is not equal, and ethical and biosafety issues have not been resolved. This section gives a critical overview of these problems based on recent research and points out ways to make things better in the future.

9.7.1 Limitations of the Dataset and Quality of Curation

Data is the basis of all AI-based research, and for plant ncRNAs, this basis is not very strong. In contrast to animals, where extensive and meticulously curated datasets have been compiled over decades, plant ncRNA resources are fragmented and frequently inconsistent (Xu et al., 2022). There are a number of effects of this uneven distribution. Most AI tools learn from data from model organisms like *A. thaliana* and *Oryza sativa*. Because of this, they are very good at these systems but have trouble recognizing ncRNA features in species that have not been studied as much.

Sequencing artifacts and technical noise add another level of difficulty. High-throughput sequencing platforms are very powerful, but they can also give you data that have incomplete reads, adapter contamination, or misaligned fragments. If these mistakes are not filtered out correctly, they get into AI models and make them less reliable for predicting things (Adjeroh et al., 2024). Different research groups and databases use different methods for annotating data, which makes things even more inconsistent. For instance, one repository might call some transcripts lncRNAs, while another might call them protein-coding, which would leave researchers with datasets that do not match up.

Benchmarking is still a problem, too. Resources like UniProt, which give widely accepted reference standards, help with predicting protein-coding genes. There is no equivalent to plant ncRNA prediction, which makes it hard to fairly compare different AI models (Xu et al., 2022). Even though ncPlantDB and other similar projects are a good start toward consolidation, with over 353,000 entries from many different species (Liu et al., 2025), this is still not enough because plant genomes are so different from each other.

Recent techniques such as MFPINC, which amalgamates various feature types including sequence composition and anticipated structure, have attained superior classification accuracy (Nie et al., 2024). PINC, an automated ML platform, has also reported accuracy levels above 92% across test datasets (Zhang et al., 2022). But these methods cannot make up for the fact that there are not enough data.

9.7.2 Cross-Species Generalization in Plants

The fast evolution of ncRNAs poses an additional difficulty. According to Mattick et al. (2023), ncRNAs have the potential to change quickly and exhibit remarkable lineage specificity, in contrast to protein-coding sequences, which frequently maintain conserved patterns across distant species. Cross-species prediction is so challenging. Despite the fact that sorghum and maize are closely related crops, a model trained on rice, for instance, may collect properties that are irrelevant to these crops.

Despite having a slightly higher degree of conservation, miRNAs are not exempt from this problem. It is difficult to create generalizable predictors since variations frequently appear in certain species or even in specific environmental circumstances (Xu et al., 2022). Though they show some promise, methods like transfer learning—which involves pre-training a model on one dataset and then fine-tuning it using smaller data from another—become less effective at high evolutionary divergence.

In order to overcome these constraints, more recent frameworks such as PlantLncBoost have been created. Using new mathematical descriptors and gradient boosting, this approach has demonstrated enhanced performance on several datasets (Tian et al., 2025). Predictive power nevertheless decreases when used for species that are phylogenetically distant. Underspecification exacerbates the problem by causing models that perform similarly well on training datasets to behave erratically on new data, eroding confidence in their resilience (D'Amour et al., 2020).

Pan-genomic models, which combine data from several species, are a promising but resource-intensive option. These would give a more comprehensive picture of ncRNA characteristics, but they need a lot of curated data and processing power.

9.7.3 Barriers to Biological Validation

It is now feasible to produce thousands of ncRNA candidates in a single investigation because of computational workflows. This surge of predictions, however, has outpaced biological validation. Wet-lab methods that are still costly and time-consuming include CRISPR-based knockouts, Northern blotting, qRT-PCR, and degradome sequencing (Xu et al., 2022).

Because of this bottleneck, a large number of anticipated ncRNAs never go past the computational step. False positives are even more prevalent. Several putative lncRNAs generated in silico have been found to reflect transcriptional noise, whereas others genuinely encode short peptides (Adjeroh et al., 2024). However, since authentic ncRNAs are only expressed at specific embryonic phases or under specific stressors, they can go undetected.

Crops lacking effective transformation mechanisms are more affected. For instance, gene-editing systems are widely used in Arabidopsis but are somewhat less dependable in cereals such as maize or wheat. Because of this, experimental validation of AI predictions is very difficult. Medium-throughput validation systems could be a solution. Synthetic biology platforms that enable the testing of several candidates at once, such as multiplex CRISPR screening, may speed up the validation process and lessen dependency on single-gene methods.

9.7.4 Interpretability and Explainability of AI Models

High prediction accuracy alone is not enough. Researchers also aim to understand why a model arrives at a given decision. Unfortunately, many DL systems serve as "black boxes," generating classifications without revealing insight into the biological basis. This restricts their effectiveness for hypothesis creation.

Explainable AI (XAI) frameworks are beginning to solve this issue. Attention processes, saliency maps, and feature attribution analyses can identify key sequence motifs or structural aspects that influence predictions (Adjeroh et al., 2024). However, such technologies are still infrequently employed in plant systems. Most XAI approaches were originally created for human medical data and may not fully capture plant-specific regulatory intricacies.

Developing interpretability tools specific to plant ncRNAs is therefore an urgent priority. Doing so would allow researchers to move beyond blind trust in model outputs instead of using forecasts as beginning points for new biological theories.

9.7.5 Integration with Multiomics Data

Bottlenecks NcRNA function rarely happens in isolation; it is altered by interactions with proteins, epigenetic changes, and metabolic pathways. For AI to capture this complexity, it must combine information across multiple omics layers.

This is easier said than done. Transcriptomic, proteomic, and epigenomic datasets differ in scale, format, and reliability. Combining them into a coherent feature collection for AI training is a big technical obstacle (Pepe et al., 2022). Noise or missing values in one dataset might also impact overall model performance.

Human systems benefit from large-scale efforts like ENCODE and GTEx, which give extensive multiomics data. Plants lack such initiatives, allowing researchers to rely on fragmented and often incomplete datasets (Adjeroh et al., 2024). Even when such data exist, they are frequently constrained to a particular tissue or time point, limiting their potential to record dynamic ncRNA function. Until complete and standardized plant-specific pipelines are built, multiomics integration will remain a weak spot in AI-based research.

9.7.6 Resource and Accessibility Constraints

The computing demands of AI models present extra impediments. Training deep networks demands high-performance GPUs, large memory, and substantial storage. Many research groups, especially in developing nations, do not have access to such infrastructure.

Accessibility is also a concern. Most AI pipelines are supplied as command-line tools requiring programming expertise (Xu et al., 2022). This presents a steep learning curve for plant molecular biologists who may lack computational knowledge. Cloud platforms offer potential solutions by giving on-demand resources, but issues of cost, internet connectivity, and data protection remain important impediments.

A less evident but equally significant concern is software sustainability. AI models and pipelines often become outdated within a few years if they are not actively maintained. Dependencies break, new versions of libraries generate incompatibilities, and reproducibility decreases (Shivashankar & Martini, 2024). Without long-term support, even promising tools risk falling into disuse.

9.7.7 Ethical and Biosafety Considerations

Although rarely highlighted, ethical and biosafety considerations are increasingly crucial in AI-driven ncRNA research. Many plant genetic datasets remain confidential, regulated by intellectual property agreements or commercial breeding projects. This hinders collaboration and slows growth.

The combination of AI predictions with genome-editing techniques such as CRISPR creates biosafety concerns. Misidentified ncRNAs could lead to unwanted off-target effects with ramifications for agricultural performance or ecological balance. Intellectual property problems may also occur if projected ncRNAs or AI models are patented, restricting their use in public sector research (Ji et al., 2018).

Establishing transparent data-sharing protocols and comprehensive biosafety guidelines will be important to ensuring that AI-driven ncRNA research proceeds ethically and equitably.

9.8 Future Perspectives of AI Technology for Studying Plant ncRNAs

The study of ncRNAs in plants, such as circular RNAs (circRNAs), lncRNAs, siRNAs, and miRNAs, has greatly benefited from AI and ML techniques. However, as sequencing prices decrease, datasets grow, and breeding applications (such as for stress tolerance, yield, and adaptation) become more pressing, there arise new demands. Thousands of potential plant ncRNAs have been identified by high-throughput sequencing, but very few of them have strong functional annotations or roles that have been confirmed by experiment. This gap between discovery and functional understanding creates an urgent need for computational approaches that can prioritize candidate ncRNAs for experimental validation and help translate molecular signals into breeding-relevant information. AI and modern ML techniques offer powerful strategies to classify, predict, and interpret ncRNA function, but progress is constrained by challenges unique to plants like diverse genomes, variable expression across tissues/environments, and complex cross-talk between molecular layers.

9.8.1 More Advanced Multiclass and Functional Predictive Frameworks

Many existing tools focus on binary classification, such as distinguishing coding from ncRNAs (e.g., tools like CPC2, CPAT, and CNCI) or identifying specific ncRNA classes (e.g., lncRNA vs mRNA). For instance, PINC is an AutoML-based tool tailored to distinguish ncRNAs from coding RNAs in plants. It shows good sensitivity across multiple species (Zhang et al., 2022).

Current AI-based methods for ncRNA research have primarily focused on binary classification. Thus, it oversimplifies the ncRNA landscape. The biological reality is more complicated as lncRNAs may act via multiple mechanisms (cis/trans regulation, miRNA sponging, and chromatin remodeling), and a single ncRNA may participate in multiple functional classes. Models of the future must therefore be able to handle multiclass and multilabel outputs, which include predicting not only that a transcript is noncoding but also whose functional category it belongs to and which tissue, developmental stage, or environmental setting it acts in. Future frameworks must progress toward multiclass classification systems that can accurately and sensitively detect several ncRNA categories (such as miRNAs, lncRNAs, siRNAs, and circRNAs) at the same time. Multiclass classifiers, such as NCodR, use sequence and secondary structure data to classify several ncRNA classes in Viridiplantae, including lncRNAs, miRNAs, pre-miRNAs, rRNAs, snRNAs, snoRNAs, and tRNAs (Nithin et al., 2022). There are also some tools that combine several features like MFPINC, which uses multisource feature fusion and sequence plus other features, to distinguish ncRNAs from coding RNAs in multiple plant species. It is among the newer tools with improved accuracy and generalization (Nie et al., 2024).

In addition, functional prediction remains a critical challenge. Many plant ncRNAs are poorly annotated, and their regulatory roles are not well defined. Emerging AI models should incorporate advanced feature learning, including structural motifs, evolutionary conservation, and expression patterns, to provide a more holistic prediction of ncRNA function. Developing such frameworks requires richer labeled datasets with function annotations, standardized ontologies for ncRNA function and phenotypes, carefully curated negative/control sets, and modeling approaches that can output calibrated probabilities for multiple labels rather than hard, single-label assignments. The majority of functional annotations for plant lncRNAs that have been experimentally confirmed are limited and biased toward model species (rice, Arabidopsis) and a small number of characteristic classes (flowering, stress). A large number of predicted lncRNAs and some functional metadata are compiled via databases and community resources such as PLncDB and species-specific repositories; nevertheless, many entries lack conventional phenotype labels or experimental confirmation. Investing in curated corpora that integrate sequencing, expression, interaction, and phenotypic labels is essential to improving prediction frameworks. Robust supervised learning and equitable model benchmarking will be made possible by aggregated datasets with standard ontologies (Domínguez-Rosas et al., 2023).

9.8.2 Integration with Multiomics

Transcriptomic information alone is insufficient to completely comprehend the intricacy of plant ncRNA biology. To create a complete picture of ncRNA-mediated regulation, future studies must concentrate on integrating multiomics datasets, including transcriptomics, proteomics, metabolomics, and epigenomics.

Transcriptomics and small RNA sequencing (miRNAomics) integration has been used to elucidate regulatory networks. For example, in rice coleoptile senescence, integration of transcriptome and small RNAome data, together with target prediction and protein-protein interaction networks, revealed modules of miRNAs, their targets, and downstream processes (e.g., ROS homeostasis) (Sasi et al., 2022). Numerous studies on plant stress response have combined proteomics and transcriptomics to show how changes in gene expression under biotic and abiotic stress translate into changes in proteins. For instance, Aslam et al. (2024) analyzed the integration of transcriptome and proteomic information in their analysis of the rice stress response. Some studies include metabolomics, environmental inputs, and even epigenomic layers (DNA methylation and histone modifications) to understand regulation often in model species. However, explicit AI models that jointly integrate all these layers, especially for ncRNA functional predictions, are relatively few.

To exploit multiomics integration fully, AI research on plant ncRNAs should develop models that accept multiple omics inputs like sequences, expression (RNA-seq), small RNA data, proteomics, epigenomics (e.g., DNA methylation, chromatin accessibility), and possibly metabolomics.

AI technologies, particularly GNNs and multimodal DL architectures, offer powerful strategies for integrating heterogeneous omics data. By combining different molecular layers, these models can uncover cross-level regulatory relationships, such as ncRNA-mediated control of protein translation or epigenetic modifications. Linking small RNA sequencing data with methylome maps could reveal how siRNAs regulate DNA methylation and transposon silencing in plants. Similarly, integrating proteomics and metabolomics with ncRNA networks could provide valuable insights into how ncRNAs influence metabolic pathways critical for stress adaptation and nutritional quality.

Such integrative approaches will not only expand fundamental knowledge but also create an understanding of ncRNA functions, enabling researchers to design targeted strategies for crop resilience and productivity under changing climates.

9.8.3 User-Friendly Tools and Better Access for Plant Breeders and Molecular Biologists

Many AI-based pipelines for ncRNA analysis are complex and unavailable to plant biologists without sophisticated computing skills. Developing user-friendly systems and interfaces that democratize the use of AI in ncRNA research is a crucial future direction.

Breeders, agronomists, and molecular biologists will be able to examine ncRNA datasets without the need for programming knowledge in web-based platforms that feature customized research pipelines, interactive visualization tools, and user-friendly dashboards. AI-driven suggestions, target prediction, automated annotation, or functional enrichment would also greatly speed up biological discovery and application. A web server for multi-class ncRNA classification in plants is included in NCodR, one of the tools that provides web interfaces (Nithin et al., 2022), and tools such as MFPINC come with usage instructions and code packaged on GitHub (Nie et al., 2024). Non-model plants benefit from

PINC's (AutoML-based) feature that allows users to upload their data (Zhang et al., 2022). The development of graphical user interfaces enables users to enter their sequences or expression/omics data, choose species or parameters, run predictions, and examine results in order to expand the effectiveness of AI for plant ncRNA research.

In agricultural contexts, this accessibility is particularly vital. Breeders could directly utilize AI-powered ncRNA analysis platforms to identify stress-responsive ncRNAs, evaluate their potential as biomarkers, and design molecular breeding strategies for crop improvement. Ultimately, the creation of such interfaces will bridge the gap between advanced computational research and practical agricultural applications, ensuring that AI innovations in ncRNA research translate into real-world impact.

9.9 Conclusion

AI has transformed the study of plant ncRNAs—including miRNAs, siRNAs, lncRNAs, and circRNAs—by accelerating their identification, classification, and functional characterization. Through advanced ML and DL models, researchers can now analyze primary sequences, structural motifs, and evolutionary patterns to predict ncRNA roles with remarkable accuracy, even in underexplored crop species. AI-driven databases such as JustRNA and ncPlantDB integrate large-scale multiomics data, enabling systemic exploration of expression profiles, regulatory networks, and ncRNA-target interactions. The fusion of AI with multiomics and NLP has enhanced the understanding of ncRNA functions in stress responses, development, and immunity. Explainable AI and graph-based models are uncovering intricate gene regulatory networks from single-cell omics, providing deeper insights into cell-specific regulation. However, challenges persist, including limited annotated datasets, inconsistent data quality, and the lack of interpretable, transparent models connecting computational predictions with experimental validation. The future of AI in plant ncRNA research depends on developing robust predictive frameworks, multimodal integration, and user-friendly tools accessible to both computational and experimental scientists. Federated learning, real-time analytics, and high-throughput phenotyping will further link molecular data with crop performance under diverse conditions. Democratizing AI technologies and establishing open-access platforms will ensure broader participation and reproducibility across disciplines. By empowering plant scientists and breeders with AI-guided insights, this field can accelerate both discovery and application—designing regulatory RNAs, identifying biomarkers, and engineering resilient, high-yield crops. Ultimately, AI-driven ncRNA research will become a cornerstone of sustainable agriculture, advancing food security and climate adaptability worldwide.

References

Adjeroh, D. A., Zhou, X., Paschoal, A. R., Dimitrova, N., Derevyanchuk, E. G., Shkurat, T. P., Loeb, J. A., Martinez, I., & Lipovich, L. (2024). Challenges in lncRNA biology: Views and opinions. *Non-coding RNA*, *10*(4), 43. https://doi.org/10.3390/ncrna10040043

Agu, P. C. (2025). Dissecting ncRNA pathways and omics integration. In *Progress in molecular biology and translational science* https://doi.org/10.1016/bs.pmbts.2025.01.001. Elsevier.

Aslam, N., Li, Q., Bashir, S., Yuan, L., Qiao, L., & Li, W. (2024). Integrated review of transcriptomic and proteomic studies to understand molecular mechanisms of rice's response to environmental stresses. *Biology*, *13*(9), 659. https://doi.org/10.3390/biology13090659

Bartel, D. P. (2004). MicroRNAs: Genomics, biogenesis, mechanism, and function. *Cell*, *116*(2), 281–297. https://doi.org/10.1016/S0092-8674(04)00045-5

Bernstein, E., Caudy, A. A., Hammond, S. M., & Hannon, G. J. (2001). Role for a bidentate ribonuclease in the initiation step of RNA interference. *Nature*, *409*(6818), 363–366. https://doi.org/10.1038/35053110

Bhasuran, B., Manoharan, S., Iyyappan, O. R., Murugesan, G., Prabahar, A., & Raja, K. (2024). Large language models and genomics for summarizing the role of microRNA in regulating mRNA expression. *Biomedicine*, *12*(7), 1535. https://doi.org/10.3390/biomedicines12071535

Bhogireddy, S., Mangrauthia, S. K., Kumar, R., Pandey, A. K., Singh, S., Jain, A., Budak, H., Varshney, R. K., & Kudapa, H. (2021). Regulatory non-coding RNAs: A new frontier in regulation of plant biology. *Functional & Integrative Genomics*, *21*(3–4), 313–330. https://doi.org/10.1007/s10142-021-00787-8

Brahim, J., Loubna, R., & Noureddine, F. (2021). RNN-and CNN-based weed detection for crop improvement: An overview. *Foods and Raw Materials*, *9*(2), 387–396. auspublishers.com.au/temp/9596bff3dd91dac401f8b326ab8d026f.pdf

Britten, R. J., & Davidson, E. H. (1969). Gene regulation for higher cells: A theory: New facts regarding the organization of the genome provide clues to the nature of gene regulation. *Science*, *165*(3891), 349–357.

Cao, L., Wang, Y., Bi, C., Ye, Q., Yin, T., & Ye, N. (2020). PreLnc: An accurate tool for predicting lncRNAs based on multiple features. *Genes*, *11*(9), 981.

Cao, Z., Zhao, T., Wang, L., Han, J., Chen, J., Hao, Y., & Guan, X. (2021). The lincRNA XH123 is involved in cotton cold-stress regulation. *Plant Molecular Biology*, *106*(6), 521–531. https://doi.org/10.1007/s11103-021-01169-1

Cao, W., Yang, L., Zhuang, M., Lv, H., Wang, Y., Zhang, Y., & Ji, J. (2024). Plant non-coding RNAs: The new frontier for the regulation of plant development and adaptation to stress. *Plant Physiology and Biochemistry*, *208*, 108435. https://doi.org/10.1016/j.plaphy.2024.108435

Chao, H., Hu, Y., Zhao, L., Xin, S., Ni, Q., Zhang, P., & Chen, M. (2022). Biogenesis, functions, interactions, and resources of non-coding RNAs in plants. *International Journal of Molecular Sciences*, *23*(7), 3695. https://doi.org/10.3390/ijms23073695

Chen, C. J., Zhou, H., Chen, Y. Q., Qu, L. H., & Gautheret, D. (2011). Plant noncoding RNA gene discovery by "single-genome comparative genomics". *RNA (New York, N.Y.)*, *17*(3), 390–400. https://doi.org/10.1261/rna.2426511

Chen, Y., Zhou, Q., Chen, W., Zhang, Y., Chen, M., & Xiang, Y. (2018). RNA sequencing reveals the long noncoding RNA landscape of the mouse CNS. *Frontiers in Genetics*, *9*, 513. https://doi.org/10.3389/fgene.2018.00513

Chitwood, D. H., Nogueira, F. T., Howell, M. D., Montgomery, T. A., Carrington, J. C., & Timmermans, M. C. (2009). Pattern formation via small RNA mobility. *Genes & Development*, *23*(5), 549–554. https://doi.org/10.1101/gad.1770009

Correia, D. C., Couto, F. M., & Martiniano, H. F. (2024). Predicting non-coding RNA function using artificial intelligence. *bioRxiv* (Cold Spring Harbor Laboratory). https://doi.org/10.1101/2024.12.30.630736

Creux, C., Zehraoui, F., Radvanyi, F., & Tahi, F. (2024). Comparison and benchmark of deep learning methods for non-coding RNA classification. *PLoS Computational Biology, 20*(9), e1012446. https://doi.org/10.1371/journal.pcbi.1012446

Csorba, T., Questa, J. I., Sun, Q., & Dean, C. (2014). Antisense COOLAIR mediates the coordinated switching of chromatin states at FLC during vernalization. *Proceedings of the National Academy of Sciences, 111*(45), 16160–16165. https://doi.org/10.1073/pnas.1419030111

D'Amour, A., Heller, K., Moldovan, D., Adlam, B., Alipanahi, B., Beutel, A., Chen, C., Deaton, J., Eisenstein, J., Hoffman, M. D., & Hormozdiari, F. (2020). Under specification presents challenges for credibility in modern machine learning. *arXiv preprint arXiv:2011.03395.*

Dhamija, S., & Menon, M. B. (2021). Measuring lncRNA expression by real-time PCR (Vol. 2348, pp. 93–111). Humana, New York, NY. https://doi.org/10.1007/978-1-0716-1581-2_6

Di, C., Yuan, J., Wu, Y., Li, J., Lin, H., Hu, L., Zhang, T., Qi, Y., Gerstein, M. B., Guo, Y., & Lu, Z. J. (2014). Characterization of stress-responsive lncRNAs in *arabidopsis thaliana* by integrating expression, epigenetic and structural features. *The Plant Journal, 80*(5), 848–861. https://doi.org/10.1111/tpj.12679

Ding, J., Lu, Q., Ouyang, Y., Mao, H., Zhang, P., Yao, J., Xu, C., Li, X., Xiao, J., & Zhang, Q. (2012). A long noncoding RNA regulates photoperiod-sensitive male sterility, an essential component of hybrid rice. *Proceedings of the National Academy of Sciences, 109*(7), 2654–2659. https://doi.org/10.1073/pnas.1121374109

Djebali, S., Davis, C. A., Angelika Merkel, Dobin, A., Lassmann, T., Mortazavi, A., Tanzer, A., Lagarde, J., Lin, W., Schlesinger, F., Xue, C., Marinov, G. K., Khatun, J., Williams, B. A., Zaleski, C., Rozowsky, J., Ro¨Der, M., Kokocinski, F., Abdelhamid, R. F., & Gingeras, T. R. (2012). Landscape of transcription in human cells. *Nature.* https://doi.org/10.1038/nature11233

Domínguez-Rosas, E., Hernández-Oñate, M. Á., Fernandez-Valverde, S. L., & Tiznado-Hernández, M. E. (2023). Plant long non-coding RNAs: Identification and analysis to unveil their physiological functions. *Frontiers in Plant Science, 14*, 1275399. https://doi.org/10.3389/fpls.2023.1275399

Fonouni-Farde, C., Ariel, F., & Crespi, M. (2021). Plant long noncoding RNAs: New players in the field of post-transcriptional regulations. *Non-coding RNA, 7*(1), 12. https://doi.org/10.3390/ncrna7010012

Gao, P., Bai, X., Yang, L., Lv, D., Pan, X., Li, Y., Cai, H., Ji, W., Chen, Q., & Zhu, Y. (2011). osa-MIR393: A salinity- and alkaline stress-related microRNA gene. *Molecular Biology Reports, 38*(1), 237–242. https://doi.org/10.1007/s11033-010-0100-8

Ge, L., Guo, H., Li, X., Tang, M., Guo, C., Bao, H., Huang, L., Yi, Y., Cui, Y., & Chen, L. (2022). OSSIDP301, a member of the DUF1644 family, negatively regulates salt stress and grain size in rice. *Frontiers in Plant Science, 13*. https://doi.org/10.3389/fpls.2022.863233

Gill, S. S., Khan, N. A., Agarwala, N., Singh, K., Sunkar, R., & Tuteja, N. (2024). ncRNAs in plant development and stress responses. *Plant Physiology and Biochemistry, 214*, 108950. https://doi.org/10.1016/j.plaphy.2024.108950

Green, A., Ribas, C. E., Ontiveros-Palacios, N., Griffiths-Jones, S., Petrov, A. I., Bateman, A., & Sweeney, B. (2025). LitSumm: Large language models for literature summarization of noncoding RNAs. *Database, 2025*. https://doi.org/10.1093/database/baaf006

Guo, J. C., Fang, S. S., Wu, Y., Zhang, J. H., Chen, Y., Liu, J., ... Zhao, Y. (2019). CNIT: A fast and accurate web tool for identifying protein-coding and long non-coding transcripts based on intrinsic sequence composition. *Nucleic Acids Research*, *47*(W1), W516–W522.

Han, S., Liang, Y., Ma, Q., Xu, Y., Zhang, Y., Du, W., Wang, C., & Li, Y. (2019). LncFinder: An integrated platform for long non-coding RNA identification utilizing sequence intrinsic composition, structural information and physicochemical property. *Briefings in Bioinformatics*, *20*(6), 2009–2027. https://doi.org/10.1093/bib/bby065

Hong, Y., Zhang, Y., Cui, J., Meng, J., Chen, Y., Zhang, C., Yang, J., & Luan, Y. (2022). The lncRNA39896–miR166b–HDZs module affects tomato resistance to Phytophthora infestans. *Journal of Integrative Plant Biology*, *64*(10), 1979–1993. https://doi.org/10.1111/jipb.13339

Jan, M., Spangaro, A., Lenartowicz, M., & Mattiazzi Usaj, M. (2024). From pixels to insights: Machine learning and deep learning for bioimage analysis. *BioEssays*, *46*(2), e2300114.

Ji, Y., Zhang, H., Wang, J., & Luo, X. (2018). Model-reuse attacks on machine learning models. *arXiv preprint arXiv:1812.00483.*

Jin, X., Wang, Z., Li, X., Ai, Q., Wong, D. C. J., Zhang, F., Yang, J., Zhang, N., & Si, H. (2024). Current perspectives of lncRNAs in abiotic and biotic stress tolerance in plants. *Frontiers in Plant Science*, *14*. https://doi.org/10.3389/fpls.2023.1334620

Kajrolkar, A. (2025). Integrating multi-omics data for plant stress response: Current advances and future directions. *Premier Journal of Plant Biology*, *3*, 100012. https://doi.org/10.70389/pjpb.100012

Keyl, P., Bischoff, P., Dernbach, G., Bockmayr, M., Fritz, R., Horst, D., Blüthgen, N., Montavon, G., Müller, K., & Klauschen, F. (2023). Single-cell gene regulatory network prediction by explainable AI. *Nucleic Acids Research*, *51*(4), e20. https://doi.org/10.1093/nar/gkac1212

Khalid, R., Naveed, H., & Khalid, Z. (2023). Computational prediction of disease related lncRNAs using machine learning. *Scientific Reports*, *13*, 806. https://doi.org/10.1038/s41598-023-27680-7

Kurihara, Y., & Watanabe, Y. (2004). Arabidopsis micro-RNA biogenesis through Dicer-like 1 protein functions. *Proceedings of the National Academy of Sciences*, *101*(34), 12753–12758. https://doi.org/10.1073/pnas.0403115101

Lee, T. F., Li, P., & Meyers, B. C. (2012). The biology and dynamics of plant small RNAs. In J. F. Wendel, J. Greilhuber, J. Dolezel, & I. J. Leitch (Eds.), *Plant Genome Diversity Volume 1: Plant Genomes, their Residents, and their Evolutionary Dynamics* (pp. 83–101). Springer Vienna. https://doi.org/10.1007/978-3-7091-1130-7

Li, M., & Liang, C. (2022). LncDC: A machine learning-based tool for long non-coding RNA detection from RNA-Seq data. *Scientific Reports*, *12*(1) https://doi.org/10.1038/s41598-022-22082-7, 19083.

Li, S., Liu, J., Liu, Z., Li, X., Wu, F., & He, Y. (2014). HEAT-INDUCED TAS1 TARGET1Mediates thermotolerance via HEAT STRESS TRANSCRIPTION FACTOR A1A–Directed pathways in arabidopsis. *The Plant Cell*, *26*(4), 1764–1780. https://doi.org/10.1105/tpc.114.124883

Liu, X., Li, D., Zhang, D., Yin, D., Zhao, Y., Ji, C., Zhao, X., Li, X., He, Q., Chen, R., Hu, S., & Zhu, L. (2018). A novel antisense long noncoding RNA,TWISTED LEAF, maintains leaf blade flattening by regulating its associated sense R2R3-MYB gene in rice. *New Phytologist*, *218*(2), 774–788. https://doi.org/10.1111/nph.15023

Liu, L., Liu, E., Hu, Y., Li, S., Zhang, S., Chao, H., Hu, Y., Zhu, Y., Chen, Y., Xie, L., Shen, Y., Wu, L., & Chen, M. (2025). ncPlantDB: A plant ncRNA database with potential ncPEP information and cell type-specific interaction. *Nucleic Acids Research*, *53*(D1), D1587–D1594. https://doi.org/10.1093/nar/gkae1017

López-Galiano, M. J., García-Robles, I., González-Hernández, A. I., Camañes, G., Vicedo, B., Real, M. D., & Rausell, C. (2019). Expression of *MIR159* is altered in tomato plants undergoing drought stress. *Plants*, *8*(7), 201. https://doi.org/10.3390/plants8070201

Lucero, L., Ferrero, L., Fonouni-Farde, C., & Ariel, F. (2021). Functional classification of plant long noncoding RNAs: A transcript is known by the company it keeps. *New Phytologist*, *229*(3), 1251–1260. https://doi.org/10.1111/nph.16903

Ma, L., Bajic, V. B., & Zhang, Z. (2013). On the classification of long non-coding RNAs. *RNA Biology*, 924–933. https://doi.org/10.4161/rna.24604

Ma, X., Zhao, F., & Zhou, B. (2022). The characters of non-coding RNAs and their biological roles in plant development and abiotic stress response. *International Journal of Molecular Sciences*, *23*(8), 4124. https://doi.org/10.3390/ijms23084124

Mancini, M., Lena, A. M., & Candi, E. (2020). Identification of long noncoding RNA by in situ hybridization approaches. *Methods in Molecular Biology*, *2154*, 175–185. https://doi.org/10.1007/978-1-0716-0648-3_15

Marin, E., Jouannet, V., Herz, A., Lokerse, A. S., Weijers, D., Vaucheret, H., Nussaume, L., Crespi, M. D., & Maizel, A. (2010). miR390, Arabidopsis TAS3 tasiRNAs, and their AUXIN RESPONSE FACTOR targets define an autoregulatory network quantitatively regulating lateral root growth. *The Plant Cell*, *22*(4), 1104–1117. https://doi.org/10.1105/tpc.109.072553

Mattick, J. S., & Makunin, I. V. (2006). Non-coding RNA. *Human Molecular Genetics*, *15*(suppl_1), R17–R29. https://doi.org/10.1093/hmg/ddl046

Mattick, J. S., Amaral, P. P., Carninci, P., Carpenter, S., Chang, H. Y., Chen, L., Chen, R., Dean, C., Dinger, M. E., Fitzgerald, K. A., Gingeras, T. R., Guttman, M., Hirose, T., Huarte, M., Johnson, R., Kanduri, C., Kapranov, P., Lawrence, J. B., Lee, J. T., & Wu, M. (2023). Long non-coding RNAs: Definitions, functions, challenges and recommendations. *Nature Reviews Molecular Cell Biology*, *24*(6), 430–447. https://doi.org/10.1038/s41580-022-00566-8

Messaoudi, I., Elloumi-Oueslati, A., & Lachiri, Z. (2014). Building specific signals from frequency chaos game and revealing periodicities using a smoothed Fourier analysis. *IEEE/ACM Transactions on Computational Biology and Bioinformatics*, *11*, 863–877. https://doi.org/10.1109/TCBB.2014.2315991

Mishra, B., Kumar, N., & Mukhtar, M. S. (2019). Systems biology and machine learning in plant–pathogen interactions. *Molecular Plant-Microbe Interactions*, *32*(1), 45–55. https://doi.org/10.1094/MPMI-08-18-0221-FI

Negri, T. D. C., Alves, W. A. L., Bugatti, P. H., Saito, P. T. M., Domingues, D. S., & Paschoal, A. R. (2019). Pattern recognition analysis on long noncoding RNAs: A tool for prediction in plants. *Briefings in Bioinformatics*, *20*(2), 682–689. https://doi.org/10.1093/bib/bby034

Nie, Z., Gao, M., Jin, X., Rao, Y., & Zhang, X. (2024). MFPINC: Prediction of plant ncRNAs based on multi-source feature fusion. *BMC Genomics*, *25*(1), 531. https://doi.org/10.1186/s12864-024-10439-3

Nithin, C., Mukherjee, S., Basak, J., & Bahadur, R. P. (2022). NCodR: A multi-class support vector machine classification to distinguish non-coding RNAs in Viridiplantae. *Quantitative Plant Biology*, *3*, e23. https://doi.org/10.1017/qpb.2022.18

Orro, A., & Trombetti, G. A. (2023). High-accuracy NCRNA function prediction via deep learning using global and local sequence information. *Biomedicine*, *11*(6), 1631. https://doi.org/10.3390/biomedicines11061631

Pandey, S., & Sharma, R. (2025a). A deep learning approach for long non-coding RNA identification in plants: DeepPlnc V2.0. *Biomedical Informatics and Smart Healthcare, 1*(2), 79–88. https://doi.org/10.62762/BISH.2025.421075

Pandey, S., & Sharma, R. (2025b). DeepPInc V2.0: A DenseNet-CNN hybrid for accurate plant long non-coding RNA identification. *Biomedical Informatics and Smart Healthcare, 1*(2), 79–88. https://doi.org/10.62762/BISH.2025.421075

Parmar, S., Gharat, S. A., Tagirasa, R., Chandra, T., Behera, L., Dash, S. K., & Shaw, B. P. (2020). Identification and expression analysis of miRNAs and elucidation of their role in salt tolerance in rice varieties susceptible and tolerant to salinity. *PLoS One, 15*(4), e0230958. https://doi.org/10.1371/journal.pone.0230958

Pepe, G., Appierdo, R., Carrino, C., Ballesio, F., Helmer-Citterich, M., & Gherardini, P. F. (2022). Artificial intelligence methods enhance the discovery of RNA interactions. *Frontiers in Molecular Biosciences, 9*, 1000205. https://doi.org/10.3389/fmolb.2022.1000205

Pierouli, K., Papageorgiou, L., Chrousos, G. P., Eliopoulos, E., & Vlachakis, D. (2025). Role of lncRNAs related to NRs in the regulation of gene expression. *International Journal of Epigenetics, 5*, 3. https://doi.org/10.3892/ije.2025.26

Prasad, A., Sharma, N., & Prasad, M. (2021). Noncoding but coding: Pri-miRNA into the action. *Trends in Plant Science, 26*(3), 204–206. https://doi.org/10.1016/j.tplants.2020.12.004

Reyes, J. L., & Chua, N. (2007). ABA induction of miR159 controls transcript levels of two MYB factors during arabidopsis seed germination. *The Plant Journal, 49*(4), 592–606. https://doi.org/10.1111/j.1365-313x.2006.02980.x

Roychowdhury, R., Ḋas, S. P., Gupta, A., Parihar, P., Chandrasekhar, K., Sarker, U., Kumar, A., Ramrao, D. P., & Sudhakar, C. (2023). Multi-omics pipeline and omics-integration approach to decipher plant's abiotic stress tolerance responses. *Genes, 14*(6), 1281. https://doi.org/10.3390/genes14061281

Sah, S. K., Kaur, G., & Kaur, A. (2014). Rapid and reliable method of high-quality RNA extraction from diverse plants. *American Journal of Plant Sciences, 05*(21), 3129–3139. https://doi.org/10.4236/AJPS.2014.521329

Saha, C., Saha, S., & Bhattacharyya, N. P. (2025). LncRNAomics: A comprehensive review of long non-coding RNAs in plants. *Genes, 16*(7), 765. https://doi.org/10.3390/genes16070765

Saranya, S. M., Rajalaxmi, R. R., Prabavathi, R., Suganya, T., Mohanapriya, S., & Tamilselvi, T. (2021). Deep learning techniques in tomato plant—A review. *Journal of Physics: Conference Series, 1767*(1), 012010. https://doi.org/10.1088/1742-6596/1767/1/012010

Sasi, J. M., VijayaKumar, C., Kukreja, B., Budhwar, R., Shukla, R. N., Agarwal, M., & Katiyar-Agarwal, S. (2022). Integrated transcriptomics and miRNAomics provide insights into the complex multi-tiered regulatory networks associated with coleoptile senescence in rice. *Frontiers in Plant Science, 13*, 985402. https://doi.org/10.3389/fpls.2022.985402

Schmal, M., Girod, C., Yaver, D., Mach, R. L., & Mach-Aigner, A. R. (2022). A bioinformatic-assisted workflow for genome-wide identification of ncRNAs. *NAR Genomics and Bioinformatics, 4*. https://doi.org/10.1093/nargab/lqac059

Schwab, R., Maizel, A., Ruiz-Ferrer, V., Garcia, D., Bayer, M., Crespi, M., Voinnet, O., & Martienssen, R. A. (2009). Endogenous TasiRNAs mediate non-cell autonomous effects on gene regulation in arabidopsis thaliana. *PLoS One, 4*(6), e5980. https://doi.org/10.1371/journal.pone.0005980

Shafiq, S., Li, J., & Sun, Q. (2016). Functions of plants long non-coding RNAs. *Biochimica et Biophysica Acta (BBA) - Gene Regulatory Mechanisms, 1859*(1), 155–162. https://doi.org/10.1016/j.bbagrm.2015.06.009

Shen, Z., Luo, T., Zhou, Y., Yu, H., & Du, P. (2021). NPI-GNN: Predicting ncRNA–protein interactions with deep graph neural networks. *Briefings in Bioinformatics, 22*(5). https://doi.org/10.1093/bib/bbab051

Shen, T., Zheng, Y., Liu, Q., Chen, C., Huang, L., Deng, S., Xu, M., & Yang, C. (2022a). Integrated SMRT and Illumina sequencing provide new insights into crocin biosynthesis of *Gardenia jasminoides. International Journal of Molecular Sciences, 23*(11), 6321. https://doi.org/10.3390/ijms23116321

Shen, X., Jiang, C., Wen, Y., Li, C., & Lu, Q. (2022b). A brief review on deep learning applications in genomic studies. *Frontiers in Systems Biology, 2*, 877717. https://doi.org/10.3389/fsysb.2022.877717

Shi, G., Wang, S., Wang, P., Zhan, J., Tang, Y., Zhao, G., Li, F., Ge, X., & Wu, J. (2022). Cotton miR393-TIR1 module regulates plant defense against verticillium dahliae via auxin perception and signaling. *Frontiers in Plant Science, 13*. https://doi.org/10.3389/fpls.2022.888703

Shivashankar, R., & Martini, B. (2024). Challenges in reproducibility of machine learning software. *arXiv preprint arXiv:2408.09196.*

Singh, U., Khemka, N., Rajkumar, M. S., Garg, R., & Jain, M. (2017). PLncPRO for prediction of long non-coding RNAs (lncRNAs) in plants and its application for discovery of abiotic stress-responsive lncRNAs in rice and chickpea. *Nucleic Acids Research, 45*(22), e183–e183.

Song, D., Yang, Y., Yu, B., Zheng, B., Deng, Z., Lu, B. L., ... Jiang, T. (2009). Computational prediction of novel non-coding RNAs in arabidopsis thaliana. *BMC Bioinformatics, 10*(Suppl 1), S36.

Tian, H., Guo, F., Zhang, Z., Ding, H., Meng, J., Li, X., Peng, Z., & Wan, S. (2020). Discovery, identification, and functional characterization of long noncoding RNAs in Arachis hypogaea L. *BMC Plant Biology, 20*(1), 308. https://doi.org/10.1186/s12870-020-02510-4

Tian, X. C., Chen, Z. Y., Nie, S., Shi, T. L., Yan, X. M., Bao, Y. T., Li, Z. C., Ma, H. Y., Jia, K. H., Zhao, W., & Mao, J. F. (2024). Plant-LncPipe: A computational pipeline providing significant improvement in plant lncRNA identification. *Horticulture Research, 11*(4), uhae041. https://doi.org/10.1093/hr/uhae041

Tian, X. C., Nie, S., Domingues, D. S., Paschoal, A. R., Jiang, L. B., & Mao, J. F. (2025). PlantLncBoost: Key features for plant lncRNA identification and significant improvement in accuracy and generalization. *The New Phytologist, 247*, 1538–1549. https://doi.org/10.1111/nph.70211

Tomari, Y., & Zamore, P. D. (2005). Perspective: Machines for RNAi. *Genes & Development, 19*(5), 517–529. https://doi.org/10.1101/gad.1284105

Trotter, M., Harris, C., Cloutier, M., Samanta, M., & Kalantry, S. (2021). Highly resolved detection of long non-coding RNAs in situ. *Methods in Molecular Biology, 2372*, 123–144. https://doi.org/10.1007/978-1-0716-1697-0_12

Tseng, K., Wu, N., Chow, C., Zheng, H., Chou, C., Yang, C., Wang, M., Chang, S., & Chang, W. (2023). JustRNA: A database of plant long noncoding RNA expression profiles and functional network. *Journal of Experimental Botany, 74*(17), 4949–4958. https://doi.org/10.1093/jxb/erad186

Vaistij, F. E., Elias, L., George, G. L., & Jones, L. (2010). Suppression of microRNA accumulation via RNA interference in arabidopsis thaliana. *Plant Molecular Biology*, *73*(4–5), 391–397. https://doi.org/10.1007/s11103-010-9625-4

Van Den Broeck, L., Gordon, M., Inzé, D., Williams, C., & Sozzani, R. (2020). Gene regulatory network inference: Connecting plant biology and mathematical modeling. *Frontiers in Genetics*, *11*. https://doi.org/10.3389/fgene.2020.00457

Vivek, A. T., Sahu, N., Kalakoti, G., & Kumar, S. (2025). ANNInter: A platform to explore ncRNA-ncRNA interactome of arabidopsis thaliana. *Computational Biology and Chemistry*, *115*, 108328. https://doi.org/10.1016/j.compbiolchem.2024.108328

Wang, L., Park, H. J., Dasari, S., Wang, S., Kocher, J. P., & Li, W. (2013). CPAT: Coding-potential assessment tool using an alignment-free logistic regression model. *Nucleic Acids Research*, *41*(6), e74. https://doi.org/10.1093/nar/gkt006

Wang, S., Sun, X., Hoshino, Y., Yu, Y., Jia, B., Sun, Z., Sun, M., Duan, X., & Zhu, Y. (2014). MicroRNA319 positively regulates cold tolerance by targeting OsPCF6 and OsTCP21 in rice (Oryza sativa L.). *PLoS One*, *9*(3), e91357. https://doi.org/10.1371/journal.pone.0091357

Wang, T., Liu, M., Zhao, M., Chen, R., & Zhang, W. (2015). Identification and characterization of long non-coding RNAs involved in osmotic and salt stress in *Medicago truncatula* using genome-wide high-throughput sequencing. *BMC Plant Biology*, *15*(1). https://doi.org/10.1186/s12870-015-0530-5

Wang, L., Jin, J., Li, L., & Qu, S. (2020). Long non-coding RNAs responsive to blast fungus infection in rice. *Rice*, *13*(1) https://doi.org/10.1186/s12284-020-00437-w, 77.

Wang, T., Li, M., Zhang, W., & Chen, J. (2021). Deeppinc: A deep learning method for plant long non-coding RNA identification using sequence and secondary structure features. *bioRxiv*. https://doi.org/10.1101/2021.12.10.472074

Wierzbicki, A. T., Blevins, T., & Swiezewski, S. (2021). Long noncoding RNAs in plants. *Annual Review of Plant Biology*, *72*, 245–271. https://doi.org/10.1146/annurev-arplant-093020-035446

Xia, K., Wang, R., Ou, X., Fang, Z., Tian, C., Duan, J., Wang, Y., & Zhang, M. (2012). OSTIR1 and OSAFB2 downregulation via OSMIR393 overexpression leads to more tillers, early flowering and less tolerance to salt and drought in rice. *PLoS One*, *7*(1), e30039. https://doi.org/10.1371/journal.pone.0030039

Xu, D., Yuan, W., Fan, C., Liu, B., Lu, M.-Z., & Zhang, J. (2022). Opportunities and challenges of predictive approaches for the non-coding RNA in plants. *Frontiers in Plant Science*, *13*, 890663. https://doi.org/10.3389/fpls.2022.890663

Yadav, A., Mathan, J., Dubey, A. K., & Singh, A. (2024). The emerging role of Non-Coding RNAs (NCRNAs) in plant growth, development, and stress response signaling. *Non-coding RNA*, *10*(1), 13. https://doi.org/10.3390/ncrna10010013

Yajnik, K. N., Singh, I. K., & Singh, A. (2024). lncRNAs and epigenetics regulate plant's resilience against biotic stresses. *Plant Physiology and Biochemistry*, *214*, 108892. https://doi.org/10.1016/j.plaphy.2024.108892

Yan, J., & Wang, X. (2022). Unsupervised and semi-supervised learning: The next frontier in machine learning for plant systems biology. *The Plant Journal*, *111*(6), 1527–1538.

Yang, C., Yang, L., Zhou, M., Xie, H., Zhang, C., Wang, M. D., & Zhu, H. (2018). LncADeep: An ab initio lncRNA identification and functional annotation tool based on deep learning. *Bioinformatics*, *34*(22), 3825–3834. https://doi.org/10.1093/bioinformatics/bty428

Yang, L., Wang, H., Zou, M., Chai, H., & Xia, Z. (2025). Artificial intelligence-driven plant bio-genomics research: A new era. *Tropical Plants*, *4*(1). https://doi.org/10.48130/tp-0025-0008

Yu, Y., Zhang, Y., Chen, X., & Chen, Y. (2019). Plant noncoding RNAs: Hidden players in development and stress responses. *Annual Review of Cell and Developmental Biology*, *35*(1), 407–431. https://doi.org/10.1146/annurev-cellbio-100818-125218

Yuan, Q., & Duren, Z. (2025). Inferring gene regulatory networks from single-cell multiome data using atlas-scale external data. *Nature Biotechnology* https://doi.org/10.1038/s41587-024-02182-7, *43*, 247–257.

Zhang, B., Pan, X., Cobb, G. P., & Anderson, T. A. (2006). Plant microRNA: A small regulatory molecule with big impact. *Developmental Biology*, *289*(1), 3–16. https://doi.org/10.1016/j.ydbio.2005.10.036

Zhang, H., Liu, Y., Wang, C., Tang, D., Xu, J., Chen, Y., & Xiang, Y. (2022). PINC: A tool for non-coding RNA identification in plants based on an automated machine learning framework. *International Journal of Molecular Sciences*, *23*(19), 11825. https://doi.org/10.3390/ijms231911825

Zhang, A., Pi, W., Wang, Y., Li, Y., Wang, J., Liu, S., Cui, X., Liu, H., Yao, D., & Zhao, R. (2024a). Update on functional analysis of long non-coding RNAs in common crops. *Frontiers in Plant Science*, *15*. https://doi.org/10.3389/fpls.2024.1389154

Zhang, X., Zhao, L., Chai, Z., Wu, H., Yang, W., Li, C., Jiang, Y., & Liu, Q. (2024b). NPI-DCGNN: An accurate tool for identifying ncRNA-Protein interactions using a dual-channel graph neural network. *Journal of Computational Biology*, *31*(8), 742–756. https://doi.org/10.1089/cmb.2023.0449

Zhao, M., Tai, H., Sun, S., Zhang, F., Xu, Y., & Li, W. X. (2012). Cloning and characterization of maize miRNAs involved in responses to nitrogen deficiency. *PLoS One*, *7*(1), e29669.

Zhao, Z., Yang, Y., Iqbal, A., Wu, Q., & Zhou, L. (2024). Biological insights and recent advances in plant long non-coding RNA. *International Journal of Molecular Sciences*, *25*(22), 11964. https://doi.org/10.3390/ijms252211964

Zhao, S., Bai, H., Fan, Z., Zhu, M., & Qiu, Z. (2025). A long non-coding RNA lncRNA18313 regulates resistance against cadmium stress in wheat. *Frontiers in Plant Science*, *16*. https://doi.org/10.3389/fpls.2025.1583758

Zhu, J., Li, Y., Lin, J., Wu, Y., Guo, H., Shao, Y., Wang, F., Wang, X., Mo, X., Zheng, S., Yu, H., & Mao, C. (2019). CRD1, an Xpo1 domain protein, regulates miRNA accumulation and crown root development in rice. *The Plant Journal*, *100*(2), 328–342. https://doi.org/10.1111/tpj.14445

Zhuo, L., Chen, Y., Song, B., Liu, Y., & Su, Y. (2022). A model for predicting ncRNA–protein interactions based on graph neural networks and community detection. *Methods*, *207*, 74–80. https://doi.org/10.1016/j.ymeth.2022.09.001

Zou, Y., Wang, S., Zhou, Y., Bai, J., Huang, G., Liu, X., Zhang, Y., Tang, D., & Lu, D. (2018). Transcriptional regulation of the immune receptor FLS2 controls the ontogeny of plant innate immunity. *The Plant Cell*, *30*(11), 2779–2794. https://doi.org/10.1105/tpc.18.00297

10

AI-Enabled Plant RNA Interference

Manoj Mani[1], Gnana Sowndariyan Gnanasekaran[1], Poovarasan Manjeeswaran[1], Kathiresan Nagaraj[1], Antony Prabhu Jeyabal Philomenathan[1], Akilandeswari Govindraj[2], Maria Jerline Babu[3], and Vijaya Anand Arumugam[1]

[1] *Department of Human Genetics and Molecular Biology, Bharathiar University, Coimbatore, Tamil Nadu, India*
[2] *Department of Biochemistry, CMS College of Science & Commerce, Coimbatore, Tamil Nadu, India*
[3] *Department of Biochemistry, Dr. R.A.N.M Arts and Science College, Erode, Tamil Nadu, India*

10.1 Introduction to Plant RNA Interference

Evolutionary origins of RNA interference (RNAi) in plants date back to the early eukaryotic past, with evidence indicating that RNAi mechanisms predated the last eukaryotic common ancestor (Rosa et al., 2018). RNAi was probably an ancient defense against RNA viruses and transposable elements, as evidenced by its universal conservation among different eukaryotic groups (Torri et al., 2022). In plants, RNAi is primarily mediated by microRNAs (miRNAs) and small interfering RNAs (siRNAs) that regulate gene expression post-transcription through sequence-specific base-pairing with target mRNAs. The preservation of these key RNAi factors, such as Dicer-like 1 (DCL1) and Hyponastic Leaves 1 (HYL1), across various plant lineages implies that these molecular factors existed early in plant evolution before the divergence of bryophytes and seed plants ca. 500 million years ago (Jia et al., 2021). Despite evolutionary replacement of many miRNA genes, others have been conserved since they are essential in controlling development and phenotypic plasticity (Gramzow & Theißen, 2019). The convergent evolution of plant and animal miRNA sets is one such case of convergent evolution, where RNAi protein machinery likely was constructed from ancient archaeal, bacterial, and phage proteins involved in RNA processing and DNA repair (Shabalina & Koonin, 2008).

The composition of RNAi in plants includes a complex network of small RNAs (sRNAs) and protein complexes, such as Argonaute (AGO) proteins, DCL enzymes, and RNA-dependent RNA polymerases (RdRPs). These proteins, in combination, enable sequence-specifically silencing of genes and genomic defense responses. The RNAi pathway mechanism is initiated by double-stranded RNA (dsRNA) molecules being cleaved by DCL proteins into siRNA or miRNA duplexes. These duplexes are subsequently incorporated into AGO-containing RNA-induced silencing complexes (RISC), which mediate the degradation or translational repression of complementary mRNAs. RdRPs amplify the silencing signal by synthesizing dsRNA from single-stranded RNA templates to generate secondary siRNAs that reinforce gene silencing (Fang & Qi, 2016; Morgado, 2020; Svoboda, 2020; Tan et al., 2024). Amplification only takes place in plants and increases the RNAi-mediated regulatory processes. Moreover, specific AGO isoforms have also evolved to gain specialized roles in post-transcriptional and transcriptional gene silencing, such as miRNA, trans-acting siRNA (ta-siRNA), and heterochromatic siRNAs (hc-siRNA) pathways. Through these paired processes, RNAi ensures genome integrity and mediates adaptive responses to abiotic and biotic stresses.

At the subfunctional level, RNAi is a critical defense and regulatory system for plants. It facilitates post-transcriptional and transcriptional silencing of genes, enabling fine control of gene expression and contributing to antiviral immunity. Long hairpin RNA-derived siRNAs direct DNA methylation and heterochromatin formation, leading to transcriptional gene silencing, and also direct mRNA degradation by the RISC complex (Pattanayak et al., 2013). In antiviral defense, RNAi serves as part of the innate immune system to identify and destroy viral RNAs with the help of virus-derived sRNAs (Deng et al., 2022). However, plant viruses have evolved RNAi suppressor proteins that inhibit various steps of this process, which shows an evolutionary arms race between host and pathogen (Szittya & Burgyán, 2013). Transcriptional regulation of stress-mediated RNAi-related genes controls dynamic modularity of silencing machinery in response to environmental and pathogenic signals (Gong et al., 2022).

Apart from its endogenous regulatory roles, RNAi has been utilized as a groundbreaking tool in agricultural biotechnology. It targets the repression of specific genes to improve crop quality, stress tolerance, and confer pathogen and pest resistance. RNAi-based techniques have been successfully employed to enhance nutritional content, delay fruit ripening, enhance shelf life, and improve plant architecture by regulating key metabolic and developmental processes (Isshiki & Kodama, 2009; Mamta & Rajam, 2018; Rajam, 2020). The cross-kingdom RNAi (ckRNAi) action also broadens the biological application of the mechanism, with plant siRNAs capable of silencing genes in interacting pathogens such as fungi, nematodes, and insects, thereby making a new choice for integrated pest management (Kaur & Talekar, 2024).

Moreover, RNAi is closely associated with plant immunity and epigenetic control. It functions as a key mediator of transcriptional gene silencing by RNA-guided DNA methylation and histone modification, facilitating dynamic epigenetic reprogramming in plant–pathogen interactions (Tao et al., 2023). In mutualism and pathogen association, too, ckRNAi is involved in bidirectional regulation of host plant-associated microbes' genes, influencing defense signaling and symbiotic stability (Qiao et al., 2023). Collectively, RNAi is a fundamental regulatory process that integrates gene expression control, stress tolerance, and immune response, emphasizing its functional and evolutionary significance in plant biology.

10.2 Computational Biology Foundations of RNAi

The application of computational biology has greatly facilitated understanding of RNAi mechanisms in plants, enabling the discovery, annotation, and functional analysis of sRNAs such as siRNAs and miRNAs. Numerous bioinformatics pipelines for the detection and analysis of these molecules have been developed, each by merging specific methodological paradigms and analytical tools. SUmir pipeline, comprised of scripts such as SUmirFind, SUmirFold, SUmirPredictor, and SUmirLocator, adopts a homology-based approach that is plant-specific, highly effective in minimizing misannotations and false positives, a typical difficulty encountered in crops with complicated, repetitive genomes such as cereals (Alptekin et al., 2017). Similarly, PmiRDiscVali integrates miRDeep-P predictive miRNA identification with degradome sequencing-based validation to enable reference-guided and de novo discovery, as shown in *Dendrobium officinale* (Yu et al., 2019). Other pipelines, such as miRTour and sRNATargetDigger, automate primary features of miRNA discovery, including precursor selection, similarity searches, and target prediction, with sRNATargetDigger also using a bidirectional search method to reveal co-regulatory sRNA-target networks (Milev et al., 2011; Ye et al., 2020). In addition, species-level pipelines, such as that for Sorghum bicolor, use PERL scripts and structure prediction tools to identify miRNAs from expressed sequence tag (EST) and genome survey sequence (GSS) data to contribute to species-level lists of miRNAs (Katiyar et al., 2012).

In addition to these pipelines, curated repositories and databases have become crucial sources of information for plant RNAi research, providing large data integration and cross-species annotation. The sRNAanno database offers a comprehensive collection of uniformly annotated sRNAs–miRNAs, phased siRNAs (phasiRNAs), and hc-siRNAs for 143 plant genomes spanning more than 22,721 PHAS loci and more than 22 million hc-siRNA loci (Chen et al., 2021). PmiREN is an encyclopedic database of plant miRNAs that captures 16,422 high-confidence novel miRNA loci in 88 species with parallel analysis of RNA ends (PARE) sequencing-supervised annotations for target validation (Guo et al., 2020). miRBase remains a reference basis for miRNA sequences, annotation, and predicted targets (Griffiths-Jones et al., 2006), while the *Arabidopsis* sRNA Project (ASRP) database is *Arabidopsis thaliana*-specific, offering well-curated datasets and computational tools for miRNA and siRNA identification (Gustafson et al., 2005). TarDB has also recently devoted a platform to the study of miRNA targets and miRNA-induced phasiRNAs with a user-friendly interface for high-confidence target and phasiRNA analyzes across different plant lineages (Liu et al., 2021).

Computational modeling further enhanced efficiency prediction in RNAi and off-target effects, addressing critical problems in crop improvement and functional genomics by RNAi. RNA string kernel algorithms, as in the instance of Qiu et al., can significantly identify base mismatches, G-U wobbles, bulges, and seed regions to enable faster siRNA target predictions than common alignment software (Qiu et al., 2007). The SIREN pipeline integrates siRNA design with RNA-RNA thermodynamic models to enable scalable off-target prediction, with high accuracy in plant and oomycete transcriptomes (Vargas-Mejía & Vega-Arreguín, 2025). Computer programs such as siRNA Scan and pssRNAit extend these approaches to genome-wide studies, providing experimental evidence of off-target risk in post-transcriptional gene silencing (PTGS) assays (Ahmed et al., 2020; Xu et al., 2006). Likewise, dsCheck simulates Dicer cleavage patterns for the identification of potential off-target sequences and therefore increases the validity and accuracy of dsRNA-mediated silencing assays (Naito et al., 2005).

Mathematical and statistical models have further facilitated improved quantitative analysis of RNAi data to enable high-throughput measurement of gene-silencing kinetics and phenotype scores. Statistical deconvolution approaches such as the gespeR model describe gene-specific effects through the correction of off-target perturbations in RNAi phenotypes, thereby improving hit accuracy in large-scale screens (Schmich, 2016). Probabilistic graphical models like netprioR integrate multidimensional data sets to rank functionally important genes and enhance the discovery of new regulatory elements within biological pathways. Similarly, regression-based statistical models have been successful in reducing false negatives and positives in RNAi screens for chemotherapeutic drug sensitivity (Ye et al., 2012). Ordinary differential equation (ODE)-based models, including the logistic Lotka–Volterra model, produce dynamic models of RNAi kinetics to enable quantitative prediction of apoptosis induced by siRNA (Ueda et al., 2013). Bayesian and Markov Chain Monte Carlo (MCMC) modeling has also been applied to model intracellular processes of RNAi, such as siRNA release from endosomes, and offer theoretical platforms for enhancing experimental efficiency and cost (Liu et al., 2019).

Despite these advances, computational methods continue to suffer from repeated setbacks upon implementation to massive and complex plant genomes. Heterogeneity among the methods of pan-genome assembly accounts for variability in the detection of gene presence–absence variants and non-reference pools of genes, usually based on variation in sequencing depth, algorithmic architecture, and input data (Baichoo & Ouzounis, 2017). The computational demand of genome assembly, sequence alignment, and short-read comparison also imposes a limitation on scalability, particularly in polyploid crops and high-repeat-content species. The effective next-generation sequencing (NGS) technologies produce short reads that are suboptimal for constructing large, complicated genomes—such as those of conifers, which exceed 30 Gbp—resulting in fragmentation of the assemblies and annotation errors (Deschamps & Llaca, 2016; Kuzmin et al., 2019). Though stepwise de novo assembly protocols have been suggested to make these easier, they find limited application. Additionally, RNA-sequencing-based expression profiling in the very complex genomes also has interpretational issues with the occurrence of gene duplications, isoform complexity, and structural variations, which distort expression quantification.

10.3 Artificial Intelligence in RNAi Research

Artificial intelligence (AI), including machine learning (ML) and deep-learning (DL) technologies, has been a prelude to a paradigm shift for RNAi investigations, primarily in plants. Such computational technologies allow for high-throughput prediction, annotation, and designing of sRNA, including siRNAs and miRNAs, and reveal complex gene regulatory RNAi-mediated silencing RNAi networks. DL models, including the Transformer architecture and convolutional neural networks (CNNs), have been employed to predict selectively accumulated siRNAs across a range of plant species. This advanced strategy integrates DL techniques, utilizing word2 and fastText-based word sequence representations within a hybrid neural model to forecast virus-derived siRNAs, thereby enhancing antiviral RNAi technologies for plants (Bu et al., 2023). Additionally, ML-based approaches for plants are increasingly being utilized for predicting miRNAs and complementing schemes of experimental identification, facilitating a high demand for prediction models for plants (Jayasundara et al., 2021). Combinational integration of AI-aided omics technologies and RNAi research offers a comprehensive framework for understanding plant defense signaling and adaptation molecular reactions against stressors of the environment and pathogens (Murmu et al., 2024).

Refinement of ML models of predicting functional miRNAs and siRNAs significantly improved RNAi studies' precision. For predicting miRNA targets, TargetNet represents a milestone by employing DL through relaxed candidate target site (CTS) criteria, a new miRNA-CTS sequence encoding format, and a deep residual network structure, acting cumulatively to improve identification of high-functional miRNA targets (Min et al., 2022). MIRZA-G suite programs employ a genome-wide, accurate, biophysical interaction model of RNA structure and sequence attributes for miRNA targets and siRNA off-target prediction, including noncanonical sites (Gumienny & Zavolan, 2015). More recently, siRNADiscovery, by employing a framework of a graph neural network (GNN), has identified higher-order RNA silencing dependencies by modeling siRNA position effects, AGO2 binding kinetics, and target mRNA structure and has achieved state-of-the-art prediction performance across validation sets (Long et al., 2024). Overall, the models have refined the RNAi mechanism of action by cumulatively employing thermodynamic, structural, and contextual sequence information.

Computational models triggered a paradigm shift for RNA structure biology prediction and play a major role in RNAi efficiency and specificity predictions. CNNs, RNNs, and transformer-based models have been exemplarily adept for RNA secondary and tertiary structure modeling. A specific case of transformer-based models is their ability to model long-range associations intrinsic to RNA sequences, a self-evident necessity for adequate structure prediction. BPfold, for instance, uses transformer and CNN models for predicting RNA secondary structures by deriving energy maps of base-pairing motive, both accuracy- and generality-wise better than traditional techniques (Zhou et al., 2024). Similarly, RNA backbone torsion angle prediction by employing a transformer encoder–decoder model simplifies tertiary structure modeling (Zhao et al., 2024). Even inspired variants of ResNet have been utilized for tertiary structure prediction by compensating current inadequacies of scoring functions for longer RNA oligonucleotides (Devata & Priyakumar, 2024). Despite all these successes, the generality of a structure prediction model for novel RNA classes is a challenge, requiring fine-tuning and its reconciliation with large RNA structure datasets (Zhu et al., 2025).

The emergence of generative AI technologies advanced frontiers of RNAi research by making it possible for the de novo designing of optimized synthetic RNA molecules for silencing for new ends. Artificial miRNAs (amiRNAs) designed through generative modeling are less specific and flexible compared to classical RNAi methods which employs natural miRNA bioprocessing pathways for nucleating of precise transcript targeting (Yadava & Mukherjee, 2011). Sequence-to-function learning paradigms-based generative models like Sequence and Network Data-driven Structure/sequence-based Tool for RNA Modeling (SANDSTORM), Generative Adversarial RNA Design Network (GARDN), are utilized for designing RNA elements of higher activity, including synthetic toehold switches of better dynamic behavior of range (Riley et al., 2025). Generative adversarial networks (GANs) have also been utilized for designing RNA molecules of selective protein-binding capability, e.g., designing piRNAs by optimization of binding to certain proteins, like inducing proteins of interacting capability for binding to the SOX2 protein, by employing GANs trained on relevant interaction data (Ozden et al., 2023). Through integration of Bayesian optimization, Transformer-based Alternative Splicing Predictor + Bayesian Optimization for Splicing (TrASPr+BOS) progresses RNA biology further by predicting and designing RNA molecules for tissue-specific splicing events, a development of potential value for designing plant and therapeutic RNA (Wu et al., 2025).

Also central to the discovery and annotation of new noncoding RNAs (ncRNAs) of plants, a field critical to understanding the regulatory complexity of RNAi systems, are ML-based programs like Plant Non-Coding RNA identification (PINC) and Plant Long Non-Coding RNA Prediction by Random Forests (PLncPRO), which apply supervised classification techniques to identify ncRNAs efficiently and accurately, separating them from protein-coding transcripts. Using an automatic ML (AutoML) approach based on the AutoGluon framework, PINC achieves accuracy ranging between 92.74% and 96.42% across several species of plants and shows better performance compared to traditional state-of-the-art approaches like Coding Potential Calculator 2 (CPC2) and Coding-Potential Assessment Tool (CPAT) (Zhang et al., 2022). PLncPRO, utilizing a random forest classifier, identifies long ncRNA (lncRNA) transcripts effectively by separating them from coding sequences, having proved true applicability across monocot and dicot plant systems, including stress-responsive lncRNA identification in rice (Khemka et al., 2020) and chickpea (Singh et al., 2017). Moreover, LncMachine offers a crop-specific, alignment-independent pipeline by bringing together thorough feature selection, coupled with ML algorithms, thereby guaranteeing suitability across plant taxa by way of scaling and robust behavior (Cagirici et al., 2021). The encompassing union of these AI programs not only expedited ncRNA discovery but also democratized computational genomics by bringing sophisticated prediction pipelines to plant researchers' desks (Narayan & Kumar, 2022).

10.4 AI-Driven RNAi Design and Optimization

With integration of thermodynamic modeling, sequence feature analysis, and advanced computational learning techniques, AI improved RNAi construct designing for increased silencing efficiency and off-target decrease for several plant species. The SIREN pipeline is a synthesis of it by utilizing thermodynamic modeling and ML for RNAi construct designing for maximized specificity as well as scalability for several plant systems (Vargas-Mejía & Vega-Arreguín, 2025). Likewise, PlantRNA-FM, a model for plant RNA specifically, utilizes DL for structure and sequence motif analysis of relevance for gene silence, increasing RNAi efficiency prediction (Yu et al., 2024). Hybrid computational systems, by interfusing the respective strengths of backpropagation neural networks and support vector machines, even better calibrate siRNA efficacy prediction by categorizing effective siRNA based on validated sets (Wang & Zhang, 2015). Additionally, multispecies RNAi reagent designing is offered by the E-RNAi web application for optimized selection of target sites based on predicted efficiency and specificity (Horn & Boutros, 2010). Even DL structures, including CNNs, have been observed reaching state-of-the-art accuracy for siRNA efficacy prediction, overcoming classical algorithms for both classification and correlation test (Han et al., 2017).

Off-target effect reduction is a chief RNAi construct fine-tuning concern. AI pipeline systems, like SIREN, combine thermodynamic and sequence-modeling protocols with user-specified transcriptomic environments to anticipate and prevent unwanted silencing, providing tunable sensitivity to exchange accuracy for computational speed (Vargas-Mejía & Vega-Arreguín, 2025). Off-target interaction molecular processes—frequently a by-product of partial sequence similarity—must be treated with utmost care by computational modeling, where successive longer-longer matches longer than 15 nucleotides or sequence identities of about 80% are known to cause unwanted gene silencing (Chen et al., 2023). ML protocol systems, like order siRNAs by supporting sequence, thermodynamic, and accessibility parameters to form the best siRNA target sites (Ladunga, 2007). Additionally, transcript expression amounts directly govern off-target potential, hence placing growing importance on prediction systems for quantitative transcriptomic information (Senthil-Kumar & Mysore, 2011).

To achieve cross-species generalizability, transfer learning was developed as a highly influential framework of RNAi modeling. Through shared genomic and regulatory patterns, models of one species are transferred for RNAi efficiency prediction across other species. Such platforms, like Deep Transcription Factor Binding Sites (deepTFBS), employ multitask and transfer learning for better predictions of transcription factor binding sites across taxa, achieving higher precision and recall rates for classical models (Zhai et al., 2025). Likewise, FloraBERT accumulates genomic attributes across several plant species to reflect taxonomic relationships, obtaining better gene expression and RNAi-related predictions (Latyshev et al., 2023). Transfer learning obtains better prediction precision for miRNA-mRNA interactions by employing conserved rules of interaction, obtaining species-species advantages for nonmodel species, for which comparatively fewer data are available (Hadad et al., 2024). Domain-adaptive neural networks also prevent prediction biases based on species-specific sequence contexts, obtaining more generalized model robustness (Cochran et al., 2022).

Reinforcement learning (RL) offers an optimization iteration layer for RNAi design, which accommodates adaptive fine-tuning of construct parameters for optimized silencing efficiency. GFlowNets with δ-Conservative Search are an illustration of that process, navigating effectively across the large RNAi design space of biological sequences by means of proxy evaluation and conservative exploration policies (Kim et al., 2024). As a component of RNAi pipelines like SIREN and pssRNAit, RL-based approaches potentially optimize siRNA design dynamically by balancing iteratively specificity, thermodynamic stability, and off-target reduction (Ahmed et al., 2020; Vargas-Mejía & Vega-Arreguín, 2025). Platforms like OligoRL reflect how RL potentially addresses multi-constraint oligonucleotide design problems, and its potential for designing context-specific RNAi constructs for plants improved for better resilience (Abdellatef et al., 2021).

Real-world implementation of AI-designed RNAi is illustrated by case studies of virus and pest management based on dsRNA. Locally administered dsRNA against the coat protein of bean common mosaic virus (BCMV) effectively suppressed aphid-mediated transmission of *Nicotiana benthamiana* and cowpea by a delivery vehicle based on clay (BioClay) (Worrall et al., 2019). Likewise, lethal phenotypes and profound downregulation of chitin synthase A (CHSA) gene of *Spodoptera litura* and *Chilo partellus* were caused by dsRNA constructs of the gene, illustrating AI-designed dsRNA's capability of controlling pests

(Rana et al., 2020). As a third illustration, virus resistance against Nicotiana benthamiana occurred by induction of dsRNA against PepMoV gene HC-Pro and NIb, revealing the versatility of AI-directed RNAi for virus suppression (Yoon et al., 2021). Such uses are representative of how RNAi technologies and AI, when complementary, are establishing precision-based, sustainable approaches to securing crops (Morozov et al., 2019).

10.5 Functional Genomics Through AI-Enhanced RNAi

Guided RNAi is a new tool of plant functional genomics of immense value for selective gene silencing for the identification of novel gene activities. RNAi allows downregulation-directed targeting of a desired transcript, and phenotypic analysis of silenced lines offers functional information on uncharacterized genes despite genome sequence data (Waterhouse & Helliwell, 2003). AI, including DL, has complemented RNAi efficiency by predicting selectively accumulating siRNAs on several plant species. Such predictors point to principal sequence motifs, which define siRNA accumulation, and hence enhance selectivity and precision of gene silencing (Enoki et al., 2024). Additionally, RNAi protocols by amiRNAs allow highly precise gene targeting with few off-target activities, and hence are of immense value for gene dissection of function, particularly in species of a polyploid character, where gene redundancy impedes classical approaches for gene knocking out (Singh & Das, 2021). Temporal, including spatial, gene-silencing regulation is also possible by utilizing inducible RNAi systems, e.g., heat-inducible constructs, which allow conditional suppression of a necessary gene function but exert no influence on entire plant viability (Masclaux & Galaud, 2011).

Integration of AI with RNAi knockdown data facilitates reconstruction of rich gene regulatory networks (GRNs) for improved gene-to-gene interaction understanding for the development of plants and stress responses. Computational systems, for example, NetREX-CF, integrate RNAi perturbation data with gene expression patterns by collaborative filtering, compensating for missing transcription factor data sets efficiently and cross-validation of inferred GRNs by large-scale RNA-sequencing (RNA-seq) and RNAi screens of *Drosophila* and human systems (Wang et al., 2022). Similarly, network transformation methods calibrate reference signaling networks to RNAi perturbation data, enabling systems of hundreds of constituents to be modeled scalably (Hashemikhabir et al., 2012). Multisource integration methodologies, including perturbation information and steady-state transcriptomes, increased causal network accuracy compared to single-source methodologies by consensus network assembly (Shojaie et al., 2014). Bayesian inference systems generalize by integrating RNAi information with genome-wide binding and expression data sets, thereby increasing regulatory network precision (Liang et al., 2018). Additionally, probabilistic Boolean threshold models employ Bayesian learning for RNAi data sets for inferring pathways' topologies, including missing observations and supporting hypothesis-guided experimental design (Kaderali et al., 2009).

Dynamic modeling methodologies universalize gene regulation of environmental stress, applicability of RNAi by deriving post-transcriptional regulation of gene silencing's dynamics, and mechanistic insight over time. Random environmental perturbation by white noise terms is incorporated into stochastic modeling, for example, by Ahmed and Elshehabey's stochastic model, for stochastic RNA silencing dynamics prediction, characterizing the regulation of environmental variability post-transcriptional regulation (Ahmed & Elshehabey, 2025). Likewise, dynamic modular systems of genotype, environment, and genotype-by-environment interaction integration by Vallejos et al. encourage phenotypic Phaseolus vulgaris shifting environmental condition prospects of modeling (Vallejos et al., 2021). ML-based models like BADDADAN recognize co-expressed gene modules, build ODE-based dynamic network, and stress-induced regulatory reaction of *Arabidopsis thaliana* incorporation (Noordijk et al., 2025). Likewise, *Arabidopsis RD29A* gene expression dynamic modeling reveals ABA-dependent gene regulation transcription kinetics and a water stress regulation of gene individual ABRE theme critical role (Ndathe et al., 2022). FIT statistical modeling framework advances transcriptomic field condition shifting predictions, supporting environmental influences of transcriptome variation analysis (Iwayama et al., 2017).

High-throughput phenotyping through AI-based computer vision has revolutionized RNAi-induced phenotyping evaluation by facilitating large-scale, automatic, and non-destructive gene silencing impacts analysis. ANN and CNN are utilized for inferring morphological and physiological traits from plant images for facilitating precise trait categorization and counting (Singhvi et al., 2024). Advanced optical character recognition (OCR) and image division schemes mechanistically automate identification of treatments and feature extraction with high fidelity; meanwhile, multispectral imagery by unmanned aerial vehicle (UAV)-integrated Internet of Things (IoT)-enabled sensors improves perceiving sub-ttif fine stress reactions and quality variations accuracy (Tripodi et al., 2020). Such AI-supported phenotyping systems enable transcriptome perturbation correlating and linking to seemingly phenotypic manifestations, thereby unveiling gene regulatory interactions and function (Mochida et al., 2019). Additionally, phenotypic large datasets analysis by AI algorithms facilitates determining complex genotype-phenotype associations and data-informed trait prediction-based precision agriculture (Sheikh et al., 2024).

10.6 AI and RNAi in Plant Stress and Disease Management

AI-guided RNAi technologies are a revolutionary way of creating plant resistance to both abiotic and biotic stresses for predictive and precise systems of gene silencing. RNAi, a universal antiviral defense strategy of plants, is a phenomenon of plants being attacked by sRNAs to complementary virus RNA sequences, leading to gene silencing and virus suppression (Agarwal & Mangrauthia, 2013). AI gives a boost to this process by studying primary molecular parameters, such as target sequence variability, RNA secondary structure, and sequence accessibility, for better RNAi construct designing and silencing efficiency (Akbarimotlagh et al., 2023). AI-guided algorithms also enable the design of amiRNAs for precise targeting of virus suppressor genes, exemplified by better resistance of virus suppressor gene silencing suppressor 2b of cucumber mosaic virus by expressing amiRNA-transgenic plants (Qu et al., 2007). Moreover, integration of AI technologies into exogenous dsRNA-based RNAi systems, such as spray-induced gene silencing (SIGS), also enables optimization of dsRNA production, delivery, efficiency, and environmental persistence, resulting in improved stability (Taliansky et al., 2021). Such technologies enable the prediction and designing of stable, sustainable RNAi technologies for off-target decrease and pathogen resistance decrease, resulting in durable virus protection (Venu et al., 2024).

RNAi technologies supplemented by AI have also been employed for the management of fungal, bacteriological, and insect pests as green alternatives to traditional chemical pesticides. RNAi-induced gene silencing, by predominantly dsRNA molecules, inhibits critical genes of target pests, thereby inhibiting their pathogenicity or survival. HIGS, virus-induced gene silencing (VIGS), and SIGS are a few of the delivery systems that have been successful against major groups of pests like lepidopterans, coleopterans, and hemipterans by knocking down critical biological processes like energy metabolism, hormone regulation, and chitin synthesis (Choudry et al., 2024; Munawar et al., 2023). RNAi's selectivity lessens environmental harm compared to chemical pesticides, but variation of silencing efficiency, along with the development of resistance by pests, are a few of its drawbacks. AI models are being utilized for fine-tuning target selection of dsRNA and also for forecasting cross-species efficacy, whereas delivery systems based on nanocarriers also optimize stability and absorption of dsRNA (Hanamasagar et al., 2024; Hasan et al., 2024; Lu et al., 2023). The integration of AI-guided RNAi and nanobiotech, therefore, facilitates the development of next-generation "smart biopesticides."

The ability of AI to predict is central to designing and implementing RNAi-based "RNA pesticides" for greatest efficiency. Random forests, besides other ML algorithms, have exhibited good prediction ability for evaluating RNAi efficiency by predicting key parameters like target gene identity, dsRNA size, dose employed, and kinetics of response (List et al., 2022). Such prediction models normalize experimental design and RNAi bioassays to a normalized format, eliminating inter-laboratory variations. Complementary to RNAi delivery are nanocarrier systems, boosting precision, environmental safety, and delivery efficiency (Qiao et al., 2024). Bioinformatics-guided designs of dsRNAs assist in selecting targets for high specificity and environmental safety (Fletcher et al., 2020), and transgenic technologies utilizing their plant-incorporated protectants for steady, in planta expression of dsRNA for long-term management of pests (Li et al., 2022b).

In addition to management of pathogens and pests, AI-directed RNAi also improves plant tolerances to abiotic stresses of water deficiency, heat, and salinity. RNAi-directed silencing of stress-related genes modulates stress-linked key biochemical and physiological processes, leading to improved resistance. RNAi-directed silencing of stress-linked gene *SlHK2*, for instance, brought about improved tolerances to water deficiency and heat by activating osmoprotectant synthesis, antioxidant defense systems, and membrane stability (Mushtaq et al., 2022). Likewise, RNAi control of miRNAs related to ABA and auxin connection signaling pathways helps plants maintain homeostasis during osmotic stress (Akash Pradhan et al., 2015). RNAi combinations with Clustered Regularly Interspaced Short Palindromic Repeats and CRISPR-associated protein (CRISPR/Cas)-based modulations like co-overexpression of AVP1 and OsSIZ1 fine-tune stress tolerance technologies further, leading to improved yields under stress (Esmaeili et al., 2019). Such combination technologies are reflective of AI-based RNAi's potential for engineering stress-tolerant, climate-resilient crops (Figure 10.1) (Mohapatra et al., 2024).

Computer vision and AI systems expand RNAi applications by automatic, high-input-output surveillance of stress reactions and suppression of diseases. CNN and other deep-learning programs allow real-time, precise recognition of phenotypic variations and RNAi-treated plants' disease symptoms. By utilizing large, annotated picture sets, the systems distinguish disease intensity and RNAi treatment effectiveness across a range of

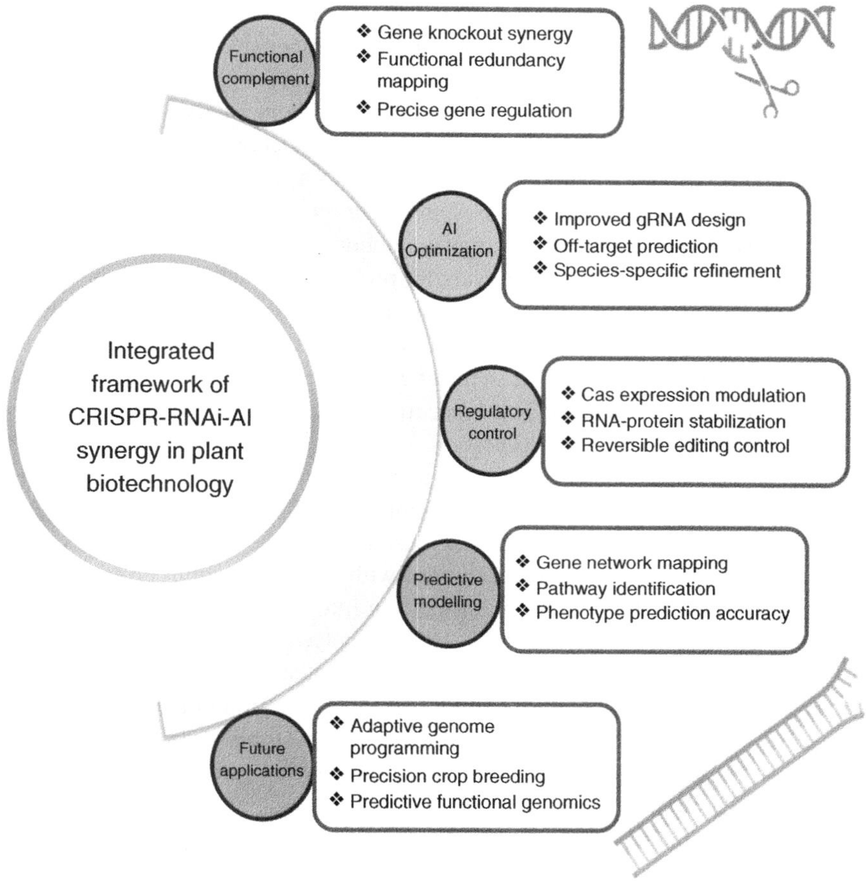

Figure 10.1 Integrated framework of CRISPR-RNAi-AI synergy in plant biotechnology.

environmental conditions (Hue et al., 2024). IoT platform integration and edge computing allow a constant, field-based evaluation, offering real-time feedback for adaptable management. New AI systems, which integrate image evaluation and natural language models, like GPT-4, expand diagnostic precision and RNAi-treated crops' decision support. Such technologies, including UAV-based pictures and diagnostics by cell phone, make AI-based phenotyping useful for RNAi endpoint analysis, guaranteeing precise, data-based disease management for sustainable agriculture (Prabha, 2021).

10.7 Integration of Multiomics Data with AI and RNAi

AI greatly eases the interpretation of short RNA-sequencing (sRNA-seq) data for RNAi target discovery by uniting sophisticated computational methodologies and ML paradigms. Such programs as Oasis 2.0 have better cross-species identification and classification of a spectrum of sRNA, including miRNA and virus-derived sRNA, for a high-throughput identification of possible RNAi targets across a spectrum of taxa at elevated speed and precision of processing (Rahman et al., 2018). Likewise, COMprehensive Platform for Small RNA-Seq data Analysis (COMPSRA) provides a broad-based analysis platform for profiling a large number of a variety of sRNA species, including circulatory sRNA, which are prospective biomarkers, pushing RNAi target discovery ahead of classical miRNA-focused studies (Li et al., 2020). ML-based paradigms like TransfoRNA use adaptable schemes of annotation for a precise classification of sRNA, even across data sets of sparse ground truth, broadening the functional set of RNAi target prediction (Taha et al., 2024). Augmentor to these, iSmaRT provides a handy set of software for profiling of small ncRNA, including piRNA hitherto broadly underappreciated class of critical importance for post-transcriptional regulation (Panero et al., 2017). Together, these AI-based programs, discussable by a review by Minhyeok Lee (Lee, 2023), foster a paradigm shift for RNAi research by increasing precise prediction of siRNA, better accuracy of designing, and elevated functional comprehension of RNAi-directed regulation.

More and more ML systems are being deployed to model sophisticated RNAi and chromatin modification interactions, extending our current comprehension of epigenetic regulation based on integrative computational schemes. One of these is AsteRIa, where nucleosome affinity purification data are intertwined with high-throughput proteomics to provide a quantitative perspective of chromatin-protein interactions and explore recruitment processes of chromatin-binding proteins such as polycomb group member CBX8 (Stadler et al., 2024). Similarly, transformer-based deep-learning schemes such as Chromoformer permit predictions of three-dimensional chromatin architectures, revealing long-distance associations of histone modifications and their transcriptional regulatory consequences (Lee et al., 2022). Similarly, multiomics integration based on DeepRCI identifies RNA-chromatin interactions accurately, providing new insight into transcriptional control processes. Experimental validation for this interaction also occurs, where RNAi-mediated *H3K9* methylation (H3K9me) for stable assembly of heterochromatin and transgenerational epigenetic inheritance has been revealed, demonstrating a cooperative involvement of RNAi and chromatin modification for gene silencing (Yu et al., 2018).

Interfusion of AI-driven plant proteomics and RNAi knockdown investigations has served as a baseline for ascertaining functional consequences of gene silencing. RNAi is a useful reverse-genetics resource for investigating loss-of-function phenotypes, notably in polyploid plants where gene redundancy complicates classical mutagenesis analysis. Two-dimensional gel electrophoresis and mass-spectrophotometric proteomic technologies enable protein-level changes following RNAi-directed silencing to be measured quantitatively, thereby affirming transcript-level consequences (Lee & Lee, 2011). Convolution-transformer neural models have also aided RNAi design by predicting selectively accumulated siRNA species and categorizing essential sequence motifs guiding silencing efficiency (Enoki et al., 2024). Targeted proteomics technologies, such as selected reaction monitoring mass spectrophotometry, enable direct measurement of specified protein downregulation after silencing (Glukhova et al., 2013), whereas amiRNA-based epitope-tagged screening systems expedite selection of highly efficient silencing constructs (Li et al., 2014). Such technologies individually enhance the precision and reliability of RNAi-based functional genomics, supporting its suitability for exploitation for crop improvement and metabolic engineering (Kumar et al., 2012).

Any systems biology approach offers a framework for integration of genomics, transcriptomics, proteomics, and metabolomics, and AI-informed RNAi predictions, for a systems-level interpretation of complex biology systems. Integrated multiomics analysis offers concurrent mapping of gene regulatory, metabolic, and signaling pathways, and disclosure of determinants of phenotype and stress reaction, at a molecular scale. Application of AI algorithms compensates for high-dimensional, noisy management of biological information, extracting informative, strong features and concise network inference (Xin et al., 2024). Combinatorial analytical approaches, like co-expression networks and metabolite-gene correlation mapping, also unveil complex regulatory patterns, lost by single-omics studies (Sanches et al., 2024). Model-based approaches, like kinetic and constraint-based modeling, also allow quantitative integration of multiple omic layers, for systems-level interpretation of cell processes (de Mas, 2018). The paradigmatic case of the INTEGRATE pipeline, for instance, integrates transcriptome and metabolome information, based on constraint-based metabolism models, predicting regulatory nodes and flux distributions, and is of great translational potential for personalized treatments (Di Filippo et al., 2022).

The intersection of RNAi, AI, and metabolomics also brought new horizons for predicting and controlling plant secondary metabolite biosynthesis. RNAi technology allows for fine-tuned modulation of secondary metabolite pathways comprised of genes, creating new chemotypes of increased pharmacological or industrial interest. This has been duly illustrated for *Centella asiatica* and *Artemisia annua* species, where RNAi revealed principal genes for the biosynthesis of bioactive compounds (Wagner & Kroumova, 2008). Predictive ML models, for example, random forest algorithms supplemented by RDKit molecular fingerprints, were also used to predict plant metabolites of desired bioactivities, for example, α-glucosidase inhibitory activity for antidiabetic therapy (Jamir & Hariprasad, 2024). Interpretation of metabolomic and NGS datasets also hastens the drafting of biosynthetic pathways, revealing genetic and regulatory foundations of metabolic diversity and guiding targeted metabolic engineering (Scossa et al., 2018).

10.8 AI-Enhanced Delivery Systems for RNAi in Plants

Classic RNAi delivery methodologies for plants, such as foliar and transgenic technologies, are handicapped by intrinsic disadvantages, which deter their effectiveness (Figure 10.2). Foliar delivery of dsRNA, although being green and safe, suffers from sluggish uptake and weak systemic mobility. It should induce an RNAi response by crossing the cuticle, surviving nuclease degradation, and penetrating both cell wall and plasma membrane interfaces, which are penetrated hardly under field conditions (Hoang et al., 2022). Moreover, systematic silencing of endogenous gene expression is yet to be achieved dependably, lowering the prospects of its application for crop protection and weed control (Bennett et al., 2020). Transgenic technologies, despite their success under some conditions, are technologically demanding and are even regulated, thereby spurring research into nontransgenic technologies.

Nanotechnology also provides a potential route to enhance RNAi delivery by increasing plant tissue cells' stability, uptake, and systemic transport of dsRNA. *Mesoporous silica* nanoparticles, for instance, have been reported to deliver effective siRNA transport and

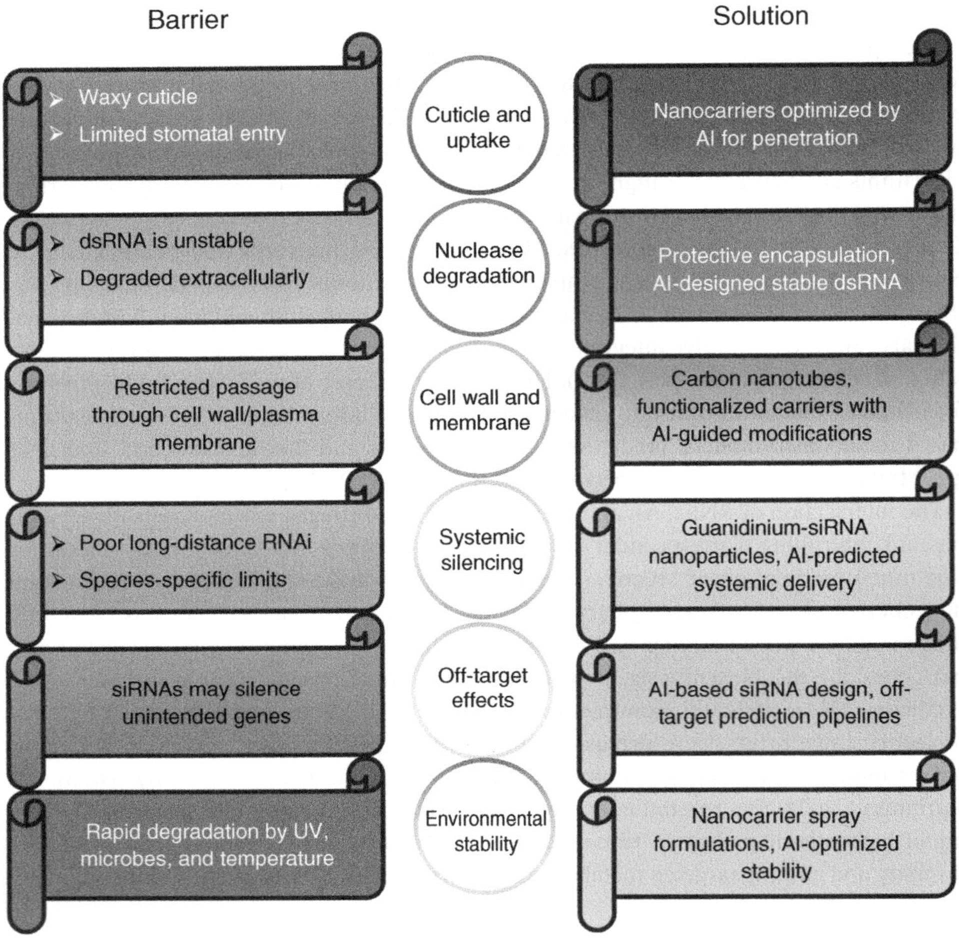

Figure 10.2 Barriers and solutions in RNAi delivery.

allow multigene silencing for weeks at a stretch without mechanical aid (De Schutter et al., 2021). Layered double hydroxide (LDH) nanoparticles have been exceptionally effective, boosting leaf surface adhesion of dsRNA and inducing its systemic transport via vasculature to act against pests like *Panonychus citri* (Cheng et al., 2024). LDH nanocarriers have also been used for delivering dsRNA into pollen of tomatoes, for which gene silencing efficiency relied on particle size and on dsRNA distinction (Yong et al., 2021). Other nanocarriers like carbon dots and Mesoporous Silica Nanoparticle (MSNs) have been efficiently utilized for counteracting plant virus pests like turnip mosaic virus and beet curly top virus, yielding better results compared to naked treatments of dsRNA (Zarrabi et al., 2024). Core-shell polymeric nanostructures also shield dsRNA from extracellular degradation and allow ATP-activated intracellular release for boosting their transfection efficiency for their delivery against pests (Bi et al., 2023).

AI adds a transformative dimension to nanoparticle delivery systems by bringing about rational design, optimization, and adaptability. ML protocols and neural networks are also

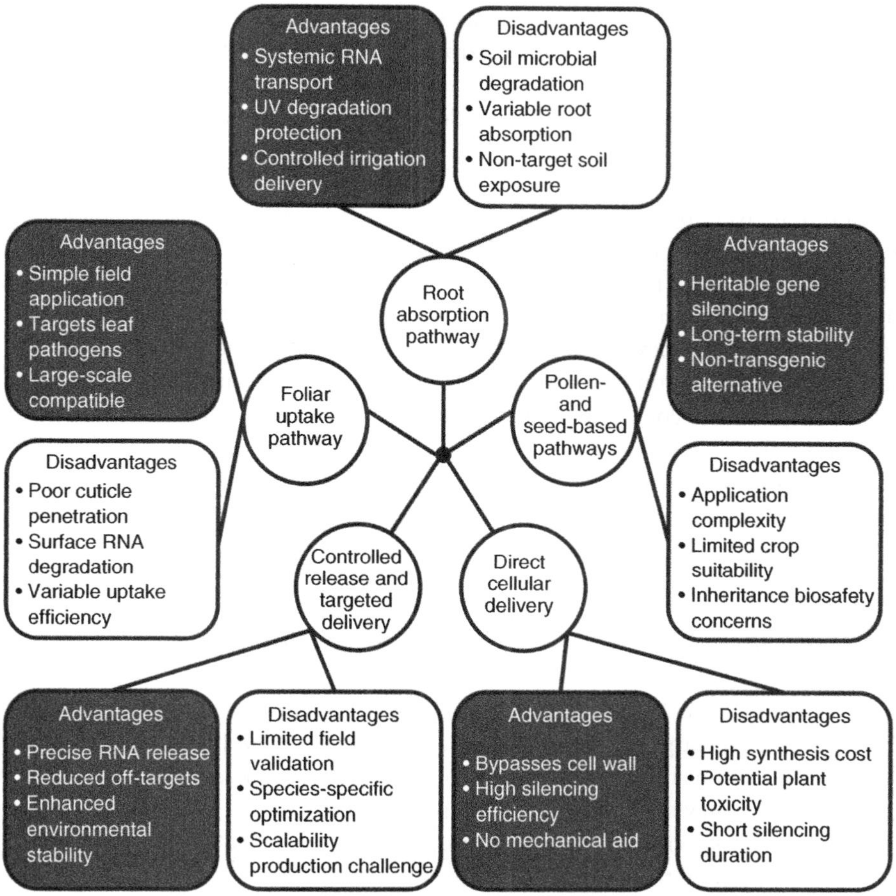

Figure 10.3 Nanocarrier-based RNAi delivery pathways.

being increasingly utilized for the prediction of desired nanocarrier geometries, thereby expediting the discovery of plant-specific RNAi delivery systems and improved stability and uptake of RNA molecules (Kapoor et al., 2024; Yong et al., 2024). Biomedical research also indicated how optimization of lipid nanoparticles for stabilization and targeted delivery of RNA can be accomplished by AI, for example, for cancer vaccine development, having direct translational applicability to plant biotechnology (Bhujel et al., 2025). Of considerable importance, AI-directed design also facilitated the development of species-independent systems, for example, of guanidinium-siRNA nanoparticles, which allow for their effective systemic gene silencing across a large palette of plant taxa (Figure 10.3) (Lin et al., 2025).

At the same time, AI models are also used to assess dsRNA uptake efficiency, persistence, and spatiotemporal dynamics of plant tissue. Systemic delivery of dsRNA using vascular pathways gives it protective integrity against attack by pathogens, setting it apart for its potential broad protection schemes based on RNAi (Brosnan et al., 2021). Nanocarriers, like functionalized graphene oxide nanoparticles, also enhance it, eliciting as high as 97.2%

of silencing efficiency for a period of a meager 24 hours, despite its short-lived effect (Li et al., 2022a). Likewise, single-walled carbon nanotubes and their oligohistidine-functionalized successors also give RNA delivery precision, bypassing cell wall limitations without inducing tissue damage, meanwhile releasing pH-sensitive controlled release of RNA molecule for targeted growth-controlling factors modulation (Jeong et al., 2025; Shehzad et al., 2025).

In addition to delivery optimization, AI also provides for spatiotemporal regulation of RNAi induction, controlling gene silencing with accuracy. Such systems are illustrated by synthetic trans-acting small interfering RNA (syn-tasiRNA) for modulating silencing efficiency by precursor positioning or base-pairing, as illustrated by *Arabidopsis thaliana* and *Nicotiana benthamiana* (Lopez-Dolz et al., 2020). Inducible expression systems, including the ethanol-inducible alc promoter, also provide for tissue-specific and temporal expression of dsRNA by heterologous plants, causing minimum pleiotropic effects and selective silencing of desired genetics, like those controlling chlorophyll synthesis (Chen et al., 2003). Very recently, syn-tasiRNA-based VIGS (syn-tasiR-VIGS) also followed, utilizing minimal RNA virus-derived precursors for large-scale and nontransgenic RNAi purposes, including antiviral defense strategy by plants (Cisneros et al., 2025).

10.9 Challenges, Limitations, and Ethical Considerations

RNAi efficiency of plants is also limited by a few biological constraints, inhibiting its useful applicability. One of its large constraints is a plant's cuticle, which hinders exogenously administered penetrations of dsRNA to deeper plant tissue layers. Even if dsRNA entities manage to penetrate, they are still significantly prone to degradation by nucleases before accessing the RNAi machinery of the cytoplasm. Additionally, dsRNA needs to penetrate both the hard cell wall and the plasma membrane to be useful, a challenge that cuts down delivery efficiency under realistic circumstances (Bennett et al., 2020). Off-target phenomena also pose a large constraint, where siRNA is also capable of knocking down off-target transcripts by partial sequence similarity, confusing experimental interpretation and also raising fears of field usefulness (Senthil-Kumar & Mysore, 2011). Another huge constraint is systemic delivery inefficiency between plant species; current technologies remain species-specific and lack the capability of long-distance gene silencing. Glimpses of new technologies, like guanidinium-siRNA nanoparticles, have also illustrated their ability to get across by ensuring systemic delivery across a variety of taxa (Lin et al., 2025). Environmental instability of naked dsRNA is also a constraint, but nanomaterial-formulated preparations are demonstrating their ability to increase stability and persistence, increasing RNAi efficiency (Venu et al., 2024).

Besides biological limitations, computational limitations also limit the application of AI to plant RNAi research. One of the major limitations is RNAi-associated dataset quality and availability. Small and weakly annotated datasets challenge traditional and DL techniques, requiring large-scale training sets for their ability to identify regulatory elements and refine siRNA designs effectively (Huang et al., 2023). Such a limitation is highest for RNAi, for which large-scale, comprehensive model-building ML datasets are scarce, leading to a delay in siRNA pipeline development (Lee, 2023). Few data are a generic challenge in plant biology, where large numbers of high-quality, curated datasets are costly and logistically challenging to generate, thereby limiting AI application (Xu et al., 2023). Additionally, multidimensional multisource dataset integration and management are unanswered, placing a focus on better integration and visualization paradigms (Williamson et al., 2023).

These computational limitations are further degraded by algorithmic biases, diminishing RNAi prediction accuracy, and generalization across species. ML algorithms trained on human or mouse data sets generally attain high predictive ability for those systems but do not generalize efficiently across plant genomes due to evolutionary divergence of miRNA seed sequences and patterns of target recognition (Ben Or & Veksler-Lublinsky, 2021). Such differences are also exacerbated by pseudo-miRNA sequences, which are common across nonmodel plants, increasing false positive rates and masking true targets. Computational software like SMIRP, based on sequence preservation and phylogenetic distance, attempts to optimize prediction accuracy and specificity across a spectrum of species (Peace et al., 2015). For most nonmodel plants, however, predictions are made purely computationally, lacking corroboration by experimental verification, and therefore risk propagation of false targets into downstream analysis like Gene Ontology enrichment, which generates confounding biological interpretation (Fridrich et al., 2019).

Through a farm scale, RNAi technologies also act ecologically and through biosafety risks by means of intervention. One of their major risks is their unintended involvement with non-target species, whereby off-target silencing might occur external to desired pest species, potentially disequilibriating ecological equilibria (Chen & De Schutter, 2024). The ecological behavior of RNAi biopesticides, such as persistence, degradation, and bioavailability, also impacts the extent of exposure of neighboring systems, for which increased ecological exposure monitoring ensues (Luo et al., 2024). Resistance development by target pests also marks a major threat, whereby adaptation might decrease long-term efficiency, leading to resistance management development (Christiaens et al., 2022). Genetic polymorphism and mutation of RNAi-transformed plants also lead to unexpected ecological consequences, justifying thorough ecological risk analysis (Auer & Frederick, 2009). Even if RNAi favors being species-specific, its selectivity largely depends on dsRNA design, which should preferentially lower chances of off-target effect, as well as of unintended cross-kingdom interactions (Christiaens et al., 2022).

These difficulties are further complicated by changing conventions of regulation and ethics for RNAi uses for agriculture, especially for the increasing number of AI-based delivery systems. Spray RNAi biopesticides are a paradigm of a nontransgenic, green lifestyle of protection for crops, but they also pose ethical, legal, and social impacts, requiring stringent regulatory approval. Existing conventions are being redesigned to meet environmental risk concerns, regulatory uncertainty, and acceptability, trust-building being a leading ethical criterion (Rinaldi et al., 2025). RNAi combination on other technologies like CRISPR/Cas exacerbates regulation, for hybrid technologies have the potency to modify or widen agronomic traits but risk off-target behavior and ecotoxicity (Touzdjian Pinheiro Kohlrausch Távora et al., 2022). Successful ERAs are also required, emphasizing routes of exposure, persistence of sRNA, and possible effects on non-target species (Auer & Frederick, 2009). Diffuse development of nanoparticle-based delivery systems is an example of how conventions of regulation themselves are called upon to submit to redesign to balance simultaneously efficiency, ecotoxicity, and ethical responsibility across agricultural biotechnologies (Hasan et al., 2024).

10.10 Future Prospects and Next-Generation Directions

AI holds great potential for increasing the precision RNAi in plants by encouraging tissue-selective and inducible systems of silencing. Such systems are desirable for studying genes critical for plant survival without inducing undesirable phenotypic impacts. Heat-inducible systems of RNAi, for instance, achieve temporal controls of gene silence, thereby averting growth defects characteristic of constitutive suppression (Masclaux & Galaud, 2011). Likewise, dexamethasone-regulated inducible RNAi vectors allow reversible, high-throughput gene silence, wherein the study of gene function leading to embryo lethality or constitutive pleiotropic effects becomes possible (Wielopolska et al., 2005). Synthetic tasiRNAs extend RNAi technology further by enabling gene silence at specified gene sites targeting at least precursors. Having been used individually with VIGS, it provides a nontransgenic, large-scale, and even antiviral exploitation (Cisneros et al., 2025). amiRNA technology joins its numbers by delivering highly precise silence and minimum off-target impacts, whereas spatiotemporal utilization of promoters allows fine regulation of gene expression across stress-related and developmental circuits (Singh & Das, 2021).

The integration of CRISPR and RNAi technologies, propelled by AI-optimization, is a multidimensional approach to plant functional genomics and trait improvement. The integration of CRISPR gene-knockout and RNAi knockdown-based gene-knockdown induces simultaneous investigation of gene function and regulatory associations, yielding a powerful platform for disentangling genetic networks (Qiao et al., 2022). Such bidirectional systems have been employed for the identification of synthetic lethal interactions, unveiling functional redundancies and potential Achilles heels of intricate networks (Housden et al., 2017). AI is central to this integration by enlarging guide RNA optimization, diminishing off-target events, and enhancing editing efficiency (Kim et al., 2025). RNAi itself, furthermore, may be employed for modulating CRISPR action by employing amiRNAs and chemical activators such as enoxacin, affording an additional precision dimension (Huang et al., 2020). The integration of CRISPR, RNAi, and AI not only refines functional genome screens but also greatly accelerates the discovery of novel cellular pathways and stress-response regulatory proteins (Mohr et al., 2014).

These digital twin technologies are also prospective for modeling plant reactions to RNAi treatments, closing computational prediction and experimental biology. By virtual models of causal networks, whereby nodes are RNA molecules and edges are regulatory interactions, digital twins are prospective predictors of gene expression change on RNAi-guided suppression. Such models, trained by RNA activity patterns from transcriptome datasets, give predictive information on cellular reactions for selective gene suppression (Kumar et al., 2012). Such technologies for agriculture are prospective for modeling SIGS for controlling pests and pathogens, pre-emptive testing of efficiency, and of ecological repercussions (Abdellatef et al., 2021; Lim et al., 2020). Computational pipelines like SIREN are improved similarly by predicting RNAi construct efficiency, whereas off-target risks are reduced (Vargas-Mejía & Vega-Arreguín, 2025). By both digital twin modeling and experimental data, scientists are better placed to efficiently designing RNAi-based systems for stress resistance improvement, disease resistance, and yield improvement.

In tandem with computational intelligence, AI is resuscitating nanocarrier and RNA spray technologies, positioning them center stage of next-generation sustainable agriculture. Controlled release and stability are improved by nanocarriers, brokering targeted stomatal uptake and root absorption delivery (Ashique et al., 2025). Compositions at a nanoscale, bioavailability, and environmental footprint are improved compared to classical chemical pesticides (Qiao et al., 2024; Rajesh, 2025). AI-aided optimization of nanocarrier structure permits real-time evaluation of their interactions with plants, pests, and, therefore, precision delivery (Ashique et al., 2025). Moreover, bioengineered nanocarriers for custom crop-pathogen systems have been proposed, ensuring performance and efficiency across a spectrum of field conditions (Mat Jalaluddin et al., 2023). Through a combination of nanotechnology and AI-based analytics, agriculture could turn a page toward nanoscale agrochemicals, balancing efficiency, selectivity, and ecological accountability (Bala et al., 2024).

In times to come, AI-fore RNAi crop breeding pipeline development shall speed up climate-resilient agriculture development. RNAi for specific manipulation of stress tolerance, increased yields, and nutrient quality traits, coupled with AI, enables large-scale integration of phenotypic and omics datasets. It assists predictive modeling of cropping behavior during circumstances of climate stress and miniaturizes breeding cycles (Thingujam et al., 2025; Rai, 2022). As part of the larger picture of “Breeding 5.0,” RNAi, speed breeding, and genome editing intersect AI-based analytics for providing climate-smart agriculture and sustainable food systems (Fang, 2024; Rai, 2022). Cobbing of multi-omics and phenotyping of complex types also assists in the production of high-yielding and resilient types (Bakala et al., 2020). The point of emphasis is where such technologies go one step further and enable farmers by providing precision agriculture solutions and equitable technology provisioning (Thingujam et al., 2025).

Accompanying these breeding technologies, edge AI and IoT sensors allow for real-time evaluation of RNAi effectiveness under field conditions. Those environmental factors of temperature, moisture, and humidity having direct impacts on RNA stability and absorption are tracked in real-time by IoT sensors (Hoang et al., 2022; Kumar et al., 2024). Edge AI onboard data processing minimizes latency, which permits quick adjustment of RNAi application (Yousuf et al., 2022). The ML evaluation of these data sets also identifies stress and disease development, and hence, dynamic adjustment of RNAi application techniques (Costa et al., 2024) is possible. This overlap of AI, IoT, and RNAi is a linchpin point for systems of real-time precision agriculture for assisting productivity under changing environmental conditions.

10.11 Conclusion

RNAi is an elusive technology for boosting plant productivity, stress resistance, and defense against pests and diseases. However, traditional delivery technologies are still plagued by inefficiencies during uptake, systemic transport, and selectivity. The confluence of nanotechnology and AI, however, holds new hopes of bypassing limitations, offering stable, efficient, and targeted RNAi delivery. AI-based technologies, such as predictive modeling, inducible systems, and CRISPR-RNAi fusions, allow for yet unknown precision of gene regulation of minimizing off-target effects. Additionally, digital twin modeling, IoT-based monitoring, and novel innovations of nanocarriers also broaden applications of RNAi from benchtop to large-scale field practice. Such benefits notwithstanding, ecological security, bias of algorithms, and regulatory acceptability are critical challenges requiring careful mitigation for eventual responsible use. With upgrading, AI-polished RNAi technologies have rich potential for upgrading plant biology, achieving climate-smart agriculture, and solving global food security in environmentally friendly ways.

Acknowledgement

All the authors are thankful to their respective Universities and Institutions for their valuable support.

Data Availability

The authors confirm that the data supporting the findings of this study are available within the book/chapter.

Abbreviations

AGO	Argonaute
AI	Artificial Intelligence
amiRNAs	Artificial microRNAs
ASRP	*Arabidopsis* Small RNA Project
CHSA	Chitin Synthase A
CNN	Convolutional Neural Network
CTS	Candidate Target Site
DCL1	Dicer-Like 1
DL	Deep Learning
dsRNA	Double-Stranded RNA
EST	Expressed Sequence Tag
GANs	Generative Adversarial Networks
GSS	Genome Survey Sequence
hc-siRNA	Heterochromatic siRNAs
HYL1	Hyponastic Leaves 1
IoT	Internet of Things
LDH	Layered Double Hydroxide
lncRNA	Long Noncoding RNA
MCMC	Markov Chain Monte Carlo
miRNAs	MicroRNAs
ML	Machine Learning
ncRNAs	Noncoding RNAs

NGS	Next-Generation Sequencing
OCR	Optical Character Recognition
ODE	Ordinary Differential Equation
PARE	Parallel Analysis of RNA Ends
phasiRNAs	Phased siRNAs
RdRPs	RNA-dependent RNA Polymerases
RISC	RNA-Induced Silencing Complexes
RL	Reinforcement Learning
RNAi	RNA Interference
siRNAs	Small Interfering RNAs
sRNAs	Small RNAs
sRNA-seq	Short RNA-Sequencing
ta-siRNA	Trans-Acting Small Interfering RNA

References

Abdellatef, E., Kamal, N. M., & Tsujimoto, H. (2021). Tuning beforehand: A foresight on RNA interference (RNAi) and *in vitro*-derived dsRNAs to enhance crop resilience to biotic and abiotic stresses. *International Journal of Molecular Sciences, 22*(14), 7687. https://doi.org/10.3390/ijms22147687

Agarwal, S., & Mangrauthia, S. K. (2013). Advances, challenges and prospects in small RNA mediated approaches of virus resistance in plants. *Journal of Genomes and Exomes, 2*, 43–62. https://doi.org/10.4137/jge.s11938

Ahmed, R., & Elshehabey, H. M. (2025). Analysis and simulation of a novel stochastic RNA silencing model: R. Ahmed, HM Elshehabey. *Scientific Reports, 15*(1), 20108. https://doi.org/10.1038/s41598-025-04142-w

Ahmed, F., Senthil-Kumar, M., Dai, X., Ramu, V. S., Lee, S., Mysore, K. S., & Zhao, P. X. (2020). pssRNAit: A web server for designing effective and specific plant siRNAs with genome-wide off-target assessment. *Plant Physiology, 184*(1), 65–81. https://doi.org/10.1104/pp.20.00293

Akash Pradhan, A. P., Nupur Naik, N. N., & Sahoo, K. K. (2015). RNAi mediated drought and salinity stress tolerance in plants. *American Journal of Plant Sciences, 6*(12), 1990–2008. https://doi.org/10.4236/ajps.2015.612200

Akbarimotlagh, M., Azizi, A., Shams-Bakhsh, M., Jafari, M., Ghasemzadeh, A., & Palukaitis, P. (2023). Critical points for the design and application of RNA silencing constructs for plant virus resistance. *Advances in Virus Research, 115*, 159–203. https://doi.org/10.1016/bs.aivir.2023.02.001

Alptekin, B., Akpinar, B. A., & Budak, H. (2017). A comprehensive prescription for plant miRNA identification. *Frontiers in Plant Science, 7*, 2058. https://doi.org/10.3389/fpls.2016.02058

Ashique, S., Raikar, A., Jamil, S., Lakshminarayana, L., Gajbhiye, S. A., De, S., & Kumar, S. (2025). Artificial intelligence integration with nanotechnology: A new frontier for sustainable and precision agriculture. *Current Nanoscience, 21*(2), 242–273. https://doi.org/10.2174/0115734137275111231206072049

Auer, C., & Frederick, R. (2009). Crop improvement using small RNAs: Applications and predictive ecological risk assessments. *Trends in Biotechnology, 27*(11), 644–651. https://doi.org/10.1016/j.tibtech.2009.08.005

Baichoo, S., & Ouzounis, C. A. (2017). Computational complexity of algorithms for sequence comparison, short-read assembly and genome alignment. *Biosystems*, *156*, 72–85. https://doi.org/10.1016/j.biosystems.2017.03.003

Bakala, H. S., Singh, G., & Srivastava, P. (2020). Smart breeding for climate resilient agriculture. In I. Y. Abdurakhmonov (Ed.), *Plant breeding-current and future views* (pp. 1–25). IntechOpen.

Bala, M., Sharma, R., & Gupta, S. (2024). Integration of hybrid nanomaterials and artificial intelligence for sustainable agriculture. In V. Khanna, P. Sharma, & P. Mahajan (Eds.), *Technological applications of nano-hybrid composites* (pp. 97–118). IGI Global Scientific Publishing.

Ben Or, G., & Veksler-Lublinsky, I. (2021). Comprehensive machine-learning-based analysis of microRNA-target interactions reveals variable transferability of interaction rules across species. *BMC Bioinformatics*, *22*(1), 264. https://doi.org/10.1186/s12859-021-04164-x

Bennett, M., Deikman, J., Hendrix, B., & Iandolino, A. (2020). Barriers to efficient foliar uptake of dsRNA and molecular barriers to dsRNA activity in plant cells. *Frontiers in Plant Science*, *11*, 816. https://doi.org/10.3389/fpls.2020.00816

Bhujel, R., Enkmann, V., Burgstaller, H., & Maharjan, R. (2025). Artificial intelligence-driven strategies for targeted delivery and enhanced stability of RNA-based lipid nanoparticle cancer vaccines. *Pharmaceutics*, *17*(8), 992. https://doi.org/10.3390/pharmaceutics17080992

Bi, F., Yu, D., Wei, Z., Wei, H., Ren, H., Wang, Y., Ren, D., Hua, Z., Huang, B., & Yang, G. (2023). Core-shell polymeric nanostructures with intracellular ATP-fueled dsRNA delivery toward genetic control of insect pests. *Journal of Agricultural and Food Chemistry*, *71*(6), 2762–2772. https://doi.org/10.1021/acs.jafc.2c05737

Brosnan, C.A., Sawyer, A., de Felippes, F.F., Carroll, B.J., Waterhouse, P. M., & Mitter, N. (2021). Intact double stranded RNA is mobile and triggers RNAi against viral and fungal plant pathogens. *bioRxiv*, pp. 2021-11. https://doi.org/10.1101/2021.11.24.469772

Bu, Y., Jia, C., Guo, X., Li, F., & Song, J. (2023). COPPER: An ensemble deep-learning approach for identifying exclusive virus-derived small interfering RNAs in plants. *Briefings in Functional Genomics*, *22*(3), 274–280. https://doi.org/10.1093/bfgp/elac049

Cagirici, H. B., Galvez, S., Sen, T. Z., & Budak, H. (2021). LncMachine: A machine learning algorithm for long noncoding RNA annotation in plants. *Functional & Integrative Genomics*, *21*(2), 195–204. https://doi.org/10.1007/s10142-021-00769-w

Chen, Y., & De Schutter, K. (2024). Biosafety aspects of *RNAi*-based pests control. *Pest Management Science*, *80*(8), 3697–3706. https://doi.org/10.1002/ps.8098

Chen, S., Hofius, D., Sonnewald, U., & Börnke, F. (2003). Temporal and spatial control of gene silencing in transgenic plants by inducible expression of double-stranded RNA. *The Plant Journal*, *36*(5), 731–740. https://doi.org/10.1046/j.1365-313X.2003.01914.x

Chen, C., Li, J., Feng, J., Liu, B., Feng, L., Yu, X., Li, G., Zhai, J., Meyers, B. C., & Xia, R. (2021). sRNAanno-a database repository of uniformly annotated small RNAs in plants. *Horticulture Research*, *8*, 45. https://doi.org/10.1038/s41438-021-00480-8

Chen, J., Sheng, C. W., Peng, Y., Wang, K., Jiao, Y., Palli, S. R., & Cao, H. (2023). Transcript level and sequence matching are key determinants of off-target effects in RNAi. *Journal of Agricultural and Food Chemistry*, *72*(1), 577–589. https://doi.org/10.1021/acs.jafc.3c07434

Cheng, X., Zhou, Q., Xiao, J., Qin, X., Zhang, Y., Li, X., Zheng, W., & Zhang, H. (2024). Nanoparticle LDH enhances RNAi efficiency of dsRNA in piercing-sucking pests by promoting dsRNA stability and transport in plants. *Journal of Nanobiotechnology*, *22*(1), 544. https://doi.org/10.1186/s12951-024-02819-4

Choudry, M. W., Nawaz, P., Jahan, N., Riaz, R., Ahmed, B., Raza, M. H., Fayyaz, Z., Malik, K., & Afzal, S. (2024). RNA based gene silencing modalities to control insect and fungal plant pests-Challenges and future prospects. *Physiological and Molecular Plant Pathology, 130*, 102241. https://doi.org/10.1016/j.pmpp.2024.102241

Christiaens, O., Sweet, J., Dzhambazova, T., Urru, I., Smagghe, G., Kostov, K., & Arpaia, S. (2022). Implementation of RNAi-based arthropod pest control: Environmental risks, potential for resistance and regulatory considerations. *Journal of Pest Science*, *95*(1), 1–15. https://doi.org/10.1007/s10340-021-01439-3

Cisneros, A. E., Alarcia, A., Llorens-Gámez, J. J., Puertes, A., Juárez-Molina, M., Primc, A., & Carbonell, A. (2025). Syn-tasiR-VIGS: Virus-based targeted RNAi in plants by synthetic trans-acting small interfering RNAs derived from minimal precursors. *Nucleic Acids Research, 53*(5), gkaf183. https://doi.org/10.1093/nar/gkaf183

Cochran, K., Srivastava, D., Shrikumar, A., Balsubramani, A., Hardison, R. C., Kundaje, A., & Mahony, S. (2022). Domain-adaptive neural networks improve cross-species prediction of transcription factor binding. *Genome Research*, *32*(3), 512–523. https://doi.org/10.1101/gr.275394.121

Costa, G., Forestiero, A., Gentile, A. F., Macrì, D., Ortale, R., Bernardi, B., & Cerruto, E. (2024, December 15–18). An ai-driven architecture for precision agriculture: Iot, machine learning, and digital twin integration for sustainable crop protection. *2024 IEEE International Conference on Big Data (BigData),* Washington, DC, USA (pp. 4685–4691). IEEE.

De Schutter, K., Christiaens, O., Taning, C. N. T., & Smagghe, G. (2021). Boosting dsRNA delivery in plant and insect cells with peptide-and polymer-based carriers: Case-based current status and future perspectives. In B. Mezzetti, J. Sweet, & L. Burgos (Eds.), *RNAi for plant improvement and protection* (pp. 102–116). CABI.

Deng, Z., Ma, L., Zhang, P., & Zhu, H. (2022). Small RNAs participate in plant-virus interaction and their application in plant viral defense. *International Journal of Molecular Sciences*, *23*(2), 696. https://doi.org/10.3390/ijms23020696

Deschamps, S., & Llaca, V. (2016). Strategies for sequence assembly of plant genomes. In I. Y. Abdurakhmonov (Ed.), *Plant genomics* (pp. 45–66). IntechOpen.

Devata, S., & Priyakumar, D. (2024). TorRNA-improved prediction of backbone torsion angles of RNA by leveraging large language models. *ChemRxiv*. https://doi.org/10.26434/chemrxiv-2024-cj4r0

Di Filippo, M., Pescini, D., Galuzzi, B. G., Bonanomi, M., Gaglio, D., Mangano, E., Consolandi, C., Alberghina, L., Vanoni, M., & Damiani, C. (2022). INTEGRATE: Model-based multi-omics data integration to characterize multi-level metabolic regulation. *PLoS Computational Biology*, *18*(2), e1009337. https://doi.org/10.1371/journal.pcbi.1009337

Enoki, N., Kuwada, E., Matsuo, S., Fujita, N., Noda, S., Matsubayashi, Y., Uchida, S., & Akagi, T. (2024). Explainable AI-guided identification of a novel protein-RNA interactive frame for selective siRNA accumulation in plants. *bioRxiv*, pp. 2024-12. https://doi.org/10.1101/2024.12.02.626406

Esmaeili, N., Yang, X., Cai, Y., Sun, L., Zhu, X., Shen, G., Payton, P., Fang, W., & Zhang, H. (2019). Co-overexpression of AVP1 and OsSIZ1 in Arabidopsis substantially enhances plant tolerance to drought, salt, and heat stresses. *Scientific Reports*, *9*(1), 7642. https://doi.org/10.1038/s41598-019-44062-0

Fang, J. (2024). Breeding 5.0: AI-driven revolution in designed plant breeding. *Molecular Plant Breeding*, *15*(1), 27–33. https://doi.org/10.5376/mpb.2024.15.0004

Fang, X., & Qi, Y. (2016). RNAi in plants: An argonaute-centered view. *The Plant Cell, 28*(2), 272–285. https://doi.org/10.1105/tpc.15.00920

Fletcher, S. J., Reeves, P. T., Hoang, B. T., & Mitter, N. (2020). A perspective on RNAi-based biopesticides. *Frontiers in Plant Science, 11*, 51. https://doi.org/10.3389/fpls.2020.00051

Fridrich, A., Hazan, Y., & Moran, Y. (2019). Too many false targets for microRNAs: Challenges and pitfalls in prediction of miRNA targets and their gene ontology in model and non-model organisms. *BioEssays, 41*(4), 1800169. https://doi.org/10.1002/bies.201800169

Glukhova, V. A., Tomazela, D. M., Findlay, G. D., Monnat, R. J., Jr., & MacCoss, M. J. (2013). Rapid assessment of RNAi-mediated protein depletion by selected reaction monitoring mass spectrometry. *Journal of Proteome Research, 12*(7), 3246–3254. https://doi.org/10.1021/pr400067k

Gong, Q., Wang, Y., Jin, Z., Hong, Y., & Liu, Y. (2022). Transcriptional and post-transcriptional regulation of RNAi-related gene expression during plant-virus interactions. *Stress Biology, 2*(1), 33. https://doi.org/10.1007/s44154-022-00057-y

Gramzow, L., & Theißen, G. (2019). Plant miRNA conservation and evolution. In S. de Folter (Ed.), *Plant microRNAs: Methods and protocols* (pp. 41–50). Humana Press.

Griffiths-Jones, S., Grocock, R. J., Van Dongen, S., Bateman, A., & Enright, A. J. (2006). miRBase: MicroRNA sequences, targets and gene nomenclature. *Nucleic Acids Research, 34*(suppl_1), D140–D144. https://doi.org/10.1093/nar/gkj112

Gumienny, R., & Zavolan, M. (2015). Accurate transcriptome-wide prediction of microRNA targets and small interfering RNA off-targets with MIRZA-G. *Nucleic Acids Research, 43*(3), 1380–1391. https://doi.org/10.1093/nar/gkv050

Guo, Z., Kuang, Z., Wang, Y., Zhao, Y., Tao, Y., Cheng, C., Yang, J., Lu, X., Hao, C., Wang, T., Cao, X., Wei, J., Li, L., & Yang, X. (2020). PmiREN: A comprehensive encyclopedia of plant miRNAs. *Nucleic Acids Research, 48*(D1), D1114–D1121. https://doi.org/10.1093/nar/gkz894

Gustafson, A. M., Allen, E., Givan, S., Smith, D., Carrington, J. C., & Kasschau, K. D. (2005). ASRP: The Arabidopsis small RNA project database. *Nucleic Acids Research, 33*(suppl_1), D637–D640. https://doi.org/10.1093/nar/gki127

Hadad, E., Rokach, L., & Veksler-Lublinsky, I. (2024). Empowering prediction of miRNA-mRNA interactions in species with limited training data through transfer learning. *Heliyon, 10*(7), e28000. https://doi.org/10.1016/j.heliyon.2024.e28000

Han, Y., He, F., Tan, X., & Yu, H. (2017, November 13–16). Effective small interfering RNA design based on convolutional neural network. *2017 IEEE International Conference on Bioinformatics and Biomedicine (BIBM),* Kansas City, MO (pp. 16–21). IEEE.

Hanamasagar, Y., Ramesha, N. M., Mahapatra, S., Panigrahi, C. K., Vidhya, C. S., Agnihotri, N., & Satapathy, S. N. (2024). Advancing RNAi for sustainable insect management: Targeted solutions for eco-friendly pest control. *Journal of Experimental Agriculture International, 46*(6), 740–775. https://doi.org/10.9734/jeai/2024/v46i62531

Hasan, W., Ramesha, N. M., Archana, B. R., Saivamsireddy, G., Choudhuri, S., Parmar, S., Charitha, K., & Pandey, S. K. (2024). Advancing RNAi-based strategies for eco-friendly and targeted insect pest management in sustainable agriculture. *Journal of Experimental Agriculture International, 46*(6), 833–863. https://doi.org/10.9734/jeai/2024/v46i62537

Hashemikhabir, S., Ayaz, E. S., Kavurucu, Y., Can, T., & Kahveci, T. (2012). Large-scale signaling network reconstruction. *IEEE/ACM Transactions on Computational Biology and Bioinformatics, 9*(6), 1696–1708. https://doi.org/10.1109/TCBB.2012.128

Hoang, B. T. L., Fletcher, S. J., Brosnan, C. A., Ghodke, A. B., Manzie, N., & Mitter, N. (2022). RNAi as a foliar spray: Efficiency and challenges to field applications. *International Journal of Molecular Sciences*, *23*(12), 6639. https://doi.org/10.3390/ijms23126639

Horn, T., & Boutros, M. (2010). E-RNAi: A web application for the multi-species design of RNAi reagents-2010 update. *Nucleic Acids Research*, *38*(suppl_2), W332–W339. https://doi.org/10.1093/nar/gkq317

Housden, B. E., Nicholson, H. E., & Perrimon, N. (2017). Synthetic lethality screens using RNAi in combination with CRISPR-based knockout in Drosophila cells. *Bio-Protocol*, *7*(3), e2119–e2119. https://doi.org/10.21769/BioProtoc.2119

Huang, X., Chen, Z., & Liu, Y. (2020). RNAi-mediated control of CRISPR functions. *Theranostics*, *10*(15), 6661–6673. https://doi.org/10.7150/thno.44880

Huang, K., Tian, J., Sun, L., Xie, P., Zhou, S., Deng, A., Mo, P., Zhou, Z., Jiang, M., Li, G., & Wang, Y. (2023). TransGeneSelector: A transformer-based approach tailored for key gene mining with small plant transcriptomic datasets. *bioRxiv*, pp. 2023-09. https://doi.org/10.1101/2023.09.26.559592

Hue, Y., Kim, J. H., Lee, G., Choi, B., Sim, H., Jeon, J., Ahn, M. I., Han, Y. K., & Kim, K. T. (2024). Artificial intelligence plant doctor: Plant disease diagnosis using Gpt4-vision. *Research in Plant Disease*, *30*, 99–102. https://doi.org/10.5423/RPD.2024.30.1.99

Isshiki, M., & Kodama, H. (2009). Plant RNAi and crop improvement. In S. M. Jain & D. S. Brar (Eds.), *Molecular techniques in crop improvement* (2nd ed., pp. 653–673). Springer Netherlands.

Iwayama, K., Aisaka, Y., Kutsuna, N., & Nagano, A. J. (2017). FIT: Statistical modelling tool for transcriptome dynamics under fluctuating field conditions. *Bioinformatics*, *33*(11), 1672–1680. https://doi.org/10.1093/bioinformatics/btx049

Jamir, L., & Hariprasad, P. (2024). Employing machine learning models to predict potential α-glucosidase inhibitory plant secondary metabolites targeting type-2 diabetes and their in vitro validation. *Journal of Chemical Information and Modeling*, *64*(24), 9150–9162. https://doi.org/10.1021/acs.jcim.4c00955

Jayasundara, S., Lokuge, S., Ihalagedara, P., & Herath, D. (2021). Machine learning for plant microrna prediction: A systematic review. *arXiv preprint arXiv:2106.15159*. https://doi.org/10.48550/arXiv.2106.15159

Jeong, Y. Y., Kwak, H., Jeong, W., Lee, H., Jeon, S. J., Lee, H. J., Seo, P. J., & Kwak, S. Y. (2025). Oligohistidine-functionalized single-walled carbon nanotube-guided RNA delivery to improve shoot regeneration efficiency in plant calli. *Advanced Functional Materials*, *36*, e10105. https://doi.org/10.1002/adfm.202510105

Jia, H., Aadland, K., Kolaczkowski, O., & Kolaczkowski, B. (2021). Direct molecular evidence for an ancient, conserved developmental toolkit controlling posttranscriptional gene regulation in land plants. *Molecular Biology and Evolution*, *38*(11), 4765–4777. https://doi.org/10.1093/molbev/msab201

Kaderali, L., Dazert, E., Zeuge, U., Frese, M., & Bartenschlager, R. (2009). Reconstructing signaling pathways from RNAi data using probabilistic Boolean threshold networks. *Bioinformatics*, *25*(17), 2229–2235. https://doi.org/10.1093/bioinformatics/btp375

Kapoor, D. U., Sharma, J. B., Gandhi, S. M., Prajapati, B. G., Thanawuth, K., Limmatvapirat, S., & Sriamornsak, P. (2024). AI-driven design and optimization of nanoparticle-based drug delivery systems. *Science, Engineering and Health Studies*, *18*, 24010003. https://doi.org/10.69598/sehs.18.24010003

Katiyar, A., Smita, S., Chinnusamy, V., Pandey, D. M., & Bansal, K. (2012). Identification of miRNAs in sorghum by using bioinformatics approach. *Plant Signaling & Behavior*, *7*(2), 246–259. https://doi.org/10.4161/psb.18914

Kaur, H., & Talekar, N. (2024). The function of RNA interference (RNAi) in crop enhancement-mechanism and applications: A review. *Journal of Experimental Agriculture International*, *46*(6), 351–370. https://doi.org/10.9734/jeai/2024/v46i62487

Khemka, N. K., Singh, U., Dwivedi, A. K., & Jain, M. (2020). Machine learning-based annotation of long noncoding RNAs using PLncPRO. In M. Jain & R. Garg (Eds.), *Legume genomics: Methods and protocols* (pp. 253–260). Springer US.

Kim, H., Kim, M., Yun, T., Choi, S., Bengio, E., Hernández-García, A., & Park, J. (2024). Improved off-policy reinforcement learning in biological sequence design. *arXiv preprint arXiv:2410.04461*. https://doi.org/10.48550/arXiv.2410.04461

Kim, M. G., Go, M. J., Kang, S. H., Jeong, S. H., & Lim, K. (2025). Revolutionizing CRISPR technology with artificial intelligence. *Experimental & Molecular Medicine*, *57*, 1–13. https://doi.org/10.1038/s12276-025-01462-9

Kumar, P., Kamle, M., & Pandey, A. (2012). RNAi: New era of functional genomics for crop improvement. In A. Goyal & P. Maheshwari (Eds.), *Frontiers on recent developments in plant science* (Vol. 1, pp. 24–38). Bentham Science Publishers.

Kumar, J.S., Rashmita, P., Pushpitha, M. S., & Deepika, K. (2024, June 21–23). SAMS: Smart agricultural monitoring system integrating edge AI and IoT for early detection of rice crop stress. *2023 4th International Conference on Intelligent Technologies (CONIT),* Bangalore, India (pp. 1–5). IEEE.

Kuzmin, D. A., Feranchuk, S. I., Sharov, V. V., Cybin, A. N., Makolov, S. V., Putintseva, Y. A., Oreshkova, N. V., & Krutovsky, K. V. (2019). Stepwise large genome assembly approach: A case of Siberian larch (Larix sibirica Ledeb). *BMC Bioinformatics*, *20*(Suppl 1), 37. https://doi.org/10.1186/s12859-018-2570-y

Ladunga, I. (2007). More complete gene silencing by fewer siRNAs: Transparent optimized design and biophysical signature. *Nucleic Acids Research*, *35*(2), 433–440. https://doi.org/10.1093/nar/gkl1065

Latyshev, P., Pavlov, F., Herbert, A., & Poptsova, M. (2023). Unsupervised domain adaptation methods for cross-species transfer of regulatory code signals. *Frontiers in Big Data*, *6*, 1140663. https://doi.org/10.3389/fdata.2023.1140663

Lee, M. (2023). Machine learning for small interfering RNAs: A concise review of recent developments. *Frontiers in Genetics*, *14*, 1226336. https://doi.org/10.3389/fgene.2023.1226336

Lee, S. Y., & Lee, K. O. (2011). Proteomic analysis of RNA interference induced knockdown plant. In H. Kodama & A. Komamine (Eds.), *RNAi and plant gene function analysis: Methods and protocols* (pp. 211–224). Humana Press.

Lee, D., Yang, J., & Kim, S. (2022). Learning the histone codes with large genomic windows and three-dimensional chromatin interactions using transformer. *Nature Communications*, *13*(1), 6678. https://doi.org/10.3389/fgene.2023.1226336

Li, J. F., Zhang, D., & Sheen, J. (2014). Epitope-tagged protein-based artificial miRNA screens for optimized gene silencing in plants. *Nature Protocols*, *9*(4), 939–949. https://doi.org/10.1038/nprot.2014.061

Li, J., Kho, A. T., Chase, R. P., Pantano, L., Farnam, L., Amr, S. S., & Tantisira, K. G. (2020). COMPSRA: A COMprehensive platform for small RNA-Seq data analysis. *Scientific Reports*, *10*(1), 4552. https://doi.org/10.1038/s41598-020-61495-0

Li, S., Li, J., Du, M., Deng, G., Song, Z., & Han, H. (2022a). Efficient gene silencing in intact plant cells using siRNA delivered by functional graphene oxide nanoparticles. *Angewandte Chemie International Edition*, *61*(40), e202210014. https://doi.org/10.1002/anie.202210.14

Li, X., Liu, X., Lu, W., Yin, X., & An, S. (2022b). Application progress of plant-mediated RNAi in pest control. *Frontiers in Bioengineering and Biotechnology*, *10*, 963026. https://doi.org/10.3389/fbioe.2022.963026

Liang, X., Young, W. C., Hung, L. H., Raftery, A. E., & Yeung, K. Y. (2018). Integration of multiple data sources for gene network inference using genetic perturbation data. *Proceedings of the 2018 ACM International Conference on Bioinformatics, Computational Biology, and Health Informatics* (pp. 601–602). https://doi.org/10.1145/3233547.3233692

Lim, F. H., Rasid, O. A., As'wad, A. W. M., Vadamalai, G., Parveez, G. K. A., & Wong, M. Y. (2020). The future is now: Revolution of RNA-mediated gene silencing in plant protection against insect pests and diseases. *Plant Biotechnology Reports*, *14*(6), 643–662. https://doi.org/10.1007/s11816-020-00640-7

Lin, S., Zhang, Q., Bai, S., Yang, L., Qin, G., Wang, L., Wang, W., Cheng, C., Zhang, D., Lu, C., & Yuan, J. (2025). Beyond species and spatial boundaries: Enabling long-distance gene silencing in plants via guanidinium-siRNA nanoparticles. *Plant Biotechnology Journal*, *23*(4), 1165–1177. https://doi.org/10.1111/pbi.14575

List, F., Tarone, A. M., Zhu-Salzman, K., & Vargo, E. L. (2022). RNA meets toxicology: Efficacy indicators from the experimental design of RNAi studies for insect pest management. *Pest Management Science*, *78*(8), 3215–3225. https://doi.org/10.1002/ps.6884

Liu, T., Pei, Y., Li, C., & Ye, M. (2019). Amount of escape estimation based on Bayesian and MCMC approaches for RNA interference. *Molecular Therapy–Nucleic Acids*, *18*, 893–902. https://doi.org/10.1016/j.omtn.2019.10.010

Liu, J., Liu, X., Zhang, S., Liang, S., Luan, W., & Ma, X. (2021). TarDB: An online database for plant miRNA targets and miRNA-triggered phased siRNAs. *BMC Genomics*, *22*(1), 348. https://doi.org/10.1186/s12864-021-07680-5

Long, R., Guo, Z., Han, D., Liu, B., Yuan, X., Chen, G., Heng, P. A., & Zhang, L. (2024). Sirnadiscovery: A graph neural network for siRNA efficacy prediction via deep RNA sequence analysis. *Briefings in Bioinformatics*, *25*(6), bbae563. https://doi.org/10.1093/bib/bbae563

Lopez-Dolz, L., Spada, M., Daròs, J. A., & Carbonell, A. (2020). Fine-tune control of targeted RNAi efficacy by plant artificial small RNAs. *Nucleic Acids Research*, *48*(11), 6234–6250. https://doi.org/10.1093/nar/gkaa343

Lu, Y., Deng, X., Zhu, Q., Wu, D., Zhong, J., Wen, L., & Yu, X. (2023). The dsRNA delivery, targeting and application in pest control. *Agronomy*, *13*(3), 714. https://doi.org/10.3390/agronomy13030714

Luo, X., Nanda, S., Zhang, Y., Zhou, X., Yang, C., & Pan, H. (2024). Risk assessment of RNAi-based biopesticides. *New Crops*, *1*, 100019. https://doi.org/10.1016/j.ncrops.2024.100019

Mamta, B., & Rajam, M. V. (2018). RNA interference: A promising approach for crop improvement. In S. S. Gosal & S. H. Wani (Eds.), *Biotechnologies of crop improvement, volume 2: Transgenic approaches* (pp. 41–65). Springer International Publishing.

de Mas, I. M. (2018). Multiomic data integration and analysis via model-driven approaches. In J. Jaumot, C. Bedia, & R. Tauler (Eds.), *Comprehensive analytical chemistry* (Vol. 82, pp. 447–476). Elsevier.

Masclaux, F., & Galaud, J. P. (2011). Heat-inducible RNAi for gene functional analysis in plants. In H. Kodama & A. Komamine (Eds.), *RNAi and plant gene function analysis: Methods and protocols* (pp. 37–55). Humana Press.

Mat Jalaluddin, N. S., Asem, M., Harikrishna, J. A., & Ahmad Fuaad, A. A. H. (2023). Recent progress on nanocarriers for topical-mediated RNAi strategies for crop protection—a review. *Molecules, 28*(6), 2700. https://doi.org/10.3390/molecules28062700

Milev, I., Yahubyan, G., Minkov, I. and Baev, V., 2011. miRTour: Plant miRNA and target prediction tool. *Bioinformation*, 6(6), p.248, 249. https://doi.org/10.6026/97320630006248

Min, S., Lee, B., & Yoon, S. (2022). TargetNet: Functional microRNA target prediction with deep neural networks. *Bioinformatics, 38*(3), 671–677. https://doi.org/10.1093/bioinformatics/btab733

Mochida, K., Koda, S., Inoue, K., Hirayama, T., Tanaka, S., Nishii, R., & Melgani, F. (2019). Computer vision-based phenotyping for improvement of plant productivity: A machine learning perspective. *GigaScience, 8*(1), giy153. https://doi.org/10.1093/gigascience/giy153

Mohapatra, R., HM, H. K., Naan, T., Chitra, M., Ashwini, R., Rout, A., & Lallawmkimi, M. C. (2024). A review on biotechnological innovations in developing stress-tolerant crops for adverse environmental conditions. *Journal of Scientific Research and Reports, 30*(7), 901–920. https://doi.org/10.9734/jsrr/2024/v30i72200

Mohr, S. E., Smith, J. A., Shamu, C. E., Neumüller, R. A., & Perrimon, N. (2014). RNAi screening comes of age: Improved techniques and complementary approaches. *Nature Reviews Molecular Cell Biology, 15*(9), 591–600. https://doi.org/10.1038/nrm3860

Morgado, L. (2020). Biogenesis of small RNA: Molecular pathways and regulatory mechanisms. In P. Guleria & V. Kumar (Eds.), *Plant small RNA* (pp. 49–64). Academic Press.

Morozov, S. Y., Solovyev, A. G., Kalinina, N. O., & Taliansky, M. E. (2019). Double-stranded RNAs in plant protection against pathogenic organisms and viruses in agriculture. *Acta Naturae (англоязычная версия), 11*(4 (43)), 13–21.

Munawar, I., Shamoon, F., Ashraf, H., Aleem, K., Imran, M., Nisar, T., Tariq, F., Ramzan, U., & Munawar, N. (2023). RNAi for the management of insect pests: A review. *Agrobiological Records, 11*, 21–27. https://doi.org/10.47278/journal.abr/2023.004

Murmu, S., Sinha, D., Chaurasia, H., Sharma, S., Das, R., Jha, G. K., & Archak, S. (2024). A review of artificial intelligence-assisted omics techniques in plant defense: Current trends and future directions. *Frontiers in Plant Science, 15*, 1292054. https://doi.org/10.3389/fpls.2024.1292054

Mushtaq, N., Wang, Y., Fan, J., Li, Y., & Ding, J. (2022). Down-regulation of cytokinin receptor gene SlHK2 improves plant tolerance to drought, heat, and combined stresses in tomato. *Plants, 11*(2), 154. https://doi.org/10.3390/plants11020154

Naito, Y., Yamada, T., Matsumiya, T., Ui-Tei, K., Saigo, K., & Morishita, S. (2005). Dscheck: Highly sensitive off-target search software for double-stranded RNA-mediated RNA interference. *Nucleic Acids Research, 33*(suppl_2), W589–W591. https://doi.org/10.1093/nar/gki419

Narayan, A., & Kumar, S. (2022). Identification of novel RNAs in plants with the help of next-generation sequencing technologies. In P. Sharma, D. Yadav, & R. K. Gaur (Eds.), *Bioinformatics in agriculture* (pp. 177–189). Elsevier.

Ndathe, R., Dale, R., & Kato, N. (2022). Dynamic modelling of ABA-dependent expression of the Arabidopsis *RD29A* gene. *Frontiers in Plant Science, 13*, 928718. https://doi.org/10.3389/fpls.2022.928718

Noordijk, B., Reinders, M., van Dijk, A. D., & de Ridder, D. (2025). BADDADAN: Mechanistic modelling of time-series gene module expression. *Quantitative Plant Biology*, *6*, e31. https://doi.org/10.1017/qpb.2025.10.17

Ozden, F., Barazandeh, S., Akboga, D., Tabrizi, S.S., Seker, U. O. S., & Cicek, A. E. (2023). RNAGEN: A generative adversarial network-based model to generate synthetic RNA sequences to target proteins. *bioRxiv*, pp. 2023-07. https://doi.org/10.1101/2023.07.11.548246

Panero, R., Rinaldi, A., Memoli, D., Nassa, G., Ravo, M., Rizzo, F., Tarallo, R., Milanesi, L., Weisz, A., & Giurato, G. (2017). iSmaRT: A toolkit for a comprehensive analysis of small RNA-Seq data. *Bioinformatics*, *33*(6), 938–940. https://doi.org/10.1093/bioinformatics/btw734

Pattanayak, D., Solanke, A. U., & Kumar, P. A. (2013). Plant RNA interference pathways: Diversity in function, similarity in action. *Plant Molecular Biology Reporter*, *31*(3), 493–506. https://doi.org/10.1007/s11105-012-0520-9

Peace, R. J., Biggar, K. K., Storey, K. B., & Green, J. R. (2015). A framework for improving microRNA prediction in non-human genomes. *Nucleic Acids Research*, *43*(20), e138–e138. https://doi.org/10.1093/nar/gkv698

Prabha, K. (2021). Disease sniffing robots to apps fixing plant diseases: Applications of artificial intelligence in plant pathology-a mini review. *Indian Phytopathology*, *74*(1), 13–20. https://doi.org/10.1007/s42360-020-00290-3

Qiao, J., Sun, W., Yin, W., Ma, L., & Liu, Y. (2022). Rational design of Cas9 ribonucleoprotein with a "gRNA-shRNA" for multidimensional genome manipulation and enhanced homology-directed repair. *bioRxiv*, pp. 2022-03. https://doi.org/10.1101/2022.03.17.484717

Qiao, S. A., Gao, Z., & Roth, R. (2023). A perspective on cross-kingdom RNA interference in mutualistic symbioses. *New Phytologist*, *240*(1), 68–79. https://doi.org/10.1111/nph.19122

Qiao, H., Chen, J., Dong, M., Shen, J., & Yan, S. (2024). Nanocarrier-based eco-friendly RNA pesticides for sustainable management of plant pathogens and pests. *Nanomaterials (Basel)*, *14*(23), 1874. https://doi.org/10.3390/nano14231874

Qiu, S., Lane, T., & Yang, C. (2007). Efficient search algorithms for RNAi target detection. *The Journal of Supercomputing*, *42*(3), 303–319. https://doi.org/10.1007/s11227-007-0121-9

Qu, J., Ye, J., & Fang, R. (2007). Artificial microRNA-mediated virus resistance in plants. *Journal of Virology*, *81*(12), 6690–6699. https://doi.org/10.1128/JVI.02457-06

Rahman, R. U., Gautam, A., Bethune, J., Sattar, A., Fiosins, M., Magruder, D. S., Capece, V., Shomroni, O., & Bonn, S. (2018). Oasis 2: Improved online analysis of small RNA-seq data. *BMC Bioinformatics*, *19*(1), 54. https://doi.org/10.1186/s12859-018-2047-z

Rai, K. K. (2022). Integrating speed breeding with artificial intelligence for developing climate-smart crops. *Molecular Biology Reports*, *49*(12), 11385–11402. https://doi.org/10.1007/s11033-022-07769-4

Rajam, M. V. (2020). RNA silencing technology: A boon for crop improvement. *Journal of Biosciences*, *45*(1), 118. https://doi.org/10.1007/s12038-020-00082-x

Rajesh, P. (2025). Nanocarriers for targeted delivering and translocation of chemical cargoes in plant cells. In J.-T. Chen (Ed.), *Nanocarriers in plant science and agriculture* (pp. 125–164). IGI Global Scientific Publishing.

Rana, S., Rajurkar, A. B., Kumar, K. K., & Mohankumar, S. (2020). Comparative analysis of chitin synthaseA dsRNA mediated RNA interference for management of crop pests of different families of Lepidoptera. *Frontiers in Plant Science*, *11*, 427. https://doi.org/10.3389/fpls.2020.00427

Riley, A. T., Robson, J. M., Ulanova, A., & Green, A. A. (2025). Generative and predictive neural networks for the design of functional RNA molecules. *Nature Communications*, *16*(1), 4155. https://doi.org/10.1038/s41467-025-59389-8

Rinaldi, A., Mat Jalaluddin, N. S., Mohd Hussain, R. B., & Abdul Ghapor, A. (2025). Building public trust and acceptance towards spray-on RNAi biopesticides: Lessons from current ethical, legal and social discourses. *GM Crops & Food*, *16*(1), 398–412. https://doi.org/10.1080/21645698.2025.2510735

Rosa, C., Kuo, Y. W., Wuriyanghan, H., & Falk, B. W. (2018). RNA interference mechanisms and applications in plant pathology. *Annual Review of Phytopathology*, *56*(1), 581–610. https://doi.org/10.1146/annurev-phyto-080417-050044

Sanches, P. H. G., de Melo, N. C., Porcari, A. M., & de Carvalho, L. M. (2024). Integrating molecular perspectives: Strategies for comprehensive multi-omics integrative data analysis and machine learning applications in transcriptomics, proteomics, and metabolomics. *Biology*, *13*(11), 848. https://doi.org/10.3390/biology13110848

Schmich, F. (2016). *Computational studies in genetic perturbation screening* [Doctoral dissertation, ETH Zurich].

Scossa, F., Benina, M., Alseekh, S., Zhang, Y., & Fernie, A. R. (2018). The integration of metabolomics and next-generation sequencing data to elucidate the pathways of natural product metabolism in medicinal plants. *Planta Medica*, *84*(12/13), 855–873. https://doi.org/10.1055/a-0630-1899

Senthil-Kumar, M., & Mysore, K. S. (2011). Caveat of RNAi in plants: The off-target effect. In H. Kodama & A. Komamine (Eds.), *RNAi and plant gene function analysis: Methods and protocols* (pp. 13–25).

Shabalina, S. A., & Koonin, E. V. (2008). Origins and evolution of eukaryotic RNA interference. *Trends in Ecology & Evolution*, *23*(10), 578–587. https://doi.org/10.1016/j.tree.2008.06.005

Shehzad, J., Bibi, S., Khan, I., Hasan, M., & Mustafa, G. (2025). Nanocarriers for delivering biomolecules in plants: Mechanisms and applications. In J.-T. Chen (Ed.), *Nanocarriers in plant science and agriculture* (pp. 165–190). IGI Global Scientific Publishing.

Sheikh, M., Iqra, F., Ambreen, H., Pravin, K. A., Ikra, M., & Chung, Y. S. (2024). Integrating artificial intelligence and high-throughput phenotyping for crop improvement. *Journal of Integrative Agriculture*, *23*(6), 1787–1802. https://doi.org/10.1016/j.jia.2023.10.019

Shojaie, A., Jauhiainen, A., Kallitsis, M., & Michailidis, G. (2014). Inferring regulatory networks by combining perturbation screens and steady state gene expression profiles. *PLoS One*, *9*(2), e82393. https://doi.org/10.1371/journal.pone.0082393

Singh, A., & Das, S. (2021). Slicing messengers by artificial designs: Artificial MicroRNA induced gene silencing in polyploid plants for functional genomics and trait modification. In G. Tang, S. Teotia, X. Tang, & D. Singh (Eds.), *RNA-based technologies for functional genomics in plants* (pp. 77–129). Springer International Publishing.

Singh, U., Khemka, N., Rajkumar, M. S., Garg, R., & Jain, M. (2017). PLncPRO for prediction of long non-coding RNAs (lncRNAs) in plants and its application for discovery of abiotic stress-responsive lncRNAs in rice and chickpea. *Nucleic Acids Research*, *45*(22), e183–e183. https://doi.org/10.1093/nar/gkx866

Singhvi, V., Lunga, L., Nidhi, P., Keum, C., & Prakash, V. (2024). High-throughput phenotyping using computer vision and machine learning. *arXiv preprint arXiv:2407.06354*. https://doi.org/10.17605/OSF.IO/R6DJQ

Stadler, M., Lukauskas, S., Bartke, T., & Müller, C. L. (2024). asteRIa enables robust interaction modelling between chromatin modifications and epigenetic readers. *Nucleic Acids Research, 52*(11), 6129–6144. https://doi.org/10.1093/nar/gkae361

Svoboda, P. (2020). RNAi and miRNA pathways in plants I—molecular mechanisms. In P. Svoboda (Ed.), *Introduction to RNAi and miRNA pathways* (pp. 215–253). Karolinum Press.

Szittya, G., & Burgyán, J. (2013). RNA interference-mediated intrinsic antiviral immunity in plants. *Intrinsic Immunity, 371*, 153–181. https://doi.org/10.1007/978-3-642-37765-5_6

Taha, Y., Jehn, J., Kahraman, M., Frank, M., Heuvelman, M., Horos, R., Yau, C., Steinkraus, B., & Sikosek, T. (2024). TransfoRNA: Navigating the uncertainties of small RNA annotation with an adaptive machine learning strategy. *bioRxiv*, pp. 2024-06. https://doi.org/10.1101/2024.06.19.599329

Taliansky, M., Samarskaya, V., Zavriev, S. K., Fesenko, I., Kalinina, N. O., & Love, A. J. (2021). RNA-based technologies for engineering plant virus resistance. *Plants, 10*(1), 82. https://doi.org/10.3390/plants10010082

Tan, H., Liu, Y., & Guo, H. (2024). The biogenesis, regulation and functions of transitive siRNA in plants: Transitive siRNA in plants. *Acta Biochimica et Biophysica Sinica, 57*(1), 131–147. https://doi.org/10.3724/abbs.2024160

Tao, Z., Yan, F., Hahn, M., & Ma, Z. (2023). Regulatory roles of epigenetic modifications in plant-phytopathogen interactions. *Crop Health, 1*(1), 6. https://doi.org/10.1007/s44297-023-00003-y

Thingujam, D., Gouli, S., Cooray, S. P., Chandran, K. B., Givens, S. B., Gandhimeyyan, R. V., Tan, Z., Wang, Y., Patam, K., Greer, S. A., & Acharya, R. (2025). Climate-resilient crops: Integrating AI, multi-omics, and advanced phenotyping to address global agricultural and societal challenges. *Plants, 14*(17), 2699. https://doi.org/10.3390/plants14172699

Torri, A., Jaeger, J., Pradeu, T., & Saleh, M. C. (2022). The origin of RNA interference: Adaptive or neutral evolution? *PLoS Biology, 20*(6), e3001715. https://doi.org/10.1371/journal.pbio.3001715

Touzdjian Pinheiro Kohlrausch Távora, F., de Assis dos Santos Diniz, F., de Moraes Rêgo-Machado, C., Chagas Freitas, N., Barbosa Monteiro Arraes, F., Chumbinho de Andrade, E., & Correa Molinari, H. B. (2022). CRISPR/Cas-and topical RNAi-based technologies for crop management and improvement: Reviewing the risk assessment and challenges towards a more sustainable agriculture. *Frontiers in Bioengineering and Biotechnology, 10*, 913728. https://doi.org/10.3389/fbioe.2022.913728

Tripodi, P., Nicastro, N., Pane, C., & Cammarano, D. (2020, December 3–5). Artificial intelligence based efficient phenotyping for agronomics. *2020 3rd International conference on intelligent sustainable systems (ICISS),* Thoothukudi, India (pp. 484–488).

Ueda, T., Tominaga, D., Araki, N., & Yoshikawa, T. (2013). Estimate hidden dynamic profiles of siRNA effect on apoptosis. *BMC Bioinformatics, 14*(1), 97. https://doi.org/10.1186/1471-2105-14-97

Vallejos, C. E., Jones, J. W., Bhakta, M. S., Gezan, S. A., & Correll, M. J. (2021). Dynamic gene-based ecophysiological models to predict phenotype from genotype and environment data. *bioRxiv*, pp. 2021-02. https://doi.org/10.1101/2021.02.07.429927

Vargas-Mejía, P., & Vega-Arreguín, J. C. (2025). SIREN: Suite for intelligent RNAi design and evaluation of nucleotide sequences. *bioRxiv*, pp. 2025-05. https://doi.org/10.1101/2025.05.26.656188

Venu, E., Ramya, A., Babu, P. L., Srinivas, B., Kumar, S., Reddy, N. K., Babu, Y. M., Majumdar, A., & Manik, S. (2024). Exogenous dsRNA-mediated RNAi: Mechanisms, applications, delivery methods and challenges in the induction of viral disease resistance in plants. *Viruses*, *17*(1), 49. https://doi.org/10.3390/v17010049

Wagner, G. J., & Kroumova, A. B. (2008). The use of RNAi to elucidate and manipulate secondary metabolite synthesis in plants. In S.-Y. Ying (Ed.), *Current perspectives in microRNAs (miRNA)* (pp. 431–459). Springer Netherlands.

Wang, X., & Zhang, F. (2015). Prediction of siRNA Efficacy using BP neural network and Support Vector Machine. *Applied Mechanics and Materials*, *701*, 214–218. https://doi.org/10.4028/www.scientific.net/AMM.644-650.5341

Wang, Y., Lee, H., Fear, J. M., Berger, I., Oliver, B., & Przytycka, T. M. (2022). NetREX-CF integrates incomplete transcription factor data with gene expression to reconstruct gene regulatory networks. *Communications Biology*, *5*(1), 1282. https://doi.org/10.1038/s42003-022-04226-7

Waterhouse, P. M., & Helliwell, C. A. (2003). Exploring plant genomes by RNA-induced gene silencing. *Nature Reviews Genetics*, *4*(1), 29–38. https://doi.org/10.1038/nrg982

Wielopolska, A., Townley, H., Moore, I., Waterhouse, P., & Helliwell, C. (2005). A high-throughput inducible RNAi vector for plants. *Plant Biotechnology Journal*, *3*(6), 583–590. https://doi.org/10.1111/j.1467-7652.2005.00149.x

Williamson, H. F., Brettschneider, J., Caccamo, M., Davey, R. P., Goble, C., Kersey, P. J., May, S., Morris, R. J., Ostler, R., Pridmore, T., Rawlings, C., Studholme, D., Tsaftaris, S. A., & Leonelli, S. (2023). Data management challenges for artificial intelligence in plant and agricultural research. *F1000Research*, *10*, 324. https://doi.org/10.12688/f1000research.52204.2

Worrall, E. A., Bravo-Cazar, A., Nilon, A. T., Fletcher, S. J., Robinson, K. E., Carr, J. P., & Mitter, N. (2019). Exogenous application of RNAi-inducing double-stranded RNA inhibits aphid-mediated transmission of a plant virus. *Frontiers in Plant Science*, *10*, 265. https://doi.org/10.3389/fpls.2019.00265

Wu, D., Maus, N., Jha, A., Yang, K., Wales-McGrath, B. D., Jewell, S., Tangiyan, A., Choi, P., Gardner, J. R., & Barash, Y. (2025). Generative modelling for RNA splicing predictions and design. *bioRxiv*. https://doi.org/10.1101.2025.01.20.633986

Xin, L., Huang, C., Li, H., Huang, S., Feng, Y., Kong, Z., Liu, Z., Li, S., Yu, C., Shen, F., & Tang, H. (2024). Artificial intelligence for central dogma-centric multi-omics: Challenges and breakthroughs. *arXiv preprint arXiv:2412.12668*. https://doi.org/10.48550/arXiv.2412.12668

Xu, P., Zhang, Y., Kang, L., Roossinck, M. J., & Mysore, K. S. (2006). Computational estimation and experimental verification of off-target silencing during posttranscriptional gene silencing in plants. *Plant Physiology*, *142*(2), 429–440. https://doi.org/10.1104/pp.106.083295

Xu, M., Kim, H., Yang, J., Fuentes, A., Meng, Y., Yoon, S., Kim, T., & Park, D. S. (2023). Embracing limited and imperfect training datasets: Opportunities and challenges in plant disease recognition using deep learning. *Frontiers in Plant Science*, *14*, 1225409. https://doi.org/10.3389/fpls.2023.1225409

Yadava, P., & Mukherjee, S. K. (2011). Artificial microRNA and its applications. In B. Mallick & Z. Ghosh (Eds.), *Regulatory RNAs: Basics, methods and applications* (pp. 505–521). Springer Berlin Heidelberg.

Ye, F., Bauer, J. A., Pietenpol, J. A., & Shyr, Y. (2012). Analysis of high-throughput RNAi screening data in identifying genes mediating sensitivity to chemotherapeutic drugs: Statistical approaches and perspectives. *BMC Genomics*, *13*(Suppl 8), S3. https://doi.org/10.1186/1471-2164-13-S8-S3

Ye, X., Yang, Z., Jiang, Y., Yu, L., Guo, R., Meng, Y., & Shao, C. (2020). sRNATargetDigger: A bioinformatics software for bidirectional identification of sRNA-target pairs with co-regulatory sRNAs information. *PLoS One*, *15*(12), e0244480. https://doi.org/10.1371/journal.pone.0244480

Yong, J., Zhang, R., Bi, S., Li, P., Sun, L., Mitter, N., Carroll, B. J., & Xu, Z. P. (2021). Sheet-like clay nanoparticles deliver RNA into developing pollen to efficiently silence a target gene. *Plant Physiology*, *187*(2), 886–899. https://doi.org/10.1093/plphys/kiab303

Yong, J., Wu, M., Carroll, B. J., Xu, Z. P., & Zhang, R. (2024). Enhancing plant biotechnology by nanoparticle delivery of nucleic acids. *Trends in Genetics*, *40*(4), 352–363. https://doi.org/10.1016/j.tig.2024.01.005

Yoon, J., Fang, M., Lee, D., Park, M., Kim, K. H., & Shin, C. (2021). Double-stranded RNA confers resistance to pepper mottle virus in *Nicotiana benthamiana*. *Applied Biological Chemistry*, *64*(1), 1. https://doi.org/10.1186/s13765-020-00581-3

Yousuf, M. F., Siddique, T., & Mahmud, M. S. (2022, May 23–26). ibug: Ai enabled iot sensing platform for real-time environmental monitoring. *2022 IEEE 31st Microelectronics Design & Test Symposium (MDTS),* Albany, NY, USA (pp. 1–7). IEEE.

Yu, R., Wang, X., & Moazed, D. (2018). Epigenetic inheritance mediated by coupling of RNAi and histone H3K9 methylation. *Nature*, *558*(7711), 615–619. https://doi.org/10.1038/s41586-018-0239-3

Yu, D., Wan, Y., Ito, H., Ma, X., Xie, T., Wang, T., Shao, C., & Meng, Y. (2019). PmiRDiscVali: An integrated pipeline for plant microRNA discovery and validation. *BMC Genomics*, *20*(1), 133. https://doi.org/10.1186/s12864-019-5478-7

Yu, H., Yang, H., Sun, W., Yan, Z., Yang, X., Zhang, H., Ding, Y., & Li, K. (2024). An interpretable RNA foundation model for exploring functional RNA motifs in plants. *Nature Machine Intelligence*, *6*(12), 1616–1625. https://doi.org/10.1038/s42256-024-00946-z

Zarrabi, S., Rangel, C., Martínez-Campos, E., Delgado-Martín, J., Arpanaei, A., Shams-bakhsh, M., & Velasco, L. (2024). Topical application of carbon dots and mesoporous silica nanoparticle-derived dsRNA-nanocomposites for the control of beet curly top virus and turnip mosaic virus. *bioRxiv*, pp. 2024-12. https://doi.org/10.1101/2024.12.29.628607

Zhai, J., Zhang, Y., Zhang, C., Yin, X., Song, M., Tang, C., Ding, P., Li, Z., & Ma, C. (2025). DeepTFBS: Improving within-and cross-species prediction of transcription factor binding using deep multi-task and transfer learning. *Advanced Science*, *12*, e03135. https://doi.org/10.1002/advs.202503135

Zhang, X., Zhou, X., Wan, M., Xuan, J., Jin, X., & Li, S. (2022). PINC: A tool for non-coding RNA identification in plants based on an automated machine learning framework. *International Journal of Molecular Sciences*, *23*(19), 11825. https://doi.org/10.3390/ijms231911825

Zhao, Q., Chen, J., Zhao, Z., Mao, Q., Shi, H., & Fan, X. (2024). ML-based RNA secondary structure prediction methods: A survey. *AI Medicine*, pp. 6-6. https://doi.org/10.53941/aim.2024.100006

Zhou, Y., Zhan, T., Wu, Y., Song, B., & Shi, C. (2024). RNA secondary structure prediction using transformer-based deep learning models. *arXiv preprint arXiv:2405.06655*. https://doi.org/10.48550/arXiv.2405.06655

Zhu, H., Tang, F., Quan, Q., Chen, K., Xiong, P., & Zhou, S. K. (2025). Deep generalizable prediction of RNA secondary structure via base pair motif energy. *Nature Communications*, *16*(1), 5856. https://doi.org/10.1038/s41467-025-60048-1

11

AI Models for Uncovering Plant–Microbiome Interactions

Mohammed Radi[1,2], *Hakima Achetoui*[1], *Ilham Dehbi*[1,3], *Hajar Zennouhi*[1,4], *and Rachid Lahlali*[1]

[1] *Department of Plant Protection, Phytopathology Unit, Ecole Nationale d'Agriculture de Meknès, Meknès, Morocco*
[2] *Environment and Valorization of Microbial and Plant Resources Unit, Faculty of Sciences, Moulay Ismail University, Meknes, Morocco*
[3] *Laboratory of Plant Biotechnology and Molecular Biology, Faculty of Sciences, Moulay Ismail University, Meknes, Morocco*
[4] *Laboratory of Functional Ecology and Environmental Engineering, Sidi Mohamed Ben Abdellah University, Fez, Morocco*

11.1 Introduction

The plant microbiome, which consists of complex bacterial, fungal, archaeal, and viral communities, is emerging as a paramount determinant of plant biology. These microbial consortia colonize separate ecological niches such as rhizosphere, phyllosphere, or endosphere and establish complex and dynamic states of interactions with their host plants. All together these microbiotas offer a variety of services including nutrient solubilization (phosphorus) and nitrogen fixing; phytohormone regulation that stimulates the growth of roots or shoots; suppression of soilborne and foliar pathogens by competition and antibiosis; and contributions to plant resilience from middle stress due to drought, saltiness, and temperature extreme (Compant et al., 2019; Mendes et al., 2013; Trivedi et al., 2020). At a broader scale than the plant, microbiome functions operate to deliver critical ecosystem services, including soil fertility, biogeochemical cycles, and ecological stability, such as those supporting global food security and climate resilience (Berg et al., 2020; Hanif et al., 2024).

Despite the obvious significance of these communities, however, it is difficult to comprehend and utilize their potential. The composition and function of the microbiome are also determined by interacting factors such as plant genotype, developmental stage of plants, soil type, environmental circumstances, and agricultural management. These influences result in high spatial and temporal variability, which can limit the generalization of findings between sites (Poudel et al., 2016; Walters et al., 2018). Moreover, inter-kingdom cross communications between bacteria, fungi, archaea, and viruses within the microbiome result in such networks exhibiting redundancy along with functional complementarity. Conventional reductionist methods are generally insufficient to encapsulate such complexity. As recently reviewed integratively, methods at the systems-level and computational sciences are the only tools that can be used to try to unravel such interactions, as well as their relationship with plant performance (Kimotho & Maina, 2024).

In parallel, the advent of next-generation sequencing (NGS) technologies and multiomics platforms (metagenomics, metatranscriptomics, metabolomics, and proteomics) has revolutionized microbiome science by generating vast amounts of high-dimensional data. However, the sheer volume, heterogeneity, and complexity of these datasets exceed the capacity of classical statistical methods to deliver meaningful insights. Here, artificial intelligence (AI) has emerged as a transformative paradigm, providing advanced analytical tools capable of identifying hidden patterns, predicting functional outcomes, and modeling plant–microbe–environment interactions with unprecedented precision (Hernández Medina et al., 2022).

Machine learning (ML) techniques, including random forests (RFs) and support vector machines (SVMs), and gradient boosting algorithms, have already been employed in microbiome studies to classify microbial taxa, associate collections of microbes with host phenotypes, and predict disease outcomes (Chang et al., 2017). On the other hand, a class of deep learning (DL) architectures, such as convolutional neural networks (CNNs) and recurrent neural networks, has achieved much higher regression accuracy by considering nonlinear relationships in high-dimensional measurements and temporal dynamics in large-scale observations (Angermueller et al., 2016; LeCun et al., 2015). These methods not only increase descriptive capability but also play predictive roles, with researchers able to predict organism stress responses from microbiome profiles and plant traits.

Applications of AI in plant–microbiome studies are not confined to data analysis but are increasingly involved in practical agricultural frameworks. For example, AI-powered predictive models are in the works as early detection systems of disease outbreaks, even before visual symptoms can be spotted, through which microbiome shift can be predicted. Conversely, computational pipelines are informing the rational design of microbial consortia customized for promoting crop productivity under defined environmental scenarios (Berruto & Demirer, 2024; D'Urso & Broccolo, 2024). When used in combination with precision or digital agriculture tools, that is, remote sensing, the Internet of Things (IoT), and near-real-time monitoring solutions, these models may also facilitate the deployment of microbiome-informed strategies directly in the field.

Nevertheless, significant challenges persist. Due to the absence of standardized methods for sampling, sequencing, and data processing, reproducibility across studies is also limited. Large enough differences between sites and agro-systems make model transferability difficult. And even as AI models have demonstrated considerable predictive power, many are "black boxes" that provide little explanation of the biological mechanisms behind their predictions. This constraint is an issue for meeting the needs of the agricultural sector, where induced resistance (IR) systems must be transparent and reliable to make use of. The novel field of explainable AI (XAI) has begun to bridge this gap by implementing algorithmic responses that not only predict but explain, therefore closing the loop for computational accuracy in light of biological ingest (Höhl et al., 2024).

The present chapter provides an overview of the ways in which AI is revolutionizing our understanding of plant–microbiome interactions. We start by reviewing the ecology and function of plant microbiomes, and then discuss AI techniques most relevant to microbiome data analysis. We further address their applications on multiomics integration, predictive modeling, and some example case studies in model plants and crops. Finally, we discuss current hurdles and future outlooks, such as XAI, microbial engineering, and the fusion of digital technology in sustainable agriculture. Connections of computational sciences with plant biology made by AI open opportunities for both fundamental understanding and practical applications to drive the transition toward resilient, microbiome-aware agriculture.

11.2 Plant Microbiomes and Their Roles

11.2.1 Niches: Rhizosphere, Phyllosphere, and Endosphere

The plant microbiome encompasses microorganisms linked together with the shoot-related rhizosphere within the plant system, involving organisms inhabiting within and on the surface of vegetative tissues.

11.2.1.1 Function Activity of Microbial Consortia in the Rhizosphere

The colonization of external and internal plants, which leads to the formation of an ecosystem, is crucial to plant metabolism (Turner et al., 2013). Microbial communities can vary and contain plant growth-promoting rhizobacteria (PGPR) that are vast and dynamic. Additionally, endophytic and epiphytic comprise a wide range of bacteria, fungi, and certain algae that can colonize plant tissues. Interactions among these microbes and between microbes and their host play a pivotal role in plant regulation and growth (Tarnita, 2017). (Figure 11.1)

The plant microorganisms are commonly categorized into three major ecological niches: the *rhizosphere*, phyllosphere, and endosphere. The endosphere comprises microbes

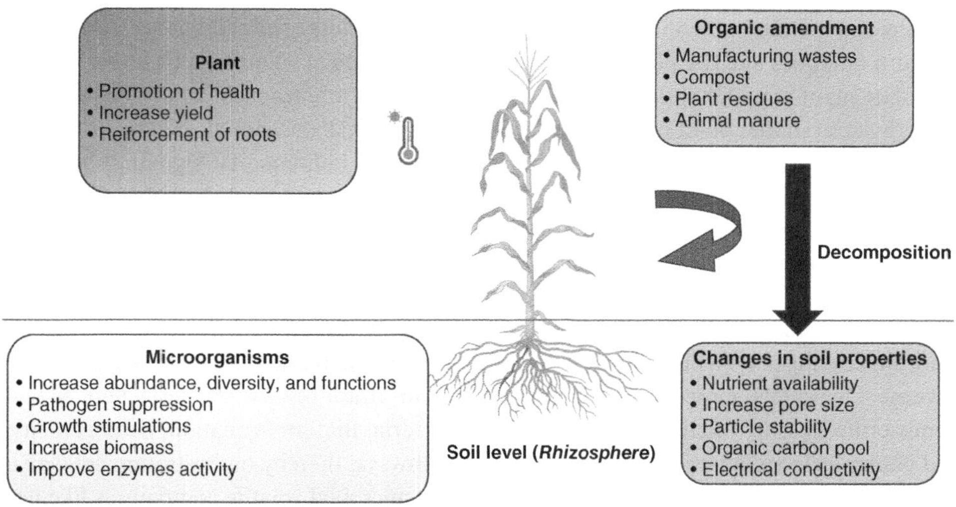

Figure 11.1 Organic amendments shape the rhizosphere: decomposition-driven changes in soil, microbiota, and plant performance.

residing within plant tissues, or intracellularly and intercellularly (Thiergart et al., 2020), and includes the vascular system and seeds. The phyllosphere refers to microbial communities inhabiting aerial parts of the plant, particularly leaf tissues with shoots. The *rhizosphere*, in contrast, is the narrow edaphic region associated with the plant system, where intense interactions occur between roots, soil, and microbes (Chaudhry et al., 2021).

Microbes within these niches may exist as endophytes (living inside the plant) or epiphytes (colonizing external surfaces) (Kumar Pandey et al., 2017). The microbial compositions are determined by numerous factors, like plant species, age, genotype, health status, and soil characteristics. And the capacity of colonization by forming the biofilm. Even members of the same species can harbor distinct microbial communities depending on environmental conditions (Xiong et al., 2021). Moreover, the differences in microbial composition, diversity, and abundance were affected across variations in nutrient availability, plant metabolite composition, and exposure to environmental stresses (Hacquard et al., 2015).

11.2.1.2 Rhizospheric Region: Dynamic Zone for Microbe-Driven

The hotspot of microbial activity is affected and represented by the richness of root-derived nutrients in bulk soil (Hartmann et al., 2008). The rhizosphere region is characterized by organic compounds of diverse molecular weights, like amino acids, organic acids, and sugars, which function as chemical signals to soil microbiota and promote the formation of the root-linked microbial communities derived from both plant exudates and microbial or faunal metabolism (Chen & Liu, 2024).

Plants distribute up to 40% of the carbon fixed by photosynthesis to the rhizosphere in the form of exudates containing amino acids, sugars, and organic acids, due to competition from environmental microbes for ecological niche, nutrients, and water. Microbial abundance in the rhizosphere can reach 10^{12} cells per gram of soil, encompassing up to 30,000 distinct prokaryotic species, which is much higher than in nonrhizospheric soil (Mendes et al., 2011).

Microbial metabolites enhance nutrient mineralization and availability, thereby benefiting both microbes and the plant host. Furthermore, microbial volatiles such as butanediol stimulate plant growth, while microbial regulation of phytohormones (auxins, cytokinins, and ethylene) shapes plant morphology and development (Parasar et al., 2024).

The chemical diversity of the rhizosphere underpins bidirectional signaling between plants and microbes, regulating biofilm formation, stress tolerance, and community stability. Communication within the rhizosphere can be broadly categorized as microorganism-to-microorganism, host-to-microbe, and microbe-to-plants crosstalk.

11.2.1.3 Microbe-to-Microbe Signaling

Quorum sensing plays a pivotal role in the synchronization of microbes and their activities, governed by signal molecules such as peptides in Gram-positive bacteria and N-acyl homoserine lactones (AHLs) in Gram-negative bacteria. Biofilm formation, pathogenicity, and chemotaxis are all affected by QS controls pathways, thereby orchestrating collective microbial activity (K. Singh et al., 2022). Moreover, microbial volatile compounds like terpenoids, alkanes, ketones, and sulfur compounds hydrogen cyanide, hydrogen sulfide, nitric oxide, ammonia, and carbon dioxide enable long-distance communication among microorganisms. However, the mechanisms underlying these microbial signaling processes remain incompletely understood. Additionally, signaling molecules such as thiamine and trehalose facilitate microbial coexistence and enhance survival under nutrient-limited conditions (Scherlach & Hertweck, 2018).

11.2.1.4 Plant-to-Microbe Signaling

The structure and function of microbial communities can be influenced by plant secretions releazed around a wide array of secondary metabolites containing phenolic compounds, signaling molecules, and carboxylic acids. For instance, carotenoid-derived phytohormones promote arbuscular mycorrhizal colonization during nutrient regimes, while flavones within legume root exudates stimulate rhizobial nodulation and nitrogen fixation (Aliche et al., 2020). Furthermore, these compounds can also influence nutrient acquisition by boosting phosphate bioavailability and microbial movement.

11.2.1.5 Microbe-to-Plant Signaling

Microorganisms produce microbe-associated molecular patterns (MAMPs), quorum sensing (QS) molecules, and phytohormones that regulate plant growth, defense, and stress responses (Palmer et al., 2014). MAMPs such as lipopolysaccharides, peptidoglycans, flagellin, and chitin trigger systemic tolerance to both abiotic and biotic stresses. Rhizosphere microbiota can modulate and affect plant physiology. Beneficial microbes can also activate induced systemic resistance (ISR), modulate hormonal balance, and improve stress resilience. Moreover, microbial volatile organic compounds (VOCs) can stimulate plant growth and nutrient uptake, though their underlying molecular mechanisms remain an active area of investigation (Egamberdieva et al., 2017).

11.2.2 Functions in Growth Promotion and Immunity

Plant-associated microorganisms play pivotal roles in regulating plant growth and health and can be broadly classified as symbiotic, pathogenic, or non-influential depending on the nature of these associations with the involved vegetal host. Within this group, plant growth-facilitating microorganisms (PGPMs) are of particular importance due to their capacity to enhance plant development and immunity through multiple mechanisms (Kaya et al., 2023).

Most beneficial microorganisms belong to the bacterial group widely identified as plant (PGPR), which establish root systems and stimulate growth via direct and indirect pathways. The term PGPR was first introduced by Nandal and Hooda (2013) and encompasses free-living, mutualistic, and endophytic bacteria that stimulate plant growth (Kumar et al., 2022). These PGPR may be subdivided and categorized, by symbiotic PGPR, like Rhizobium and Frankia, which develop symbiotic relationships with their hosts, forming nodules or other specialized structures, and free-living PGPR, including Bacillus and Pseudomonas, which inhabit the rhizosphere and exert their effects without forming structural symbioses (Kishore et al., 2005).

In addition to bacterial PGPR, arbuscular mycorrhizal fungi (AMF) and rhizobia participate significantly in planting resilience through boosting macronutrient incorporation, enhancing stress tolerance, and stimulating root development (Acharya et al., 2024). AMF inoculants are widely used as commercial biofertilizers due to their efficiency in improving phosphorus and water acquisition. Microbial interactions within these communities may range from mutualism to antagonism, with some species showing positive co-occurrence patterns, while others display competitive or exclusionary relationships (Pacheco et al., 2019).

The growth of vegetative crops can promote growth through multiple direct pathways of plant-beneficial rhizobacteria, like microbial P solubilization, plant hormone biosynthesis, such as auxins, gibberellins, cytokinins, and nitrogen fixation, which regulate plant development. Certain PGPR also release siderophores, high-affinity iron-chelating compounds to improve iron assimilation by plants. The root system architecture can be influenced by these microbes, which facilitate water absorption, tolerance to abiotic stresses, including salt stress, water deficit, and metal toxicity, by synthesizing osmolytes and antioxidative proteins (Hyder et al., 2023). Well-studied examples provide mycorrhizal symbiosis with Rhizobium–legume and associations with nitrogen fixation can enhance phosphorus assimilation. The efficiency of these beneficial interactions depends on multiple environmental factors, like soil humidity, root exudate composition, pH, mineral availability, and climatic temperature, all of which influence rhizobacterial colonization and activity (Philippot et al., 2013).

In addition, plant immunity can be provided directly by plant growth-promoting, and also via indirect pathways, via association with other plant beneficial microorganisms, like pathogen suppression and induction of host defense systems like resistance (ISR) in plants (Mhlongo et al., 2020). Certain strains of rhizobacteria produce antimicrobial peptides and antibiotics that inhibit pathogen proliferation or block quorum-sensing networks, which are critical for pathogen regulation. Moreover, PGPR-produced VOCs can enhance plant defense signaling and promote systemic resistance throughout the plant (Mon et al., 2021; Yang et al., 2024).

11.3 AI Approaches in Microbiome Research

AI increasingly supports end-to-end microbiome workflows from sequence processing and quality control to phenotype prediction and biological interpretation. In supervised learning, RF and SVM remain reliable baselines for sparse, zero-inflated, and compositional abundance tables, while gradient-boosted trees such as XGBoost often reach top accuracy when paired with leakage-proof evaluation (e.g., cohort/site-blocked cross-validation) and transparent feature handling (Johansson, 2023). Recent studies have shown that model selection under cross-cohort distribution shifts may contribute less to predictive success than rigorous data preprocessing, explicit control of confounding factors (e.g., batch effects, site, and diet), and robust validation across independent cohorts, particularly since models that perform well in intra-cohort tests often fail to generalize across datasets (Papoutsoglou et al., 2023).

A recurring lesson from recent benchmarking is that preprocessing and transformation choices can reshape the set of selected "biomarkers" without dramatically changing headline accuracy. In some cases, proportion-based normalizations or minimal preprocessing match or even slightly outperform composition-aware transforms (Centered Log-Ratio

[CLR]/Additive Log-Ratio [ALR]/ Isometric Log-Ratio transformations [ILR]) for classification, particularly with tree-based learners that tolerate monotone transformations and sparsity. This does not negate the value of composition-aware methods; rather, it motivates sensitivity analyses that report model stability across multiple reasonable transformations, and full transparency in methods about every preprocessing decision (Amin & Ganatra, 2024).

Beyond accuracy, explainability is essential to translate predictions into biological insight. Post-hoc tools such as permutation importance and SHAP (SHapley Additive Explanations) help rank taxa or pathways that drive model decisions at global and individual levels, enabling hypothesis generation and subgroup discovery (e.g., patient-specific microbial signatures) (Novielli et al., 2024). Recent studies has illustrated SHAP's utility for microbiome models but also warn that explanations should be interpreted alongside rigorous validation to avoid confounding and artifact over-interpretation; practically, explanations should be combined with cross-cohort testing and confounder audits (Walsh et al., 2024a, 2024b).

DL is most compelling when data are high-dimensional, structured, or multiomic. CNNs can encode k-Mer spectra or transform profiles into image-like representations; recurrent and attention-based models capture long-range dependencies across reads or contigs, facilitating phenotype prediction, functional annotation, and multiomic integration (e.g., 16S/shotgun with metabolomics) (Papoutsoglou et al., 2023). However, several practical constraints persist, including limited labeled samples, class imbalance, cohort shifts, and the need for external validation and model documentation. Nevertheless, the evidence supporting the application of DL in microbiome research continues to grow (Przymus et al., 2024).

A particularly promising line of work is taxonomy-aware neural networks that embed phylogenetic structure directly into the architecture. Instead of treating features as exchangeable, these models propagate information across related clades, improving trait prediction while yielding hierarchy-aligned latent representations that are easier to interpret biologically. The MIOSTONE framework exemplifies this approach, narrowing the gap between black-box performance and mechanism-oriented insights, especially valuable when lineage-level processes are central to the research question (Jiang et al., 2025).

Principal Component Analysis (PCA)/Multidimensional Scaling (MDS) offer a stable global structure and denoising consistent with the chosen ecological distance, like Bray–Curtis, for unsupervised analysis, clustering, and low-dimensional (DR) remain essential, making them well-suited for quality control, batch assessment, and trend discovery (Whited, 2022). t-distributed Stochastic Neighbor Embedding (t-SNE) and Uniform Manifold Approximation and Projection (UMAP) can highlight local manifolds and rare taxa groups, but are so sensitive to hyperparameters (perplexity) and may induce structure if tuning is not guided by a quantitative criterion. Current guidance suggests showing complementary projections (PCA or MDS alongside UMAP/t-SNE), validating clusters with internal indices and orthogonal biological evidence (metadata and metabolites), and resisting overinterpretation of scatterplot geometry (Whited, 2022).

Taken together, a practical AI pipeline for microbiome research typically includes: (1) rigorous sample/feature QC and contamination checks; (2) principled normalization/transform with sensitivity analyses (contrast a simple scaling against a composition-aware option); (3) confounder assessment and leakage-proof evaluation design (group-blocked

CV; strict train/test separation by cohort or site); (4) model selection emphasizing robustness (RF/SVM/XGBoost as baselines; DL when data volume/structure justify it); (5) calibrated explainability (permutation importance, SHAP) with stability checks; (6) external validation across cohorts/labs; and (7) full code/model sharing for reproducibility. Pipelines built this way generalize better and produce explanations that align with known or testable biology (Li et al., 2023).

Nutrition, ecological, and edaphic systems are covert by inter-cohort evaluations, especially in clinical diagnostics. Performance degradation is frequently observed when predictive models trained on intra-cohort data are applied to external datasets, as reflected by reduced areas under the curve (AUCs), highlighting the importance of data harmonization and domain adaptation in clinical applications (Novielli et al., 2025). In environmental microbial communities, uncovered drivers of ecosystem functions (soil respiration sensitivity) demonstrate that ML can interpret how XAI can yield process hypotheses rather than simple classification output. For plant-associated with soil microbiomes, similar limitations, including site heterogeneity, changes in taxonomic composition, and protocol variability, can dominate the signal, so blocked structure and multisite external tests are necessary (Li et al., 2023).

Two general cautions are timely. First, transformation choices can introduce artifacts or interact subtly with modeling assumptions; recent work asked when and how Centered Log-Ratio (CLR)/ Isometric Log-Ratio transformations (ILR) should be applied in predictive pipelines and advocated transparency plus robustness checks (Austin et al., 2025). Second, the striking visuals produced by manifold learning can be seductive; Dimensionality Reduction (DR) plots should be treated as exploratory aids, not hypothesis tests. In response, the field is moving toward registered pipelines, data/model cards, and prespecified evaluation plans to limit researchers' degrees of freedom and strengthen the credibility of AI-based microbiome findings (Zhang et al., 2024b).

11.4 The Omics Toolkit: Mapping the Plant–Microbiome Interface

Interactions between hosts and microbiomes are known to be essential for maintaining host health. A better understanding of the code that governs these interactions will make it easier to predict the physical condition of hosts and modify it as deemed appropriate. For this reason, it needs to learn more about the complex interactions within the plant holobiont using various high-throughput omics technologies (Xu et al., 2021).

Combining different omics data streams provides a comprehensive overview of the state of the host and its partners in this situation (Ferrocino et al., 2023). These are different types of data streams that show the different layers of work needed to explain biological connections. Due to the complexity, high creation, and differences between multiomic data types, complex AI and ML techniques are needed to discover new biological points and combine different types of data (Yetgin, 2025). The identification, through AI, of the complexity and nonlinear connections between microbial groups (metagenomics) and the functional response of the host (metabolomics and transcriptomics) is essential to better predict plant characteristics such as shape, stress resistance, and resilience to infections (Zhao et al., 2023).

11.4.1 Metagenomics and the Functional Potential of Microbial Communities

Metagenomics is the study of the entire genetic makeup of microbes attached to plants. This helps to understand the genetic mechanisms of the entire system (Rajguru et al., 2024). They study the DNA makeup of this group to find out some microbes and the two types called functional microbes that are tied to beneficial or antagonist plant–microbe relationships (Guttman et al., 2014). Knowing how the parts closer to the tools are encoded can help in actual ways, like recipes for microbes, to help plants grow and stay healthy (Batista & Singh, 2021). Metagenomes have a lot more worth when current types of data collection or genome sequencing involve the use of long-read metagenomic sequencing (Kim et al., 2024). This approach allows us to get a full picture of chromosomes, plasmids, and phages of batch size and quality in real-world samples. It also opens the door to give us better knowledge on the size and form of microbes during their movement (Marbouty et al., 2017).

11.4.2 Transcriptomics, Proteomics, and Metabolomics: Decoding Functional Expression

While metagenomics provides a map of genetic potential, transcriptomics, proteomics, and metabolomics show how this potential is realized in real time (Singh et al., 2021). Transcriptomics, often performed using RNA sequencing (RNA-seq), measures gene expression and therefore shows how stored genetic information expresses itself into functional outcome and how to find genes that are important for specific plant traits (Wolf, 2013). In horticulture research, RNA-seq is used to observe transcriptome changes during a number of timings, such as fruit ripening or in response to pathogenic stress (Martinelli et al., 2012). Proteomics and metabolomics show a more immediate window into what is happening at the molecular level within the plant–microbe system and reflect what biochemical mechanisms are taking place (Hong et al., 2016). More specifically, metabolomics deciphers the small-molecule interactions that define the chemical dialogue between host and microbiota and which are essential in understanding active interaction networks under stress (Kulkarni et al., 2025). In woody plant research, RNA-seq points out the change of the transcriptome during important stages of woody plant development, such as fruit ripening or when a plant is under pathogen stress (Martinelli et al., 2012). Proteomics and metabolomics provide a more immediate window into how molecules in the plant–microbe system work and reveal the actual biochemical processes that are in progress (Hong et al., 2016). In this context, metabolomics helps elucidate how small molecules mediate biochemical communication between the host plant and its associated microbiota (Kulkarni et al., 2025).

11.4.3 AI Pipelines for Feature Extraction and Data Harmonization

Integrating and analyzing complex high-dimensional multi-omics data remain a major challenge. Artificial intelligence (AI) approaches provide essential computational tools for transforming raw biological data into meaningful biological knowledge (Vitorino, 2024).

Large-scale omics datasets are often interconnected, incomplete, and noisy (Mathé et al., 2018; Vitorino, 2024). A major challenge is to preserve relevant biological signals while integrating heterogeneous data types (Tan et al., 2022).

AI workflows frequently start with a feature to eliminate noise and find the most useful parts of the data, such as major taxa of microbes, genes that express differently, or facts about the structure of metabolites, that drive changes seen in bio data (Gupta et al., 2024). Classic ways often do not work with this data, which is like that at all, and results in many false findings, under from getting good results in predictions (Caminero et al., 2025). ML and chance making, even deeper learning (DL), come with solutions for these needs by finding patterns that are strong and have real meaning in the field across many omics sets of data (Yetgin, 2025).

DL, a type of ML that uses neural nets, is good at changing raw omics data into meaningful forms of representations (Tsimenidis et al., 2022). These types of models act more like a brain doing work in the way data is meant to operate, which are complex and not straight lines as seen in our bodies (Ahmed et al., 2023). Systems that work like CNNs and autoencoders can pick out deep parts, building small culture spaces for data that keep the basis of biological signals and take out some noise (Pintelas et al., 2021). This prediction produces a standard way of looking at large inputs into matrices that are used by models of ML, which allow for prediction and explanation (Tsimenidis et al., 2022).

Getting the right ML system for the job of combining many types of omic data depends on the number of facts masked in the types of data, how the types relate to each other, and how much data will be fed through the setup (Eicher et al., 2020; Ferrocino et al., 2023). Careful choice of the ML system is needed to see if the different data types are different and if they are statistically related (Eicher et al., 2020; Satya et al., 2024). Several integration strategies and machine learning approaches used for multi-omics data analysis are summarized in Table 11.1.

Unsupervised models find common biological processes, or groups of things, with no thought about the end goal, such as having a disease or phenotype (Mishra et al., 2019). When data contain some but not all of each type of data, the multiomics factor analysis (MOFA2) comes in and finds a window (latent factors) that joins the types of data, which is the way the data is brought together even when data are missed, and brings together the map of the microbes to the body's metabolism (Hayes et al., 2024). When all data are complete, integrative clusters such in the way of the iCluster and iClusterBayes can find areas that are commonly seen in the data, and iClusterBayes is better when there is much data to work with, and iCluster when there is less to test on (Tyler et al., 2021).

When omics layers are not alike, the joint and individual variation explained (JIVE) way splits up the changes into parts that are shared and parts that are new to each layer. This tells us about working together in plants or animals. This step is very important when making sense of complicated systems (Lock et al., 2013). After having cut down the size and made the data alike, high-end models are used to find what the data means. These might indicate how much a crop will grow or how sick a plant might be (Misra et al., 2019). For big sets of data, fairer ways that use many models make things difficult to break and do well (Rokach, 2010). Techs like least absolute shrinkage and selection operator (LASSO) are good for this as they help to keep the data from running wild and put more weight on what counts well (Boulesteix & Strimmer, 2007; Hastie, 2009; Huang et al., 2021). With a smaller

Table 11.1 Key AI architectures for multiomics feature extraction and integration in plant systems.

Integration strategy	Machine learning/deep learning model	Data overlap/ distribution criteria	Primary function in plant-microbiome research	References
Unsupervised factor analysis	MOFA2 (multiomics factor analysis)	Partial overlap (missing data across modalities)	Identifying latent factors that explain variation across different omics datasets (e.g., linking microbiome structure to host metabolism).	Argelaguet et al. (2018)
Unsupervised clustering	iCluster/ iClusterBayes	Complete Overlap (iClusterBayes for large data)	Integrative clustering to identify novel plant or microbial subtypes based on combined molecular signatures.	Khurana et al. (2018)
Unsupervised variance decomposition	JIVE (joint and individual variation explained)	Different data distributions (heterogeneous)	Separating shared variance (joint host–microbe interaction) from modality-specific variation (individual omics noise).	Lock et al. (2013)
Supervised prediction/ selection	LASSO (Least absolute shrinkage and selection operator)	Large datasets; complete overlap	Robust feature selection and building predictive models for specific traits (e.g., disease resistance prediction).	Huang et al. (2021, Tibshirani, (1996)
Feature extraction/ representation	Deep learning (CNNs, autoencoders)	Image, spectral, or sequential omics data	Automated feature learning and nonlinear dimensionality reduction for highly complex data.	Angermueller et al. (2016), Eraslan et al. (2019)

set of data or fewer transfers in each set, other models like SVMs, Decision Trees, and Naive Bayes can be very good at ruling out the last answer (Breiman, 2001).

11.5 Predictive Applications

11.5.1 Biomarker Discovery: From Single Molecules to Functional Pathways

The strong links between AI and ML are removing the traditional way of finding biomarkers in labs. These new technologies make it easy to see multiple matching information, which is more dependable and better at predicting than single tests (Yetgin, 2025).

Conventional ways of finding biomarkers picked out fake wild genes, only caught some reproducibility, and did not bring together the multiple information that makes biological happenings so hard to pin down (Theodorakis et al., 2024). Powerful artificial intelligence (AI) approaches, including neural networks, large predictive models, and advanced feature selection techniques, enable the analysis of complex datasets and facilitate the identification of informative biomarkers for diagnosis, prediction, and treatment planning (Sartori et al., 2025). They work in many fields of medicine and especially the plant disease (Prelaj et al., 2024).

Another big change in how we use AI is that we no longer focus just on simple molecular matches, but rather on signs of functions of different pathways signatures (Zhang et al., 2025). And, the best example is the biosynthetic gene clusters (BGCs). BGCs are key functional parts of microbes (by studying unauthenticated microbes) that make new secondary metabolites (Zhu et al., 2025). These secondary metabolites may be new drugs to fight cancers and other viruses, and also help host–microbe interactions that can be used to diagnose or treat conditions and help discover new treatments (Dhaarani & Reddy, 2025). So, instead of just seeing what is wrong, AI helps us find info on what can be done to good effect (predict functional and therapeutic use), thus making sure we hit the right crop or drug at the right time (Zhang et al., 2025).

11.5.2 Disease Prediction and Early-Warning Systems (EWS)

AI use has made a big change in how we deal with plant disease, helping us move from just watching for signs to tell us ahead what disease may be and how soon (Mohamedikbal et al., 2025). Key AI-driven predictive applications in plant–microbiome research are summarized in Table 11.2.

AI and ML models can be well trained on data from many sources, such as genes, gene activity, proteins, and chemicals, which show what is inside the plant, to tell if a plant may or may not get certain diseases (Mohamedikbal et al., 2025; Murmu et al., 2024; Upadhyaya et al., 2024). By identifying relationships between molecular traits and phenotypic outcomes, these approaches enable faster and more precise selection of desirable traits compared with traditional field-based phenotyping (Cai et al., 2025; Thingujam et al., 2025). This enables further improvement of the breeding models through testing the AI predictions made (Cai et al., 2025).

Early-warning systems (EWS) are a key real-world example of how predictive AI can work (Mohamedikbal et al., 2025; Upadhyaya et al., 2024). These systems use very advanced forecasting models, Google that put together all kinds of facts, including other environmental factors (weather and ground conditions) and molecular/spectral data of the crop (Neik et al., 2025; Ngomle et al., 2025). Combining data this way can give people time to act more than several weeks before the problem is found in the field. This way of predicting the future has a big strategic value because it saves plants and the use of plant protector chemicals (Neik et al., 2025). Using AI and data from many sources is very helpful because it can find small changes or even presymptoms that show after the plant has started to get sick and before any signs of the problem are seen, which causes the earliest warning (Mohamedikbal et al., 2025; Murmu et al., 2024).

In reactive disease control, where the goal is a quick diagnosis with visual data or spectral data, deep CNNs have been shown to have very high accuracy. These techniques usually report achieving accuracies that are above 95% and can have classification scores of 0.95 to 0.99 for the disease of fungus for grains (Kumar et al., 2025; Rahman et al., 2025; Sutaji & Rosyid, 2022). A device called DenseNet-121 has been found to give very high accuracy 213, which is good for use on a mobile device (Trivedi et al., 2024).

Table 11.2 AI-driven predictive applications in plant–microbiome research.

Application focus	AI/ML technique class	Primary data inputs integrated	Achieved output/ metric	Context and significance
Early disease forecasting (EWS)	Forecasting models (ML/DL ensembles)	Weather, soil, crop data, environmental variables	Predicts spatial-temporal outbreak risks weeks in advance with 81–95% precision.	Enables preemptive intervention, optimizing resource use and minimizing crop losses.
Disease diagnosis (visual)	Deep CNNs (e.g., DenseNet-121, Faster R-CNN)	High-resolution image data, spectral data	Classification accuracy ranging from 95% to 99% in identifying fungal diseases and pests.	Automated, rapid in-field disease diagnosis, exceeding typical human visual assessment accuracy.
Functional biomarker discovery	Neural networks, transformers, feature selection	Genomics, transcriptomics, proteomics, metabolomics	Identification of complex signatures, including biosynthetic gene clusters (BGCs).	Advances in precision breeding and pharmaceutical discovery by locating novel functional units.
Hologenome resistance prediction	Multiomics ML algorithms (general)	Genomics (host), metagenomics (microbe), metabolomics (host)	Prediction of plant resilience and disease resistance traits based on integrated molecular profiles.	Facilitates faster and more precise selection in breeding programs based on integrated host–microbe function.

For planning to see if there is likely to be an outbreak a few weeks in the future, in which multiple variables in the environment and the crop are put together in a model that predicts risk, the typical accuracy rate is about 81–95%. This is sometimes a little bit lower than the accuracy seen in reactive visual diagnosis, but it is offset by the huge strategic advantage of having the extra lead time, which makes it possible to make good use of time and other resources to act before the outbreak (Neik et al., 2025; Ngomle et al., 2025). Putting together many different types of data shows other disease patterns that would not be noticed by people and makes it possible for people to make better decisions in the field (Murmu et al., 2024; Neik et al., 2025). The great power of using data from many omics, especially on the plant microbiome, is shown to work by studies that looked at how the host resists disease. This shows that the idea of the hologenome works, which says the plant and the microbes that live on or in it are one very changing unit of evolution. An extreme example is the work on tomato plant wilt caused by *Ralstonia solanacearum (R. solanacearum)*. Employing a multiomics method that combined microbiomics and metagenomics of the root zone, with host metabolomics (Yin et al., 2022). It found big foundational differences in the root-zone microbial groups of tomato plants that resisted wilt and those that did not (Kwak et al., 2018). Importantly, overcoming the wilt by moving the rhizosphere microbiomes of a plant that resisted the disease to one that could not, greatly lessened the wilt disease (Choi et al., 2020). It strongly verified the molecular basis for the way plants get or do not get wilt depends on the DNA of the host plant, and the host's interactions with microbes. Because of the captured dynamic between omics approaches, there needs to be a way to use big data, using AIs to enable simple actions and policies, for example, microbiome design to make crops better (Yin et al., 2022).

11.6 Case Studies

The use of AI to decipher plant–microbiome interactions has been studied using plant models such as Arabidopsis and crops such as maize and tomatoes.

Studies on the model species *Arabidopsis thaliana* have revealed host genes that shape the structure and function of the leaf microbiota, especially in terms of resistance to abiotic stress (Mulaudzi et al., 2025). The microbial network and the Genome-Wide Association Studies (GWAS) approach have detected many loci of agronomic interest related to plant nutrition and defense. ML models have also helped elucidate the regulators of the root microbiota composition. For instance, RF models applied to 16S rDNA sequencing have allowed prediction of the frequency of distinct bacteria in root microbiota according to plant host genotype and environment. It is also thanks to this approach that it has been possible to identify a small number of agronomically interesting genes linked with the immune system or root exudation that affect the composition and dynamics of the microbiota (Fitzpatrick et al., 2020). A new computer application has been developed as part of a project dedicated to advancing our description of one link between an *A. thaliana* root model and its microbiota. This application is that using CNNs allows for an intuitive display of associations between root structure and the existence of microbial groups inferred from correlations within the plant, enabling nondestructive prediction of the status of a microbiome in plants (Przymus et al., 2025).

DL applied to metagenomic, transcriptomic, and metabolomic data enables us to model interactions within the *Arabidopsis phytobiome.* In that aspect, the positive regulation in some defense genes and multiplication of some beneficial bacteria have already been demonstrated, revealing an improved understanding of how this phytomicrobial system functions (Yoshida et al., 2024). Recent work in this area has resulted in the creation of a categorization model based on an RF analysis that predicts the sensitivity of *A. thaliana* to *Pseudomonas syringae* strains by its leaf microbiome. This type of model is a new addition to programs carried out in this area. It can classify bacterial groups corresponding to the resistance of the plants studied with an identification accuracy of up to 92%, which helps prevent and manage this disease (Chaudhry et al., 2021).

In a more recent integrated AI and reinforcement learning study of improving maize plants' resistance to water stress. Microbial communities that have a water-retaining property were detected in the AI for their genomic sequences, and numerous bacteria–fungi pairs were found to be potential ones. These compounds were identified in the greenhouse to be promoting root number and the physiological status of the plant (Tariq et al., 2025). Metabolomic analyses by ML also revealed divergent functionality concerning microbial consortia using biostimulants and were visualized with both RF and XGBoost models in three distinct microbe-mediated mechanisms of action: photoprotection, metabolic reprogramming, and hormonal regulation (Fitzpatrick et al., 2020). A recent study conducted on maize also showed that combining metabolomics and AI is a useful tool for understanding the interactions between plants and their microbiome under natural conditions (Mulaudzi et al., 2025). Three bacillus-based microbial consortia induced distinct metabolomic signatures, characterized by the accumulation of flavonoids, hydroxycinnamic acids, and lipids (Gomes et al., 2024). Supervised analyses identified Hierarchical Cluster Analysis (HCA) derivatives as key discriminating metabolites, revealing differentiated mechanisms: photoprotection and structural reinforcement for C1, immuno-metabolic activation for C2, and hormonal modulation by oxylipins for C3 (Sanadhya et al., 2021). This integrative approach illustrates the potential of AI to map plant metabolism in detail and guides the design of sustainable biostimulants. Another study of 129 maize accessions showed that genetic variation in the host strongly influences the composition of the root microbiome, particularly under water and nutrient stress. By integrating GWAS and predictive modeling, the researchers identified the genus Massilia as a key taxon recruited in cases of nitrogen deficiency due to the secretion of flavones by the plant, which in turn leads to the formation of lateral roots and nutrient absorption (He et al., 2024). In addition, studies show a DL model using a CNN to detect gray leaf spot disease in maize caused by gray leaf spot disease. This model uses leaf microbiome sequencing data collected before symptoms appear. The model makes it possible to recognize the subtle microbial signatures that precede the outbreak of the disease, thus providing early warning systems that enable targeted and more effective interventions against the disease (Qian et al., 2022).

For tomatoes, domestication and intensive selective breeding have led to an imbalance in the composition and function of the symbiotic microbiome, leaves, and fruits. Some parts of these plants are colonized by bacteria that are considered beneficial, such as those of the genus Bacillus. By integrating plant localization data with AI-driven simulation algorithms of computer models, it is possible to indicate the resistance of the tomatoes to the various diseases that may appear on the aerial portions of these plants, roots, and leaves (F. Li et al., 2024). With the SVM and gradient boosting models used by Zhang et al. (Liu et al., 2025), it was possible to predict the resistance of the tomato varieties to *R. solanacearum* based on their microbial composition. The models were able to successfully identify the microbial indicators, such as Bacillus subtilis, able to forecast the resistance of the plants. Inquiring more, these authors used AI-based mechanisms to predict diseases and the resistance mechanisms caused by the rhizosphere microbiomes. The neural network models were trained on the metagenomic profiles and were able to predict the susceptibility of the plant to *R. solanacearum* with an accuracy superior to 85% (Khan et al., 2024). Unsupervised classification algorithms were also used to identify diseased and healthy plant microbiomes. This led to the discovery of beneficial species such as Pseudomonas and Bacillus, which are associated with the elimination of pathogens (Tracanna, 2022).

11.7 Toward Predictive Microbiome Engineering

Predictive microbiome engineering is a truly disruptive innovation that not only combines systems biology, biotechnology, and AI to design, model, and improve microbial communities associated with plants to optimize their growth, health, and resistance to multiple stresses, and to leverage these interactions for implementation, but also, by leveraging synthetic biology and functional genomics, enables the complexity of the living world to be transformed by AI into sustainable solutions (Du et al., 2025; Sahoo et al., 2025).

Synthetic microbial communities (SynComs) are a key tool for predictive engineering. By carefully reassembling the interactions within a natural microbiota, they contribute to understanding the specific functions of different types of microbes and increase resistance to stress factors (Berendsen et al., 2018). For example, the removal of *Enterobacter cloacae* from a SynCom shed light on its central role in maize mildew resistance (Herrera Paredes et al., 2018). Microbe–plant interactions are then coupled with a design-build-test-learn (DBTL) model to improve the composition of SynComs developed based on the interaction (Yusuf et al., 2025). ML models such as RFs, SVMs, XGBoost, and neural networks (CNN and RNN) can also provide information on the best combinations likely to promote plant growth or stress tolerance, based on microbial strains (Tariq et al., 2025). Recent work has shown that neural models trained on meta-tagged names, transcriptomes, and phenotypes predict the performance of microbial consortia even before their experimental validation (Zhang et al., 2024a).

Predictive engineering also applies synthetic biology and genetic manipulation approaches to optimize modified microbial activities. The Clustered Regularly Interspaced Short Palindromic Repeats / CRISPR-associated protein 9 (CRISPR/Cas9) systems can generate beneficial microbes to improve their competitiveness or stability of effects in the rhizosphere. For instance, *Bacillus velezensis* has been improved for biofilm formation, for producing antifungal metabolites and root colonization activities, and is known to promote wheat growth (Guo et al., 2025). In addition, the rapid growth in the development of decision support systems (DSS) primed by AI is also making microbiome management more immediate and predictive. These are data-driven platforms that integrate inputs via soil and geophysical sensors, genomics databases, and soil microbial profiles for geographically specific recommendations for real-time microbiological inoculation considerations on a farm-to-farm basis (Li et al., 2025). The future in this field will be created through a convergence of XAI, systems biology, multiomics, and digital technologies, facilitating the analysis of microbe-related plant traits that are the most significant to plant health, giving models more interpretability of results for decision-making and increased trust in those results (Gunina & Kuzyakov, 2022). Thus, connected objects (IoT) and smart soil sensors give real-time data on temperature, humidity, pH, and microbiome structure. In conjunction with AI platforms and digital twins of microbiomes, the capacity exists for real-time simulation models of plant–microbe–environment interactions to be formulated to facilitate the prediction of best intervention strategies (Sahoo et al., 2025). These microbiome digital twins are the logical extension of predictive engineering: they enable different microbial configurations to be tested virtually before they are implemented experimentally, ecological imbalances to be anticipated, and microbial formulations to be adjusted, according to local conditions (Yusuf et al., 2025).

11.8 Challenges and Future Prospects

The AI revolution's penetration in plant–microbiome research provides new opportunities, but also poses obstacles to overcome on a large-scale application in agriculture. The heterogeneity of microbiome data itself poses one of the major challenges. Variations in sampling techniques, DNA extraction methods, sequencing depth, and bioinformatic pipelines frequently give rise to bias that makes it difficult to compare studies and reproduce those of other environments (Knight et al., 2018). Furthermore, the nonexistence of standardized metadata on soil properties, climatic parameters, and agricultural practices makes it difficult to have trust in predictive models. In practice, integrating multiple layers beyond metagenomic data (transcriptomics, metabolomics, and proteomics) remains a formidable obstacle stat-wise as these datasets are generally at different scales and levels of noise and interpretation, which necessitates the existence of multimodal learning techniques yet to be developed (Noecker et al., 2016).

The interpretability becomes another crucial challenge. Even though ML and DL models can provide a high prediction accuracy, they are often criticized for being black boxes that do not convey added value to plant scientists and agronomists regarding the biological interpretation of the predictions. Such opacity undermines trust in AI-based outcomes, particularly when used for as high-stakes decisions as those of crop management. Recent advances around XAI have started to address this by creating models that can detect the important features or provide an understandable output (Höhl et al., 2024), but their application in microbiome data is emerging. Furthermore, the amounts of computational resources necessary to train and validate models on a large scale are still out of reach for many research institutions around the world (especially those in low- and middle-income countries lacking high-performance computing access and bioinformatics expertise). This results in inequalities on the global research map and might delay the democratization of AI-powered microbiome research.

Apart from technical issues, it continues to be a considerable challenge to translate insights obtained by AI into applied decision-making on the ground. To date, most models have been trained with data derived from controlled laboratory or greenhouse conditions, while plant–microbiome interactions in natural ecosystems are influenced by variable environmental factors that are challenging to experimentally reproduce. Inter-field variation, including variation in soil microbiota, weather, and management, is often too great to allow accurate predictions from models when they are applied beyond the region into which they were trained (Busby et al., 2017). Closing this gap will require large-scale field validation, iterative updating of systems via model improvements, and the streaming in real time of data generated by digital agriculture technologies such as remote sensing, IoTs, and precision instruments such as sensors (Finger, 2023).

Despite these challenges, the future for AI-enabled plant–microbiome research is very promising. Progress will also be achieved by interpretable AI algorithms that increase the interpretability and usability of models, such that computational predictions are biologically relevant and experimentally valid (Rudin, 2019). Intelligent analysis of microbiome with AI and digital Agriculture platform may also revolutionise precision farming to monitor the changing microbiome on-field in real time, integrate online decision-making dynamics for crop- and ecosystem-specific management strategies. From the less limiting processes for vector generation to those more permeable ones, including synthetic biology and systems biotechnology approaches and computational strategies, the development of process-oriented AI is enabling rational design of microbial communities establishing synergy-driven consortia with increasingly improved performances, be such beneficial biofertilizers or efficient biocontrol agents in a double-folded approach: increase crop productivity while saving resources from chemical inputs (Berruto & Demirer, 2024). In the end, this progress will not only be beneficial to crop production and disease resistance but also to solving global problems, including food security and climate change. In leveraging plant–microbiome relationships, AI could contribute to resilient, sustainable agricultural systems that are better positioned for the challenges posed by an uncertain future (Hanif et al., 2024; Trivedi et al., 2020).

11.9 Conclusion

Currently, technical advances in AI methods are revolutionizing plant–microbiome science by enabling complex integration and analysis of multiomics datasets using advanced ML and DL methods. This innovative combination allows scientists to discern underlying patterns in microbiomes, develop validated predictive biomarkers, design optimal plant beneficial microbes, and model predictive disease outbreaks long before the appearance of symptoms, all while facilitating initiatives in microbiome-motivated precision agriculture. Beyond these exciting possibilities, there remain several critical challenges facing the use of swarm robotic methods: the data heterogeneity, the lack of standardized use of protocols, low levels of interpretability of predictive models, and the need to validate their utility in real-world field settings over larger scales. Addressing these challenges will involve the use of key advances in technology, such as interpretable XAI, developing more robust standards for data, and a better relationship between swarm robotics and digital agriculture technologies (e.g., IoT sensors, remote sensing, and digital twins). In the longer term, AI and microbiome science could profoundly impact crop management, improve plant resilience and yield, and contribute to the pressing issues facing the world, such as food insecurity, sustainable resource use, and climate adaptation.

Acknowledgments

This study was financially supported by the Phytopathology Unit within the Department of Plant Protection and Environment at the Ecole Nationale d'Agriculture de Meknès, Morocco.

References

Acharya, B. R., Gill, S. P., Kaundal, A., & Sandhu, D. (2024). Strategies for combating plant salinity stress: The potential of plant growth-promoting microorganisms. *Frontiers in Plant Science, 15*, 1–15. https://doi.org/10.3389/fpls.2024.1406913

Ahmed, S. F., Alam, M. S., Hassan, M., Rozbu, M. R., Ishtiak, T., Rafa, N., Mofijur, M., Shawkat Ali, A. B. M., & Gandomi, A. H. (2023). Deep learning modelling techniques: Current progress, applications, advantages, and challenges. *Artificial Intelligence Review, 56*(11), 13521–13617. https://doi.org/10.1007/s10462-023-10466-8

Aliche, E. B., Screpanti, C., De Mesmaeker, A., Munnik, T., & Bouwmeester, H. J. (2020). Science and application of strigolactones. *New Phytologist, 227*(4), 1001–1011. https://doi.org/10.1111/nph.16489

Amin, L. B., & Ganatra, R. B. (2024). Peripheral administration of norepinephrine: A prospective observational study. *Chest*, *166*(2), e69–e70. https://doi.org/10.1016/j.chest.2024.03.035

Angermueller, C., Pärnamaa, T., Parts, L., & Stegle, O. (2016). Deep learning for computational biology. *Molecular Systems Biology*, *12*(7), 878. https://doi.org/10.15252/msb.20156651

Argelaguet, R., Velten, B., Arnol, D., Dietrich, S., Zenz, T., Marioni, J. C., Buettner, F., Huber, W., & Stegle, O. (2018). Multi-omics factor analysis—a framework for unsupervised integration of multi-omics data sets. *Molecular Systems Biology*, *14*(6). https://doi.org/10.15252/msb.20178124

Austin, G. I., Korem, T., Austin, G. I., & Korem, T. (2025). Phenotype-associated values into sparse features. 10(5). 10.1128/mBio.01607-23.

Batista, B. D., & Singh, B. K. (2021). Realities and hopes in the application of microbial tools in agriculture. *Microbial Biotechnology*, *14*(4), 1258–1268. https://doi.org/10.1111/1751-7915.13866

Berendsen, R. L., Vismans, G., Yu, K., Song, Y., de Jonge, R., Burgman, W. P., Burmølle, M., Herschend, J., Bakker, P. A. H. M., & Pieterse, C. M. J. (2018). Disease-induced assemblage of a plant-beneficial bacterial consortium. *The ISME Journal*, *12*(6), 1496–1507. https://doi.org/10.1038/s41396-018-0093-1

Berg, G., Rybakova, D., Fischer, D., Cernava, T., Vergès, M.-C. C., Charles, T., Chen, X., Cocolin, L., Eversole, K., & Corral, G. H. (2020). Microbiome definition re-visited: Old concepts and new challenges. *Microbiome*, *8*(1), 103.

Berruto, C. A., & Demirer, G. S. (2024). Engineering agricultural soil microbiomes and predicting plant phenotypes. *Trends in Microbiology*, *32*(9), 858–873.

Boulesteix, A.-L., & Strimmer, K. (2007). Partial least squares: A versatile tool for the analysis of high-dimensional genomic data. *Briefings in Bioinformatics*, *8*(1), 32–44.

Breiman, L. (2001). Random forests. *Machine Learning*, *45*(1), 5–32.

Busby, P. E., Soman, C., Wagner, M. R., Friesen, M. L., Kremer, J., Bennett, A., Morsy, M., Eisen, J. A., Leach, J. E., & Dangl, J. L. (2017). Research priorities for harnessing plant microbiomes in sustainable agriculture. *PLoS Biology*, *15*(3), e2001793.

Cai, S., Sun, C., & Tian, J. (2025). Towards smart agriculture: AI-driven prediction of key genes for revolutionizing crop breeding. *Planta*, *262*(5), 1–12.

Caminero, A., Tropini, C., Valles-Colomer, M., Shung, D. L., Gibbons, S. M., Surette, M. G., Sokol, H., Tomeo, N. J., & Tarr, P. I. (2025). Credible inferences in microbiome research: Ensuring rigour, reproducibility and relevance in the era of AI. *Nature Reviews Gastroenterology & Hepatology*, *22*, 1–16.

Chang, H.-X., Haudenshield, J. S., Bowen, C. R., & Hartman, G. L. (2017). Metagenome-wide association study and machine learning prediction of bulk soil microbiome and crop productivity. *Frontiers in Microbiology*, *8*, 519.

Chaudhry, V., Runge, P., Sengupta, P., Doehlemann, G., Parker, J. E., & Kemen, E. (2021). Shaping the leaf microbiota: Plant-microbe-microbe interactions. *Journal of Experimental Botany*, *72*(1), 36–56. https://doi.org/10.1093/jxb/eraa417

Chen, L., & Liu, Y. (2024). The function of root exudates in the root colonization by beneficial soil rhizobacteria. *Biology*, *13*(2), 1–18. https://doi.org/10.3390/biology13020095

Choi, K., Choi, J., Lee, P. A., Roy, N., Khan, R., Lee, H. J., Weon, H. Y., Kong, H. G., & Lee, S.-W. (2020). Alteration of bacterial wilt resistance in tomato plant by microbiota transplant. *Frontiers in Plant Science*, *11*, 1186.

Compant, S., Samad, A., Faist, H., & Sessitsch, A. (2019). A review on the plant microbiome: Ecology, functions, and emerging trends in microbial application. *Journal of Advanced Research, 19*, 29–37.

D'Urso, F., & Broccolo, F. (2024). Applications of artificial intelligence in microbiome analysis and probiotic interventions—an overview and perspective based on the current state of the art. *Applied Sciences, 14*(19), 8627.

Dhaarani, R., & Reddy, M. K. (2025). Progressing microbial genomics: Artificial intelligence and deep learning driven advances in genome analysis and therapeutics. *Intelligence-Based Medicine, 11*, 100251. https://doi.org/10.1016/j.ibmed.2025.100251

Du, Y., Han, X., & Tsuda, K. (2025). Microbiome-mediated plant disease resistance: Recent advances and future directions. *Journal of General Plant Pathology, 91*(1), 1–17. https://doi.org/10.1007/s10327-024-01204-1

Egamberdieva, D., Wirth, S. J., Alqarawi, A. A., Abd-Allah, E. F., & Hashem, A. (2017). Phytohormones and beneficial microbes: Essential components for plants to balance stress and fitness. *Frontiers in Microbiology, 8*(OCT), 1–14. https://doi.org/10.3389/fmicb.2017.02104

Eicher, T., Kinnebrew, G., Patt, A., Spencer, K., Ying, K., Ma, Q., Machiraju, R., & Mathé, E. A. (2020). Metabolomics and multi-omics integration: A survey of computational methods and resources. *Metabolites, 10*(5), 202.

Eraslan, G., Avsec, Ž., Gagneur, J., & Theis, F. J. (2019). Deep learning: New computational modelling techniques for genomics. *Nature Reviews Genetics, 20*(7), 389–403. https://doi.org/10.1038/s41576-019-0122-6

Ferrocino, I., Rantsiou, K., McClure, R., Kostic, T., De Souza, R. S. C., Lange, L., FitzGerald, J., Kriaa, A., Cotter, P., & Maguin, E. (2023). The need for an integrated multi-OMICs approach in microbiome science in the food system. *Comprehensive Reviews in Food Science and Food Safety, 22*(2), 1082–1103.

Finger, R. (2023). Digital innovations for sustainable and resilient agricultural systems. *European Review of Agricultural Economics, 50*(4), 1277–1309.

Fitzpatrick, C. R., Salas-González, I., Conway, J. M., Finkel, O. M., Gilbert, S., Russ, D., Teixeira, P. J. P. L., & Dangl, J. L. (2020). The plant microbiome: From ecology to reductionism and beyond. *Annual Review of Microbiology, 74*(1), 81–100. https://doi.org/10.1146/annurev-micro-022620-014327

Gomes, P. W. P., Mannochio-Russo, H., Schmid, R., Zuffa, S., Damiani, T., Quiros-Guerrero, L.-M., Caraballo-Rodríguez, A. M., Zhao, H. N., Yang, H., Xing, S., Charron-Lamoureux, V., Chigumba, D. N., Sedio, B. E., Myers, J. A., Allard, P.-M., Harwood, T. V., Tamayo-Castillo, G., Kang, K. Bin, Defossez, E., ... Dorrestein, P. C. (2024). PlantMASST—community-driven chemotaxonomic digitization of plants. https://doi.org/10.1101/2024.05.13.593988

Gunina, A., & Kuzyakov, Y. (2022). From energy to (soil organic) matter. *Global Change Biology, 28*(7), 2169–2182. https://doi.org/10.1111/gcb.16071

Guo, N., Wang, S., Whitfield, C. T., Batchelor, W. D., Wang, Y., Blersch, D., Higgins, B. T., Feng, Y., Liles, M. R., De-Bashan, L. E., Wang, Y., & Ma, Y. (2025). High-efficiency CRISPR-Cas9 genome editing unveils biofilm insights and enhances antimicrobial activity in *Bacillus velezensis* FZB42. *Biotechnology and Bioengineering, 122*(4), 983–994. https://doi.org/10.1002/bit.28933

Gupta, S., Kashyap, S., & Kumar, D. (2024). AI-driven analysis of environmental metagenomic data. In S. Gupta, D. Kumar, R. Negi, R. Singh, M. J. Ansari, S. Kashyap, & S. Mehrotra (Eds.), *Genomic Intelligence* (pp. 188–206). CRC Press.

Guttman, D. S., McHardy, A. C., & Schulze-Lefert, P. (2014). Microbial genome-enabled insights into plant–microorganism interactions. *Nature Reviews Genetics*, *15*(12), 797–813.

Hacquard, S., Garrido-Oter, R., González, A., Spaepen, S., Ackermann, G., Lebeis, S., McHardy, A. C., Dangl, J. L., Knight, R., Ley, R., & Schulze-Lefert, P. (2015). Microbiota and host nutrition across plant and animal kingdoms. *Cell Host & Microbe*, *17*(5), 603–616. https://doi.org/10.1016/j.chom.2015.04.009

Hanif, M. S., Tayyab, M., Baillo, E. H., Islam, M. M., Islam, W., & Li, X. (2024). Plant microbiome technology for sustainable agriculture. *Frontiers in Microbiology*, *15*, 1500260.

Hartmann, A., Rothballer, M., & Schmid, M. (2008). Lorenz Hiltner, a pioneer in rhizosphere microbial ecology and soil bacteriology research. *Plant and Soil*, *312*(1–2), 7–14. https://doi.org/10.1007/s11104-007-9514-z

Hastie, T. (2009). *The elements of statistical learning: Data mining, inference, and prediction.* Springer.

Hayes, C. N., Nakahara, H., Ono, A., Tsuge, M., & Oka, S. (2024). From omics to multi-omics: A review of advantages and tradeoffs. *Genes*, *15*(12), 1551.

He, X., Wang, D., Jiang, Y., Li, M., Delgado-Baquerizo, M., McLaughlin, C., Marcon, C., Guo, L., Baer, M., Moya, Y. A. T., von Wirén, N., Deichmann, M., Schaaf, G., Piepho, H.-P., Yang, Z., Yang, J., Yim, B., Smalla, K., Goormachtig, S., ... Yu, P. (2024). Heritable microbiome variation is correlated with source environment in locally adapted maize varieties. *Nature Plants*, *10*(4), 598–617. https://doi.org/10.1038/s41477-024-01654-7

Hernández Medina, R., Kutuzova, S., Nielsen, K. N., Johansen, J., Hansen, L. H., Nielsen, M., & Rasmussen, S. (2022). Machine learning and deep learning applications in microbiome research. *ISME Communications*, *2*(1), 98.

Herrera Paredes, S., Gao, T., Law, T. F., Finkel, O. M., Mucyn, T., Teixeira, P. J. P. L., Salas González, I., Feltcher, M. E., Powers, M. J., Shank, E. A., Jones, C. D., Jojic, V., Dangl, J. L., & Castrillo, G. (2018). Design of synthetic bacterial communities for predictable plant phenotypes. *PLoS Biology*, *16*(2), e2003962. https://doi.org/10.1371/journal.pbio.2003962

Höhl, A., Obadic, I., Fernandez-Torres, M.-A., Najjar, H., Oliveira, D. A. B., Akata, Z., Dengel, A., & Zhu, X. X. (2024). Opening the black box: A systematic review on explainable artificial intelligence in remote sensing. *IEEE Geoscience and Remote Sensing Magazine*, *12*, 261–304.

Hong, J., Yang, L., Zhang, D., & Shi, J. (2016). Plant metabolomics: An indispensable system biology tool for plant science. *International Journal of Molecular Sciences*, *17*(6), 767.

Huang, J., Pallotti, S., Zhou, Q., Kleber, M., Xin, X., King, D. A., & Napolioni, V. (2021). PERHAPS: Paired-end short reads-based HAPlotyping from next-generation sequencing data. *Briefings in Bioinformatics*, *22*(4), bbaa320.

Hyder, S., Rizvi, Z. F., los Santos-Villalobos, S., de Santoyo, G., Gondal, A. S., Khalid, N., Fatima, S. N., Nadeem, M., Rafique, K., & Rani, A. (2023). Applications of plant growth-promoting rhizobacteria for increasing crop production and resilience. *Journal of Plant Nutrition*, *46*(10), 2551–2580. https://doi.org/10.1080/01904167.2022.2160742

Jiang, Y., Aton, M., Zhu, Q., & Lu, Y. Y. (2025). Modeling microbiome-trait associations with taxonomy-adaptive neural networks. *Microbiome*, *13*(1) https://doi.org/10.1186/s40168-025-02080-3, 87.

Johansson, Å. (2023). Comment on Li et al. (2023): A dynamic 2000—540 ma earth history: From cratonic amalgamation to the age of supercontinent cycle. *Earth-Science Reviews, 241*(May), 104457. https://doi.org/10.1016/j.earscirev.2023.104457

Kaya, C., Uğurlar, F., & Adamakis, I. D. S. (2023). Epigenetic and hormonal modulation in plant–plant growth-promoting microorganism symbiosis for drought-resilient agriculture. *International Journal of Molecular Sciences, 24*(22). https://doi.org/10.3390/ijms242216064

Khan, M., Gulan, F., Arshad, M., Zaman, A., & Riaz, A. (2024). Early and late blight disease identification in tomato plants using a neural network-based model to augmenting agricultural productivity. *Science Progress, 107*(3). https://doi.org/10.1177/00368504241275371

Khurana, S., Rawi, R., Kunji, K., Chuang, G.-Y., Bensmail, H., & Mall, R. (2018). DeepSol: A deep learning framework for sequence-based protein solubility prediction. *Bioinformatics, 34*(15), 2605–2613. https://doi.org/10.1093/bioinformatics/bty166

Kim, C., Pongpanich, M., & Porntaveetus, T. (2024). Unraveling metagenomics through long-read sequencing: A comprehensive review. *Journal of Translational Medicine, 22*(1), 111.

Kimotho, R. N., & Maina, S. (2024). Unraveling plant–microbe interactions: Can integrated omics approaches offer concrete answers? *Journal of Experimental Botany, 75*(5), 1289–1313.

Kishore, G. K., Pande, S., & Podile, A. R. (2005). Biological control of late leaf spot of peanut (Arachis hypogaea) with chitinolytic bacteria. *Phytopathology, 95*(10), 1157–1165. https://doi.org/10.1094/PHYTO-95-1157

Knight, R., Vrbanac, A., Taylor, B. C., Aksenov, A., Callewaert, C., Debelius, J., Gonzalez, A., Kosciolek, T., McCall, L.-I., McDonald, D., Melnik, A. V., Morton, J. T., Navas, J., Quinn, R. A., Sanders, J. G., Swafford, A. D., Thompson, L. R., Tripathi, A., Xu, Z. Z., ... Dorrestein, P. C. (2018). Best practices for analysing microbiomes. *Nature Reviews Microbiology, 16*(7), 410–422.

Kulkarni, A. S., Carrara, G. M. P., Jin, J., Laro, J., Peramuna, T., McCall, L. -I., & Garg, N. (2025). *Mass spectrometry-based metabolomics approaches to interrogate host–microbiome interactions in mammalian systems*. Natural Product Reports.

Kumar Pandey, P., Singh, S., Singh, M. C., Singh, A. K., Pandey, P., Pandey, A. K., Pathak, M., Kumar, M., Shakywar, R. C., & Patidar, R. K. (2017). Inside the plants: Bacterial endophytes and their natural products. *International Journal of Current Microbiology and Applied Sciences, 6*(6), 33–41. 10.20546/ijcmas.2017.606.003

Kumar, R., Swapnil, P., Meena, M., Selpair, S., & Yadav, B. G. (2022). Plant growth-promoting rhizobacteria (PGPR): Approaches to alleviate abiotic stresses for enhancement of growth and development of medicinal plants. *Sustainability (Switzerland), 14*(23). https://doi.org/10.3390/su142315514

Kumar, S., Aravinth, S. S., Kollem, S., Kumar, M., Naveed, Q. N., Mubarakali, A., Bhattacherjee, A., & Emma, A. F. (2025). Spatial attention-guided pre-trained networks for accurate identification of crop diseases. *Scientific Reports, 15*(1), 23213.

Kwak, M.-J., Kong, H. G., Choi, K., Kwon, S.-K., Song, J. Y., Lee, J., Lee, P. A., Choi, S. Y., Seo, M., Lee, H. J., Jung, E. J., Park, H., Roy, N., Kim, H., Lee, M. M., Rubin, E. M., Lee, S. W., & Kim, J. F. (2018). Rhizosphere microbiome structure alters to enable wilt resistance in tomato. *Nature Biotechnology, 36*(11), 1100–1109.

LeCun, Y., Bengio, Y., & Hinton, G. (2015). Deep learning. *Nature, 521*(7553), 436–444.

Li, M., Liu, J., Zhu, J., Wang, H., Sun, C., Gao, N. L., Zhao, X. M., & Chen, W. H. (2023). Performance of gut microbiome as an independent diagnostic tool for 20 diseases: Cross-cohort validation of machine-learning classifiers. *Gut Microbes*, *15*(1). https://doi.org/10.1080/19490976.2023.2205386

Li, F., Lyu, H., Li, H., Xi, K., Yi, Y., & Zhang, Y. (2024). Domestication and genetic improvement alter the symbiotic microbiome structure and function of tomato leaf and fruit pericarp. *Microorganisms*, *12*(7), 1351. https://doi.org/10.3390/microorganisms12071351

Li, F., Gao, W., Lin, Y., Liu, J., Su, Y., Zhang, Y., & Cai, K. (2025). Soil volatile organic compounds: Source-sink, function, mechanism, detection, and application analysis in environmental ecology. *TrAC Trends in Analytical Chemistry*, *184*, 118125. https://doi.org/10.1016/j.trac.2024.118125

Liu, S., Li, H.-Y., Zhang, T., Li, Y.-K., Liao, X.-D., & Xing, S.-C. (2025). Organic fertilizer enhances the secretion of microRNAs from tomato roots to facilitate beneficial rhizosphere microorganism expansion and suppress Ralstonia solanacearum proliferation. *Microbiome*, *13*(1), 159. https://doi.org/10.1186/s40168-025-02137-3

Lock, E. F., Hoadley, K. A., Marron, J. S., & Nobel, A. B. (2013). Joint and individual variation explained (JIVE) for integrated analysis of multiple data types. *The Annals of Applied Statistics*, *7*(1) https://doi.org/10.1214/12-AOAS597, 523–542.

Marbouty, M., Baudry, L., Cournac, A., & Koszul, R. (2017). Scaffolding bacterial genomes and probing host-virus interactions in gut microbiome by proximity ligation (chromosome capture) assay. *Science Advances*, *3*(2), e1602105.

Martinelli, F., Uratsu, S. L., Albrecht, U., Reagan, R. L., Phu, M. L., Britton, M., Buffalo, V., Fass, J., Leicht, E., Zhao, W., Lin, D., D'Souza, R., Davis, C. E., Bowman, K. D., & Dandekar, A. M. (2012). Transcriptome profiling of citrus fruit response to huanglongbing disease. *PLoS One*, *7*(5), e38039.

Mathé, E., Hays, J. L., Stover, D. G., & Chen, J. L. (2018). The omics revolution continues: The maturation of high-throughput biological data sources. *Yearbook of Medical Informatics*, *27*(01), 211–222.

Mendes, R., Kruijt, M., De Bruijn, I., Dekkers, E., Van Der Voort, M., Schneider, J. H. M., Piceno, Y. M., DeSantis, T. Z., Andersen, G. L., Bakker, P. A. H. M., & Raaijmakers, J. M. (2011). Deciphering the rhizosphere microbiome for disease-suppressive bacteria. *Science*, *332*(6033), 1097–1100. https://doi.org/10.1126/science.1203980

Mendes, R., Garbeva, P., & Raaijmakers, J. M. (2013). The rhizosphere microbiome: Significance of plant beneficial, plant pathogenic, and human pathogenic microorganisms. *FEMS Microbiology Reviews*, *37*(5), 634–663.

Mhlongo, M. I., Piater, L. A., Steenkamp, P. A., Labuschagne, N., & Dubery, I. A. (2020). Metabolic profiling of PGPR-treated tomato plants reveal priming-related adaptations of secondary metabolites and aromatic amino acids. *Metabolites*, *10*(5). https://doi.org/10.3390/metabo10050210

Mishra, B., Kumar, N., & Mukhtar, M. S. (2019). Systems biology and machine learning in plant–pathogen interactions. *Molecular Plant-Microbe Interactions*, *32*(1), 45–55.

Misra, B. B., Langefeld, C., Olivier, M., & Cox, L. A. (2019). Integrated omics: Tools, advances and future approaches. *Journal of Molecular Endocrinology*, *62*(1), R21–R45.

Mohamedikbal, S., Al-Mamun, H. A., Bestry, M. S., Batley, J., & Edwards, D. (2025). Integrating multi-omics and machine learning for disease resistance prediction in legumes. *Theoretical and Applied Genetics*, *138*(7), 1–21.

Mon, Y. Y., Bidabadi, S. S., Oo, K. S., & Zheng, S. J. (2021). The antagonistic mechanism of rhizosphere microbes and endophytes on the interaction between banana and *Fusarium oxysporum* f. sp. *cubense*. *Physiological and Molecular Plant Pathology*, *116*, 101733. https://doi.org/10.1016/j.pmpp.2021.101733

Mulaudzi, M. S., Nephali, L. P., & Tugizimana, F. (2025). AI-integrated metabolomics maps functional divergence of microbial consortia in field-grown maize. *Plant Cell Reports*, *44*(10) https://doi.org/10.1007/s00299-025-03600-z, 211.

Murmu, S., Sinha, D., Chaurasia, H., Sharma, S., Das, R., Jha, G. K., & Archak, S. (2024). A review of artificial intelligence-assisted omics techniques in plant defense: Current trends and future directions. *Frontiers in Plant Science*, *15*, 1292054.

Nandal, M., & Hooda, R. (2013). Plant growth promoting rhizobacteria: A review article. *International Journal of Current Research*, *5*(12), 3863–3871.

Neik, T. X., Dolatabadian, A., Danilevicz, M. F., Upadhyaya, S. R., Zhang, F., Batley, J., & Edwards, D. (2025). Plant disease epidemiology in the age of artificial intelligence and machine learning. *Agriculture Communications*, *3*, 100089.

Ngomle, S., Haokip, S. W., Singh, Y. D., Sheikh, K. H. A., Singh, S. R., & Katare, M. B. (2025). Omics approaches in plant disease management: An insightful review. *Physiological and Molecular Plant Pathology*, *140*, 102916.

Noecker, C., Eng, A., Srinivasan, S., Theriot, C. M., Young, V. B., Jansson, J. K., Fredricks, D. N., & Borenstein, E. (2016). Metabolic model-based integration of microbiome taxonomic and metabolomic profiles elucidates mechanistic links between ecological and metabolic variation. *MSystems*, *1*(1), 10–1128.

Novielli, P., Romano, D., Magarelli, M., Di Bitonto, P., Diacono, D., Chiatante, A., Lopalco, G., Sabella, D., Venerito, V., Filannino, P., Bellotti, R., De Angelis, M., Iannone, F., & Tangaro, S. (2024). Explainable artificial intelligence for microbiome data analysis in colorectal cancer biomarker identification. *Frontiers in Microbiology*, *15*. https://doi.org/10.3389/fmicb.2024.1348974

Novielli, P., Magarelli, M., Romano, D., Di Bitonto, P., Stellacci, A. M., Monaco, A., Amoroso, N., Bellotti, R., & Tangaro, S. (2025). Leveraging explainable AI to predict soil respiration sensitivity and its drivers for climate change mitigation. *Scientific Reports*, *15*(1), 1–13. https://doi.org/10.1038/s41598-025-96216-y

Pacheco, A. R., Moel, M., & Segrè, D. (2019). Costless metabolic secretions as drivers of interspecies interactions in microbial ecosystems. *Nature Communications*, *10*(1) https://doi.org/10.1038/s41467-018-07946-9, 103.

Palmer, A. G., Senechal, A. C., Mukherjee, A., Ané, J. M., & Blackwell, H. E. (2014). Plant responses to bacterial N-acyl l-homoserine lactones are dependent on enzymatic degradation to l-homoserine. *ACS Chemical Biology*, *9*(8), 1834–1845. https://doi.org/10.1021/cb500191a

Papoutsoglou, G., Tarazona, S., Lopes, M. B., Klammsteiner, T., Ibrahimi, E., Eckenberger, J., Novielli, P., Tonda, A., Simeon, A., Shigdel, R., Béreux, S., Vitali, G., Tangaro, S., Lahti, L., Temko, A., Claesson, M. J., & Berland, M. (2023). Machine learning approaches in microbiome research: Challenges and best practices. *Frontiers in Microbiology*, *14*(September). https://doi.org/10.3389/fmicb.2023.1261889

Parasar, B. J., Sharma, I., & Agarwala, N. (2024). Root exudation drives abiotic stress tolerance in plants by recruiting beneficial microbes. *Applied Soil Ecology*, *198*(May 2023), 105351. https://doi.org/10.1016/j.apsoil.2024.105351

Philippot, L., Raaijmakers, J. M., Lemanceau, P., & Van Der Putten, W. H. (2013). Going back to the roots: The microbial ecology of the rhizosphere. *Nature Reviews Microbiology*, *11*(11), 789–799. https://doi.org/10.1038/nrmicro3109

Pintelas, E., Livieris, I. E., & Pintelas, P. E. (2021). A convolutional autoencoder topology for classification in high-dimensional noisy image datasets. *Sensors (Basel)*, *21*(22), 7731.

Poudel, R., Jumpponen, A., Schlatter, D. C., Paulitz, T. C., Gardener, B. B. M., Kinkel, L. L., & Garrett, K. A. (2016). Microbiome networks: A systems framework for identifying candidate microbial assemblages for disease management. *Phytopathology*, *106*(10), 1083–1096.

Prelaj, A., Miskovic, V., Zanitti, M., Trovo, F., Genova, C., Viscardi, G., Rebuzzi, S. E., Mazzeo, L., Provenzano, L., Kosta, S., Favali, M., Spagnoletti, A., Castelo-Branco, L., Dolezal, J., Pearson, A. T., Lo Russo, G., Proto, C., Ganzinelli, M., Giani, C., ... Pedrocchi, A. L. G. (2024). Artificial intelligence for predictive biomarker discovery in immuno-oncology: A systematic review. *Annals of Oncology*, *35*(1), 29–65. https://doi.org/10.1016/j.annonc.2023.10.125

Przymus, P., Rykaczewski, K., Martín-Segura, A., Truu, J., De Santa, C., Pau, E., Kolev, M., Naskinova, I., Gruca, A., Sampri, A., Frohme, M., & Nechyporenko, A. (2024). Deep learning in microbiome analysis: A comprehensive review of neural network models. *Frontiers in Microbiology*, *15*. https://doi.org/10.3389/fmicb.2024.1516667

Przymus, P., Rykaczewski, K., Martín-Segura, A., Truu, J., De Santa, C., Pau, E., Kolev, M., Naskinova, I., Gruca, A., Sampri, A., Frohme, M., & Nechyporenko, A. (2025). Deep learning in microbiome analysis: A comprehensive review of neural network models. *Frontiers in Microbiology*, *15*. https://doi.org/10.3389/fmicb.2024.1516667

Qian, X., Zhang, C., Chen, L., & Li, K. (2022). Deep learning-based identification of maize leaf diseases is improved by an attention mechanism: Self-attention. *Frontiers in Plant Science*, *13*. https://doi.org/10.3389/fpls.2022.864486

Rahman, K. N., Banik, S. C., Islam, R., & Al Fahim, A. (2025). A real time monitoring system for accurate plant leaves disease detection using deep learning. *Crop Design*, *4*(1), 100092.

Rajguru, B., Shri, M., & Bhatt, V. D. (2024). Exploring microbial diversity in the rhizosphere: A comprehensive review of metagenomic approaches and their applications. *3 Biotech*, *14*(10), 224.

Rokach, L. (2010). Ensemble-based classifiers. *Artificial Intelligence Review*, *33*(1), 1–39.

Rudin, C. (2019). Stop explaining black box machine learning models for high stakes decisions and use interpretable models instead. *Nature Machine Intelligence*, *1*(5), 206–215.

Sahoo, A., Yadav, G., Mehta, T., Meena, M., & Swapnil, P. (2025). Omics-driven insights into plant growth-promoting microorganisms for sustainable agriculture. *Discover Sustainability*, *6*(1), 659. https://doi.org/10.1007/s43621-025-01582-2

Sanadhya, P., Kumar, A., Bucki, P., Fitoussi, N., Carmeli-Weissberg, M., Borenstein, M., & Brown-Miyara, S. (2021). Tomato divinyl ether-biosynthesis pathway is implicated in modulating of root-knot nematode meloidogyne javanica's parasitic ability. *Frontiers in Plant Science*, *12*. https://doi.org/10.3389/fpls.2021.670772

Sartori, F., Codicè, F., Caranzano, I., Rollo, C., Birolo, G., Fariselli, P., & Pancotti, C. (2025). A comprehensive review of deep learning applications with multi-omics data in cancer research. *Genes*, *16*(6), 648. https://doi.org/10.3390/genes16060648

Satya, S., Sharma, S., Choudhary, G., & Kaushik, G. (2024). Advances in environmental microbiology: A multi-omic perspective. In M. Kesheri, S. Kanchan, T. B. Salisbury, & R. P. Sinha (Eds.), *Microbial omics in environment and health* (pp. 175–204). Springer.

Scherlach, K., & Hertweck, C. (2018). Mediators of mutualistic microbe-microbe interactions. *Natural Product Reports, 35*(4), 303–308. https://doi.org/10.1039/c7np00035a

Singh, D., Geat, N., Mehriya, M., Rajawat, M. V. S., Prasanna, R., Kumar, A., Kumari, G., & Jha, M. N. (2021). Omics (genomics, proteomics, metabolomics, etc.) tools to study the environmental microbiome and bioremediation. In B. K. Kashyap, M. K. Solanki, D. V. Kamboj, & A. K. Pandey (Eds.), *Waste to energy: Prospects and applications* (pp. 235–260). Springer.

Singh, K., Chandra, R., & Purchase, D. (2022). Unraveling the secrets of rhizobacteria signaling in rhizosphere. *Rhizosphere, 21*(February), 100484. https://doi.org/10.1016/j.rhisph.2022.100484

Sutaji, D., & Rosyid, H. (2022). Convolutional neural network (CNN) models for crop diseases classification. Kinetik : Game Technology, Information System, Computer Network, Computing, Electronics, and Control.

Tan, A. W., Malik, D. M., Bottalico, L. N., Sengupta, A., & Weljie, A. M. (2022). A primer for circadian metabolic profile analysis using multi-platform metabolomics. In T. Hirota, M. Hatori, & S. Panda (Eds.), *Circadian clocks* (pp. 333–357). Springer.

Tariq, A., Guo, S., Farhat, F., & Shen, X. (2025). Engineering synthetic microbial communities: Diversity and applications in soil for plant resilience. *Agronomy, 15*(3), 513. https://doi.org/10.3390/agronomy15030513

Tarnita, C. E. (2017). The ecology and evolution of social behavior in microbes. *Journal of Experimental Biology, 220*(1), 18–24. https://doi.org/10.1242/jeb.145631

Theodorakis, N., Feretzakis, G., Tzelves, L., Paxinou, E., Hitas, C., Vamvakou, G., Verykios, V. S., & Nikolaou, M. (2024). Integrating machine learning with multi-omics technologies in Geroscience: Towards personalized medicine. *Journal of Personalized Medicine, 14*(9), 931. https://doi.org/10.3390/jpm14090931

Thiergart, T., Durán, P., Ellis, T., Vannier, N., Garrido-Oter, R., Kemen, E., Roux, F., Alonso-Blanco, C., Ågren, J., Schulze-Lefert, P., & Hacquard, S. (2020). Root microbiota assembly and adaptive differentiation among European arabidopsis populations. *Nature Ecology & Evolution, 4*(1), 122–131. https://doi.org/10.1038/s41559-019-1063-3

Thingujam, D., Gouli, S., Cooray, S. P., Chandran, K. B., Givens, S. B., Gandhimeyyan, R. V., Tan, Z., Wang, Y., Patam, K., & Greer, S. A. (2025). Climate-resilient crops: Integrating AI, multi-omics, and advanced phenotyping to address global agricultural and societal challenges. *Plants, 14*(17), 2699.

Tibshirani, R. (1996). Regression shrinkage and selection via the lasso. *Journal of the Royal Statistical Society, Series B: Statistical Methodology, 58*(1), 267–288. https://doi.org/10.1111/j.2517-6161.1996.tb02080.x

Tracanna, V. (2022). *Dissecting disease-suppressive rhizosphere microbiomes using metagenomics* [Wageningen University]. https://doi.org/10.18174/561495

Trivedi, P., Leach, J. E., Tringe, S. G., Sa, T., & Singh, B. K. (2020). Plant–microbiome interactions: From community assembly to plant health. *Nature Reviews Microbiology, 18*(11), 607–621.

Trivedi, P., Narayan, Y., Ravi, V., Kumar, P., Kaur, P., Tabianan, K., Singh, P., & Diwakar, M. (2024). Plant leaf disease detection and classification using segmentation encoder techniques. *The Open Agriculture Journal*, *18*(1).

Tsimenidis, S., Vrochidou, E., & Papakostas, G. A. (2022). Omics data and data representations for deep learning-based predictive modeling. *International Journal of Molecular Sciences*, *23*(20), 12272.

Turner, T. R., James, E. K., & Poole, P. S. (2013). The plant microbiome. *Genome Biology*, *14*(6), 1–10. https://doi.org/10.1186/gb-2013-14-6-209

Tyler, S. R., Chun, Y., Ribeiro, V. M., Grishina, G., Grishin, A., Hoffman, G. E., Do, A. N., & Bunyavanich, S. (2021). Merged affinity network association clustering: Joint multi-omic/clinical clustering to identify disease endotypes. *Cell Reports*, *35*(2), 108975.

Upadhyaya, S. R., Danilevicz, M. F., Dolatabadian, A., Neik, T. X., Zhang, F., Al-Mamun, H. A., Bennamoun, M., Batley, J., & Edwards, D. (2024). Genomics-based plant disease resistance prediction using machine learning. *Plant Pathology*, *73*(9), 2298–2309.

Vitorino, R. (2024). Transforming clinical research: The power of high-throughput omics integration. *Proteomes*, *12*(3), 25.

Walsh, C., Stallard-Olivera, E., & Fierer, N. (2024a). Nine (not so simple) steps: A practical guide to using machine learning in microbial ecology. *MBio*, *15*(2). https://doi.org/10.1128/mbio.02050-23

Walsh, S., Wallace, L., Kuhn, I., Mytton, O., Lafortune, L., Wills, W., Mukadam, N., & Brayne, C. (2024b). Population-level interventions for the primary prevention of dementia: A complex evidence review. *EClinicalMedicine*, *70*(ii), 102538. https://doi.org/10.1016/j.eclinm.2024.102538

Walters, W. A., Jin, Z., Youngblut, N., Wallace, J. G., Sutter, J., Zhang, W., González-Peña, A., Peiffer, J., Koren, O., Shi, Q., Knight, R., Glavina del Rio, T., Tringe, S. G., Buckler, E. S., Dangl, J. L., & Ley, R. E. (2018). Large-scale replicated field study of maize rhizosphere identifies heritable microbes. *Proceedings of the National Academy of Sciences*, *115*(28), 7368–7373.

Whited, T. M. (2022). Parallels between structural estimation and causal inference: A discussion of Armstrong et al. (2022). *Journal of Accounting and Economics*, *74*(2–3), 101541. https://doi.org/10.1016/j.jacceco.2022.101541

Wolf, J. B. W. (2013). Principles of transcriptome analysis and gene expression quantification: An RNA-seq tutorial. *Molecular Ecology Resources*, *13*(4), 559–572.

Xiong, C., Singh, B. K., He, J. Z., Han, Y. L., Li, P. P., Wan, L. H., Meng, G. Z., Liu, S. Y., Wang, J. T., Wu, C. F., Ge, A. H., & Zhang, L. M. (2021). Plant developmental stage drives the differentiation in ecological role of the maize microbiome. *Microbiome*, *9*(1), 1–15. https://doi.org/10.1186/s40168-021-01118-6

Xu, L., Pierroz, G., Wipf, H. M.-L., Gao, C., Taylor, J. W., Lemaux, P. G., & Coleman-Derr, D. (2021). Holo-omics for deciphering plant-microbiome interactions. *Microbiome*, *9*(1), 69.

Yang, L., Qian, X., Zhao, Z., Wang, Y., Ding, G., & Xing, X. (2024). Mechanisms of rhizosphere plant-microbe interactions: Molecular insights into microbial colonization. *Frontiers in Plant Science*, *15*, 1–12. https://doi.org/10.3389/fpls.2024.1491495

Yetgin, A. (2025). Revolutionizing multi-omics analysis with artificial intelligence and data processing. *Quantitative Biology*, *13*(3), e70002.

Yin, J., Zhang, Z., Zhu, C., Wang, T., Wang, R., & Ruan, L. (2022). Heritability of tomato rhizobacteria resistant to *Ralstonia solanacearum*. *Microbiome*, *10*(1), 227.

Yoshida, T., Mergner, J., Yang, Z., Liu, J., Kuster, B., Fernie, A. R., & Grill, E. (2024). Integrating multi-omics data reveals energy and stress signaling activated by abscisic acid in arabidopsis. *The Plant Journal*, *119*(2), 1112–1133. https://doi.org/10.1111/tpj.16765

Yusuf, A., Li, M., Zhang, S.-Y., Odedishemi-Ajibade, F., Luo, R.-F., Wu, Y.-X., Zhang, T.-T., Yunusa Ugya, A., Zhang, Y., & Duan, S. (2025). Harnessing plant–microbe interactions: Strategies for enhancing resilience and nutrient acquisition for sustainable agriculture. *Frontiers in Plant Science*, *16*. https://doi.org/10.3389/fpls.2025.1503730

Zhang, G., Bai, J., Zhai, Y., Jia, J., Zhao, Q., Wang, W., & Hu, X. (2024a). Microbial diversity and functions in saline soils: A review from a biogeochemical perspective. *Journal of Advanced Research*, *59*, 129–140. https://doi.org/10.1016/j.jare.2023.06.015

Zhang, Y., Schluter, J., Zhang, L., Cao, X., Jenq, R. R., Feng, H., Haines, J., & Zhang, L. (2024b). Review and revamp of compositional data transformation: A new framework combining proportion conversion and contrast transformation. *Computational and Structural Biotechnology Journal*, *23*(November), 4088–4107. https://doi.org/10.1016/j.csbj.2024.11.003

Zhang, X., Rahnavard, A., & Crandall, K. (2025). Machine learning enhances biomarker discovery: From multi- omics to functional genomics. *Medical Research Archives*, *13*(8). 10.18103/mra.v13i8.6888

Zhao, L., Walkowiak, S., & Fernando, W. G. D. (2023). Artificial intelligence: A promising tool in exploring the phytomicrobiome in managing disease and promoting plant health. *Plants*, *12*(9), 1852.

Zhu, S., Xu, H., Liu, Y., Hong, Y., Yang, H., Zhou, C., & Tao, L. (2025). Computational advances in biosynthetic gene cluster discovery and prediction. *Biotechnology Advances*, *79*, 108532. https://doi.org/10.1016/j.biotechadv.2025.108532

12

AI-Assisted Omics Tools for Predicting Functions of Plant RNAs

Ather Manzoor[1], Umer Fayaz[2], Nelofar Lone[1], Asma Majid[1], Showkat A. Waza[1,2], and A. K. M. Aminul Islam[3]

[1] *Division of Genetics and Plant Breeding, Sher-e-Kashmir University of Agricultural Sciences & Technology of Kashmir, Jammu and Kashmir, India*
[2] *Accelerated Crop Improvement, International Crops Research Institute for the Semi-Arid Tropics, Patancheru, Telangana, India*
[3] *Department of Genetics and Plant Breeding, Gazipur Agricultural University, Gazipur, Bangladesh*

12.1 Introduction

Ribonucleic acid (RNA) plays an important role in plants. RNA also plays many other important roles beyond being the carrier of genetic information. Messenger RNA acts as an essential link between DNA and ribosomes, transferring the genetic information encoded in DNA and regulating the synthesis of proteins (Bai et al., 2024). Long noncoding RNAs (lncRNAs) modulate DNA structure and control gene expression, while microRNAs (miRNAs) are short RNA fragments that regulate post-transcription gene activity by guiding RNA-induced silencing complexes to specific mRNA targets (Chekulaeva & Rajewsky, 2019; Yadav et al., 2024). Associates with specific mRNAs inhibit translation, helping plants maintain their growth and developmental processes. Circular RNAs (circRNAs) represent another group of noncoding RNAs that are involved in the regulation of gene transcription by engaging with RNA-binding proteins to regulate the activity of cellular genes. Transfer RNAs (tRNAs), ribosomal RNAs (rRNAs), and mRNA carry genetic information into proteins (Gao et al., 2025). In this process, the tRNA supplies the amino acids, and the rRNA facilitates peptide bond formation. The variation and flexibility in the structure of these RNAs enable them to play an important role in the growth, adaptation, and survival of the plant. Based on complexes of omics data, advanced artificial intelligence (AI) algorithms predict the structures and functions of RNA, thus helping in the identification of new regulatory networks. Various platforms like genomics, transcriptomics, and epigenomics help in the development of better models of gene regulation, which play an essential role in developing stress-tolerant and high-yielding crop varieties. The different methods, which include single-molecule fluorescence microscopy and cryo-electron microscopy, provide the RNA localization and three-dimensional structure, facilitating important findings into their functional mechanisms.

12.2 Omics Data in Plant RNA Studies

Omics studies the entire set of biological molecules, including DNA, RNA, proteins, and metabolites within a cell or tissue of an organism and reveals the functions of the molecules in different biological processes (Samarskaya et al., 2025). Omics is based on advanced instruments and high-throughput platforms, which include genome-wide sequencing and mass spectrometry. Genomics studies the entire DNA sequences of the plant and identifies the various genetic variations present that result from major chromosomal rearrangements to single nucleotide polymorphism (SNPs). The variations present in the genetic basis of many potential traits include disease resistance, yield, or stress adaptation and are an important source of genetic diversity. Transcriptomics studies the genes that are actively expressed under specific developmental stages and are actively transcribed into RNA. The transcriptomic profiling is done using RNA sequencing (RNA-*seq*), which reads millions of RNA fragments to measure gene expression in the whole plant or in specific tissues at different developmental stages (Table 12.1). Other tools, including single-cell RNA-*seq*, provide a detailed view of gene expression in each cell, helping in the characterization of rare cell types and processes (Sanches et al., 2024).

Epigenomics studies heritable changes in gene expression that occur without alterations in the DNA sequence, primarily through mechanisms such as DNA methylation and histone modifications. These changes improve the molecular stability and functional competence of RNA (Chen et al., 2024; Wei 2024), therefore significantly contributing to the process of controlling plants' growth, developmental patterns, and adaptation to environmental stress factors. Proteomics deals with the study of interactions of proteins with different kinds of RNA, which include messenger RNA, transfer RNA, and noncoding RNAs, to regulate gene expression and provide a detailed map of these molecular interactions across different tissues (Hafner et al., 2021; Kapral et al., 2022). The integration of AI with multilayered omics datasets, which include genomics, transcriptomics, proteomics, epigenomics, and phenomics, helps in the efficient prediction of the role of RNAs in complex biological processes like drought resistance and pathogen defense, enabling early detection (Murmu et al., 2024; Yu et al., 2024).

12.3 AI Methodologies in Omics Analysis

AI has emerged as a powerful tool that helps scientists decipher how the complex mechanisms of plant RNA work, especially when dealing with unorganized huge datasets that resemble the chaos of some unsorted library (Murmu et al., 2024; Yu et al., 2024). The supervised learning approach has become a widely adopted methodology in this domain, in which computational models are trained using well-characterized, for example, RNA molecules that have an important role in the mechanism of protein synthesis. These models can identify underlying patterns autonomously following this training phase, demonstrating the capacity to analyze novel RNA sequences and make predictions about their functional characteristics based on previously processed/acquired knowledge (Murmu et al., 2024). Deep learning has proven transformative, particularly in this field, by employing sophisticated algorithms that easily replicate the different aspects of neural processing

Table 12.1 Omics tools, focus areas, technological platforms, data output, and applications in plant RNA studies.

Omics tools	Focus area	Technologies or platform	Data output	Applications	Examples
Phenomics	HTP-trait phenotyping	Imaging, sensors, robotics	Quantitative traits	Integrating transcriptomic and phenotypic response to stress	Leaf morphology in rice
Genomics	Whole-genome sequence	WGS, PacBio, Illumina Sequencing	Gene model, exon-intron architecture, promoter elements	Mapping ncRNA, RNA gene annotation	Identification of miRNA in maize
Transcriptomics	RNA expression profiles	RNA-seq, Microarray, cDNA-seq	Transcript abundance, alternative splicing	lncRNA, miRNA identification, mRNA expression under stress	Drought-responsive lncRNA in rice
Proteomics	Protein expression, post-translational modifications	LC–MS/MS, 2D-GE, iTRAQ	Protein abundance and modification profiles	RNA–protein interaction networks	RNA-binding protein in RNA stability
Metabolomics	Metabolite profile linked to RNA	GC–MS, LC–MS, NMR	Metabolite abundance, flux profiles	RNA expression under stress	Metabolic pathways in maize under drought
Interactomics	RNA–protein, RNA–DNA, RNA–RNA networks	RIP-seq, CLIP-seq, RNA assay	Interaction networks	Characterization of lncRNA–protein and miRNA–mRNA interactions	RNA–protein interaction modules
Epigenomics	DNA methylation, histone modification	Bisulfate sequence, ChIP-seq, ATAC-seq	Methylation maps, histone marks profiles	Epigenetic marks, ncRNA expression, RNA gene silencing	lncRNA expression in soybean under salt stress

that occurs in the human brain through interconnected layers that can process information sequentially. Convolutional neural networks (CNNs) and recurrent neural networks (RNNs) are excellent tools for analyzing both the sequential composition and structural architecture of RNA molecules with accurate precision comparable to examining text through high-powered magnification (Kajrolkar, 2025). These models excel at pointing out the hidden patterns that are embedded within complex datasets that can escape human observation, thereby facilitating new scientific insights (Kajrolkar, 2025; Murmu et al., 2024). Transfer learning offers another strategic advantage by enabling researchers to leverage existing knowledge from characterized model organisms, for example, *Arabidopsis thaliana*, whose whole biology has been documented thoroughly over decades, and helping them apply these insights to less-studied species (Cao et al., 2025). But when data availability is limited for particular plant taxa, this approach helps in solving this problem by attempting specimen identification from incomplete or degraded species by allowing cross-application of biological principles to overcome data scarcity in under-represented species.

12.4 AI-Assisted Omics Tools for Predicting Plant RNA Functions

12.4.1 Plant Long Noncoding RNA Prediction by Random Forests9 (PLncPRO)

PLncPRO was initially presented in the journal *Plant Physiology* in 2019, which provides an effective way to identify lncRNAs in plants, similar to identifying a single green thread in a tangle of roots. Despite never producing proteins, these lncRNAs are a unique type of RNA that play a crucial role in controlling gene activity, such as turning certain genes on or off at the right moment (Palos et al., 2023). These regulatory switches do not set anything up on their own, but they do control when and how other sections of the plant's instruction manual, the genome, get put to work, like deciding exactly when a leaf should start to unfurl (Mansuri et al., 2022). Spotting lncRNAs is not easy, especially in plants, where RNA can mimic protein-coding strands in several ways, even though key differences set them apart (Palos et al., 2023). PLncPRO relies on a powerful deep learning method called bidirectional long short-term memory (BiLSTM) (Cheng et al., 2021). Modern deep learning models for the classification of RNA are utilizing bidirectional sequence scanning, such as the Bidirectional Anchored Generation (RNA-BAnG) model, which has efficiently classified coding and noncoding RNAs. It utilizes geometric attention and cross-modal learning. The model has been trained on sequence data and predicted protein structures (AlphaFold2) to reveal the sequence pattern for the RNA–protein interaction. This helps in supporting the advanced scientific progress in RNA biology and synthetic biology by overcoming the limitations of earlier models (Cheng et al., 2021). lncRNAs play an important role in plants' development, stress resilience, and adaptation to diverse environmental conditions. lncRNAs coordinate gene expression through complex regulatory networks controlling growth and flowering transitions in plants and also regulating responses to drought and nutrient deficiency (Mansuri et al., 2022; Palos et al., 2023). The PLncPRO tool has been evaluated on model plants, such as *A. thaliana*, and can identify lncRNAs with an accuracy of 95% outperforming many already well-known established models (Cheng et al., 2021). PLncPRO is an effective tool for the identification and analysis of lncRNAs in plants. PreLnc2 is also a commonly used tool that improves lncRNA detection across many species. DeepPlnc utilizes deep learning models, including LSTM and CNN, to identify lncRNA with high accuracy. LncFinder is a sequence-based system that facilitates the classification of lncRNAs. PLncDB and CANTATAdb provide a large dataset of plant lncRNAs from many species, with annotation tools that help in studying the functional and evolutionary aspects of lncRNAs. PlantLncBoost and Plant-LncPipe provide ML-based tools that help in the identification and analysis of plant lncRNA datasets (Singh et al., 2017).

12.4.2 Plant Target Prediction for microRNAs (P-TarPmiR)

MicroRNAs are small noncoding RNAs about 20–24 nucleotides long, playing an important role in gene regulation and expression post-transcription in plants. These small RNA strands regulate the expression of messenger RNA (mRNA) controls, effectively controlling the activity of target genes. miRNAs associate with plant hormone signaling pathways, affect plant hormone responses, and help the plant adapt to diverse environmental changes (Ajila et al., 2023). Understanding plants' defense responses to biotic and abiotic stresses, growth control, and evolutionary adaptation over time mainly depends on accurately identifying miRNA targets (Ajila et al., 2023). Methods such as ITS-CLIP, CLASH, and degradome sequencing provide accurate miRNA identification, but these methods are often costly and time-consuming. Unlike other methods, bioinformatic tools facilitated by AI, such as P-TarPmiR, are fast and reliable, trained on plant data to predict miRNA targets quickly without compromising accuracy (Ding et al., 2016). P-TarPmiR is an advanced machine learning (ML) tool designed to predict miRNA–mRNA interactions specific to plants, identifying matches much like finding the right key for a single lock (Ajila et al., 2023). This updated TarPmiR model builds on the original design, first created for animal studies (Ding et al., 2016), much like refining a tool that once worked only in a single lab. We retrained the model with a plant-focused dataset from TarDB that holds tens of thousands of confirmed plant miRNA–mRNA interactions, enough to fill page after page of dense, annotated sequences (Ajila et al., 2023). Instead of the older, rule-heavy approaches that depend on fixed parameters such as seed pairing or target-site accessibility, P-TarPmiR uses a random forest classifier that draws on a wide mix of biological traits folding energy, AU content, seed-match quality, and more like reading the subtle twists in a strand of RNA (Ajila et al., 2023; Ding et al., 2016). Experiments conducted using P-TarPmiR, a plant microRNA target prediction method in four diverse plant species, soybean, brachypodium, rice, and poplar, showed that models trained on plant interaction networks produced better results than models trained on human or mixed-species data. Using only plant-specific data and avoiding animal-driven results provides more precision in predicting plant miRNA targets (Ajila et al., 2023). The plant-specific P-TarPmiR showed better performance in predicting mRNA targets, reaching a high precision score ranging from 0.984 to 1.0 and recall values between 0.993 and 0.997. The scores obtained were better than the already known established models, including psRNATarget and TargetFinder, showing the superiority of plant-specific models in predicting plant miRNA targets. Unlike conventional prediction tools, P-TarPmiR is an important bioinformatics tool that is mainly focused on sequence matching. Besides P-TarPmiR, the other tools available, psRNATarget and TargetFinder, rely more on classic sequence-pairing rules to predict plant miRNA–mRNA interactions or identify noncoding RNAs.

Advanced ML tools DeepTarget and PlantMirTar utilize ML algorithms evaluated under experimental datasets in order to improve sensitivity and specificity. Combining predictions from multiple sources, including datasets such as psRNATarget and databases like PMTED and PNRD, provides a more comprehensive understanding of miRNA regulatory networks. The P-TarPmiR tool is highly effective in enhancing our understanding of plant gene regulation, particularly in soybean (*Glycine max*). The P-TarPmiR model accurately identified a large number of miRNA targets, which were further examined for gene

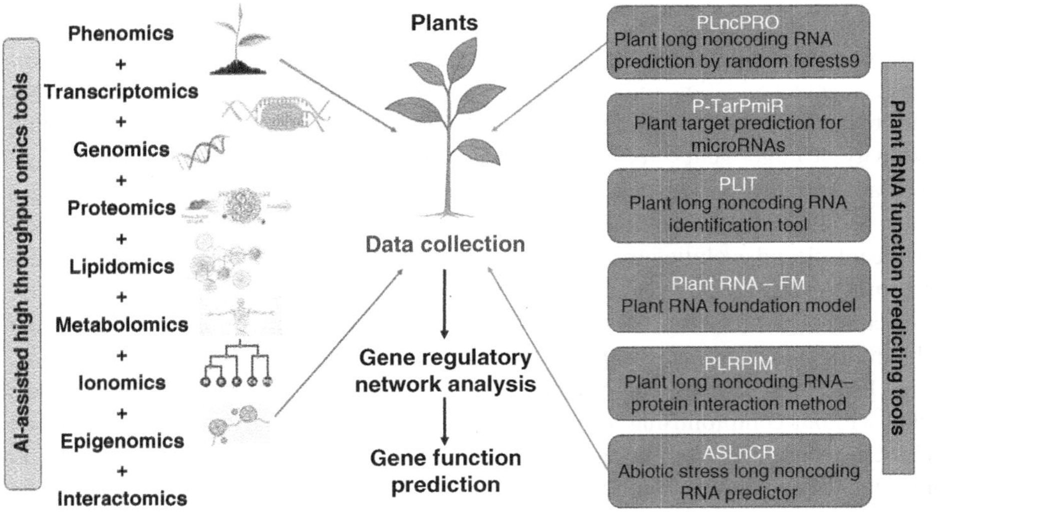

Figure 12.1 AI-assisted omics tools for analyzing gene regulatory networks and predicting plant gene functions.

ontology enrichment patterns. This method enables the process of identification of various biological processes, including stress responses and developmental processes. Analyzing complex data could transform crop improvement strategies to identify the regulatory pathways that control plant growth, adaptation, and yield (Figure 12.1).

12.4.3 Plant Long Non-coding RNA Identification Tool (PLIT)

The Plant lncRNA Identification Tool, or PLIT, represents an important development in computational biology that uses ML algorithms to find and study lncRNAs within plant genomes with great accuracy (Deshpande et al., 2019). Although these lncRNAs do not encode proteins, they do perform important regulatory functions by modulating gene expression patterns, which effectively controls which genes are to be activated and which ones are to remain dormant. This regulatory capacity is crucial, particularly during the growth and development of plants, and even when plants face stresses such as drought or nutrient-poor soil. Conventional techniques/tools were designed using an animal dataset, which resulted in their poor performance when they were applied to plant data, flooding RNA-seq results with false positives (Deshpande et al., 2019). PLIT solves this problem by using its plant-specific design to distinguish between protein-coding and noncoding RNA transcripts by utilizing ML at its best (Deshpande et al., 2019). The framework examines 73 distinct sequence attributes for each RNA sequence, including parameters such as transcript length, open reading frame (ORF) dimensions, Guanine–Cytosine (GC) content, and codon usage patterns, including the relative frequency of GC versus AT nucleotides (Deshpande et al., 2019). PLIT uses a random forest classifier to classify RNAs into coding or noncoding groups once the most informative features are identified and selected. Researchers can choose between two operational modes—an expedited analysis (rapid one) that uses 7 key features or a comprehensive evaluation (detailed mode) which incorporates 31 features—allowing researchers to optimize the tradeoff between computational efficiency and classification accuracy (Deshpande et al., 2019). PLIT performed a lot better than tools such as CPC2, CPAT, PLEK, and lncScore in tests on plants including *Arabidopsis*, rice, maize, and tomato, obtaining over 96% accuracy in maize (Deshpande et al., 2019). Avoiding slow sequence alignments like BLAST and quickly identifying novel lncRNAs, even those hidden in a noisy read file, are some advantages of PLIT. It can reveal lncRNAs directly from raw RNA-seq data (Deshpande et al., 2019). PLIT utilizes AI and advanced ML models in an accurate and efficient way in order to identify important noncoding RNAs that play an essential role in plant development and tolerance to biotic and abiotic stresses. This approach not only facilitates studying the gene function of regulatory RNAs but also plays a major role in crop improvement efforts, particularly in the face of changing climatic conditions. It also uses AI to help scientists quickly and precisely identify important noncoding RNAs that promote plant development and resilience, facilitating crop improvement even under the harsh conditions of climate change (Deshpande et al., 2019).

12.4.4 Plant RNA-FM (Plant RNA Foundation Model)

The plant RNA foundation model is an ML tool that can process the difficult language in plant RNA sequences and their 3D structural configurations. It has been trained on an extensive dataset of RNA that includes a dataset covering more than a thousand plant species, making it capable of capturing diversity in plant transcriptomes, including the small variations in gene expression patterns that occur during different physiological states (Yu et al., 2024). One of the features of this tool that sets it apart from others is its integrated approach, combining sequence information with three-dimensional structural data that improves predictions of gene regulatory properties and enhances translation efficiency across major eukaryotic lineages. Because of this, the predictions about gene regulation and protein synthesis get more accurate as both aspects are considered together across many groups of organisms (Yu et al., 2024). This is also of great value as it results in the analysis of minor alterations in RNA strands that can result in significant functional consequences (Yu et al., 2024). The strength of this framework lies in its ability to interpret data, enabling researchers to identify important regulatory RNA motifs governing gene expression (Yu et al., 2024). Plant RNA-FM has successfully identified and analyzed fresh RNA motifs that are linked to gene regulation. These include well-classified conserved sequences that are found near the initiation site of translation. The model uncovers patterns that would usually not be observed through conventional approaches by systematically identifying and mapping these regulatory elements (Yu et al., 2024). Plant RNA-FM is a methodology that efficiently maps all the structural features, such as stem loops and hairpins, that traditional methods are usually not able to detect. This design contributes to both the accuracy and reliability of its predictions (Yu et al., 2024). This tool has shown great success in identifying novel RNA regulatory motifs that play important roles in plant development and environmental responses, including reactions to light and changes in temperature. Such discoveries are important for agricultural enhancement as they contribute greatly toward the development of crop varieties having more tolerance to abiotic stresses such as drought and nutrient deficiency through the identification of previously unstable regulatory sequences (Yu et al., 2024). Plant RNA-FM is thus an important development in plant molecular biology that has provided researchers with new ways to understand the regulatory mechanisms encoded within plant transcriptomes across diverse species (Yu et al., 2024).

12.4.5 Plant Long Noncoding RNA–Protein Interaction Method (PLRPIM)

Understanding the functions of RNA molecules in plants has never been easy, particularly IncRNAs; it has always been a major scientific challenge. Though RNAs do not encode proteins, they do play a critical regulatory role in plants by modulating gene expression, functioning like molecular switches within plant cells. To solve this difficulty of predicting the interaction of such lncRNAs with cellular proteins, PLRPIM, an AI-assisted framework, was developed for plants. This approach enables rapid prediction of the RNA–protein interaction by integrating advanced deep learning methods with classical learning algorithms, such as random forests and gradient boosting (Wekesa et al., 2019). Transforming RNA and protein into compressed digital representations, capturing essential structural and sequence-based features, it converts key biological information into simplified, learnable patterns as its first step. It then uses an autoencoder network to find small and complex associations that many tools typically overlook. Instead of working on a single-model approach, decisions aggregated by PLRPIM are from multiple predictive models, which enhances reliability and sensitivity in identifying authentic RNA–protein interactions even within highly complex or unarranged datasets (Wekesa et al., 2019). Researchers compared PLRPIM to conventional tools such as IPMiner and RPISeq-RF after testing it on experimental datasets from familiar plants like Arabidopsis, which have small seeds and tall corn stalks. PLRPIM predicted interactions in maize with an accuracy of over 90%. What makes PLRPIM particularly valuable is its adaptability, as it performs reliably across different genomic contexts, whether one is working with a well-annotated genome or one that is only partially mapped. With this flexibility, scientists can easily explore RNA–protein interactions even in non-model plants or orphan plants that usually lack comprehensive genomic data. By doing so, PLRPIM provides researchers with new avenues for investigating how genes are regulated in wild or represented species that they have not been able to do research on and have limited research attention (Wekesa et al., 2019). Most regulatory RNAs in plants are still not defined with their roles, leaving a major gap in the understanding of their significance in biological aspects. Tools like PLRPIM provide a breakthrough by acting as a bridge between vast genomic data and functional insights in plant science. It enables fast and accurate prediction of RNA–protein interactions, thus helping researchers in understanding raw sequencing outputs by translating them into meaningful biological knowledge (Wekesa et al., 2019). This development of PLRPIM is bridging the gap between sequencing technologies and biological insights, with great potential to impact future agricultural practices by supporting developments in crop improvement, sustainability, and adaptation to environmental stresses, such as identifying genes that could enhance frost tolerance. PLRPIM simplifies complex datasets, allowing researchers to analyze leaf morphology in the laboratory or monitor soil dynamics in the field, making molecular data more accessible and interpretable (Wekesa et al., 2019).

12.4.6 Abiotic Stress Long Non-coding RNA Predictor (ASLnCR)

ASLnCR is a cutting-edge AI-based tool that is helpful in predicting how lncRNAs in plants help them cope with tough conditions like drought, salinity (salt water), heat, and cold conditions, including frost. ASLnCR uses advanced ML-supporting vector machines trained on carefully selected plant RNA-sequencing data, letting it classify/differentiate stress-responsive lncRNAs apart from those that are not stress-responsive (Pradhan et al., 2023). ASLnCR processes by analyzing and identifying small sequence details and expression patterns that are unique to stress-related lncRNAs, hence overcoming the limitations of previous models that only relied on animal-derived or generalized datasets. During its development, ASLnCR was trained on a dataset that had an equal number of stress-responsive and non-responsive lncRNAs, which helps it in identifying important sequence motifs and structural details, like a unique loop in the RNA that plays a vital role in stress adaptation. By removing information that is not useful, feature selection identifies the most promising k-mer patterns, which is similar to finding rare sequences within genes, which helps improve model accuracy and prediction reliability (Pradhan et al., 2023). The final classifier that we get performs better than traditional IncRNA prediction tools, providing plant researchers with a promising source for investigating gene regulation under stress conditions such as drought. ASLnCR is a web-based platform that allows researchers to upload transcriptomic data and helps in quick prediction of regulatory noncoding RNAs (IncRNAs) associated with abiotic stress tolerance in plants, improving the process of identification of major bio-molecules that include drought-resistant proteins, making it valuable for crop improvement because it can rapidly detect functionally relevant IncRNAsn (Murmu et al., 2024; Pradhan et al., 2023). In addition to supporting basic research, this platform helps us clarify and understand lncRNAs that are stress-sensitive and regulate the expression of genes, thus providing information and allowing for the development of crop cultivars that could adapt to changing climatic conditions and extreme environmental conditions. This platform is an example of how a tool can combine multiomics data with AI and, as such, offers accurate, species-specific results that can provide us with information on the regulatory complexities of lncRNAs. These developments/advances are crucial for decoding the molecular responses occurring in plant tissue under different stress conditions, ultimately contributing to improved and stress-tolerant crops with better performance.

12.4.7 Alternative Splicing and microRNA Interaction Resource (ASmiR)

ASmiR is an AI-based system developed to predict the function of plant microRNAs. These miRNAs help plants adapt to various biotic and abiotic stresses, which include high temperature, extremely cold weather, drought, and salinity. ASmiR is one of the first ML-based approaches for predicting miRNAs related to plant stresses and uses advanced computer tools to detect patterns in biological data. This model uses pseudo-K-tuple nucleotide composition (PseKNC) to convert miRNA sequences into numerical features and then helps in the identification of the most informative variant by feature selection. Several ML methods have been tested, with the support vector machine (SVM) performing the best and having a prediction accuracy of 90% for all the stresses (cold, drought, heat, and salt). It is easily accessible through an online web server (https://iasri-sg.icar.gov.in/asmir/). This method helps in the prediction of possible stress relationships based on miRNA sequences. Post-transcriptional plant miRNAs precisely regulate gene expression, enabling plants to acclimatize to different environmental conditions, including extreme weather conditions, such as the transition from heat to cold or drought to high humidity. Noncoding RNA molecules, which are about 21 nucleotides long, bind to the messenger RNA (mRNA) of target genes in order to suppress translation or initiate mRNA degradation, thereby regulating the production of proteins that are important for survival under stress (Pradhan et al., 2023). Transcription factors integrating development pathways and stress response are controlled by many plant microRNAs. The miR156-SPL and miR172-AP2 modules control plant growth and floral transitions, and miR396-GRF and miR398-CSD control leaf and root development and response to drought, salinity, and oxidative stress. In contrast to earlier tools developed for animal miRNAs or lacking a stress-specific focus, AsmiR specifically identifies sequence patterns associated with plant responses under stress conditions, like the small genetic changes after a sudden frost. It taps into advanced feature selection to zero in on the most telling sequence traits, like a sudden spike in signal boosting both accuracy and confidence in its predictions. In order to make technology accessible to all, even those outside computational biology, scientists built a user-friendly web platform where researchers can quickly upload RNA sequences for fast analysis and prediction (Pradhan et al., 2023). Experiments show that ASmiR consistently outperforms many other tools in terms of accuracy; it stays accurate and reliable under a wide range of abiotic stresses like heat, drought, and even high salinity. Addressing the challenges of climate change requires tools that can quickly and accurately identify microRNAs (miRNAs) involved in key biological processes in plants. ASmiR is capable of integrating multiomics datasets (genomics, transcriptomics, and epigenomics) and building a map of detailed miRNA regulatory networks in plants. These models can examine large sets of RNA-sequencing data, thus promoting the computation of effective miRNAs and their individual targets (Murmu et al., 2024; Pradhan et al., 2023).

12.4.8 miRNA Finder/MicroRNA Finder

MicroRNAs or miRNAs are short RNA molecules that play an important role in controlling gene activity, particularly when plants face stress from drought, chilling cold, or salty soil. Previous methods of prediction often have problems accounting for plant miRNA sequence intricacy and diversity. The miRNAFinder tool utilizes an advanced method of computation to analyze a large scale of RNA-sequencing data with high precision. This tool analyzes even the smallest of molecular features and thermodynamic variations, helping distinguish genuine miRNAs from other small RNA fragments (Lokuge et al., 2021). Researchers can upload plant RNA data quickly, with the help of its user-friendly interface, and obtain accurate predictions for previously known as well as novel miRNAs in a short duration of time, even without requiring programming expertise. This tool's high accessibility helps increase research efficiency and expands participation in miRNA discovery (Lokuge et al., 2021). The tool not only identifies the role of each miRNA but also differentiates them from other small RNAs, and assists in mapping gene networks that are responsible for plants' growth and response to stresses, including drought and heat. The miRNA Finder's impact goes beyond theoretical research, connecting discoveries from labs to practical farmer applications. By uncovering miRNAs that help in regulating important growth and survival pathways, it contributes directly to agricultural innovation, increasing crop resilience and yield (Lokuge et al., 2021; Murmu et al., 2024). In today's time, climate changes are intensifying, and tools like miRNA Finder are proving to be helpful in making a transition toward more sustainable agricultural practices by increasing our understanding of gene regulation at the molecular level, even at the level of a single strand of RNA.

12.5 Challenges

Through the use of AI, researchers are efficiently analyzing massive, complex datasets like millions of tangled strands of genetic code to uncover exact plant RNA functions. Supervised learning is a common method in which AI is trained on known examples of RNA functions, such as sequences linked to specific cellular roles, until it recognizes similar patterns on its own. Once training is done, the AI can apply the information that it has learned to predict the roles of unfamiliar RNAs, for example, identifying one that is responsible for regulating cell growth (Murmu et al., 2024). Deep learning methods such as CNNs and RNNs are good examples of providing excellent results in analyzing RNA sequences and structures, identifying even subtle patterns that experienced researchers might not detect, like a slight twist in a folded strand (Yu et al., 2024). One effective method is transfer learning, which uses insights from well-known plants like *Arabidopsis* to study lesser-known species. It is valuable especially when data on certain crops is limited, as in the case of many limited-researched legumes or minor cereal crops (rare bean variety) (Pradhan et al., 2023). Even with notable progress, many problems still persist. One of the main problems is that we do not have large, well-labeled datasets with known results for many key crops, especially those that have been neglected or those with minimal research such as sorghum. Since AI model results mainly depend upon the quality of training data provided, insufficient datasets can result in inconsistent predictions that affect model reliability. Another problem faced by scientists is AI's black box; while these models may generate precise results, the process behind the predictions is often opaque. In agriculture, this lack of transparency makes scientists reluctant to rely on AI-based decisions, as there always exists the risk of crop mismanagement or yield loss (Yu et al., 2024). Furthermore, models optimized for one plant species, like *Arabidopsis,* may not work when applied to other species because each species carries its unique RNA structure and genetic composition. Therefore, to enhance AI's performance in plant RNA research, there is a need for broader and more varied datasets, improvements in interpretability mechanisms that explain how models arrive at their given results, and design frameworks that are flexible enough to maintain consistent performance across different crop types and growing conditions. Only by addressing these barriers can the full potential of AI-based RNA research be realized, paving the way for notable advances in crop enhancement, including sturdier stems, better yields, and greater tolerance to environmental stressors.

12.6 Conclusion

With AI-based omics tools, scientists are quickly changing their approach to exploring plant RNA by spotting possible functions in minutes rather than days. Recent developments in RNA biology have highlighted the role of small noncoding RNAs in helping plants regulate their stress-responsive pathways, offering a clear understanding of the molecular pathways that control plant growth, drought tolerance, and nutrient absorption. Post-transcription, miRNAs and other noncoding RNAs (ncRNAs) precisely control gene expression by activating or silencing stress-related genes and help plants adapt their growth and metabolism when facing severe environmental conditions like heat, cold, drought, and salinity. These miRNAs regulate the key pathways in plants like root development, photosynthesis, and hormone signaling, coupling environmental signaling with plant physiological processes (Murmu et al., 2024; Yu et al., 2024). Although major progress has been made in AI-integrated smart farming systems, various significant challenges continue to limit the full utilization of these technologies. The major problems include incomplete datasets, underlying molecular complexities, and large variations across plant species, which prevent the generalization of these models to crops other than major crops like rice and wheat, leading to the poor performance of these models when applied to the minor crops. Besides this, the complex nature and nonlinear dynamics of biological systems remain a major challenge for the accurate performance of these current data-driven models. Addressing these gaps requires collaboration among researchers from diverse fields and also highlights the need for more reliable data and combined methods. Computational models, when combined with the knowledge of plant biology, enhance the efficiency of model training, the comprehension of biological features, and model performance validation under field conditions. Collaboration across different disciplines strengthens the data accuracy and reliability of models while enabling the development of efficient workflows that can handle the whole process, starting from data collection to interpretation, and also ensuring the consistent production of good results. This collaborative system is making a promising change in the direction of precision agriculture, in which the combination of different sensors, AI, and genomic information is making major progress in precision agriculture, helping us make better decisions on planting time, water use, fertilization, and harvesting. By ensuring data connectivity, addressing species challenges, and encouraging open data sharing, it is important to build a sustainable agriculture system powered by AI. Together, all these strategies can greatly contribute toward higher yields and improved tolerance to environmental stress and provide a stable global food supply (Pradhan et al., 2023).

References

Ajila, V., Colley, L., Ste-Croix, D. T., Nissan, N., Golshani, A., Cober, E. R., Mimee, B., Samanfar, B., & Green, J. R. (2023). P-TarPmiR accurately predicts plant-specific miRNA targets. *Scientific Reports*, *13*, 332. https://doi.org/10.1038/s41598-022-27283-8

Bai, Q., Shi, L., Li, K., Xu, F., & Zhang, W. (2024). The construction of lncRNA/circRNA-miRNA-mRNA networks reveals functional genes related to growth traits in *Schima superba*. *International Journal of Molecular Sciences*, *25*(4), 2171. https://doi.org/10.3390/ijms25042171

Cao, G., Chao, H., Zheng, W., Lan, Y., Lu, K., Wang, Y., Chen, M., Zhang, H., & Chen, D. (2025). scPlantLLM: a foundation model for exploring single-cell expression atlases in plants. *Genomics, Proteomics & Bioinformatics, 23*(3), qzaf024.

Chekulaeva, M., & Rajewsky, N. (2019). Roles of long noncoding RNAs and circular RNAs in translation. *Cold Spring Harbor Perspectives in Biology, 11*(6), a032680. https://doi.org/10.1101/cshperspect.a032680

Chen, T., Greene, G. H., Motley, J., Zhang, Y., Liu, L., Wang, W., Wu, J., Liu, Z., Shi, H., & Dong, X. (2024). m6A modification plays an integral role in mRNA stability and translation during pattern-triggered immunity. *Proceedings of the National Academy of Sciences, 121*(33), e2411100121. https://doi.org/10.1073/pnas.241110.121

Cheng, X., Wang, J., Li, Q., & Liu, T. (2021). BiLSTM-5mC: A bidirectional long short-term memory-based approach for predicting 5-methylcytosine sites in genome-wide DNA promoters. *Molecules, 26*(24), 7414. https://doi.org/10.3390/molecules26247414

Deshpande, S., Shuttleworth, J., Yang, J., Taramonli, S., & England, M. (2019). PLIT: An alignment-free computational tool for identification of long non-coding RNAs in plant transcriptomic datasets. *Computers in Biology and Medicine, 105*, 169–181.

Ding, J., Li, X., & Hu, H. (2016). TarPmiR: a new approach for microRNA target site prediction. *Bioinformatics (Oxford, England), 32*(18), 2768–2775. https://doi.org/10.1093/bioinformatics/btw318

Gao, Y., Takenaka, K., Xu, S.-M., Cheng, Y., & Janitz, M. (2025). Recent advances in investigation of circRNA/lncRNA-miRNA-mRNA networks through RNA sequencing data analysis. *Briefings in Functional Genomics, 24*, elaf005. https://doi.org/10.1093/bfgp/elaf005

Hafner, M., Katsantoni, M., Köster, T., Marks, J., Mukherjee, J., Staiger, D., Ule, J., & Zavolan, M. (2021). CLIP and complementary methods. *Nature Reviews Methods Primers, 1*, 20. https://doi.org/10.1038/s43586-021-00018-1

Kajrolkar, A. (2025). Deep learning applications in plant biology: Bridging genotype and phenotype. *Premier Journal of Plant Biology, 3*, 100011.

Kapral, T. H., Farnhammer, F., Zhao, W., Lu, Z. J., & Zagrovic, B. (2022). Widespread autogenous mRNA–protein interactions detected by CLIP-seq. *Nucleic Acids Research, 50*(17), 9984–9999. https://doi.org/10.1093/nar/gkac756

Lokuge, S., Jayasundara, D., Wickramasinghe, I., & Tennakoon, C. (2021). Mirnafinder: An AI-based tool for accurate identification of plant micrornas from small RNA sequencing data. *Frontiers in Genetics*. https://doi.org/10.3389/fgene.2021.674695

Mirdar Mansuri, R., Azizi, A. H., Sadri, A. H., & Shobbar, Z. S. (2022). Long non-coding RNAs as the regulatory hubs in rice response to salt stress. *Scientific Reports, 12*, 21696. https://doi.org/10.1038/s41598-022-26133-x

Murmu, S., Sinha, D., Chaurasia, H., Sharma, S., Das, R., Jha, G. K., & Archak, S. (2024). A review of artificial intelligence-assisted omics techniques in plant defense: Current trends and future directions. *Frontiers in Plant Science, 15*, 1292054. https://doi.org/10.3389/fpls.2024.1292054

Palos, K., Yu, L., Railey, C. E., Nelson Dittrich, A. C., & Nelson, A. D. L. (2023). Linking discoveries, mechanisms, and technologies to develop a clearer perspective on plant long noncoding RNAs. *The Plant Cell, 35*(6), 1762–1786. https://doi.org/10.1093/plcell/koad027

Pradhan, U. K., Thakur, A. K., & Singh, A. (2023). ASLnCR: a novel computational tool for prediction of abiotic stress-responsive long non-coding RNAs in plants. *Functional & Integrative Genomics, 23*(1), 113–127. https://doi.org/10.1007/s10142-023-00928-7

Samarskaya, V., Spechenkova, N., Kalinina, N. O., Love, A. J., & Taliansky, M. (2025). The emerging role of omics-based approaches in plant virology. *Viruses*, *17*(7), 986. https://doi.org/10.3390/v17070986

Sanches, P. H. G., de Melo, N. C., Porcari, A. M., & de Carvalho, L. M. (2024). Integrating molecular perspectives: Strategies for comprehensive multi-omics integrative data analysis and machine learning applications in transcriptomics, proteomics, and metabolomics. *Biology*, *13*(11), 848. https://doi.org/10.3390/biology13110848

Singh, U., Khemka, N., Rajkumar, M. S., Garg, R., & Jain, M. (2017). PLncPRO for prediction of long non-coding RNAs (lncRNAs) in plants and its application for discovery of abiotic stress-responsive lncRNAs in rice and chickpea. *Nucleic Acids Research*, *45*(22), e183. https://doi.org/10.1093/nar/gkx866

Wei, G. (2024). RNA m6A modification, signals for degradation or stabilisation? *Biochemical Society Transactions*, *52*(2), 707–717. https://doi.org/10.1042/BST20230574

Wekesa, J. S., Luan, Y., Chen, M., & Meng, J. (2019). A hybrid prediction method for plant lncRNA-protein interaction. *Cells*, *8*(6), 521. https://doi.org/10.3390/cells8060521

Yadav, A., Mathan, J., Dubey, A. K., & Singh, A. (2024). The emerging role of non-coding RNAs (ncRNAs) in plant growth, development, and stress response signaling. *Non-coding RNA*, *10*(1), 13. https://doi.org/10.3390/ncrna10010013

Yu, H., Yang, H., Sun, W., Yan, Z., Yang, X., Zhang, H., Ding, Y., & Li, K. (2024). An interpretable RNA foundation model for exploring functional RNA motifs in plants. *Nature Machine Intelligence*, *6*, 1616–1625. https://doi.org/10.1038/s42256-024-00946-z

13

The Integration of Artificial Intelligence and Big Data in Plant Epigenetics

Isha Sharma[1], *Varucha Misra*[2], *and A.K. Mall*[2]

[1] *Department of Microbiology, Dr. Harisingh Gour Vishwavidyalaya (A Central University), Sagar, Madhya Pradesh, India*
[2] *Division of Crop Improvement, ICAR-Indian Institute of Sugarcane Research, Lucknow, Uttar Pradesh, India*

13.1 Introduction

A crucial factor of our understanding of plant gene control is epigenetics. It explains how gene expression can be modified without changing the underlying DNA sequence, thus regulating phenotype–environment interactions. Epigenetics also plays an important role in stress adaptation, development, and transgenerational inheritance (Sampson et al., 2024; Singh et al., 2023). Unlike genetic mutations, which are permanent, epigenetic changes like DNA methylation (DM), histone modifications, changes in chromatin structure, and regulation by noncoding RNAs (ncRNAs) are flexible and reversible (Agarwal et al., 2020; Ashapkin et al., 2020). This gives plants a powerful way to fine-tune their gene activity in response to both environmental stress and developmental signals. The progressions in sequence technology have opened new avenues in the study of plant epigenomes as researchers are now able to explore chromatin architecture, methylation profiles, and regulation of gene expression across diverse tissues and environmental conditions (Balazova et al., 2022; Kumar & Mohapatra, 2021) because of the high-throughput approaches like whole-genome bisulfite sequencing (WGBS), Hi-C, assay for transposase-accessible chromatin using sequencing (ATAC-seq), chromatin immunoprecipitation followed by sequencing (ChIP-seq), and RNA-seq (Shen et al., 2025; Wu et al., 2025). Huge datasets that have been produced through these technologies provide in-depth knowledge about the epigenetic profile of various plant species and assist the scientists in comprehending how gene regulation changes over time and space.

Despite these technological trends, the interpretation of plant epigenomic data remains computationally rigorous and multifaceted. The multidimensional nature of these datasets creates a bottleneck in data integration, modeling, and hypothesis generation using traditional bioinformatics approaches, as it is coupled with tissue specificity, developmental heterogeneity, and genotype–environment interactions (Zhao et al., 2020). The above-discussed factors invoked a paradigm shift toward the integration of big data methodologies and artificial intelligence (AI) into plant epigenetics research.

AI, specifically through machine learning (ML) and deep learning (DL), is transforming the means of understanding complex biological data. In recent times, some of the ML techniques, such as support vector machines (SVMs), random forests (RFs), k-nearest neighbors (kNNs), and ensemble models, are being extensively used to categorize chromatin states, identify key regulatory regions in noncoding DNA, and map out gene regulatory networks (GRNs), and majorly to predict gene expression based on epigenomic information (Tahir et al., 2024; Turkoglu et al. 2023).

Moreover, in DL, models like convolutional neural networks (CNNs) and recurrent neural networks (RNNs) could use the complicated sequential features of genomic data, making predictions not only more accurate but also more biologically meaningful (Angermueller et al., 2017). Further, tools such as DeepCpG and Basset have effectively predicted the patterns of DM and chromatin accessibility directly from raw DNA sequences and their associated epigenomic data (Angermueller et al., 2017; Kelley et al., 2016). These developments pinpoint the role of AI in reforming huge, complicated biological data into meaningful insights, aiding researchers to better comprehend the underlying mechanisms that govern gene regulation (Eraslan et al., 2019). Moreover, in modern plant epigenetics, big data technologies such as Hadoop and Spark and cloud-based platforms such as Google Earth Engine and Terra play a crucial role. Bio (https://terra.bio) is used to store, manage, and connect massive, diverse datasets from different types of biological research (Gorelick et al., 2017).

Furthermore, predictive phenotyping is one of the most compelling areas where AI and big data merge through the linking of epigenetic markers (Washburn et al., 2019). This method is opening a new avenue for improving the key crop traits such as drought tolerance, flowering time, nutrient efficiency, and resistance to pests and diseases (Varshney et al., 2021). Nowadays, modeling epigenetic memory, i.e., how the exposure of plants to stress can influence not just their own traits but also their offspring, is one of the most fascinating areas (Springer & Schmitz, 2017). Using these platforms and methodologies, researchers are developing new methods and tools to breed more robust and adaptable crops for the future.

The integration of AI and big data is reforming plant epigenetics from a largely descriptive discipline into analytical and translational science. As computational tools continue to evolve and data around the world become more accessible to researchers, we are entering into a period of notable discoveries in the field of plant epigenetics. These advancements hold great promise for the future of agriculture, helping the researchers to uncover how plants store, interpret, and pass on epigenetic information to develop crops that are more resistant to climate change, more efficient in their use of resources, and better suited for sustainable farming and ecological conservation (Springer & Schmitz, 2017; Varshney et al., 2021; Washburn et al., 2019).

Thus, the integration of big data and AI in plant epigenetics is covered in this chapter. A framework of the mechanism and function of epigenetic processes in plant growth was provided at the outset. This chapter explains the emergence of big data in plant research, typical AI technologies and how they are used to examine regulatory elements, chromatin states, and methylation profiles. Integrative methods for locating epigenetic biomarkers and important resources, including tools, databases, and analytic pipelines, have also been discussed.

13.2 Fundamentals of Plant Epigenetics

13.2.1 Overview of Epigenetic Mechanisms

Epigenetics refers to heritable changes in gene activity that occur without changing the DNA code. In plants, epigenetic regulation is crucial for the development, stress adaptation, and transgenerational inheritance. These changes include modifications mainly of DM, histone modifications, chromatin remodeling, and ncRNAs that interact dynamically to orchestrate gene expression patterns.

DM is possibly the most studied epigenetic mark. In plants, it occurs in symmetrical (CG, CHG) as well as asymmetrical (CHH) contexts, inducing stability to the genome and gene expression. Methyltransferases, including MET1, CMT2, and CMT3, ensure maintenance and proper establishment of methylation patterns. A recent study reported how stress-driven methylation changes through the regulation of methyltransferase enzymes (CMT2 and CMT3), elevating stress resistance in *Arabidopsis* (Jogam et al., 2022; Kawakatsu et al., 2017). Also, to control stress responses and development, plants sporadically employ N6-methyladenine (6mA), i.e., epitranscriptomic modifications. Further, histone modifications such as acetylation, methylation, ubiquitination, and phosphorylation act as adaptive chromatin signatures (dynamic chromatin markers) (Ashapkin et al., 2020). The balance between H3K9ac, H3K4me3, i.e., active marks, and H3K27me3, H3K9me2, i.e., repressive marks, controls the outcomes of the gene expression (Ueda & Seki, 2019). Recent research on tomato has linked both histone processes and ncRNAs to ripening and quality of the fruit, signifying how epigenetics can modify developmental trajectories and traits. Moreover, SWI/SNF, ISWI, CHD, and INO80 (ATP-dependent complexes) change the composition and positioning of nucleosomes, thus mediating chromatin remodeling. Also, recent studies reported the roles of lncRNAs (long noncoding RNAs) as molecular guides or scaffolds in both stress response and transgenerational epigenetic memory (Park et al., 2017). Specific ncRNAs have been associated with fruit maturation through their specific role in hormone signaling; thus, both ncRNAs and lncRNAs coordinate epigenetic modifications (Tonosaki et al. 2022). These mechanisms do not work in isolation; they form interrelated links. For instance, histone readers, writers, and erasers work in tandem with DM enzymes and chromatin remodelers, while ncRNAs fine-tune both processes. To fully understand the significance of these interactions, researchers have studied these correlated networks at single-cell and tissue-specific levels.

13.2.2 Function in Adaptation, Stress Response, and Plant Development

DNA methyltransferases target dinucleotides CpG, resulting in chromatin compaction and transcriptional repression (Rehman & Tanti, 2020). In *A. thaliana*, by silencing genes responsible for flowering, DM regulates defined traits such as delay in the onset and modulating leaf morphology, leading to the development of smaller leaves. Likewise, infection of *Pseudomonas syringae* activates the DM, resulting in the activation of hormonal and defense pathways, which in turn elevated the resistance to *Hyaloperonospora arabidopsidis*, *Plectosphaerella cucumerina*, and *Alternaria brassicicola*, thus contributing to plant immunity (Ferrari et al., 2020). Also, differences in the levels of DM between salt-tolerant and salt-sensitive rice varieties significantly influence the plant's response to salt stress and changes in methylation within promoters and genes under high salinity and play a vital role in controlling gene expression. Similarly, *Oryza sativa* induces both hyper- and hypo-methylation in specific genomic regions, which causes alterations in gene expression and improved salt tolerance (Kovalchuk, 2021; Pathak et al., 2021). Furthermore, in *A. thaliana*, drought stress causes hypomethylation in plants, which is exclusively found in repetitive and mobile genomic components, which in turn activate the DRG (drought-responsive genes). Plants under environmental stress may have variations in DM; some of these variations are transgenerational heritable (Yaish, 2017). For instance, methylation reactions to UV, heat, and cold in maize were rarely heritable (Eichten & Springer, 2015). In contrast, laser-induced activation of micro-inverted-repeat transposable elements (MITEs) was persistent, and such heritable alterations in methylation have been seen in rice (Li et al., 2017).

Research indicates that the genes related to stress reactions were activated by altering the patterns of histone acetylation (Hu et al., 2019; Kumar et al., 2021). For instance, treatment with trichostatin A (histone deacetylase inhibitor) demonstrated the increased resistance of *A. thaliana* to the fungus *Botrytis cinerea* (Hu et al., 2019). Also, plants undergo dynamic changes in histone alterations because of complex signaling pathways being activated by abiotic stressors such as drought, salt, temperature extremes, and heavy metals (Nunez-Vazquez et al., 2022). Likewise, recent studies have demonstrated the function of stress-responsive short RNAs, such as miRNAs and lncRNAs, in regulating plant stress tolerance (Zhan & Meyers, 2023). These RNAs can alter the expression of genes during stress by either preventing translation or causing the destruction of mRNA. For example, in *A. thaliana*, miR159 regulates genes associated with drought resistance (Wang & Qiao, 2020), whereas miR398 down-regulates the CSD transcripts against oxidative stress (Singroha et al., 2021). Furthermore, short RNAs were produced by stress-activated mobile genetic elements, which aid in regulating genes in unfavorable conditions (Zhan & Meyers, 2023). Besides, chromatin remodeling complexes help plants adapt to stress by increasing the accessibility of important genes required for expression by utilizing ATP to modify nucleosomes (He et al., 2023). This method permits rapid and reversible changes in gene activity, which are essential for plant survival under stress without changing the DNA sequence itself (Pecinka et al., 2019). Current studies have revealed that environmental signals may directly affect the behavior of these remodelers by linking internal regulatory pathways to external stress signals (Kim, 2021). The high-throughput epigenetic profiling methods were also discussed in Table 13.1.

13.3 From Field to Cloud: Big Data Transforming Plant Research

Plant science has been transformed with the emergence of next-generation sequencing (NGS) and high-throughput epigenomic profiling. These helped to understand the change in gene expression in response to various stimuli such as stress, development, and varied environmental changes without altering their DNA sequences and by making big, comprehensive datasets that record gene regulation across genomes accessible.

WGBS is a commonly used standard technique for revealing 5-methylcytosine (5-mc) across the whole genome with per-base precision. It provides a fair insight into the

Table 13.1 A summary of tools, software, databases, and pipelines commonly used in plant epigenetics research.

Tools/Software/Framework/Database	Description	Web link	References
(i) DNA methylation			
Bismark	Aligns bisulfite-treated sequencing reads (WGBS, RRBS) and performs methylation calling for CpG, CHG, and CHH contexts; widely used in plant methylation studies	https://www.bioinformatics.babraham.ac.uk/projects/bismark/	Gong et al. (2022)
MethylKit	Performs statistical analysis of differentially methylated regions, clustering, Principal Component Analysis, and visualizations	https://bioconductor.org/packages/release/bioc/html/methylKit.html	Akalin et al. (2012)
MethGo	A pipeline for WGBS analysis that provides visualization tools for methylation level, gene body methylation, and DMRs	http://bioinfo.nwafu.edu.cn/software/methgo	Liao et al. (2015)
CGmapTools	Efficient post-processing and visualization of methylation maps from WGBS data, supports plant genome formats	https://cgmaptools.github.io	Guo et al. (2017)
nf-core/methylseq	Community-maintained WGBS pipeline with quality checks, alignment, and methylation calling	https://nfco.re/methylseq	—
EpiDiverse WGBS Pipeline	Developed specifically for non-model plants; supports low-coverage, complex genomes and generates DMRs and quality reports	—	Nunn et al. (2021)
DMRichR	R pipeline for DMR analysis using dmrseq and bsseq, with rich visualization options	https://github.com/ben-laufer/DMRichR	Salam et al. (2025)
CpG_Me	Workflow for converting raw WGBS reads into CpG count matrices with quality checks	https://github.com/ben-laufer/CpG_Me	Xu et al. (2007)
MEDIPS	R package for analyzing MeDIP-seq enrichment data	https://bioconductor.org/packages/release/bioc/html/MEDIPS.html	Lienhard et al. (2013)
MethylScore	Unsupervised DMR calling in plants using context-aware segmentation	https://github.com/weigelworld/methylscore	Huther et al. (2022)

(*Continued*)

Table 13.1 (Continued)

Tools/Software/Framework/Database	Description	Web link	References
(ii) Histone modification			
ChIPseeker	Annotates ChIP-seq peaks, links them to genes, and helps in downstream functional enrichment analysis	https://bioconductor.org/packages/release/bioc/html/ChIPseeker.html	Yu et al. (2015)
deepTools	Enables quality control, normalization, and visualization (e.g., heatmaps, profiles) of ChIP-seq and ATAC-seq data	https://deeptools.readthedocs.io/en/develop/	Ramírez et al. (2014)
SICER	Identifies broad enrichment regions typical of histone modifications (e.g., H3K27me3); useful in plants for heterochromatin analysis	https://home.gwu.edu/~wpeng/Software.htm	Xu et al. (2014)
(iii) Chromatin accessibility			
nf-core/atacseq	A reproducible pipeline for processing ATAC-seq data, including quality control, alignment, and peak calling	https://nf-co.re/atacseq	—
Genrich	Performs peak calling optimized for ATAC-seq and DNase-seq data; can handle biological replicates and paired-end reads	https://github.com/jsh58/Genrich	—
MACS2	One of the most widely used peak callers for ChIP-seq and ATAC-seq; works well for narrow and broad peaks	https://github.com/macs3-project/MACS	Zhang et al. (2024)
ChromHMM	Study chromatin states from multiple ChIP-seq datasets using hidden Markov models	http://compbio.mit.edu/ChromHMM/	Ernst and Kellis (2017)
PCSD	Plant Chromatin State Database for Arabidopsis, rice, and maize	http://systemsbiology.cau.edu.cn/chromstates/	Liu et al. (2017)
PlantCADB	Database of open chromatin regions from ATAC/DNase-seq across 37 plant species	http://systemsbiology.cau.edu.cn/PlantCADB/	Ding et al. (2022)
(iv) Noncoding RNA (epigenetic regulation)			
ShortStack	Specialized in detecting and annotating miRNA and siRNA clusters from small RNA-seq data, validated on plants	https://github.com/MikeAxtell/ShortStack	—
psRNATarget	Predicts plant miRNA targets based on transcriptome sequences using user-uploaded or public datasets	https://www.zhaolab.org/psRNATarget/	Dai et al. (2018)

TAPIR	Predicts miRNA targets in plants using sequence complementarity and target accessibility models	http://bioinformatics.psb.ugent.be/webtools/tapir/	Bonnet et al. (2010)
iSRAP	Integrated pipeline for sRNA-seq processing and miRNA analysis	https://github.com/twkieh/isarap	Quek et al. (2015)
miRDeep-P	Tool for identifying plant miRNAs from sRNA-seq data	https://github.com/zhangtaolab/miRDeep-P	Yang and Li (2011)
(v) Databases (epigenome)			
Plant Methylome DB	Provides methylome data across numerous plant species and tissues; useful for comparative studies	http://epigenome.genetics.uga.edu/PlantMethylome/	Song et al. (2013)
EPICO	Web portal for cross-species comparison of DM patterns in plants	http://epico.genetics.uga.edu/	—
ePlant	Allow users to visualize multi-omics data (transcriptome, epigenome, and protein interactions) for Arabidopsis and crops	http://bar.utoronto.ca/eplant/	Waese et al. (2017)
RiceENCODE	Integrates DM, histone marks, and transcriptomics in rice	http://glab.hzau.edu.cn/RiceENCODE/	—
(vi) Integrative omics and epigenomics			
Galaxy	Offers user-friendly GUI with tools for DM, ChIP-seq, RNA-seq, and WGBS analysis without programming	https://usegalaxy.org/	McGowan et al. (2020)
SnakePipe	Automates analysis for RNA-seq, ATAC-seq, WGBS, and ChIP-seq, with modular support for downstream steps	https://github.com/maxplanck-ie/snakepipe	Bhardwaj et al. (2019)
EpiMiner	Integrative analysis of epigenetic marks and gene expression	https://doi.org/10.1093/nar/gkaa365	Shang et al. (2013)
Integrative Genomics Viewer (IGV)	Visualizing genomic and epigenomic data	https://software.broadinstitute.org/software/igv/	Robinson et al. (2011)

ATAC-seq, assay for transposase-accessible chromatin using sequencing; ChIP-seq, chromatin immunoprecipitation followed by sequencing; DM, DNA methylation; DMR, differentially methylated region; DNase-seq, DNase I hypersensitivity sequencing; WGBS, whole-genome bisulfite sequencing.

stress-induced DM profile based on tissue-specific datasets. WGBS datasets were created for *A. thaliana*, *O. sativa*, *Zea mays*, *Brassica napus*, and *Glycine max* (Agius et al., 2023; To et al., 2022; Rao et al., 2024). The datasets are available on the public databases, including the European Nucleotide Archive (ENA), the Gene Expression Omnibus (GEO) of the National Center for Biotechnology Information (NCBI), and crop-specific websites. Similarly, ChIP-seq offered information on genome-wide maps of histone modifications and transcription factor binding. This method is helpful for identifying the regulatory areas and chromatin states under different developmental and stress situations and exposes the patterns of active and repressive marks, such as H3K4me3 and H3K27me3, respectively (Ueda & Seki, 2019). Recently, two new alternatives to ChIP have been introduced in plant research, i.e., CUT and RUN and CUT and Tag.

Techniques such as ATAC-seq and DNase-seq (DNase I hypersensitivity sequencing) are generally used to evaluate chromatin accessibility. The major application of these datasets is to identify open chromatin regions that reveal active functional regulatory sites such as promoters and enhancers, mainly in tissues that are developing or under stress (Shen et al., 2025; Wu et al., 2025). RNA-seq and small RNA-seq, such as transcriptomic datasets, enhance the epigenomic profile by providing information on regulatory RNA expressions. Furthermore, to investigate epigenetic regulation at several phases, these data types are frequently integrated with methylome and histone datasets. Precisely, siRNA clusters and RNA-directed DNA methylation processes in crops, including rice, maize, and tomato, have been recognized through short RNA libraries (Chow & Mosher, 2023).

Plant epigenomic datasets have facilitated streamlined integration and organization in recent years because of platforms like MethBank 4.0, Plant Regulomics, and Plant Chromatin State Database (https://plantepigenetics.oist.jp/). These platforms provide comprehensive and predictive research support by facilitating well-documented information, genome browsers, and visualization tools in an easily accessible manner (Ran et al., 2019; Zhang et al., 2022). Furthermore, they enable the investigation of epigenomic changes across several omics layers, including DNA, RNA, and proteins. PlantEMS is a freely accessible, user-friendly database that is the first of its type, gathering more than 13 million modification sites from 54 plant species, including model organisms like *A. thaliana* and *Fragaria vesca*. It is different from the widely used tools such as PRMD and AraENCODE by combining transcriptomics, proteomics, and genomic data into a single platform. Users may study the alteration trends and direct particular loci using an integrated genome browser. Notably, PlantEMS has an ML-driven tool that allows users to predict epigenetic alterations and modifications in a variety of plant species with high-precision annotation. It is an advanced platform for investigating plant epigenetics, development, and stress adaptation through the integration of sophisticated visualization systems and ML-based prediction models (Dao et al., 2024).

The advent of epigenomic techniques such as scATAC-seq and scRNA-seq (single-cell sequencing) provides epigenetic profiles at single-cell resolution, opening new avenues to explore the tissue-specific regulatory processes in different plants. Altogether, these varied data sources enable exploration of plant epigenomes, facilitating useful applications of AI and ML aimed at modeling GRNs, identifying stress-responsive markers, and guiding crop improvement strategies. Advanced tools, databases, software, and pipelines involved in big data analysis are highlighted in Table 13.2.

Table 13.2 Advanced computational tools, databases, and pipelines used in large-scale plant epigenetics and big data analysis.

Tools/software/ framework/database	Description	Web link	References
BioWardrobe	Web-based GUI for ChIP-seq, RNA-seq, and WGBS analyses with built-in QC and visualization	https://scidap.com/	Kartashov and Barski (2015)
GobyWeb	Scalable HTS (RNA-seq, small RNA, and WGBS) analysis on clusters	http://gobyweb.campagnelab.org	Dorff et al. (2013)
Plant Phenomics and Genomics PGP	Multi-omics (epigenomic data)	https://edal-pgp.ipk-gatersleben.de/	—
NCBI Epigenomics	WGBS and ATAC across rice accessions	https://commonfund.nih.gov/epigenomics	—
methylPipe / compEpiTools	Integrative analysis of WGBS and ChIP-seq at base-resolution; scalable and parallelized	https://www.bioconductor.org/packages/release/bioc/html/compEpiTools.html	Kishore et al. (2015)
MethBase	High-resolution methylomes (including Arabidopsis) track	http://smithlabresearch.org/software/methbase	—
DeepSignal-Plant	ML-based nanopore model optimized for plant cytosine methylation; scales to population-level big data	https://github.com/PengNi/deepsignal-plant	Ni et al. (2021)
PhytoOracle	Phenomic and epigenomic pipeline	https://phytooracle.readthedocs.io/en/latest/	Gonzalez et al. (2023)
PlantCADB	ATAC/DNase-seq aggregated for 37+ species; supports big epigenomic queries	http://systemsbiology.cau.edu.cn/PlantCADB/	Ding et al. (2022)
ChIP-Hub	>10,000 datasets including ChIP-seq and accessibility across 40+ species	https://biobigdata.nju.edu.cn/ChIPHub	Fu et al. (2022)
EXPath 2.0	Integrates high-throughput transcriptome data with pathway contexts (e.g., rice)	https://academic.oup.com/pcp/article/61/10/1818/5902840	Tseng et al. (2020)
Arabidopsis/Rice ENCODE	Large-scale epigenomic mapping (ChIP, ATAC, WGBS, Hi-C) for foundational model plants	—	—

ATAC, assay for transposase-accessible chromatin; ChIP-seq, chromatin immunoprecipitation followed by sequencing; DNase-seq, DNase I hypersensitivity sequencing; ML, machine learning.

13.4 AI Unraveled: Techniques for Biological Pattern Discovery

The use of AI in plant science has immensely improved our capability to estimate phenotypic traits, decode complex biological processes, and crop improvement. For example, Misra and Mail (2024) showed that the use of image processing helps in identifying diseases earlier and helps in the early management of the sugar beet crop. To systematically comprehend the vast and multilayered datasets produced by contemporary advanced technologies, AI techniques, precisely ML and DL, have become indispensable tools. AI techniques are altering the way one studies plants, from genome annotation to phenotypic prediction and epigenetic control (Angermueller et al., 2017; Tahir et al., 2024; Turkoglu et al., 2023).

13.4.1 ML Approaches for Biological Data Analysis

A subfield of AI is ML, which enables computer systems to learn and make predictions from data without needing to be precisely encoded for each task. Certainly, ML is particularly suitable in the biological sciences, where datasets are often huge, varied, and complex. ML has emerged as a pivotal analytical tool in plant research for the interpretation of the large-scale datasets that originate from several sources, like environmental inputs, molecular profiles, and phenotypic trait assessments. Such models are frequently employed in tasks including pattern recognition, data classification, unsupervised clustering, and forecasting, allowing researchers to unravel the connotations between genetic, epigenetic, and phenotypic traits under various developmental or stress-related conditions (Springer & Schmitz, 2017; Varshney et al., 2021; Washburn et al., 2019). In this context, SVM, a frequently used ML algorithm in plant epigenetics, is supervised algorithms that work by drawing the best possible line or margin between different groups in the data to achieve accurate classification (Kok et al., 2021). They are predominantly effective for binary classifications, for example, marking differences between methylated and unmethylated cytosines in CpG dinucleotides. For the classification of methylation profiles, based on nucleic base composition, motifs, and genomic framework, it has been extensively used in plant systems for providing valuable insights toward the mechanism of gene regulation in varied environmental conditions (Rehman & Tanti, 2020).

RF is a popular ML technique used in biological data analyses; it creates a "forest" of decision trees proficient in different data samples, and the combined output leads to stable, robust, and accurate predictions. For noisy and unfiltered biological data, RF models are especially suited and they have been widely used for envisaging differential gene expression, distinguishing the methylated regions, and selecting informative features in multiomics datasets (Angermueller et al., 2017). Also, RF highlights the most informative variables, allowing the investigators to locate significant genes or epigenetic markers involved in traits, including flowering or environmental stress adaptation. Moreover, kNN is a straightforward and powerful technique; it classifies the most predominant class from the datasets among its neighbors. It has been widely used in investigations such as plant phenotyping and gene expression profiling (N'Diaye et al., 2020).

In addition, naive Bayes and logistic regression, the data-driven probabilistic approaches, are often employed in genomic prediction pipelines and are helpful in smaller scale classifications (N'Diaye et al., 2020).

These ML methods together are changing the landscape of plant epigenetics and systems biology. They enable scientists/researchers to approach beyond descriptive analyses toward extrapolative, data-driven inference models that deepen our understanding of the mechanisms of plants' adaptation, growth, and evolution in response to their surroundings. AI tools and algorithms applied in plant epigenetics research are discussed in Table 13.3.

13.4.2 DL Versus ML in Biological Data Analysis

The landscape of life sciences research has transformed due to the explosive growth of biological data driven by high-efficiency sequencing platforms. This data revolution has generated an immediate requirement for machine-based analysis systems proficient in processing and comprehending complex, multiparametric datasets. Among the powerful tools emerging from this age are AI techniques, particularly ML and DL. These two subfields are progressively being applied across all the biological domains, like genomics, transcriptomics, proteomics, epigenomics, and phenomics. Though ML and DL pursue aligned objectives, they significantly vary in research strategy, data requirements, large-scale application, and interpretability. To utilize them effectively in biological research, it is important to understand and pinpoint their individual fortes and limitations.

In the context of biological sciences, ML is defined as the family of algorithms that study the biological profiles from structured data and classify without being manually coded. The ML algorithms have proven instrumental in solving problems such as prediction of gene function, classification of proteins, phenotype-genotype associations, and disease outcome modeling. Some of the extensively used ML techniques are SVMs, RFs, kNN, Decision Trees, and Gradient Boosting (ensemble methods) (Angermueller et al., 2017; Kok et al., 2021; Mahamat et al., 2024; N'Diaye et al., 2020). Based on domain knowledge, these models typically require manually crafted features, for instance, GC content in genomic sequences or expression fold changes in transcriptomic data. Despite its complexity, feature engineering plays an intensive role in making models more comprehensible and in determining their learning process (Montesinos-López et al., 2024).

Conventionally, ML methods have certain boundaries in spite of their efficacy when applied to huge unstructured data such as raw DNA or RNA sequences, histopathology images, or single-cell omics data. ML algorithms may struggle to capture complicated dependences or hierarchical profiles as data becomes more complicated and nonlinear; in such circumstances, DL offers a highly adaptable and extensible approach. DL models are based on complex artificial neural networks that can originally derive abstract patterns from raw data, eliminating the requirement for manual feature engineering. Also, these are highly appropriate for biological applications that demand great resolution and understanding, as these models can recognize the subtle patterns and extended-range correlations in large-scale datasets (Alharbi & Rashid, 2022; Yang et al., 2025).

In life sciences, various kinds of DL architectures have been adapted for use where CNNs outshine in handling and processing spatial data that have been applied in predicting sites of transcription factor binding, identifying epigenomic markers, and classifying cellular

Table 13.3 Artificial Intelligence (AI) tools and algorithms applied in plant epigenetics research.

Tools/software/framework/database	Description	Web link	References
Support vector machine	Study methylated/unmethylated cytosines, RNA-seq subtype classification	https://scikit-learn.org/stable/modules/svm.html	Kok et al. (2021)
Random forest	Predict gene expression, identify key epigenetic markers, and predict methylation state	https://scikit-learn.org/stable/modules/generated/sklearn.ensemble.RandomForestClassifier.html	Angermueller et al. (2017)
K-Nearest Neighbors	Gene expression profiling, phenotyping classification	https://scikit-learn.org/stable/modules/neighbors.html	N'Diaye et al. (2020)
Naive Bayes	Genomic predictions in small-scale datasets	https://scikit-learn.org/stable/modules/naive_bayes.html	N'Diaye et al. (2020)
Logistic Regression	Classification tasks in genomic pipelines	https://scikit-learn.org/stable/modules/linear_model.html#logistic-regression	N'Diaye et al. (2020)
Gradient Boosting / Decision Trees	Feature selection and classification in omics data	https://xgboost.readthedocs.io/en/stable/	Mahamat et al. (2024)
DeepCpG	Predict DNA methylation from sequence and single-cell bisulfite data	https://github.com/cangermueller/deepcpg	Angermueller et al. (2017)
Basset	Predict chromatin accessibility from raw sequence	https://github.com/davek44/Basseth	Kelley et al. (2016)
D-GEX	Gene expression inference from landmark genes	https://github.com/uci-cbcl/D-GEX	Chen et al. (2016)
AlphaFold2	Predict 3D protein structures, applicable in epigenetic protein analysis	https://www.deepmind.com/research/highlighted-research/alphafold	Yang et al. (2023)
SHAP / LIME	Model interpretation tools for DL/ML models in epigenetics	https://github.com/shap/shap	Lundberg and Lee (2017)
ChronoRoot	Image-based time-series root growth tracking	https://arxiv.org/abs/2504.14736	Gaggion et al. (2021)

DL, deep learning; LIME, Local Interpretable Model-Agnostic Explanations; ML, machine learning; SHAP, SHapley Additive exPlanations.

morphology in microscopy images (Yang et al., 2025). RNNs and their progressive variants like Long Short-Term Memory (LSTM) systems are optimally designed for investigating sequential data such as time-dependent courses of gene expression patterns or amino acid sequences (Yang et al., 2025). Furthermore, for dimensionality reduction, denoising biological signals, and integrating multiomics data, unsupervised DL models such as Autoencoders are used (Ballard et al., 2024). Lately, in tools like AlphaFold2, which have revolutionized the field of 3D protein conformations directly from sequences, the transformer-based architectures governed by self-attention mechanisms have notably shown extraordinary performance in the prediction of protein structure (Yang et al., 2023).

ML and DL differ significantly in terms of their data demands and the complexity of tools needed to implement them. ML algorithms frequently perform well with small to moderate datasets and can be trained rapidly with limited computational resources. They also deliver interpretability, which is crucial in terms of biology, where the knowledge of the decision-making process of the model could provide a mechanistic understanding. DL models generally require huge, complex, annotated datasets and substantial computational resources. They outdo the ML models in terms of precision and scalability. Analyzing NGS data, high-content imaging, or spatial transcriptomics and their ability to learn directly from raw, unstructured inputs without predefined features makes it ideal (Alharbi & Rashid, 2022; Yang et al., 2025).

From an epigenetic perspective, ML and DL have both made prominent contributions. Based on local sequence context and chromatin traits, RF, a traditional ML model, has been used to predict the methylation states of DNA. Likewise, to identify potent regulatory elements, these models also provide feature importance scores to the researchers. On the contrary, DeepCpG, a DL model, uses integration of convolutional and recurrent architectures, allowing models to learn from both DNA sequence patterns and methylation signals in single-cell bisulfite data, resulting in reliable and robust predictions. Also, chromatin accessibility from raw sequence data was predicted by Basset-like DL models, an operation that traditional ML algorithms would struggle to handle efficiently (Angermueller et al., 2017; Kelley et al., 2016).

In RNA-seq–based investigations, to classify disease subtypes and infer gene networks from microarray or RNA-seq data, SVMs, an ML technique, have been used. In contrast, DL models go a step further by identifying nonlinear patterns and capturing context-specific regulatory signals. For instance, D-GEX (a deep neural network model) used a small set of landmark genes to infer the expression of thousands of target genes, significantly outdoing linear regression and other ML baselines. DL has surpassed ML in the discipline of protein bioinformatics because of its ability to model structural and functional properties directly from amino acid sequences (Chen et al., 2016).

In addition, the advantage of DL is substantial; its major drawback remains in its interpretability. Since neural networks often operate as "black boxes," it could be difficult to understand the mechanism behind the predictions made. This lack of transparency is often challenging in research, where conviction depends heavily on mechanistic understanding. Nevertheless, growing interest in explainable AI (XAI) has led to the development of tools like SHapley Additive exPlanations (SHAP) that aim to make DL models more transparent (Zhou et al., 2023). However, to address this challenge, the field of XAI has introduced tools such as SHAP, Local Interpretable Model-Agnostic Explanations, and attention heatmaps, which aid researchers in better understanding the decision-making processes of DL models (Lundberg & Lee, 2017).

An alternative promising approach is the upsurge of hybrid models that bring together the best of both ML and DL. In these contexts, DL can first be used to automatically extract complex features from raw data, which are then passed on to more interpretable models, including RFs or logistic regression for final classification. This approach helps to create a balance between high accuracy and transparency in predictions, making it valuable in biological research, where understanding both the results and the reasoning behind them is crucial (Zhao et al., 2023).

Moving forward, DL models are accessible and effective when the advancements in transfer and self-supervised learning are made, even in cases of low-data biological investigations, including rare diseases or non-model organisms. These methods reprocess the knowledge from unlabeled data to improve learning efficacy, addressing one of the chief limitations of DL algorithms (Mavaie et al., 2023).

To conclude, ML and DL are transforming the way biological data are examined and studied. For problems where interpretability and efficacy are priorities, ML acts as a powerful tool, whereas DL could be an asset when it comes to modeling and handling complex massive datasets; it opens new avenues in systems biology and personalized medicine. The use of ML and DL depends on the nature of the research, data complexity, and the availability of the tools and resources. These methods will continue to reveal prevailing understandings of complex biological systems.

13.5 Integrative Approaches Using AI and Big Data

13.5.1 AI-driven Identification of Epigenetic Biomarkers for Breeding

With the advancements in sequencing and multiomics approaches, researchers could examine complex epigenetic instances, such as methylation profiles, histone markers, and ncRNAs across a range of developmental and environmental responses, thanks to enhanced analytical capabilities. Deciphering the epigenomic complexity for crop development is still quite difficult, though. To uncover epigenetic markers linked to characteristics like stress and drought tolerance, adaptation, and crop production, AI, in particular, ML and DL, is essential. This allows for targeted and effective plant breeding techniques (Herath et al., 2020; Tahir et al., 2024).

The genes related to flowering, drought and stress tolerance, disease resistance, and nutrient absorption are regulated by epigenetic biomarkers, which are heritable but reversible changes not encoded in the DNA sequence. A layer of control is provided by epigenetic markers that might be useful in adaptive breeding; they represent dynamic reactions to environmental stimuli. AI systems can effectively explore multidimensional information to identify trends and rank potential biomarkers (Rajpal et al., 2022).

RF, SVMs, and LASSO regression ML models have been successfully applied to study DM sites or histone markers based on their prediction for phenotypic variation (Angermueller et al., 2017; Kok et al., 2021; N'Diaye et al., 2020). These models not only deal with the intricated datasets but also provide feature importance scores that are used in the interpretation of biological applications. For instance, ML has been used to foresee stress-resistant epigenetic markers in rice and maize by integrating methylome and transcriptome data (N'Diaye et al., 2020; Wang et al., 2021).

By epigenetic alterations, CNNs and DL models such as LSTM can model three-dimensional and sequential patterns within the genome, which enables the identification of regulatory elements. Recently, it was reported that alterations in chromatin and tissue-specific methylation sites were directly related to the breeding methods directing the developmental traits and stress resilience (Yang et al., 2025).

Platforms like Terra (https://terra.bio), Galaxy, and Google Cloud provide scalable frameworks to handle and examine multiomics data, facilitating comparative analysis across varied environments and plant genotypes. Such analysis enables the identification of consistent, environment-specific, robust epigenetic markers conserved among different plant populations (Koppad et al., 2021; McGowan et al., 2020). AI-enabled identification of epigenetic markers improves crop features through genomic selection and marker-assisted breeding. With increasing alignment and compatibility with biological workflows, AI has the potential to shift plant breeding from trial-and-error to a precision-guided process based on epigenetic information. AI-enabled framework for interpreting plant epigenetic responses to environmental stress has been represented in Figure 13.1.

13.5.2 Network-Based Models for Epigenetic Regulation

A comprehensive understanding of epigenetic complexity involves transitioning from isolated genomic elements toward a broader system biology framework. Network-based modeling serves as a robust method for intricately visualizing the relationships between regulatory elements, epigenetic signatures, and gene activity. Researchers use GRNs, co-methylation patterns, and chromatin interaction frameworks to dissect how DM, histone markers, and ncRNAs function cooperatively to modify gene expression under specific developmental and environmental responses (Davila-Velderrain et al., 2018; McCoy et al., 2021).

These network models help identify crucial epigenetic regulators by pinpointing key regulatory nodes or "epigenetic hubs" that drive extensive variations in gene expression. For example, fusing ATAC-seq with DM and gene expression data uncovers stress-specific regulatory networks in crops like rice and maize. Graph-based approaches and ML tools like graph neural networks (GNNs), ARACNe, and GENIE3 are widely used to infer regulatory networks and predict their biological outcomes (Ahmed et al., 2024). Integrating multiomics data into network models enables researchers to visualize dynamic regulatory interactions while also assisting in the selection of promising genes for validation or agronomic enhancement. As AI-powered network models grow more advanced and scalable, they offer new opportunities to understand plant adaptability and accelerate the search for epigenetic levers in crop improvement strategies.

13.5.3 Integrating Epigenomic, Transcriptomic, and Phenotypic Data

The fusion of comprehensive omics information from epigenetic, gene expression, and trait-level observations has advanced the field of plant systems biology, facilitating a comprehensive perspective on genotypic to phenotypic relationships. Regulatory mechanisms involving DM, histone tail alterations, and nucleosome remodeling reversibly modulate gene expression without changing the DNA code. Evaluating these epigenetic features in combination

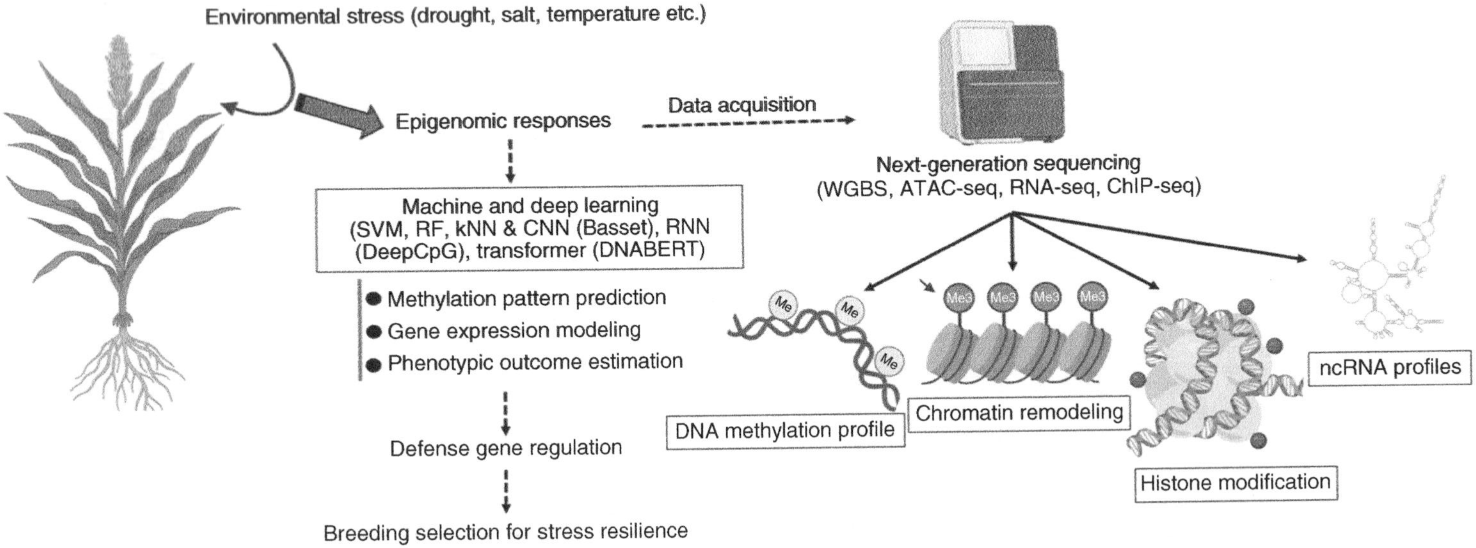

Figure 13.1 A schematic representation of artificial intelligence (AI)–enabled framework for interpreting plant epigenetic responses to environmental stress. The integration of high-throughput omics data (e.g., whole-genome bisulfite sequencing, assay for transposase-accessible chromatin using sequencing, RNA-seq) with AI models, including machine learning and deep learning, to decode plant epigenetic mechanisms. AI tools process input data to predict the methylation profile, histone alterations, and noncoding RNA regulation, leading to gene expression changes, defense priming, and long-term stress memory. [Created in BioRender. Sharma, I. (2025) https://BioRender.com]

with gene expression datasets reflecting dynamic transcriptional activity and correlated with morphological or physiological characteristics, under varying developmental contexts or ecological stimuli, enables the dissection of sophisticated gene expression systems that govern agronomic characteristics (Ali & Georgiev, 2023; Fan et al., 2025). Comprehensive multilayered omics profiling supports the detection of epigenetic focal points and transcriptional interaction networks that are heritable epigenetic indicators correlated with vital agricultural traits, including drought/salinity resistance, biomass gain, and flowering schedule. By linking variable methylation profiles of transcriptomic readouts, researchers have identified major regulators and signaling components that contribute to acclimation mechanisms connecting omics profiles to trait expressions such as foliar drooping, decreased plant mass, or shifted blooming period, offering mechanistic insights instead of surface-level correlations (Ali & Georgiev, 2023; Balazova et al., 2022; Fan et al., 2025).

The application of data-driven models connecting predictive models and interaction-based frameworks has enabled effective investigation of these high-dimensional omics data. These models not only uncover relevant data linkages but also pinpoint key regulatory genes for selective breeding and gene function studies. Omics-combined frameworks, including iDEP, Multiomics Factor Analysis, and DL-based predictors, aid in harmonizing and understanding multiomics information (Argelaguet et al., 2018; Ge et al., 2018). Furthermore, this holistic data integration offers significant potential in foresight-based phenotype analysis, facilitating the discovery of early-stage omics markers that anticipate phenotypic variations. This is highly useful in plant breeding techniques where preliminary identification of intricate agronomic traits has the potential to improve the pace of breeding. Eventually, merging multiomics and phenotypic insights has redefined the comprehension of biological complexity from individual factors to holistic regulatory networks promoting computationally guided crop management and robust plant breeding (Argelaguet et al., 2018; Yang et al., 2021).

13.5.4 Case Study: AI-Based Identification of Rice Epigenetic Markers Responsive to Drought

Climate-induced drought stress poses a major challenge to global rice (*O. sativa*) productivity, which requires advanced techniques for pinpointing molecular traits associated with stress tolerance. Recently, computational AI approaches have been applied to examine drought-induced molecular profiles in rice varieties. Using advanced learning algorithms, scientists can detect drought-associated regulatory patterns with functional importance (Kaur et al., 2023).

A widely cited application employed ML models, RF, and SVM algorithms to distinguish differentially methylated regions (DMRs) in rice exposed to water deficit. The algorithms were developed using high-resolution WGBS and RNA-seq data derived from drought-resistant (N22) and vulnerable (IR64) rice strains (Gour et al., 2021; Kaur et al., 2023).

Additionally, explicable AI approaches like SHAP were applied to underline the main contributors to drought resilience, facilitating better gene expression in breeding pipelines. These investigations highlight the transformative role of AI in the discovery of inheritable methylation patterns associated with drought resilience (Smet et al., 2023).

13.5.5 Case Study: DL for *Arabidopsis* Chromatin Accessibility Prediction

Understanding the mechanism of gene regulation in plants is crucial for studying the chromatin basis, as it controls the binding of regulatory proteins to DNA, affecting the levels of gene expression. Differentiating the free DNA regions in *A. thaliana* revealed the patterns of transcriptional regulation during different environmental stresses. Conventional methods like ATAC-seq and DNase-seq help to detect genome-wide open chromatin domains, but these techniques are labor-intensive, costly, and constrained by tissue-specific and condition-dependent variabilities (Shen et al., 2025; Wu et al., 2025). To address these barriers, DL-based frameworks are being employed to build predictive systems that estimate DNA accessibility from raw nucleotide data and multidimensional omics data based on genomic sequences with crucial epigenetic signatures like methylation profiles, histone marks, and nucleosome positioning. The study was performed from combined data of ATAC-seq, methylome mapping (WGBS), and histone profiling (ChIP-seq) across different physiological and environmental conditions (Balazova et al., 2022; Kumar & Mohapatra, 2021; Shen et al., 2025; Wu et al., 2025). By optimizing the CNN designs, the model was tailored to identify spatial configurations in DNA that correlate with signature patterns characteristic of chromatin accessibility (Angermueller et al., 2017).

Significantly, the predictive framework attained more than 90% accuracy in identifying the accessible DNA areas, demonstrating robust adaptability across varied plant tissues and experimental conditions. The model was enhanced by integrating attention layers into the CNN models, which pinpoint the nucleotide patterns and epigenetic markers involved in chromatin dynamics. Researchers were able to extract mechanistic insights through this framework, such as the overrepresentation of DREB2A and WRKY motifs within open chromatin in stress-induced samples (Banerjee & Roychoudhury, 2015).

The incorporation of DL techniques with open chromatin datasets provides a robust and cost-effective alternative to traditional experimental strategies. It paves the way for annotating large-scale mapping of cis-regulatory elements throughout the genome, even where wet-lab assays are constrained. Additionally, this approach is versatile, and its implementation requires minimal recalibration and optimization when backed by appropriate data resources (Hu et al., 2025).

This case demonstrates the efficacy of DL-based techniques in recreating detailed epigenetic frameworks and assessing DNA accessibility profiles. This advancement significantly boosts the pace of functional genomic mapping and critical understanding of regulatory mechanisms under hostile conditions, nurturing a way toward breeding pipelines and molecular crop development.

13.6 From Code to Crops: AI Applications in Plant Epigenetics

13.6.1 Prediction of Histone Modification Sites and DM

AI-driven detection of epigenetic marks has become a key priority of research in plant epigenomic biology. It plays critical functions in modulating gene activity, developmental signaling, and adaptive responses. These molecular signatures, such as methylated cytosines (CpG, CHG, and CHH contexts) and key histone tail modifications (H3K4me3, H3K27ac, or H3K9me2), act as regulatory toggles that govern gene expression. However, wet-lab profiling via chromatin immunoprecipitation, methylation sequencing, and WGBS protocols is often resource-intensive, laborious, costly, and constrained (Kawakatsu et al., 2017; Ueda & Seki, 2019).

To address these experimental bottlenecks, advanced AI techniques such as ML and DL are commonly used to infer DNA and epigenetic patterns. ML methods, including RF, SVM, and kNN, have been used to differentiate methylated and non-methylated cytosine residues or detect significant histone signal intensities from conserved DNA patterns and genomic environment (Angermueller et al., 2017; Kok et al., 2021; N'Diaye et al., 2020).

However, DL structures using CNNs and temporal RNNs offer enhanced precision and accuracy in modeling distant interactions and local–global features, from unprocessed genomic cues. These models minimize the reliance on resource-heavy experimental work and promote the recognition of new epigenetic regions in various growth stages and environmental contexts. Structured multiomics repositories are likely to be pivotal in our understanding of epigenetic signaling circuits driving data-targeted crop improvement and agricultural practices.

13.6.2 Recognizing Regulatory Elements and Chromatin State

Unraveling the transcriptional blueprint of plant genomes is fundamental for interpreting how gene expressions are regulated under developmental signals and external stress factors. Cis-regulatory modules comprising initiation, boosting, silencing, and insulating elements are major regulatory intergenic elements that influence upregulation and silencing of genes. Their regulation is modulated by epigenetic conformation, which is determined by methyl marks, histone variants, and chromatin packaging. Identifying them offers structural context into regulatory architecture and uncovers major key levers for adaptation, stress response, and agronomic performance.

Conventionally, researchers used assays such as DNase hypersensitivity, ChIP-seq, and transposase assays; despite their precision, they require large sample sizes and suffer from scalability and material constraints. Cutting-edge developments in computational intelligence, including methods like ML/DL, enable high-throughput annotation from sequence and mark data. DL systems, chiefly CNN-based architectures, can detect topological correlations from histone marks and DNA patterns, allowing fine-scale mapping of gene-accessible zones and modified histone sites (Yang et al., 2025).

13.6.3 Categorizing Epigenetic Patterns Under Stress

Stress-induced epigenetic changes function as responsive switches that influence the regulation of plant gene expression. Classifying these modifications is essential to reveal the patterns of how plants cope with drought, salt stress, heat, cold, and diseases. The combination of methyl groups, histone alteration, structural chromatin shifts, and ncRNAs shapes the stress-induced transcriptional landscape (Tonosaki et al. 2022). Findings indicate that stress triggers region-specific methylation to control stress-adaptive gene activity. For example, water scarcity commonly results in demethylation of gene-start regions that occur in regulatory sequences, leading to transcriptional activation. Transcription is enhanced by acetylation and methylation at specific histone sites, like H3K4me3, whereas repression uses trimethylation at H3K27 (Ueda & Seki, 2019). These epigenetic changes follow stress-specific methylation changes in spatial and sequential patterns. Stress-induced epigenetic traits are memorized and transferred to progeny, forming epigenetic stress memory. WGBS, ATAC-seq, ChIP sequencing, and chromatin accessibility assays are some advanced techniques that facilitate precise epigenetic profiling under stress (Shen et al., 2025; Wu et al., 2025). Differentiating stress epigenetic patterns helps define resilience pathways and key molecular markers for stress-tolerant crops.

13.6.4 Utilizing AI to Predict Genotypes to Epigenotypes

Processes such as methylation, histone tagging, and the accessibility of chromatin induce the genetic variation that shapes epigenomic features. Understanding this dynamic is key to underpinning our knowledge of regulatory mechanisms, development, and stress resilience in plant biology. However, predicting such modifications from genome data alone is challenging as epigenetic mechanisms are highly variable and environment-dependent. Modern AI methods, like ML and DL, have proven effective in associating genetic motifs with downstream epigenetic traits (Tahir et al., 2024; Turkoglu et al. 2023). Particularly, CNNs and GNN architectures, the AI techniques, are effective at predicting exact positions of methylation marks and histone changes, including CpG methylation and histone alterations, under different biological and environmental stimuli (Ahmed et al., 2024). For example, DL frameworks built on bisulfite sequencing data have been shown to effectively predict cytosine methylation directly from unprocessed DNA, demonstrating sensitivity to nearby nucleotide patterns, CpG composition, and nearby regulatory motifs (Rehman & Tanti, 2020). These AI-driven frameworks exhibit robust scalability and cross-species flexibility, helping researchers analyze epigenetic states in species that lack comprehensive methylation or histone modification maps (Agarwal et al., 2020; Ashapkin et al., 2020). Additionally, tools from explainable AI, like SHAP values and attention maps, provide biological understanding by pinpointing sequence patterns or single-nucleotide polymorphisms critical to epigenomic fluctuation (Lundberg & Lee, 2017). Such prediction tools hold great value in regulating epigenetic malleability for modern crop improvement methods, enabling fine-tuned breeding and site-specific agricultural planning. The growing fusion of AI technologies with integrated omics datasets promises advancements in modeling genetic-to-epigenetic connections, facilitating deeper decoding of gene regulatory blueprints, and hastening phenotype-centered breeding.

13.7 Future Perspectives

The convergence of AI and big data in plant epigenetics is poised to revolutionize our understanding of gene regulation and accelerate precision breeding. As multiomics datasets continue to expand, future efforts will focus on developing interpretable and transferable AI models capable of predicting epigenetic modifications across species, tissues, and stress conditions. Explainable AI tools such as SHAP and attention-based mechanisms will enhance biological insights by identifying key sequence motifs, SNPs, and regulatory regions linked to epigenetic variability. The fusion of computational intelligence and massive datasets in omics is expected to reshape insights to overhaul gene expression research and refine crop selection. With the rise of complex omics data, future research will aim at designing understandable and generalizable computational frameworks to map epigenetic alterations across genotypes and conditions. Advancing research will increasingly rely on integrating single-cell epigenetic datasets and temporal epigenomic maps with AI-driven and network-based approaches to uncover more nuanced patterns in chromatin activity and gene expression. These developments will empower phenotypic prediction, identification of heritable epigenetic markers, genetic linkage studies, and biomarker detection. As ML systems grow more explainable and become standard in breeding strategies, these frameworks will help breed stress-tolerant and yield-optimized crops.

13.8 Conclusions

This chapter elucidates that the transformative role of AI-based synthesis of epigenomic and multidimensional omics datasets is reshaping plant genomics by elucidating regulatory architecture. Predictive analytics and molecular marker identification allow epigenetic biomarker discovery, trait-based crop improvement, and enhanced stress resistance in cultivars. With better comprehensibility and growth potential, AI is set to transform breeding pipelines and promote epigenetically guided agricultural innovation.

References

Agarwal, G., Kudapa, H., Ramalingam, A., Choudhary, D., Sinha, P., Garg, V., Singh, V. K., Patil, G. B., Pandey, M. K., Nguyen, H. T., Guo, B., Sunkar, R., Niederhuth, C. E., & Varshney, R. K. (2020). Epigenetics and epigenomics: Underlying mechanisms, relevance, and implications in crop improvement. *Functional & Integrative Genomics*, *20*(6), 739–761. https://doi.org/10.1007/S10142-020-00756-7

Agius, D. R., Kapazoglou, A., Avramidou, E., Baranek, M., Carneros, E., Caro, E., Lieberman-Lazarovich, M., & Vassileva, V. (2023). Exploring the crop epigenome: A comparison of DNA methylation profiling techniques. *Frontiers in Plant Science*, *14*. https://doi.org/10.3389/fpls.2023.1181039

Ahmed, S. S., Shabab, N., Samee, M. A. H., & Rahman, M. S. (2024). Graphage: Unleashing the power of graph neural network to decode epigenetic aging. *arXiv (Cornell University)*. https://doi.org/10.48550/arxiv.2408.00984

Akalin, A., Kormaksson, M., Li, S., Garrett-Bakelman, F. E., Figueroa, M. E., Melnick, A., & Mason, C. E. (2012). methylKit: A comprehensive R package for the analysis of genome-wide DNA methylation profiles. *Genome Biology*, *13*(10) https://doi.org/10.1186/gb-2012-13-10-r87, R87.

Alharbi, W. S., & Rashid, M. (2022). A review of deep learning applications in human genomics using next-generation sequencing data. *Human Genomics*, *16*(1) https://doi.org/10.1186/s40246-022-00396-x, 26.

Ali, K., & Georgiev, M. I. (2023). Editorial: Omics and its integration: A systems biology approach to understanding plant physiology. *Frontiers in Plant Science, 14*. https://doi.org/10.3389/fpls.2023.1324901

Angermueller, C., Lee, H. J., Reik, W., & Stegle, O. (2017). DeepCpG: Accurate prediction of single-cell DNA methylation states using deep learning. *Genome Biology, 18*(1). https://doi.org/10.1186/s13059-017-1189-z, 67.

Argelaguet, R., Velten, B., Arnol, D., Dietrich, S., Zenz, T., Marioni, J. C., Buettner, F., Huber, W., & Stegle, O. (2018). Multi-omics factor analysis—a framework for unsupervised integration of multi-omics data sets. *Molecular Systems Biology, 14*(6). https://doi.org/10.15252/msb.20178124

Ashapkin, V. V., Kutueva, L. I., Aleksandrushkina, N. I., & Vanyushin, B. F. (2020). Epigenetic mechanisms of plant adaptation to biotic and abiotic stresses. *International Journal of Molecular Sciences, 21*(20), 7457. https://doi.org/10.3390/ijms21207457

Balazova, E., Balazova, A., & Oblozinsky, M. (2022). Epigenetic modifications in plants-impact on phospholipid signaling and secondary metabolism. *Chemicke Listy, 116*, 416–422.

Ballard, J. L., Wang, Z., Li, W., Shen, L., & Long, Q. (2024). Deep learning-based approaches for multi-omics data integration and analysis. *Biodata Mining, 17*(1) https://doi.org/10.1186/s13040-024-00391-z, 38.

Banerjee, A., & Roychoudhury, A. (2015). WRKY proteins: Signaling and regulation of expression during abiotic stress responses. *The Scientific World Journal, 2015*(1). https://doi.org/10.1155/2015/807560

Bhardwaj, V., Heyne, S., Sikora, K., Rabbani, L., Rauer, M., Kilpert, F., Richter, A. S., Ryan, D. P., & Manke, T. (2019). snakePipes: Facilitating flexible, scalable and integrative epigenomic analysis. *Bioinformatics, 35*(22), 4757–4759. https://doi.org/10.1093/bioinformatics/btz436

Bonnet, E., He, Y., Billiau, K., & Van De Peer, Y. (2010). TAPIR, a web server for the prediction of plant microRNA targets, including target mimics. *Bioinformatics, 26*(12), 1566–1568. https://doi.org/10.1093/bioinformatics/btq233

Chen, Y., Li, Y., Narayan, R., Subramanian, A., & Xie, X. (2016). Gene expression inference with deep learning. *Bioinformatics, 32*(12), 1832–1839. https://doi.org/10.1093/bioinformatics/btw074

Chow, H. T., & Mosher, R. A. (2023). Small RNA-mediated DNA methylation during plant reproduction. *The Plant Cell, 35*(6), 1787–1800. https://doi.org/10.1093/plcell/koad010

Dai, X., Zhuang, Z., & Zhao, P. X. (2018). psRNATarget: A plant small RNA target analysis server (2017 release). *Nucleic Acids Research, 46*(W1), W49–W54. https://doi.org/10.1093/nar/gky316

Dao, F., Xie, X., Zhang, H., Guan, Z., Wu, C., Su, W., Wei, Y., Hong, F., Luo, X., Xie, S., & Lai, H. (2024). PlantEMS: A comprehensive database of epigenetic modification sites across multiple plant species. *Plant Communications*, 101228. https://doi.org/10.1016/j.xplc.2024.101228

Davila-Velderrain, J., Caldu-Primo, J. L., Martinez-Garcia, J. C., & Alvarez-Buylla, E. R. (2018). Modeling the epigenetic landscape in plant development. *Methods in Molecular Biology*, 357–383. https://doi.org/10.1007/978-1-4939-8618-7_17

Ding, K., Sun, S., Luo, Y., Long, C., Zhai, J., Zhai, Y., & Wang, G. (2022). PlantCADB: A comprehensive plant chromatin accessibility database. *Genomics, Proteomics & Bioinformatics, 21*(2), 311–323. https://doi.org/10.1016/j.gpb.2022.10.005

Dorff, K. C., Chambwe, N., Zeno, Z., Simi, M., Shaknovich, R., & Campagne, F. (2013). GobyWeb: Simplified management and analysis of gene expression and DNA methylation sequencing data. *PLoS One*, *8*(7), e69666. https://doi.org/10.1371/journal.pone.0069666

Eichten, S. R., & Springer, N. M. (2015). Minimal evidence for consistent changes in maize DNA methylation patterns following environmental stress. *Frontiers in Plant Science*, *6*. https://doi.org/10.3389/fpls.2015.00308

Eraslan, G., Avsec, Z., Gagneur, J., & Theis, F. J. (2019). Deep learning: New computational modelling techniques for genomics. *Nature Reviews Genetics*, *20*(7), 389–403. https://doi.org/10.1038/s41576-019-0122-6

Ernst, J., & Kellis, M. (2017). Chromatin-state discovery and genome annotation with ChromHMM. *Nature Protocols*, *12*(12), 2478–2492. https://doi.org/10.1038/nprot.2017.124

Fan, B., Chen, L., Chen, L., & Guo, H. (2025). Integrative multi-omics approaches for identifying and characterizing biological elements in crop traits: Current progress and future prospects. *International Journal of Molecular Sciences*, *26*(4), 1466. https://doi.org/10.3390/ijms26041466

Ferrari, M., Torelli, A., Marieschi, M., & Cozza, R. (2020). Role of DNA methylation in the chromium tolerance of Scenedesmus acutus (Chlorophyceae) and its impact on the sulfate pathway regulation. *Plant Science*, *301*, 110680. https://doi.org/10.1016/j.plantsci.2020.110680

Fu, L., Zhu, T., Zhou, X., Yu, R., He, Z., Zhang, P., Wu, Z., Chen, M., Kaufmann, K., & Chen, D. (2022). ChIP-hub provides an integrative platform for exploring plant regulome. *Nature Communications*, *13*(1) https://doi.org/10.1038/s41467-022-30770-1, 3413.

Gaggion, N., Ariel, F., Daric, V., Lambert, É., Legendre, S., Roulé, T., Camoirano, A., Milone, D. H., Crespi, M., Blein, T., & Ferrante, E. (2021). ChronoRoot: High-throughput phenotyping by deep segmentation networks reveals novel temporal parameters of plant root system architecture. *GigaScience* https://doi.org/10.1093/gigascience/giab052, *10*.

Ge, S. X., Son, E. W., & Yao, R. (2018). iDEP: An integrated web application for differential expression and pathway analysis of RNA-Seq data. *BMC Bioinformatics*, *19*(1). https://doi.org/10.1186/s12859-018-2486-6, 534.

Gong, W., Pan, X., Xu, D., Ji, G., Wang, Y., Tian, Y., Cai, J., Li, J., Zhang, Z., & Yuan, X. (2022). Benchmarking DNA methylation analysis of 14 alignment algorithms for whole genome bisulfite sequencing in mammals. *Computational and Structural Biotechnology Journal*, *20*, 4704–4716. https://doi.org/10.1016/j.csbj.2022.08.051

Gonzalez, E. M., Zarei, A., Hendler, N., Simmons, T., Zarei, A., Demieville, J., Strand, R., Rozzi, B., Calleja, S., Ellingson, H., Cosi, M., Davey, S., Lavelle, D. O., Truco, M. J., Swetnam, T. L., Merchant, N., Michelmore, R. W., Lyons, E., & Pauli, D. (2023). PhytoOracle: Scalable, modular phenomics data processing pipelines. *Frontiers in Plant Science*, *14*. https://doi.org/10.3389/fpls.2023.1112973

Gorelick, N., Hancher, M., Dixon, M., Ilyushchenko, S., Thau, D., & Moore, R. (2017). Google earth engine: Planetary-scale geospatial analysis for everyone. *Remote Sensing of Environment*, *202*, 18–27. https://doi.org/10.1016/j.rse.2017.06.031

Gour, P., Kansal, S., Agarwal, P., Mishra, B. S., Sharma, D., Mathur, S., & Raghuvanshi, S. (2021). Variety-specific transcript accumulation during reproductive stage in drought-stressed rice. *Physiologia Plantarum*, *174*(1). https://doi.org/10.1111/ppl.13585

Guo, W., Zhu, P., Pellegrini, M., Zhang, M. Q., Wang, X., & Ni, Z. (2017). CGmapTools improves the precision of heterozygous SNV calls and supports allele-specific methylation detection and visualization in bisulfite-sequencing data. *Bioinformatics, 34*(3), 381–387. https://doi.org/10.1093/bioinformatics/btx595

He, M., Ren, T. X., Jin, Z. D., Deng, L., Liu, H. J., Cheng, Y. Y., Li, Z. Y., Liu, X. X., Yang, Y., & Chang, H. (2023). Precise analysis of potassium isotopic composition in plant materials by multi-collector inductively coupled plasma mass spectrometry. *Spectrochimica Acta, Part B: Atomic Spectroscopy, 209*, 106781. https://doi.org/10.1016/j.sab.2023.106781

Herath, H. M. S. N., Rafii, M. Y., Ismail, S. I., Jj, N., & Ramlee, S. I. (2020). Improvement of important economic traits in chilli through heterosis breeding: A review. *The Journal of Horticultural Science and Biotechnology, 96*(1), 14–23. https://doi.org/10.1080/14620316.2020.1780162

Hu, Y., Lu, Y., Zhao, Y., & Zhou, D. (2019). Histone acetylation dynamics integrates metabolic activity to regulate plant response to stress. *Frontiers in Plant Science, 10*. https://doi.org/10.3389/fpls.2019.01236

Hu, Y., Horlbeck, M. A., Zhang, R., Ma, S., Shrestha, R., Kartha, V. K., Duarte, F. M., Hock, C., Savage, R. E., Labade, A., Kletzien, H., Meliki, A., Castillo, A., Durand, N. C., Mattei, E., Anderson, L. J., Tay, T., Earl, A. S., Shoresh, N., & Buenrostro, J. D. (2025). Multiscale footprints reveal the organization of cis-regulatory elements. *Nature* https://doi.org/10.1038/s41586-024-08443-4, *638*, 779–786.

Huther, P., Hagmann, J., Nunn, A., Kakoulidou, I., Pisupati, R., Langenberger, D., Weigel, D., Johannes, F., Schultheiss, S. J., & Becker, C. (2022). MethylScore, a pipeline for accurate and context-aware identification of differentially methylated regions from population-scale plant whole-genome bisulfite sequencing data. *Quantitative Plant Biology, 3*. https://doi.org/10.1017/qpb.2022.14, e19.

Jogam, P., Sandhya, D., Alok, A., Peddaboina, V., Allini, V. R., & Zhang, B. (2022). A review on CRISPR/Cas-based epigenetic regulation in plants. *International Journal of Biological Macromolecules, 219*, 1261–1271. https://doi.org/10.1016/j.ij.iomac.2022.08.182

Kartashov, A. V., & Barski, A. (2015). BioWardrobe: An integrated platform for analysis of epigenomics and transcriptomics data. *Genome Biology, 16*(1) https://doi.org/10.1186/s13059-015-0720-3, 158.

Kaur, S., Seem, K., Duhan, N., Kumar, S., Kaundal, R., & Mohapatra, T. (2023). Transcriptome and physio-biochemical profiling reveals differential responses of rice cultivars at reproductive-stage drought stress. *International Journal of Molecular Sciences, 24*(2), 1002. https://doi.org/10.3390/ijms24021002

Kawakatsu, T., Nery, J. R., Castanon, R., & Ecker, J. R. (2017). Dynamic DNA methylation reconfiguration during seed development and germination. *Genome Biology, 18*(1). https://doi.org/10.1186/s13059-017-1251-x, 171.

Kelley, D. R., Snoek, J., & Rinn, J. L. (2016). Basset: Learning the regulatory code of the accessible genome with deep convolutional neural networks. *Genome Research, 26*(7), 990–999. https://doi.org/10.1101/gr.200535.115

Kim, J. (2021). Multifaceted chromatin structure and transcription changes in plant stress response. *International Journal of Molecular Sciences, 22*(4), 2013. https://doi.org/10.3390/ijms22042013

Kishore, K., De Pretis, S., Lister, R., Morelli, M. J., Bianchi, V., Amati, B., Ecker, J. R., & Pelizzola, M. (2015). methylPipe and compEpiTools: A suite of R packages for the integrative analysis of epigenomics data. *BMC Bioinformatics*, *16*(1). https://doi.org/10.1186/s12859-015-0742-6

Kok, Z. H., Shariff, A. R. M., Alfatni, M. S. M., & Khairunniza-Bejo, S. (2021). Support vector machine in precision agriculture: A review. *Computers and Electronics in Agriculture*, *191*, 106546. https://doi.org/10.1016/j.compag.2021.106546

Koppad, S. B. A., Gkoutos, G. V., & Acharjee, A. (2021). Cloud computing enabled big multi-omics data analytics. *Bioinformatics and Biology Insights*, *15*. https://doi.org/10.1177/11779322211035921

Kovalchuk, I. (2021). Role of DNA methylation in genome stability. In Kovalchuk, I. and Kovalchuk, O. (eds.) *Genome Stability (Second Edition): From Virus to Human Application*, Volume 26 in Translational Epigenetics. *Academic Press*. https://doi.org/10.1016/b978-0-323-85679-9.00023-4

Kumar, S., & Mohapatra, T. (2021). Dynamics of DNA methylation and its functions in plant growth and development. *Frontiers in Plant Science*, *12*. https://doi.org/10.3389/fpls.2021.596236

Kumar, V., Thakur, J. K., & Prasad, M. (2021). Histone acetylation dynamics regulating plant development and stress responses. *Cellular and Molecular Life Sciences*, *78*(10), 4467–4486. https://doi.org/10.1007/s00018-021-03794-x

Li, S., Xia, Q., Wang, F., Yu, X., Ma, J., Kou, H., Lin, X., Gao, X., & Liu, B. (2017). Laser irradiation-induced DNA methylation changes are heritable and accompanied with transpositional activation of mPing in rice. *Frontiers in Plant Science*, *8*. https://doi.org/10.3389/fpls.2017.00363

Liao, W., Yen, M., Ju, E., Hsu, F., Lam, L., & Chen, P. (2015). Methgo: A comprehensive tool for analyzing whole-genome bisulfite sequencing data. *BMC Genomics*, *16*(S12). https://doi.org/10.1186/1471-2164-16-s12-s11

Lienhard, M., Grimm, C., Morkel, M., Herwig, R., & Chavez, L. (2013). MEDIPS: Genome-wide differential coverage analysis of sequencing data derived from DNA enrichment experiments. *Bioinformatics*, *30*(2), 284–286. https://doi.org/10.1093/bioinformatics/btt650

Liu, Y., Tian, T., Zhang, K., You, Q., Yan, H., Zhao, N., Yi, X., Xu, W., & Su, Z. (2017). PCSD: A plant chromatin state database. *Nucleic Acids Research*, *46*(D1), D1157–D1167. https://doi.org/10.1093/nar/gkx91

Lundberg, S. M., & Lee, S. (2017). A unified approach to interpreting model predictions. *arXiv (Cornell University)*. https://doi.org/10.48550/arxiv.1705.07874

Mahamat, A. A., Boukar, M. M., Leklou, N., Celino, A., Obianyo, I. I., Bih, N. L., Stanislas, T. T., & Savastanos, H. (2024). Decision tree regression vs. Gradient boosting regressor models for the prediction of hygroscopic properties of *Borassus* fruit fiber. *Applied Sciences*, *14*(17), 7540. https://doi.org/10.3390/app14177540

Mavaie, P., Holder, L., & Skinner, M. K. (2023). Hybrid deep learning approach to improve classification of low-volume high-dimensional data. *BMC Bioinformatics*, *24*(1). https://doi.org/10.1186/s12859-023-05557-w, 419.

McCoy, R. M., Julian, R., Kumar, S. R. V., Ranjan, R., Varala, K., & Li, Y. (2021). A systems biology approach to identify essential epigenetic regulators for specific biological processes in plants. *Plants*, *10*(2), 364. https://doi.org/10.3390/plants10020364

McGowan, T., Johnson, J. E., Kumar, P., Sajulga, R., Mehta, S., Jagtap, P. D., & Griffin, T. J. (2020). Multi-omics Visualization Platform: An extensible galaxy plug-in for multi-omics data visualization and exploration. *GigaScience*, *9*(4). https://doi.org/10.1093/gigascience/giaa025

Misra, V., & Mall, A. K. (2024). Harnessing image processing for precision disease diagnosis in sugar beet agriculture. *Crop Design*, *3*(4), 100075. https://doi.org/10.1016/j.cropd.2024.100075

Montesinos-López, O. A., Crespo-Herrera, L., Pierre, C. S., Cano-Paez, B., Huerta-Prado, G. I., Mosqueda-González, B. A., Ramos-Pulido, S., Gerard, G., Alnowibet, K., Fritsche-Neto, R., Montesinos-López, A., & Crossa, J. (2024). Feature engineering of environmental covariates improves plant genomic-enabled prediction. *Frontiers in Plant Science*, *15*. https://doi.org/10.3389/fpls.2024.1349569

N'Diaye, A., Byrns, B., Cory, A. T., Nilsen, K. T., Walkowiak, S., Sharpe, A., Robinson, S. J., & Pozniak, C. J. (2020). Machine learning analyses of methylation profiles uncovers tissue-specific gene expression patterns in wheat. *The Plant Genome*, *13*(2). https://doi.org/10.1002/tpg2.20027

Ni, P., Huang, N., Nie, F., Zhang, J., Zhang, Z., Wu, B., Bai, L., Liu, W., Xiao, C., Luo, F., & Wang, J. (2021). Genome-wide detection of cytosine methylations in plant from nanopore data using deep learning. *Nature Communications*, *12*(1) https://doi.org/10.1038/s41467-021-26278-9, 5976.

Nunez-Vazquez, R., Desvoyes, B., & Gutierrez, C. (2022). Histone variants and modifications during abiotic stress response. *Frontiers in Plant Science*, *13*. https://doi.org/10.3389/fpls.2022.984702

Nunn, A., Can, S. N., Otto, C., Fasold, M., Rodríguez, B. D., Fernández-Pozo, N., Rensing, S. A., Stadler, P. F., & Langenberger, D. (2021). EpiDiverse Toolkit: A pipeline suite for the analysis of bisulfite sequencing data in ecological plant epigenetics. *NAR Genomics and Bioinformatics*, *3*(4). https://doi.org/10.1093/nargab/lqab106

Park, J., Dempewolf, E., Zhang, W., & Wang, Z. (2017). RNA-guided transcriptional activation via CRISPR/dCas9 mimics overexpression phenotypes in arabidopsis. *PLoS One*, *12*(6), e0179410. https://doi.org/10.1371/journal.pone.0179410

Pathak, H., Kumar, M., Molla, K. A., & Chakraborty, K. (2021). Abiotic stresses in rice production: Impacts and management. *ORYZA- an International Journal on Rice*, *58*(Special), 103–125. https://doi.org/10.35709/ory.2021.58.spl.4

Pecinka, A., Chevalier, C., Colas, I., Kalantidis, K., Varotto, S., Krugman, T., Michailidis, C., Vallés, M., Muñoz, A., & Pradillo, M. (2019). Chromatin dynamics during interphase and cell division: Similarities and differences between model and crop plants. *Journal of Experimental Botany*, *71*(17), 5205–5222. https://doi.org/10.1093/jxb/erz457

Quek, C., Jung, C., Bellingham, S. A., Lonie, A., & Hill, A. F. (2015). ISRAP—a one-touch research tool for rapid profiling of small RNA-seq data. *Journal of Extracellular Vesicles*, *4*(1). https://doi.org/10.3402/jev.v4.29454

Rajpal, V. R., Rathore, P., Mehta, S., Wadhwa, N., Yadav, P., Berry, E., Goel, S., Bhat, V., & Raina, S. N. (2022). Epigenetic variation: A major player in facilitating plant fitness under changing environmental conditions. *Frontiers in Cell and Developmental Biology*, *10*. https://doi.org/10.3389/fcell.2022.1020958

Ramírez, F., Dündar, F., Diehl, S., Grüning, B. A., & Manke, T. (2014). deepTools: A flexible platform for exploring deep-sequencing data. *Nucleic Acids Research*, *42*(W1), W187–W191. https://doi.org/10.1093/nar/gku365

Ran, X., Zhao, F., Wang, Y., Liu, J., Zhuang, Y., Ye, L., Qi, M., Cheng, J., & Zhang, Y. (2019). Plant regulomics: A data-driven interface for retrieving upstream regulators from plant multi-omics data. *The Plant Journal, 101*(1), 237–248. https://doi.org/10.1111/tpj.14526

Rao, X., Yang, S., Lü, S., & Yang, P. (2024). DNA methylation dynamics in response to drought stress in crops. *Plants, 13*(14), 1977. https://doi.org/10.3390/plants13141977

Rehman, M., & Tanti, B. (2020). Understanding epigenetic modifications in response to abiotic stresses in plants. *Biocatalysis and Agricultural Biotechnology, 27*, 101673. https://doi.org/10.1016/j.bcab.2020.101673

Robinson, J. T., Thorvaldsdóttir, H., Winckler, W., Guttman, M., Lander, E. S., Getz, G., & Mesirov, J. P. (2011). Integrative genomics viewer. *Nature Biotechnology, 29*(1), 24–26. https://doi.org/10.1038/nbt.1754

Salam, W., Wojewodzic, M. W., & Frisch, D. (2025). FAIRification of the DMRICHR pipeline: Advancing epigenetic research on environmental and evolutionary model organisms. *Bioinformatics Advances* https://doi.org/10.1093/bioadv/vbaf024, *5*.

Sampson, C., Ikenwugwu, T. H., Okagu, I. U., Yahaya, I. I., Odoh, C. K., & Eze, C. N. (2024). Epigenetics: Toward improving crop disease resistance and agronomic characteristics. *Plant Biotechnology Reports, 18*(1), 1–20. https://doi.org/10.1007/S11816-023-00876-Z

Shang, J., Zhang, J., Sun, Y., & Zhang, Y. (2013). EpiMiner: A three-stage co-information based method for detecting and visualizing epistatic interactions. *Digital Signal Processing, 24*, 1–13. https://doi.org/10.1016/j.dsp.2013.08.007

Shen, L., Qi, Z., Ai, Y., Zhang, J., Chao, Y., Han, L., & Xu, L. (2025). Integration of ATAC-seq and RNA-seq reveals the dynamics of chromatin accessibility and gene expression in zoysiagrass response to drought. *Plant Cell Reports, 44*(4) https://doi.org/10.1007/s00299-025-03469-y, 92.

Singh, B. K., Delgado-Baquerizo, M., Egidi, E., Guirado, E., Leach, J. E., Liu, H., & Trivedi, P. (2023). Climate change impacts on plant pathogens, food security and paths forward. *Nature Reviews Microbiology, 21*(10), 640–656. https://doi.org/10.1038/s41579-023-00900- 7. 2023 21:10

Singroha, G., Sharma, P., & Sunkur, R. (2021). Current status of microRNA-mediated regulation of drought stress responses in cereals. *Physiologia Plantarum, 172*(3), 1808–1821. https://doi.org/10.1111/ppl.13451

Smet, D., Opdebeeck, H., & Vandepoele, K. (2023). Predicting transcriptional responses to heat and drought stress from genomic features using a machine learning approach in rice. *Frontiers in Plant Science, 14*. https://doi.org/10.3389/fpls.2023.1212073

Song, Q., Decato, B., Hong, E. E., Zhou, M., Fang, F., Qu, J., Garvin, T., Kessler, M., Zhou, J., & Smith, A. D. (2013). A reference methylome database and analysis pipeline to facilitate integrative and comparative epigenomics. *PLoS One, 8*(12), e81148. https://doi.org/10.1371/journal.pone.0081148

Springer, N. M., & Schmitz, R. J. (2017). Exploiting induced and natural epigenetic variation for crop improvement. *Nature Reviews Genetics, 18*(9), 563–575. https://doi.org/10.1038/nrg.2017.45

Tahir, M., Norouzi, M., Khan, S. S., Davie, J. R., Yamanaka, S., & Ashraf, A. (2024). Artificial intelligence and deep learning algorithms for epigenetic sequence analysis: A review for epigeneticists and AI experts. *Computers in Biology and Medicine, 183*, 109302.

To, T. K., Yamasaki, C., Oda, S., Tominaga, S., Kobayashi, A., Tarutani, Y., & Kakutani, T. (2022). Local and global crosstalk among heterochromatin marks drives DNA methylome patterning in arabidopsis. Nature. *Communications, 13*(1). https://doi.org/10.1038/s41467-022-28468-5

Tonosaki, K., Fujimoto, R., Dennis, E. S., Raboy, V., & Osabe, K. (2022). Will epigenetics be a key player in crop breeding? *Frontiers in Plant Science, 13*. https://doi.org/10.3389/fpls.2022.958350

Tseng, K., Li, G., Hung, Y., Chow, C., Wu, N., Chien, Y., Zheng, H., Lee, T., Kuo, P., Chang, S., & Chang, W. (2020). EXPath 2.0: An updated database for integrating high-throughput gene expression data with biological pathways. *Plant and Cell Physiology, 61*(10), 1818–1827. https://doi.org/10.1093/pcp/pcaa115

Turkoglu, A., Haliloğlu, K., Demirel, F., Aydin, M., Çiçek, S., Yiğider, E., ... Niedbała, G. (2023). Machine learning analysis of the impact of silver nitrate and silver nanoparticles on wheat (Triticum aestivum L.): Callus induction, plant regeneration, and DNA methylation. *Plants, 12*(24), 4151.

Ueda, M., & Seki, M. (2019). Histone modifications form epigenetic regulatory networks to regulate abiotic stress response. *Plant Physiology, 182*(1), 15–26. https://doi.org/10.1104/pp.19.00988

Varshney, R. K., Bohra, A., Yu, J., Graner, A., Zhang, Q., & Sorrells, M. E. (2021). Designing future crops: Genomics-Assisted breeding comes of age. *Trends in Plant Science, 26*(6), 631–649. https://doi.org/10.1016/j.tplants.2021.03.010

Waese, J., Fan, J., Pasha, A., Yu, H., Fucile, G., Shi, R., Cumming, M., Kelley, L. A., Sternberg, M. J., Krishnakumar, V., Ferlanti, E., Miller, J., Town, C., Stuerzlinger, W., & Provart, N. J. (2017). EPlant: Visualizing and exploring multiple levels of data for hypothesis generation in plant biology. *The Plant Cell, 29*(8), 1806–1821. https://doi.org/10.1105/tpc.17.00073

Wang, L., & Qiao, H. (2020). Chromatin regulation in plant hormone and plant stress responses. *Current Opinion in Plant Biology, 57*, 164–170. https://doi.org/10.1016/j.pbi.2020.08.007

Wang, Y., Zhang, P., Guo, W., Liu, H., Li, X., Zhang, Q., Du, Z., Hu, G., Han, X., Pu, L., Tian, J., & Gu, X. (2021). A deep learning approach to automate whole-genome prediction of diverse epigenomic modifications in plants. *New Phytologist, 232*(2), 880–897. https://doi.org/10.1111/nph.17630

Washburn, J. D., Mejia-Guerra, M. K., Ramstein, G., Kremling, K. A., Valluru, R., Buckler, E. S., & Wang, H. (2019). Evolutionarily informed deep learning methods for predicting relative transcript abundance from DNA sequence. *Proceedings of the National Academy of Sciences, 116*(12), 5542–5549. https://doi.org/10.1073/pnas.1814551116

Wu, Q., Yang, M., Yang, Y., Iqbal, A., & Zhou, L. (2025). Assessment of bisulfite sequencing alignment tools for whole genome analysis in plants. *International Journal of Biological Macromolecules*, 140940. https://doi.org/10.1016/j.ij.iomac.2025.140940

Xu, Y., Manoharan, H. T., & Pitot, H. C. (2007). CPG patternfinder: A Windows®-based utility program for easy and rapid identification of the CPG methylation status of DNA. *BioTechniques, 43*(3), 334–342. https://doi.org/10.2144/000112537

Xu, S., Grullon, S., Ge, K., & Peng, W. (2014). Spatial clustering for identification of CHIP-Enriched Regions (SICER) to map regions of histone methylation patterns in embryonic stem cells. *Methods in Molecular Biology*, 97–111. https://doi.org/10.1007/978-1-4939-0512-6_5

Yaish, M. W. (2017). Editorial: Epigenetic modifications associated with abiotic and biotic stresses in plants: An implication for understanding plant evolution. *Frontiers in Plant Science*, *8*. https://doi.org/10.3389/fpls.2017.01983

Yang, X., & Li, L. (2011). miRDeep-P: A computational tool for analyzing the microRNA transcriptome in plants. *Bioinformatics*, *27*(18), 2614–2615. https://doi.org/10.1093/bioinformatics/btr430

Yang, Y., Saand, M. A., Huang, L., Abdelaal, W. B., Zhang, J., Wu, Y., Li, J., Sirohi, M. H., & Wang, F. (2021). Applications of multi-omics technologies for crop improvement. *Frontiers in Plant Science*, *12*. https://doi.org/10.3389/fpls.2021.563953

Yang, Z., Zeng, X., Zhao, Y., & Chen, R. (2023). AlphaFold2 and its applications in the fields of biology and medicine. *Signal Transduction and Targeted Therapy*, *8*(1) https://doi.org/10.1038/s41392-023-01381-z, 115.

Yang, L., Wang, H., Zou, M., Chai, H., & Xia, Z. (2025). Artificial intelligence-driven plant bio-genomics research: A new era. *Tropical Plants*, *4*(1). https://doi.org/10.48130/tp-0025-0008

Yu, G., Wang, L., & He, Q. (2015). ChIPseeker: An R/Bioconductor package for ChIP peak annotation, comparison and visualization. *Bioinformatics*, *31*(14), 2382–2383. https://doi.org/10.1093/bioinformatics/btv145

Zhan, J., & Meyers, B. C. (2023). Plant small RNAs: Their biogenesis, regulatory roles, and functions. *Annual Review of Plant Biology*, *74*(1), 21–51. https://doi.org/10.1146/annurev-arplant-070122-035226

Zhang, M., Zong, W., Zou, D., Wang, G., Zhao, W., Yang, F., Wu, S., Zhang, X., Guo, X., Ma, Y., Xiong, Z., Zhang, Z., Bao, Y., & Li, R. (2022). MethBank 4.0: An updated database of DNA methylation across a variety of species. *Nucleic Acids Research*, *51*(D1), D208–D216. https://doi.org/10.1093/nar/gkac969

Zhang, X., Marand, A. P., Yan, H., & Schmitz, R. J. (2024). scifi-ATAC-seq: Massive-scale single-cell chromatin accessibility sequencing using combinatorial fluidic indexing. *Genome Biology*, *25*(1) https://doi.org/10.1186/s13059-024-03235-5, 90.

Zhao, J., Lu, Z., Wang, L., & Jin, B. (2020). Plant responses to heat stress: Physiology, transcription, noncoding RNAs, and epigenetics. *International Journal of Molecular Sciences*, *22*(1), 117.

Zhao, Z., Alzubaidi, L., Zhang, J., Duan, Y., & Gu, Y. (2023). A comparison review of transfer learning and self-supervised learning: Definitions, applications, advantages and limitations. *Expert Systems with Applications*, *242*, 122807. https://doi.org/10.1016/j.eswa.2023.122807

Zhou, Z., Hu, M., Salcedo, M., Gravel, N., Yeung, W., Venkat, A., Guo, D., Zhang, J., Kannan, N., & Li, S. (2023). XAI meets biology: A comprehensive review of explainable AI in bioinformatics applications. *arXiv (Cornell University)*. https://doi.org/10.48550/arxiv.2312.06082

14

AI-Omics-Epigenetics Integration in Plants: Highlighting the Study of MicroRNAs

Aditya Pratap Singh[1], *Sanjivani Karki*[2], *Ashutosh Sawarkar*[3], *and Siddhartha Singh*[4]

[1] *Department of Genetics and Plant Breeding, Bidhan Chandra Krishi Viswavidyalaya, Nadia, West Bengal, India*
[2] *Department of Agronomy, Uttar Banga Krishi Viswavidyalaya, Pundibari, West Bengal, India*
[3] *Department of Genetics and Plant Breeding, IRDM Faculty Centre, Ramakrishna Mission Vivekananda Educational and Research Institute (RKMVERI), Narendrapur, West Bengal, India*
[4] *College of Horticulture & Forestry, Pasighat, Arunachal Pradesh, Central Agricultural University, Imphal, Manipur, India*

14.1 Introduction

MicroRNAs (miRNAs) are endogenous, non-coding RNAs approximately 20–24 nucleotides in length that play central roles in regulating gene expression in plants by directing ARGONAUTE (AGO)-containing complexes to complementary transcripts, primarily resulting in precise messenger RNA (mRNA) slicing and, in certain contexts, translational repression (Bartel, 2004; Brodersen & Voinnet, 2009). Transcribed by RNA polymerase II as stem-loop forming primary miRNAs (pri-miRNAs), these transcripts are processed by DICER-LIKE 1 (DCL1) in concert with HYPONASTIC LEAVES 1 (HYL1) and SERRATE (SE), yielding precursor and mature miRNAs; one strand is then selectively loaded into AGO to execute silencing (Rogers & Chen, 2013; Voinnet, 2009). Because certain miRNA pathways intersect with DNA methylation and chromatin modifications, miRNAs serve as molecular bridges linking post-transcriptional regulation with epigenetic control mechanisms (Chen, 2009). The identification of plant miRNAs fundamentally redefined gene-regulatory biology (Figure 14.1) by revealing a class of small RNAs that coordinate developmental processes and stress responses. Early work in *Arabidopsis thaliana* first demonstrated small RNAs as potent post-transcriptional regulators influencing plant development and environmental adaptation (Reinhart et al., 2002). Comparative genomic analyses later revealed that many miRNAs are evolutionarily conserved across plant taxa, underscoring their functional and phylogenetic significance (Bartel, 2004). Distinct miRNA families fulfill specific regulatory roles, miR156/miR172 orchestrate phase transitions, miR160/miR167 modulate auxin signaling, and stress-responsive miRNAs such as miR398 and miR393 fine-tune defense and acclimation responses (Chen, 2009; Allen et al., 2005., Sunkar & Zhu, 2004). These functions extend beyond direct mRNA repression to encompass chromatin-level regulation, thereby influencing developmental plasticity and adaptive

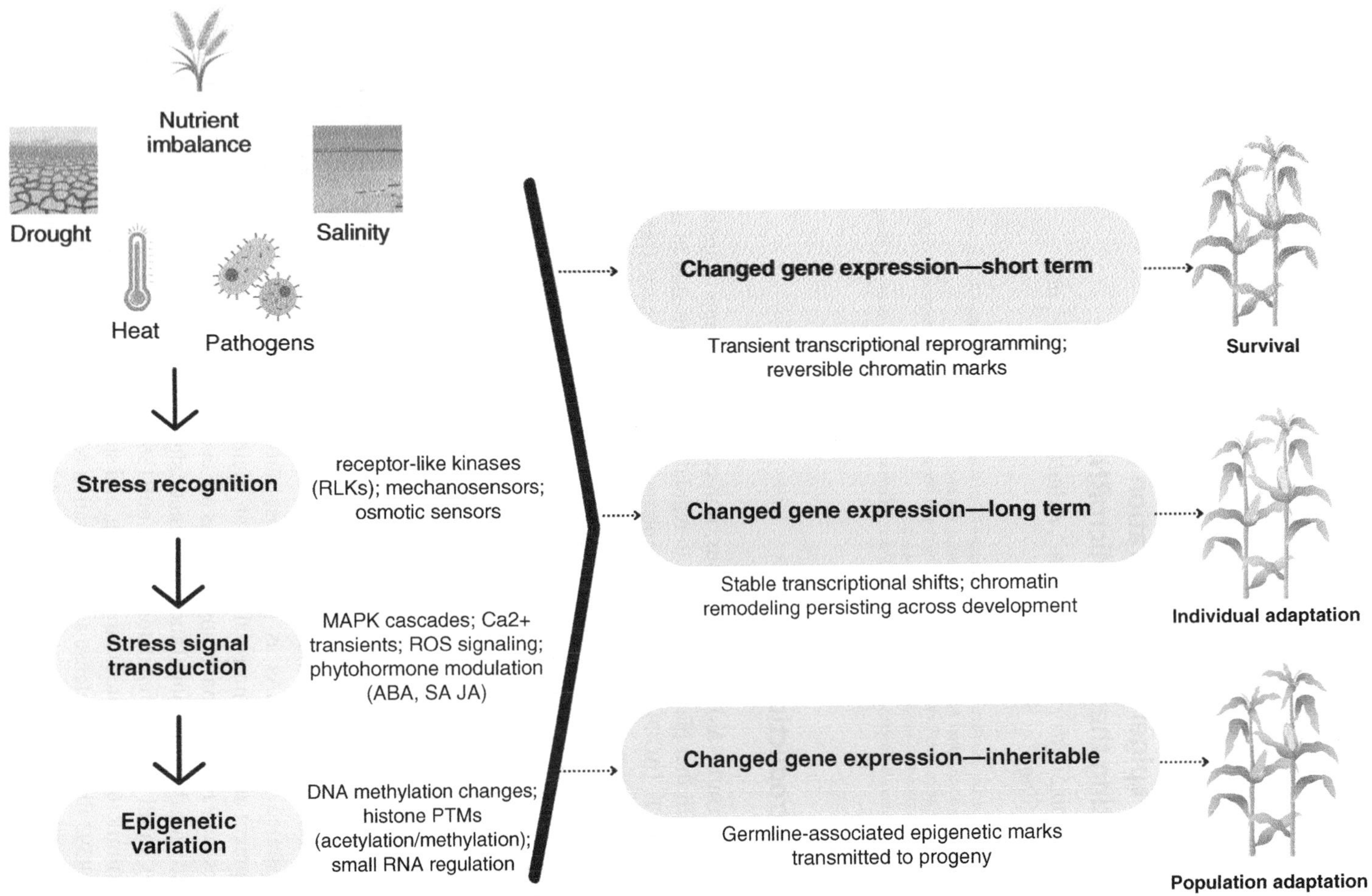

Figure 14.1 Schematic representation of how abiotic and biotic stresses are perceived and transduced into epigenetic changes that alter gene expression across short, long and heritable timescales leading to survival, individual adaptation, or population-level adaptation.

evolution. Further studies have reinforced that miRNAs are central regulators of plant development, physiology, and stress tolerance (Axtell, 2013a, 2013b; Wang et al., 2019). In major model species such as *A. thaliana*, *Oryza sativa* (rice), and *Zea mays* (maize), miRNAs modulate transcription factor networks, nutrient homeostasis, and defense signaling pathways through near-perfect complementarity with target mRNAs, facilitating AGO-mediated cleavage or translational inhibition (Jones-Rhoades et al., 2006).

Despite their biological importance, the experimental discovery and validation of plant miRNAs remain laborious and technically demanding; high-throughput small RNA sequencing (sRNA-seq) followed by stringent computational and molecular validation is required. Moreover, the extensive diversity of plant genomes, hairpin precursor structures, and species-specific targeting rules complicates cross-taxonomic generalization (Wang et al., 2019; Yu et al., 2017). In view of these challenges, computational and statistical approaches have become indispensable to accelerate plant miRNA research. Traditional rule-based pipelines exploited sequence and structural features of miRNA precursors such as minimum free energy, hairpin stability, and conservation to identify novel candidates, but they often suffered from high false-positive rates and limited sensitivity in non-model taxa (Jayasundara et al., 2021). In response, artificial intelligence (AI) technologies particularly machine learning (ML) and deep learning (DL) methods are transforming plant miRNA discovery and target prediction. These models enable automated pattern recognition directly from raw sequences or sRNA-seq datasets, minimizing human bias and improving predictive accuracy. Classical ML algorithms such as support vector machine (SVM) and random forests (RFs) have been used to discriminate true miRNA precursors, while advanced architectures including convolutional neural networks (CNNs), graph neural networks (GNNs), and self-supervised embedding frameworks are now capable of modeling complex sequence–structure relationships and plant-specific complementarity rules (Kuang et al., 2023; Zhang et al., 2024a). The success of AI-driven discovery pipelines, however, relies on high-quality, species-relevant datasets. Annotated miRNA loci from databases such as miRBase, along with plant-specific repositories and genome-wide sRNA maps from model species, provide critical training and validation data for ML and DL frameworks. Yet, considerable obstacles remain, including inconsistencies in library-preparation protocols, high genome complexity (notably in maize), and a scarcity of experimentally verified miRNA targets in non-model species; together, these factors hinder cross-taxon generalization and limit reproducibility persists (Guo et al., 2022; Martin et al., 2023). Furthermore, benchmarking across available tools remains inconsistent due to heterogeneous negative datasets, variable evaluation metrics, and limited cross-taxon performance testing. Overcoming these challenges will be key to realizing the full potential of AI in elucidating plant miRNA biology and advancing crop improvement.

14.2 Molecular Basis of Plant miRNA-Mediated Regulation

Plant miRNAs are produced from RNA polymerase II transcripts (pri-miRNAs) that fold into imperfect stem-loop structures and are sequentially processed in the nucleus by a conserved microprocessor complex composed of DCL1, HYL1, and SE to release miRNA/miRNA duplexes; one strand is then loaded into an AGO effector to assemble the

RNA-induced silencing complex (RISC) (Figure 14.2) (Baulcombe, 2004; Li & Yu, 2021). The precision of processing is influenced by accessory factors and post-translational modifications of the microprocessor components, which in plants can be further modulated by environmental signals and protein partners that affect localization and activity (Gonzalo et al., 2024; Li & Yu, 2021). Once loaded into AGO (commonly AGO1 in many species), plant miRNA-RISC recognizes target transcripts primarily through extensive sequence complementarity, which frequently results in AGO-catalyzed endonucleolytic cleavage of the target and subsequent exonucleolytic degradation (Jones-Rhoades et al., 2006). In addition to slicing, plant miRNAs can repress translation mechanistically distinct from animal miRNA repression by inhibiting translation initiation or elongation when complementarity and binding site context permit; these repression modes may precede or act in parallel with mRNA decay (Iwakawa & Tomari, 2013; Naeli et al., 2023). The catalytic activation of slicing-competent AGOs involves metal–ion coordination and conformational rearrangements that position the scissile phosphate, and differences among AGO paralogs influence target outcome (Nakanishi, 2024). High-throughput degradome (PARE) sequencing has

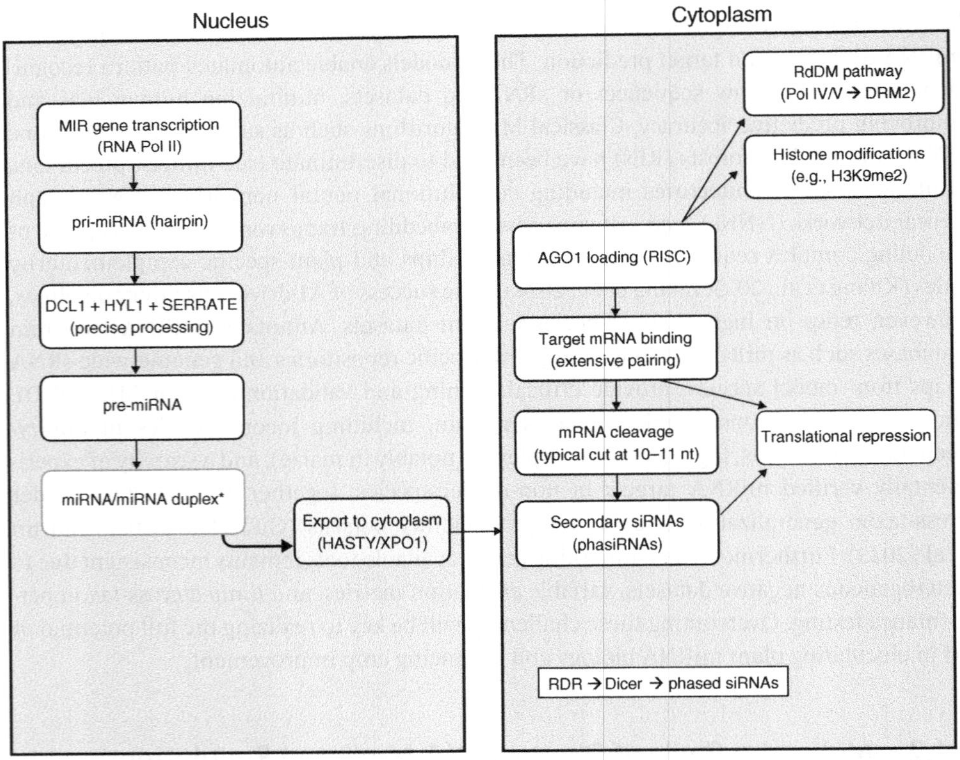

Figure 14.2 Transcription of primary miRNAs by RNA polymerase II; DCL1-HYL1-SE (DICER-LIKE 1–HYPONASTIC LEAVES 1–SERRATE) processing to pre- and mature miRNAs; AGO1-RISC (ARGONAUTE–RNA-induced silencing complex) loading; target mRNA cleavage/translational repression. Arrows indicate miRNA links to RdDM (DNA methylation) and histone-modifying pathways, emphasizing the bridge between post-transcriptional silencing and chromatin regulation.

provided transcriptome-wide evidence for miRNA-directed cleavage in diverse plant species, enabling experimental validation and the creation of training sets for computational predictors (Addo-Quaye et al., 2009; Thody et al., 2020; Zhou et al., 2022). Coupling sRNA-seq expression profiles with degradome data increases confidence in predicted miRNA–target pairs and helps disambiguate primary regulatory interactions from secondary effects. Importantly, the positional context of the target site (coding sequence versus untranslated region), local RNA structure, and target accessibility modulate the efficiency of slicing versus translational repression, complicating purely sequence-based prediction approaches (Jin et al., 2020; Thody et al., 2020).

14.3 Epigenetic Roles of miRNAs in Plants

Plant miRNAs directly and indirectly affect epigenetic modifications. In the RdDM pathway, small RNAs can guide AGO complexes to homologous genomic loci to promote de novo cytosine methylation via Pol IV/V and DNA methyltransferases, stabilizing silencing at transposons and certain stress-responsive genes (Matzke & Mosher, 2014). Indirectly, miRNAs influence histone modification and chromatin accessibility by targeting transcription factors or chromatin-associated enzymes. Representative miRNA families and their regulatory roles are summarized in Table 14.1. For example, miR156's regulation of SPL (SQUAMOSA Promoter Binding Protein-Like) transcription factors contributes to reprogramming during the vegetative-to-reproductive transition; miR393's repression of auxin receptors (TIR1/AFBs) fine-tunes chromatin responsiveness at auxin-regulated loci (Wu et al., 2009; Navarro et al., 2006). (Table 14.1)

An additional dimension is stress memory, where regulatory states persist and can influence responses upon re-exposure. Stress-inducible miRNAs such as miR398 participate in regulatory loops that can coincide with persistent changes in DNA methylation or histone marks, enabling plants to mount faster or stronger responses upon re-exposure (Crisp et al., 2016; Sunkar & Zhu, 2004). While the extent and mechanisms of transgenerational inheritance vary, these observations collectively position miRNAs as precision regulators that bridge immediate post-transcriptional control and longer-term chromatin programming.

Biological and technical hurdles continue to complicate miRNA studies. First, sequence variability and lineage-specific emergence of miRNAs (on backgrounds of polyploidy and transposable elements) complicate annotation and cross-species inference (Meyers et al., 2008). Second, spatio-temporal specificity means many miRNAs accumulate in restricted tissues or developmental phases or transiently under stress—demanding expensive, finely resolved datasets to capture true dynamics (Zhang, 2015). Third, data integration remains difficult: researchers must reconcile small RNA-seq with degradome, transcriptome, methylome, and chromatin immunoprecipitation sequencing (ChIP-seq) datasets each with its own noise profile and biases (Yu et al., 2021). Finally, false positives persist in both de novo discovery (hairpins that are not genuine miRNAs) and target prediction, particularly in non-model species with incomplete annotations (Rogers & Chen, 2013). Collectively, these issues motivate approaches that learn subtle signals from noisy, mixed datasets and generalize beyond model species.

Table 14.1 Representative plant miRNA families and roles.

miRNA family	Primary role	Epigenetic/regulatory impact	Example species	References
miR156	Phase transition	SPL TFs; chromatin remodeling during vegetative→reproductive shift	*Arabidopsis*, rice	Wu et al. (2009), Chen (2009)
miR172	Flowering control	Tuning of AP2-like TFs; floral organ identity	Maize, barley	Zhu and Helliwell (2011)
miR168	AGO regulation, stress	AGO1 feedback loop; RISC homeostasis	Rice, maize	Vaucheret et al. (2006)
miR393	Auxin and stress	TIR1/AFB repression; chromatin accessibility at auxin-responsive loci	Wheat, soybean	Navarro et al. (2006)
miR398	Oxidative stress	Cu/Zn-SOD repression; stress memory circuits	*Arabidopsis*, tobacco	Sunkar and Zhu (2004)
miR399	Phosphate homeostasis	Crosstalk with DNA methylation under Pi starvation	*Arabidopsis*, rice	Chiou et al. (2006)
miR160	Auxin regulation	ARF targeting; effects on histone marks at auxin loci	Rice, soybean	Wang et al. (2019)
miR164	Morphogenesis, senescence	NAC TFs; developmental and stress programs	*Arabidopsis*, maize	Mallory et al. (2004)
miR395	Sulfate assimilation	ATP sulfurylase and transporters; nutrient epigenetics	*Arabidopsis*, wheat	Kawashima et al. (2009)
miR827	Pi/N economy	NLA targeting; nutrient homeostasis	Rice, maize	Kant et al. (2011)
miR444	Monocot development, stress	MADS-box TFs; integration of stress/development	Rice, barley	Sunkar et al. (2008)
miR482/2118	Pathogen defense	phasiRNAs from NBS-LRR loci; reinforced defense	Tomato, soybean	Shivaprasad et al. (2012)

AGO, ARGONAUTE; ARF, auxin response factor; LRR, leucine-rich repeat; MADS, MCM1/AGAMOUS/DEFICIENS/SRF transcription factor family; NAC, NAM/ATAF/CUC transcription factor family; NBS, nucleotide-binding site; NLA, nitrogen limitation adaptation; RISC, RNA-induced silencing complex; SOD, superoxide dismutase; SPL, SQUAMOSA Promoter Binding Protein-Like; TF, transcription factor.

14.4 Artificial Intelligence in Biological Data Science

AI encompassing ML and DL infers predictive structure directly from examples rather than fixed rules. DL breakthroughs such as AlphaFold's protein structure prediction illustrate how sequence-to-function relationships can be learned at scale (Ching et al., 2018; Jumper et al., 2021). In genomics and epigenomics, AI has advanced single-cell analysis, regulatory element discovery, and image-based phenotyping by extracting nonlinear patterns across high-dimensional measurements (Zou et al., 2019). For plant sciences, AI already supports genome annotation, yield forecasting, and stress phenotyping, demonstrating value from molecules to fields.

14.4.1 Relevance of AI to Plant miRNA Research

14.4.1.1 De Novo Discovery

ML classifiers (e.g., SVMs and RFs) and DL models (e.g., CNNs) can separate true miRNA precursors from spurious hairpins by learning from features such as stem-loop stability, minimum free energy, k-mer composition, and cross-species conservation, reducing dependence on hard thresholds and improving generalization across datasets (Yousef et al., 2016). Deep models extract hierarchical features automatically, a practical advantage where sequence diversity and noise would otherwise mask signal.

14.4.2 Target Prediction and Interaction Modeling

Classical tools (e.g., TargetFinder and psRNATarget) rely on complementarity and thermodynamics but often overpredict. AI models refine these calls by incorporating structure accessibility, degradome support, evolutionary signals, and even epigenetic context (e.g., methylation states near-target loci), improving precision while providing confidence estimates (Patel et al., 2025).

14.4.3 Pattern Recognition in Multiomics

Autoencoders and related representation-learning methods compress heterogeneous omics layers (small RNA, transcriptome, methylome, ChIP-seq) into latent factors that preserve biological variance, enabling correlation of miRNA expression with chromatin and transcriptional outcomes. Ensemble learning can integrate outputs from independent models (e.g., structure-based and expression-based predictors), yielding robust miRNA–gene network hypotheses resilient to dataset noise. In rice and soybean, integrative AI frameworks have helped identify stress-inducible or tissue-specific miRNAs and prioritize biologically validated targets relevant to drought tolerance, salinity response, or nitrogen fixation (Li et al., 2019). Case studies indicate that AI both accelerates discovery and improves the quality of hypotheses, reaching experimental validation.

14.4.4 Translational Potential: From Basic Research to Precision Agriculture

AI-assisted miRNA discovery enables biomarker development for resilience (e.g., miR398 for oxidative stress, miR393 for biotic stress, and miR169 for drought), guiding selection or engineering efforts and informing management decisions (Song et al., 2019; Kuang et al., 2023). Coupled with genome and epigenome editing (e.g., CRISPR-based tuning of miRNA loci or targets), these insights support fine control over growth–defense trade-offs (Raza et al., 2021). Nutrient-responsive miRNAs provide further levers: manipulating the miR399–PHO2 axis can improve phosphate use efficiency, aligning molecular control with resource-smart agronomy (Chiou et al., 2006). In a precision agriculture framework, AI models predicting miRNA dynamics under variable environments can inform the timing and intensity of interventions from irrigation to nutrient regimes, linking molecular signals with field operations. This pathway from molecular insights to field practice is where the promise of AI-enabled miRNA research becomes tangible.

14.5 AI and ML Techniques for MiRNA Discovery and Prediction

The advent of next-generation sequencing technologies has resulted in an unprecedented accumulation of genomic and transcriptomic data from a wide array of plant species. This data deluge has created both opportunities and challenges for miRNA research. On one hand, the availability of large-scale datasets enables comprehensive exploration of miRNA repertoires across diverse taxa and conditions. On the other hand, the sheer volume and complexity of the data necessitate robust, scalable, and accurate computational tools for effective analysis. AI and ML, with their capacity to learn complex patterns from large datasets, have emerged as powerful approaches for addressing these challenges. By automating the identification of miRNA candidates and their targets, these technologies have the potential to revolutionize plant miRNA research, making it more efficient and accessible.

ML, a subset of AI, encompasses a range of algorithms that can learn from data and make predictions or decisions without being explicitly programmed for specific tasks. In the context of plant miRNA discovery, ML algorithms are trained on known miRNA sequences and their features—such as sequence composition, secondary structure, and thermodynamic stability—to distinguish true miRNAs from other small RNAs or random sequences. Commonly used ML techniques include SVMs, RFs, artificial neural networks (ANNs), and, more recently, DL models such as CNNs and recurrent neural networks (RNNs). These models can integrate diverse types of data, including sequence, structural, and evolutionary information, to improve the accuracy of miRNA prediction.

The application of AI and ML in plant miRNA research extends beyond the identification of novel miRNAs. These technologies are also employed in the prediction of miRNA targets, elucidation of miRNA-mediated regulatory networks, and functional annotation of miRNAs. Target prediction is particularly challenging in plants due to the high degree of complementarity required between miRNAs and their target mRNAs as well as the presence of multiple potential binding sites. ML-based approaches can incorporate sequence features, binding energy calculations, and evolutionary conservation to enhance the specificity and sensitivity of target prediction. Furthermore, AI-driven network analysis tools can help unravel the complex interactions between miRNAs, their targets, and other regulatory molecules, providing insights into the broader functional roles of miRNAs in plant systems.

Despite the significant progress made, several challenges remain in the application of AI and ML to plant miRNA discovery and prediction. One major issue is the limited availability of high-quality, experimentally validated miRNA datasets for many plant species, which can hinder the training and validation of ML models. Additionally, the heterogeneity of plant genomes, the presence of species-specific miRNAs, and the dynamic nature of miRNA expression under different developmental and environmental conditions add layers of complexity to the prediction task. Addressing these challenges requires the development of more sophisticated models that can generalize across species and conditions as well as the integration of multiomics data to capture the full spectrum of miRNA-mediated regulation.

Recent advances in DL, a branch of ML characterized by multilayered neural networks capable of learning hierarchical representations, hold promise for overcoming some of these challenges. DL models can automatically extract relevant features from raw sequence data, reducing the reliance on manual feature engineering and potentially uncovering novel patterns associated with miRNA biogenesis and function. Moreover, the integration of AI with cloud computing and high-performance computing resources enables the analysis of large-scale datasets in a timely and cost-effective manner, further accelerating the pace of discovery.

The integration of AI and ML into plant miRNA discovery and prediction represents a transformative shift in the field, offering new avenues for the systematic identification and functional characterization of miRNAs. As these technologies continue to evolve, they are expected to play an increasingly central role in unraveling the complexities of plant gene regulation, with far-reaching implications for plant biology, agriculture, and biotechnology. This review aims to provide a comprehensive overview of the current state of AI and ML applications in plant miRNA research, highlighting recent advances, ongoing challenges, and future directions in this rapidly developing area.

The increasing volume of plant genomic and transcriptomic data has prompted the adoption of computational approaches that leverage AI and ML to accelerate miRNA discovery, structural prediction, and functional annotation. These technologies offer scalable, efficient alternatives to experimental pipelines and have become integral to modern plant genomics.

14.6 Computational Approaches in miRNA Discovery

Computational identification of miRNAs typically begins with the extraction of precursor sequences (pre-miRNAs) from plant genomes. These precursors are characterized by hairpin-loop structures with thermodynamically stable configurations. Early algorithms, such as miRDeep and miRPlant, utilized rule-based systems and thermodynamic modeling to distinguish true miRNAs from pseudo hairpins. However, these approaches depended on manually engineered features such as GC (guanine–cytosine) content, minimum free energy (MFE), nucleotide composition, and sequence motifs.

With the advent of ML, data-driven models began automating feature extraction and classification. By converting RNA sequences into numerical representations—often using k-mer frequencies, secondary-structure metrics, and free energy parameters—ML algorithms can differentiate miRNAs from non-miRNAs more accurately. These computational frameworks have laid the foundation for integrating AI into large-scale plant miRNA analysis.

14.7 ML Models for Plant miRNA Prediction

Traditional ML algorithms such as SVMs, RFs, and naive Bayes classifiers were among the earliest to be applied for miRNA prediction. SVMs, for instance, classify sequences based on kernel functions that map input features into high-dimensional spaces, efficiently separating miRNAs from pseudo hairpins (Gálvez et al., 2012). RFs, composed of ensemble decision trees, offer higher tolerance to noise and better generalization for diverse plant datasets.

Feature-based ML models have achieved notable accuracy in plants such as *A. thaliana*, rice, and maize, but their performance is often constrained by the need for carefully designed feature sets. Consequently, research has shifted toward DL architectures capable of automatic feature extraction directly from raw sequences, enabling broader generalization across plant species.

14.8 DL Applications and Target Prediction

DL approaches, particularly CNNs and RNNs, have transformed miRNA prediction by learning spatial and sequential dependencies within RNA sequences (Murmu et al., 2024). CNNs capture local sequence motifs and base-pairing patterns, while RNNs—especially Long Short-Term Memory networks—model sequential relationships, enabling context-aware feature learning. These architectures outperform classical ML models in both sensitivity and specificity, especially for novel or low-conservation miRNAs.

Target prediction, another vital aspect of miRNA research, benefits significantly from AI integration. Conventional tools such as TargetFinder and psRNATarget rely on complementarity and energy parameters to predict miRNA–mRNA binding sites. AI-based models enhance this process by learning interaction patterns from large datasets, improving accuracy for plant-specific pairing rules.

Recent developments, such as GNNs, model miRNA–mRNA interactions as bipartite graphs. For instance, the kmerPMTF framework integrates graph topology and sequence information to accurately predict plant miRNA–mRNA relationships (Kumar & Mukhtar, 2025). These models incorporate network context—such as co-expression and gene ontology (GO) associations—providing a systems-level understanding of miRNA regulation.

14.8.1 Comparative Analysis of Computational Tools for Plant miRNA–Target Prediction

A comparative summary of computational tools is presented in Table 14.2. The table summarizes the performance metrics of four distinct computational tools used for miRNA–target prediction in plant genomics. These tools range from foundational algorithms to modern DL architectures, demonstrating an evolutionary improvement in predictive capability.

The comparison clearly illustrates the impact of transitioning from traditional computational methods to DL approaches in bioinformatics.

14.9 Conventional Prediction Tools and Their Limitations

Early computational approaches provided the first practical methods for identifying miRNA targets in plants and remain important baselines for method development. Tools such as TargetFinder and psRNATarget implement explicit sequence complementarity

Table 14.2 Comparative summary of computational tools for plant miRNA–target prediction.

Tool	Approach	Input	Primary output	Strengths	Key limitations	References
TargetFinder	Rule-based complementarity scoring	miRNA + transcript sequences	Candidate target sites + score	Fast; transparent scoring	Overprediction; limited contextual features	Dai and Zhao (2011)
psRNATarget	Rule-based + accessibility considerations (server implementation)	miRNA + transcript sequences	Candidate targets + expectation score	Widely used; user-friendly	Threshold-sensitive; species/annotation dependence	Dai and Zhao (2011)
PAREameters	Data-driven inference of targeting rules using degradome	sRNA-seq + degradome + transcripts	Calibrated targeting parameters; validated cleavage support	Leverages degradome to reduce false positives	Requires degradome; condition/tissue dependence	Thody et al. (2020)
P-TarPmiR	ML-based plant-specific target prediction	miRNA + transcript features (model-dependent)	Ranked targets + confidence	Improved plant-specific prediction	Model transferability depends on training data	Ajila et al. (2023)

rules, mismatch penalties, gap scoring, and simple energetic metrics to nominate miRNA–mRNA pairings (Dai & Zhao, 2011; Srivastava et al., 2014., Ajila et al. 2023). These methods have the advantage of transparency (their scoring rules are interpretable) and low computational cost, enabling genome-wide scans with modest resources. In practice, conventional tools typically achieve moderate predictive performance (reported in earlier comparisons at ~75–82% accuracy depending on datasets and metrics) and are useful for initial candidate filtering in well-annotated genomes. However, several intrinsic limitations reduce their utility as standalone discovery engines. First, rigid sequence-matching rules fail to capture contextual determinants such as target site accessibility, local RNA secondary structure, and co-expression dynamics that modulate effective binding in vivo (Jones-Rhoades et al., 2006). Second, many early validation datasets and parameter choices were derived from model species (e.g., *A. thaliana*) and a handful of major crops (rice, maize), imparting a training bias that reduces sensitivity and specificity when applied to non-model taxa or polyploid genomes (Meyers et al., 2008). Third, conventional tools do not readily incorporate orthogonal experimental evidence (e.g., degradome/PARE, expression correlation, and methylome status), limiting their ability to distinguish functional interactions from spurious complementarity. Finally, the use of fixed thresholds for penalty scores tends to generate either many false positives (if thresholds are loose) or miss true interactions (if thresholds are strict), providing little guidance about prediction confidence for downstream experimental prioritization. Because of these shortcomings, conventional tools are best viewed as fast, interpretable first passes that must be complemented by expression evidence, degradome validation, or more sophisticated statistical or ML rescoring to identify high-confidence, biologically meaningful miRNA targets (Srivastava et al., 2014).

14.10 DL Advancements

The past decade saw a rapid transition from feature-engineered classifiers (SVM, RF) to end-to-end DL systems that automatically learn predictive representations from raw sequence or combined feature sets (Zou et al., 2019). DL approaches brought multiple advantages: hierarchical feature extraction, tolerance to noisy training labels, and the ability to incorporate heterogeneous modalities (sequence, structure, and expression) within a single model.

14.10.1 Convolutional Neural Networks

CNNs learn spatially local motifs that are analogous to position-specific sequence features (e.g., conserved seed regions and structural signature patterns). DeepMirPlant applies convolutional filters to encoded nucleotide sequences (and optionally pairing/structure encodings) to detect discriminative patterns distinguishing bona fide miRNA precursors or binding sites from negatives. Trained on expanded crop datasets (rice, wheat, and soybean) rather than narrowly on *Arabidopsis*, DeepMirPlant reported substantially improved accuracy (~90%) and better cross-species generalization because the convolutional filters capture biologically meaningful motifs that recur across taxa (Zhang et al., 2024c). CNNs also integrate secondary-structure encodings (dot-bracket or pairing probability matrices), enabling the model to attend jointly to sequence and local pairing features that are critical for plant miRNA processing and target recognition.

14.10.2 Graph Neural Networks

GNNs represent biological entities (miRNAs, mRNAs, and proteins) and interactions as nodes and edges in graphs, allowing message passing that aggregates local and global network contexts. The miRNA-GNN framework treats miRNA–mRNA systems as bipartite graphs augmented with multiomics node features (expression levels, methylation scores, conservation metrics, and protein interactions). This enables the model to learn how an interaction's network neighborhood (co-expression partners, chromatin context, functional modules) informs binding likelihood or regulatory potency. GNNs trained on cross-species multiomics datasets achieve state-of-the-art performance (~92% accuracy) because they exploit both molecular sequence signals and systems-level context (Chang et al., 2024). GNNs are especially powerful when the biological problem depends on network topology (e.g., miRNAs that regulate hubs or belong to conserved regulatory modules).

Many successful pipelines combine deep feature extractors (CNNs or transformers) with classical classifiers (e.g., gradient boosting and RF) as ensemble learners. Deep models provide rich representations; tree-based ensemble learners convert those representations into calibrated probability estimates and are robust to label noise. Self-supervised pretraining (masked nucleotide prediction and k-mer embeddings) also improves downstream performance when labeled examples are limited, by enabling transfer learning from large unlabeled sequence corpora (Zou et al., 2019).

14.11 Integrating Multiomics and Epigenetic Data

miRNA regulation is embedded within layered epigenetic and transcriptional landscapes; therefore, integrating methylome, histone modification, chromatin accessibility, and transcriptome data substantially improves predictive and mechanistic inferences.

14.11.1 Data Fusion Strategies

For the data fusion, two broad strategies are commonly used: feature-level fusion and model-level fusion. Feature-level fusion concatenates normalized features from different omics (e.g., read counts, methylation beta values, and histone peak scores) before input to an ML model; model-level fusion trains separate submodels per modality and merges their latent representations (late fusion) (Murmu et al., 2024). Autoencoders and variational autoencoders (VAEs) are effective for denoising and compressing high-dimensional omics into compact latent factors that capture coordinated regulatory variation.

14.11.2 Epigenetic Features That Matter

DNA methylation around MIR loci or target genes, histone marks indicative of active (H3K4me3) or repressed (H3K27me3) chromatin, and Assay for Transposase-Accessible Chromatin using sequencing (ATAC-seq)–derived accessibility scores provide orthogonal evidence that influences miRNA transcription and target availability. For example, methylation of MIR promoters can downregulate miRNA expression, while hypomethylation near-target loci can increase access for RISC binding (Matzke & Mosher, 2014). Including these features helps AI models distinguish conditionally active interactions from constitutive complementarity.

14.11.3 Applications to Stress Biology

Integrating sRNA-seq with methylome and histone profiles has enabled the discovery of stress-responsive miRNAs in maize and soybean; the observed co-variation between miRNA induction and locus-specific demethylation suggests mechanistic coupling that AI models can exploit to prioritize regulatory relationships for experimental validation (Agius et al., 2023; Li et al., 2019).

14.12 Challenges and Future Directions

A persistent set of challenges constrains the deployment and generalizability of AI for plant miRNA research, but each also points to clear technical remedies and research directions. Foremost among these is the scarcity and imbalance of high-quality training data: experimentally validated miRNAs and verified miRNA–target pairs remain limited for many crops and nearly absent for most wild taxa, producing severe class-imbalance problems in which authentic positive examples are vastly outnumbered by negatives; practical mitigations therefore include data augmentation strategies (for example, in silico mutagenesis to synthetically expand positive classes), semi-supervised learning that leverages large volumes of unlabeled sequencing data, and transfer-learning schemes that pretrain models on large, related datasets before fine-tuning on smaller, species-specific collections (Meyers et al., 2008; Zou et al., 2019). A second, related obstacle is the extraordinary biological diversity among plants: genome size, polyploidy, repeat content, and the relative abundance of different small RNA classes vary widely between lineages, which undermines naive model transfer. Addressing this requires careful data harmonization and normalization, explicit domain-adaptation methods, and deliberately phylogenetically diverse training sets; rigorous cross-species evaluation (for example, leave-one-species-out benchmarking) should become standard to quantify true generalization. Equally important is interpretability and biological plausibility: many high-performing deep models behave as "black boxes," which impedes experimental uptake and regulatory confidence. Explainable AI (XAI) techniques including saliency mapping, SHAP (SHapley Additive exPlanations) values, and integrated gradients can expose the nucleotide motifs, structural signals, or epigenetic features that drive predictions, thereby guiding mechanistic follow-up and increasing trust among bench scientists (Das et al., 2017). Integration complexity poses further practical barriers: heterogeneity in library preparation, sequencing depth, and platform-specific biases across sRNA-seq, degradome, methylome, and ChIP/ATAC datasets creates confounders that will distort models unless standardized preprocessing, normalization, and well-defined negative-set construction practices are widely adopted; community benchmarking datasets and a common set of evaluation metrics are therefore essential to enable fair comparisons and reproducible results (Srivastava et al., 2014). Finally, several methodological advances offer promising routes forward: transfer learning and self-supervised pretraining on massive plant sequence corpora can provide robust initializations; multimodal transformers and graph-based multimodal fusion models can better jointly model sequence, structure, and regulatory context; and causal-inference frameworks can help move from correlation to testable mechanistic hypotheses. Complementing these computational advances with experimental designs that close the prediction–validation loop will be critical for accelerating model refinement and for translating AI predictions into validated biological insights and, ultimately, practical crop improvements (Zou et al., 2019; Murmu et al., 2024).

14.12.1 AI-Driven Functional Analysis and Epigenetic Integration

Plants exist in a dynamic, often harsh environment, facing relentless environmental fluctuations. These challenges span biotic stresses, such as attacks from pathogens, and abiotic stresses, including severe drought, high soil salinity, or temperature extremes. To survive and thrive, plants have developed intricate regulatory mechanisms that govern gene expression, dictate protein dynamics, and establish complex epigenetic configurations. These elements are the fundamental determinants of a plant's ability to respond to and successfully adapt to stress.

For decades, foundational research has relied on traditional molecular and genomic methods to explore these biological processes. While these techniques have yielded invaluable data, they often face significant limitations. They are frequently limited in scalability, making it challenging to study responses across vast numbers of samples or conditions. Crucially, they lack robust integrative capability—the power to synthesize the massive, high-dimensional datasets produced by modern "omics" technologies.

In stark contrast, AI has emerged as a genuinely transformative tool in modern biology. AI, particularly its subsets of ML and DL, excels at the very tasks that overwhelm traditional methods: processing, integrating, and interpreting enormous volumes of multiomics data. This includes information from genomics, transcriptomics, proteomics, metabolomics, and epigenomics. By leveraging AI, researchers can now achieve a comprehensive, systems-level understanding of how plants operate and respond under stress conditions (Agius et al., 2023; Zhang et al., 2024b).

The adoption of AI and ML/DL paradigms has dramatically enhanced the capacity to predict crucial biological phenomena, such as gene function, the architecture of regulatory networks, and the intricacies of the epigenetic landscape. These predictive models do more than just process data; they facilitate hypothesis generation, significantly accelerate the pace of precision breeding, and drive overall crop improvement (Cao & Chen, 2024). The convergence of powerful computational techniques with vast biological data streams is generating novel insights into how plants establish stress memory, precisely regulate gene expression, and even transmit adaptive characteristics across successive generations (Cao & Chen, 2024).

14.12.2 Advancing Functional Analysis Through Artificial Intelligence

A core application of AI in plant biology lies in functional analysis, which aims to determine the specific biological roles of genes and their corresponding proteins. Diverse AI algorithms have proven highly effective in this domain, including RFs, SVMs, and DNNs. These models are now routinely employed to infer the functions of genes based on the complex patterns found within expression datasets (Zhang et al., 2024b).

ML classifiers, for instance, are exceptionally good at identifying key stress-responsive genes. They achieve this by recognizing nonlinear patterns in gene expression profiles—patterns that would be nearly invisible to human analysis—across a range of environmental conditions (Li & Vakilian, 2025). Classic examples from staple crops illustrate this success: in *A. thaliana* and *O. sativa* (rice).

The comprehension of cellular communication is also being advanced by Graph-based AI approaches. These models are particularly well-suited for mapping and analyzing protein–protein interaction networks. By treating proteins and their interactions as nodes and edges in a vast network, AI can effectively identify the central regulators—or "hub" proteins—that are critical for mediating a plant's overall stress response (Chang et al., 2024). Furthermore, DL models are adeptly applied to forecast the intricate relationships between miRNAs and their target mRNAs. This capability is essential for uncovering the small, non-coding RNA regulatory mechanisms that govern the complex process of stress adaptation.

The integrated use of AI in functional analysis not only dramatically improves standard gene annotation but also provides powerful assistance in the finer details of molecular biology. This includes the identification of functional domains within complex proteins and mapping out the rules governing their essential post-translational regulation.

14.12.3 Decoding the Epigenetic Landscape with AI

Epigenetic modifications—changes to the genome that do not alter the underlying DNA sequence—are fundamental to gene regulation and, critically, to the establishment of stress memory in plants. These modifications include DNA methylation, various histone modifications, and dynamic chromatin remodeling. Given the complexity and sheer volume of data involved, AI models have become truly indispensable for decoding these crucial regulatory layers.

Advanced DL architectures, such as CNNs and specialized recurrent architectures, are highly effective at analyzing high-throughput sequencing data. For example, they can precisely detect methylation motifs, accurately predict histone binding sites, and classify chromatin accessibility from data generated by techniques like ChIP-seq or ATAC-seq.

Recent, AI-driven methylome studies focusing on model species like *Arabidopsis* and major crops like maize have successfully correlated stress-induced methylation changes with subsequent, long-term transcriptional reprogramming (Agius et al., 2023). This provides compelling evidence that AI is uniquely positioned to uncover the causal relationships between specific epigenetic modifications and observable adaptive phenotypes.

The scope of AI in epigenetics extends beyond DNA and histones. It is now used to predict the complex physical and regulatory associations between long non-coding RNAs (lncRNAs), miRNAs, and various chromatin modifiers (Zhang et al., 2024b). This provides critical insights into how RNA-mediated epigenetic regulation is a major factor contributing to a plant's ability to adapt to stress.

A rapidly developing and critical area is XAI. XAI seeks to move beyond the traditional "black-box" nature of many DL models to enhance biological interpretability. By identifying which epigenetic features are the primary drivers of an AI's predictions, XAI allows researchers to precisely pinpoint stress memory markers (Das et al., 2017). This transparency is not just academically valuable; it is technically crucial for precision engineering. For instance, it can guide the design of bespoke CRISPR-dCas9 systems fused with epigenetic modifiers, allowing for highly targeted, dynamic gene regulation in crops (Cao & Chen, 2024).

14.12.4 Holistic Insights Through Multiomics Integration

A truly complete picture of plant stress responses requires the simultaneous, holistic analysis of multiple omics layers. AI-based multiomics integration is the key technology enabling this level of complexity. It employs sophisticated mathematical and computational methods to effectively merge heterogeneous datasets from genomics, transcriptomics, proteomics, and epigenomics.

Key integration techniques include tensor decomposition, VAEs, and GNNs (Chang et al., 2024). By using these tools, AI models can reveal the intricate interdependencies between various cellular components—for example, how changes in a specific methylation status correlate with alterations in gene expression and subsequent shifts in metabolite accumulation.

Advanced frameworks, such as Bayesian DL, have been successfully implemented to infer causal relationships between regulatory features and important phenotypes, such as a plant's measurable tolerance to drought or salinity. Autoencoders are vital for this process; they compress the high-dimensionality of multiomics data into a set of latent variables that inherently capture co-regulated modules and key features. To aid human comprehension, specialized visualization techniques like t-distributed stochastic neighbor embedding and uniform manifold approximation and projection are used to translate these complex latent patterns into intuitive visual maps, clearly revealing underlying patterns in stress responses.

Graph-based approaches, particularly graph neural networks (GNNs), are especially powerful for modeling complex, interconnected relationships among miRNAs, mRNAs, and epigenetic factors; for example, multisource feature fusion combined with GNNs has been used to predict abiotic stress-responsive miRNAs in key crop species (Chang et al., 2024). In parallel, extensive evidence shows that miRNAs regulate plant stress responses across diverse crops (Sunkar et al., 2012).

14.12.5 Practical Applications in Modern Plant Science

The profound capabilities of AI-driven functional and epigenetic analysis translate directly into numerous practical applications for modern plant biology and agriculture:

Enhanced Stress Adaptation: AI enables the precise identification of key genes and miRNAs whose expression is tightly regulated by epigenetic mechanisms, thereby conferring enhanced resilience against critical threats like drought, salinity, and high heat (Li & Vakilian, 2025).

Targeted Crop Improvement: Computational models can predict stable epigenetic variants and desirable, favorable alleles that are strongly linked to critical agronomic traits such as yield, resource use efficiency, and developmental timing.

Precision Epigenetic Breeding: AI guides the engineering of plants to possess desirable, inheritable epigenetic traits that demonstrably boost productivity, even under challenging and adverse environmental conditions (Cao & Chen, 2024).

Synthetic Biology Design: AI can be used to model and guide the construction of novel gene circuits that incorporate epigenetic switches. This allows for the dynamic, real-time control of stress responses within a plant system (Das et al., 2017).

Beyond the molecular lab, AI tools are also being implemented directly in precision agriculture. By integrating omics data from plant tissues with real-time environmental sensor readings (e.g., soil moisture and temperature), AI can generate sophisticated forecasts of a plant's stress trajectories. This capability facilitates the real-time adjustment of essential management strategies—such as irrigation scheduling, optimizing nutrient supply, and targeted pest management—to maximize the final yield and sustainability.

14.12.6 Dynamic Modeling: Agent-Based and Reinforcement Learning

The complexity of biological systems often requires dynamic modeling techniques. Agent-Based Models (ABMs) are computational frameworks that simulate the behavior of individual biological entities—such as individual cells, molecules, or genes—in a spatial and temporal context, thereby capturing the dynamic and often stochastic nature of plant systems (Das et al., 2017).

In epigenetic modeling, ABMs can represent sophisticated concepts, like how stochastic gene expression or the movement of small molecules like miRNAs contributes to a heterogeneous stress response across different tissues within a single plant.

Reinforcement learning (RL) represents another paradigm, focused on optimizing actions based on environmental feedback. RL can significantly enhance ABM simulations by allowing virtual agents (e.g., a gene-regulatory pathway) to "learn" the optimal, most adaptive responses to specific environmental stimuli through a system of rewards and penalties (Anjali et al., 2019).

The integration of ABM and RL provides a powerful, forward-looking framework for simulating complex molecular decision-making, such as the process of transcriptional reprogramming that occurs under stress. This coupling is invaluable for generating novel hypotheses about localized epigenetic regulation and directly guiding the design of highly focused, high-impact experiments. For example, RL models can be used to iteratively optimize the design of sgRNAs in CRISPR-based epigenome editing tools, aiming to achieve desired methylation changes with the highest possible specificity and efficiency.

Real-world examples clearly demonstrate the transformative impact of these technologies:

CRISPR-Enhanced Resilience: AI-enhanced CRISPR editing has been successfully applied to identify and strategically modify key stress-regulatory miRNAs and their epigenetic targets in critical crops like rice and maize, leading to measurable improvements in drought tolerance (Cao & Chen, 2024). Furthermore, DL and GNNs have played a crucial role in computationally prioritizing miRNA–target sites to amplify crop resilience.

Spatio-temporal Stress Classification: Sophisticated spatio-temporal DL frameworks that combine CNNs and RNNs have been successfully used to classify the severity of nitrogen stress when it occurs under combined, multi-stress condition.

Biomarker Discovery and Management: Specific miRNAs, notably miR393 and miR172, have been identified as robust biomarkers for the real-time stress management of tomato plants. This innovation integrates AI-based prediction with field-level decision support systems, allowing for proactive and optimized management (Li & Vakilian, 2025).

14.12.7 Ongoing Challenges and Future Trajectories

Despite this remarkable progress, the full, widespread implementation of AI in plant epigenetics still faces several significant hurdles. A major obstacle is data heterogeneity across different omics platforms, which makes seamless integration challenging. There is also a distinct limitation in the availability of rich, fully annotated datasets for the vast number of non-model plant species. Finally, the difficulty in transferring predictive models across different species remains a major technical challenge (Agius et al., 2023). Furthermore, the inherent "black-box" nature of many state-of-the-art DL models continues to limit biological interpretability and user trust.

Explainable AI: Developing and implementing XAI methods is paramount to increasing the trust and utility of complex models by providing biological explanations for their predictions.

Single-Cell Multi-omics Integration: Integrating AI with the emerging, high-resolution data from single-cell multiomics will offer unprecedented insights into tissue-specific and cell-to-cell regulatory heterogeneity.

Cross-species Learning Frameworks: Developing models that can effectively learn from one species and generalize predictions to others will enable much broader applicability across the plant kingdom.

The ultimate, most promising avenue for the future involves the direct coupling of AI with transformative genome-editing technologies. Combining AI-driven prediction with tools like CRISPR-dCas9 fused with epigenetic modifiers holds immense potential for achieving precision epigenetic engineering and paving the way for truly sustainable crop improvement (Cao & Chen, 2024).

AI-driven approaches are fundamentally revolutionizing the fields of functional genomics and epigenetics in plants. By providing the tools to effectively integrate massive, heterogeneous multiomics datasets, AI is delivering unprecedented insights into the complexity of gene-regulatory networks, the mechanisms of stress memory, and the dynamics of heritable epigenetic traits. These computational advancements are actively accelerating the transition in plant biology from merely descriptive to fully predictive and prescriptive science. The final, critical step—coupling robust AI predictions with rigorous experimental validation—will enable the rational design of resilient crop varieties, a necessary and urgent step toward ensuring sustainable global agriculture in the face of accelerating climate change.

14.13 Conclusion

The study of miRNAs, the small non-coding RNAs critical for governing gene expression, plant development, and stress adaptation, has been fundamentally transformed by the integration of AI. Traditional experimental techniques for identifying and characterizing these vital regulatory molecules are typically slow, require extensive labor, and are often restricted to specific species. AI-powered computational methods, however, introduce unparalleled precision, scalability, and predictive capability to the field, fostering a more holistic understanding of complex plant regulatory networks (Zhang et al., 2024b).

ML and DL models are now essential tools for tasks ranging from miRNA discovery to target prediction. Algorithms like SVMs, RFs, CNNs, and RNNs are used to distinguish genuine miRNA precursors from pseudo hairpins. These methods achieve high accuracy by analyzing subtle sequence-structural motifs and contextual dependencies often missed by older heuristic models (Zhang et al., 2024a).

Furthermore, AI has drastically improved miRNA–target prediction. Modern AI-enhanced tools go beyond simple sequence matching, integrating multiple biological parameters, including sequence complementarity, binding energy, site accessibility, and evolutionary conservation. This multifactorial approach ensures higher specificity and sensitivity than conventional algorithms, greatly advancing the mapping of crucial miRNA–mRNA interaction landscapes across various plant species (Zhang et al., 2024a).

AI's impact extends to interpreting the functional consequences of miRNA regulation. NLP models, trained on vast scientific corpora, can automatically extract intricate connections between miRNAs, their target genes, and resulting phenotypes, building dynamic, data-enriched knowledge graphs. When combined with databases like Gene Ontology or KEGG pathways, these systems provide mechanistic insights into how miRNAs regulate growth, stress resilience, and even epigenetic processes (Zhang et al., 2024b).

A major leap is the use of AI for multiomics integration. Deep autoencoders and GNNs model high-dimensional data, linking miRNA activity directly to transcriptomics, proteomics, and epigenomic layers like DNA methylation and histone modifications. This unified analysis enables robust predictive modeling of a plant's response to environmental stress. Moreover, RL and ABMs simulate the complex, dynamic behavior of miRNA-mediated epigenetic feedback loops, providing virtual ecosystems for testing systems-level hypotheses (Singh et al., 2024).

AI is directly fueling translational applications in plant biotechnology. For example, AI algorithms integrated with CRISPR-Cas systems for miRNA editing can precisely identify optimal target sites and boost genome-editing efficiency while minimizing unwanted off-target effects (Sherkatghanad, 2023). These breakthroughs are vital for the rational design of high-yielding, climate-smart, and stress-tolerant crops.

Despite this progress, challenges remain. The scarcity of high-quality annotated datasets for non-model plants, the lack of standardized benchmarking protocols, and the "black-box" nature of DL models must be addressed. Future efforts must focus on establishing open-access databases, ensuring transparent model evaluation, and adopting XAI methodologies to guarantee biological interpretability and reliability. The continued convergence of AI with plant molecular and epigenetic sciences is set to rapidly accelerate innovation in crop improvement and sustainable agriculture.

References

Addo-Quaye, C., Miller, W., & Axtell, M. J. (2009). CleaveLand: A pipeline for using degradome data to find cleaved small RNA targets. *Bioinformatics, 25*(1), 130–131. https://doi.org/10.1093/bioinformatics/btn604

Agius, D. R., Kapazoglou, A., Avramidou, E., Baranek, M., Carneros, E., Caro, E., Castiglione, S., Cicatelli, A., Radanovic, A., Ebejer, J. P., Gackowski, D., Guarino, F., Gulyás, A., Hidvégi, N., Hoenicka, H., Inácio, V., Johannes, F., Karalija, E., Lieberman-Lazarovich, M., ... Vassileva, V. (2023). Exploring the crop epigenome: A comparison of DNA methylation profiling techniques. *Frontiers in Plant Science, 14*, 1181039. https://doi.org/10.3389/fpls.2023.1181039

Ajila, V., Colley, L., Ste-Croix, D. T., Nissan, N., Golshani, A., Cober, E. R., Mimee, B., Samanfar, B., & Green, J. R. (2023). P-TarPmiR accurately predicts plant-specific miRNA targets. *Scientific Reports, 13*, 332. https://doi.org/10.1038/s41598-022-27283-8

Allen, E., Xie, Z., Gustafson, A. M., & Carrington, J. C. (2005). microRNA-directed phasing during trans-acting siRNA biogenesis in plants. *Cell, 121*(2), 207–221. https://doi.org/10.1016/j.cell.2005.04.004

Anjali, N., Nadiya, F., Thomas, J., & Sabu, K. K. (2019). Identification and characterization of drought responsive microRNAs and their target genes in cardamom (*Elettaria cardamomum* Maton). *Plant Growth Regulation, 87*, 201–216. https://doi.org/10.1007/s10725-018-0462-9

Axtell, M. J. (2013a). Classification and comparison of small RNAs from plants. *Annual Review of Plant Biology, 64*, 137–159. https://doi.org/10.1146/annurev-arplant-050312-120043

Axtell, M. J. (2013b). ShortStack: Comprehensive annotation and quantification of small RNA genes. *RNA, 19*(6), 740–751. https://doi.org/10.1261/rna.035279.112

Bartel, D. P. (2004). MicroRNAs: Genomics, biogenesis, mechanism, and function. *Cell, 116*(2), 281–297. https://doi.org/10.1016/s0092-8674(04)00045-5

Baulcombe, D. (2004). RNA silencing in plants. *Nature, 431*(7006), 356–363. https://doi.org/10.1038/nature02874

Brodersen, P., & Voinnet, O. (2009). Revisiting the principles of microRNA target recognition and mode of action in plants. *Nature Reviews Genetics, 10*(2), 141–148. https://doi.org/10.1038/nrm2619

Cao, S., & Chen, Z. J. (2024). Transgenerational epigenetic inheritance during plant evolution and breeding. *Trends in Plant Science, 29*(11), 1203–1223. https://doi.org/10.1016/j.tplants.2024.04.007

Chang, L., Jin, X., Rao, Y., & Zhang, X. (2024). Predicting abiotic stress-responsive miRNAs in plants based on multi-source features and graph neural networks. *Plant Methods, 20*, 33. https://doi.org/10.1186/s13007-024-01158-7

Chen, X. (2009). Small RNAs and their roles in plant development. *Annual Review of Cell and Developmental Biology, 25*, 21–44. https://doi.org/10.1146/annurev.cellbio.042308.113417

Ching, T., Himmelstein, D. S., Beaulieu-Jones, B. K., Kalinin, A. A., Do, B. T., Way, G. P., ... Greene, C. S. (2018). Opportunities and obstacles for deep learning in biology and medicine. *Journal of the Royal Society Interface, 15*(141), 20170387. https://doi.org/10.1098/rsif.2017.0387

Chiou, T.-J., Aung, K., Lin, S.-I., Wu, C.-C., Chiang, S.-F., & Su, C.-L. (2006). Regulation of phosphate homeostasis by microRNA399 and PHO2 in *Arabidopsis*. *The Plant Cell, 18*(2), 412–421. https://doi.org/10.1105/tpc.105.038943

Crisp, P. A., Ganguly, D. R., Eichten, S. R., Borevitz, J. O., & Pogson, B. J. (2016). Reconsidering plant memory: Intersections between stress recovery, RNA turnover, and epigenetics. *Science Advances, 2*(2), e1501340. https://doi.org/10.1126/sciadv.1501340

Dai, X., & Zhao, P. X. (2011). psRNATarget: A plant small RNA target analysis server. *Nucleic Acids Research, 39*(suppl_2), W155–W159. https://doi.org/10.1093/nar/gkr319

Das, A. A., Ajayakumar Darsana, T., & Jacob, E. (2017). Agent-based re-engineering of ErbB signalling: A modelling pipeline for integrative systems biology. *Bioinformatics, 33*(5), 726–734. https://doi.org/10.1093/bioinformatics/btw709

Gálvez, F. J., Baghour, M., Hao, G., Cagnac, O., Rodríguez-Rosales, M. P., & Venema, K. (2012). Expression of LeNHX isoforms in response to salt stress in salt sensitive and salt tolerant tomato species. *Plant Physiology and Biochemistry, 51*, 109–115. https://doi.org/10.1016/j.plaphy.2011.10.012

Gonzalo, L., Giudicatti, A. J., & Manavella, P. A. (2024). HYL1's multiverse: A journey through miRNA biogenesis and beyond canonical and non-canonical functions of HYL1. *Current Opinion in Plant Biology, 80*, 102546. https://doi.org/10.1016/j.pbi.2024.102546

Guo, Z., Kuang, Z., Zhao, Y., Deng, Y., He, H., Wan, M., ... Yang, X. (2022). PmiREN2.0: From data annotation to functional exploration of plant microRNAs. *Nucleic Acids Research, 50*(D1), D1475–D1482. https://doi.org/10.1093/nar/gkab811

Iwakawa, H. O., & Tomari, Y. (2013). Molecular insights into microRNA-mediated translational repression in plants. *Molecular Cell, 52*(4), 591–601. https://doi.org/10.1016/j.molcel.2013.10.033

Jayasundara, S., Lokuge, S., Ihalagedara, P. and Herath, D. (2021). Machine learning for plant microRNA prediction: A systematic review. *arXiv preprint* arXiv:2106.15159. https://doi.org/10.48550/arXiv.2106.15159

Jin, J., Xu, Y., Lu, P., Chen, Q., Liu, P., Wang, J., ... Cao, P. (2020). Degradome, small RNAs and transcriptome sequencing of a high-nicotine cultivated tobacco uncovers miRNA's function in nicotine biosynthesis. *Scientific Reports, 10*(1), 11751. https://doi.org/10.1038/s41598-020-68691-y

Jones-Rhoades, M. W., Bartel, D. P., & Bartel, B. (2006). MicroRNAs and their regulatory roles in plants. *Annual Review of Plant Biology, 57*, 19–53. https://doi.org/10.1146/annurev.arplant.57.032905.105218

Jumper, J., Evans, R., Pritzel, A., Green, T., Figurnov, M., Ronneberger, O., ... Hassabis, D. (2021). Highly accurate protein structure prediction with AlphaFold. *Nature*, *596*(7873), 583–589. https://doi.org/10.1038/s41586-021-03819-2

Kant, S., Peng, M., & Rothstein, S. J. (2011). Genetic regulation by NLA and miR827 for maintaining nitrate-dependent phosphate homeostasis in *Arabidopsis*. *PLoS Genetics*, *7*(3), e1002021. https://doi.org/10.1371/journal.pgen.1002021

Kawashima, C. G., Yoshimoto, N., Maruyama-Nakashita, A., Tsuchiya, Y. N., Saito, K., Takahashi, H., & Dalmay, T. (2009). Sulphur starvation induces the expression of microRNA-395 and one of its target genes in *Arabidopsis*. *The Plant Journal*, *57*(2), 313–321. https://doi.org/10.1111/j.1365-313X.2008.03690.x

Kuang, Z., Zhao, Y., & Yang, X. (2023). Machine learning approaches for plant miRNA prediction: Challenges, advancements, and future directions. *Agriculture Communications*, *1*(2), 100014. https://doi.org/10.1016/j.agrcom.2023.100014

Kumar, N., & Mukhtar, M. S. (2025). An NLP-based method to mine gene and function relationships from published articles. *Scientific Reports*, *15*, 7503. https://doi.org/10.1038/s41598-025-91809-z

Li, M., & Yu, B. (2021). Recent advances in the regulation of plant miRNA biogenesis. *RNA Biology*, *18*(12), 2087–2096. https://doi.org/10.1080/15476286.2021.1899491

Li, Z., & Asefpour Vakilian, K. (2025). Detecting the type and severity of mineral nutrient deficiency in rice plants based on an intelligent microRNA biosensing platform. *Sensors*, *25*(16), 5189. https://doi.org/10.3390/s25165189

Mallory, A. C., Dugas, D. V., Bartel, D. P., & Bartel, B. (2004). Microrna regulation of NAC-domain targets is required for proper formation and separation of adjacent embryonic, vegetative, and floral organs. *The Plant Cell*, *16*(4), 935–955. https://doi.org/10.1016/j.cub.2004.06.022

Martin, G. T., Solares, E., Guadardo-Mendez, J., Muyle, A., Bousios, A., & Gaut, B. S. (2023). miRNA-like secondary structures in maize (*Zea mays*) genes and transposable elements correlate with small RNAs, methylation, and expression. *Genome Research*, *33*(11), 1932–1946. https://doi.org/10.1101/gr.277459.122

Matzke, M. A., & Mosher, R. A. (2014). RNA-directed DNA methylation: An epigenetic pathway of increasing complexity. *Nature Reviews Genetics*, *15*(6), 394–408. https://doi.org/10.1038/nrg3683

Murmu, S., Sinha, D., Chaurasia, H., Sharma, S., Das, R., Jha, G.K. and Archak, S. (2024) A review of artificial intelligence-assisted omics techniques in plant defense: Current trends and future directions. Frontiers in Plant Science, 15, 1292054. https://doi.org/10.3389/fpls.2024.1292054Naeli, P., Winter, T., Hackett, A. P., Alboushi, L., & Jafarnejad, S. M. (2023). The intricate balance between microRNA-induced mRNA decay and translational repression. *The FEBS Journal*, *290*(10), 2508–2524. https://doi.org/10.1111/febs.16422

Nakanishi, K. (2024). When Argonaute takes out the ribonuclease sword. *Journal of Biological Chemistry*, *300*(1), 105499. https://doi.org/10.1016/j.jbc.2023.105499

Navarro, L., Dunoyer, P., Jay, F., Arnold, B., Dharmasiri, N., Estelle, M., ... Jones, J. D. G. (2006). A plant miRNA contributes to antibacterial resistance by repressing auxin signaling. *Science*, *312*(5772), 436–439. https://doi.org/10.1126/science.1126088

Patel, J., Patel, D., & Raval, A. (2025). Artificial intelligence for predictive modeling in CRISPR/Cas9 gene editing: A survey of methods and design strategies. *The Journal of Gene Medicine*, *27*(12), e70061. https://doi.org/10.1002/jgm.70061

Raza, A., Charagh, S., Karikari, B., Sharif, R., Yadav, V., Mubarik, M. S., Habib, M., Zhuang, Y., Zhang, C., Chen, H., Varshney, R. K., & Zhuang, W. (2023). miRNAs for crop improvement. *Plant Physiology and Biochemistry, 201*, 107857. https://doi.org/10.1016/j.plaphy.2023.107857

Reinhart, B. J., Weinstein, E. G., Rhoades, M. W., Bartel, B., & Bartel, D. P. (2002). MicroRNAs in plants. *Genes & Development, 16*(13), 1616–1626. https://doi.org/10.1101/gad.1004402

Rogers, K., & Chen, X. (2013). Biogenesis, turnover, and mode of action of plant microRNAs. *The Plant Cell, 25*(7), 2383–2399. https://doi.org/10.1105/tpc.113.113159

Sherkatghanad, Z., Abdar, M., Charlier, J., & Makarenkov, V. (2023). Using traditional machine learning and deep learning methods for on- and off-target prediction in CRISPR/Cas9: A review. *Briefings in Bioinformatics, 24*(3), bbad131. https://doi.org/10.1093/bib/bbad131

Shivaprasad, P. V., Chen, H.-M., Patel, K., Bond, D. M., Santos, B. A. C. M., & Baulcombe, D. C. (2012). A microRNA superfamily regulates nucleotide-binding site–leucine-rich repeats and other mRNAs. *Proceedings of the National Academy of Sciences of the United States of America, 109*(8), 2790–2795. https://doi.org/10.1105/tpc.111.095380

Singh, K., Balyan, H. S., & Gupta, P. K. (2024). Bioinformatics for analyzing the role of epigenetics in plant disease resistance. In J.-T. Chen (Ed.), *Bioinformatics for Plant Research and Crop Breeding* (pp. 351–390). John Wiley & Sons, Ltd. (Wiley). https://doi.org/10.1002/9781394209965.ch15

Song, X., Li, Y., Cao, X., & Qi, Y. (2019). MicroRNAs and their regulatory roles in plant–environment interactions. *Annual Review of Plant Biology, 70*, 489–525. https://doi.org/10.1146/annurev-arplant-050718-100334

Srivastava, P. K., Moturu, T. R., Pandey, P., Baldwin, I. T., & Pandey, S. P. (2014). A comparison of performance of plant miRNA target prediction tools and the characterization of features for genome-wide target prediction. *BMC Genomics, 15*(1), 348. https://doi.org/10.1186/1471-2164-15-348

Sunkar, R., Li, Y. F., & Jagadeeswaran, G. (2012). Functions of microRNAs in plant stress responses. *Trends in Plant Science, 17*(4), 196–203. https://doi.org/10.1016/j.tplants.2012.01.010

Sunkar, R., Zhou, X., Zheng, Y., Zhang, W., & Zhu, J.-K. (2008). Identification of novel and candidate miRNAs in rice by high throughput sequencing. *BMC Plant Biology, 8*, 25. https://doi.org/10.1186/1471-2229-8-25

Sunkar, R., & Zhu, J.-K. (2004). Novel and stress-regulated microRNAs and other small RNAs from *Arabidopsis*. *The Plant Cell, 16*(8), 2001–2019. https://doi.org/10.1105/tpc.104.022830

Thody, J., Moulton, V., & Mohorianu, I. (2020). PAREameters: A tool for computational inference of plant miRNA–mRNA targeting rules using small RNA and degradome sequencing data. *Nucleic Acids Research, 48*(5), 2258–2270. https://doi.org/10.1093/nar/gkz1234

Vaucheret, H., Mallory, A. C., & Bartel, D. P. (2006). AGO1 homeostasis entails coexpression of MIR168 and AGO1 and preferential stabilization of miR168 by AGO1. *Molecular Cell, 22*(1), 129–136. https://doi.org/10.1016/j.molcel.2006.03.011

Voinnet, O. (2009). Origin, biogenesis, and activity of plant microRNAs. *Cell, 136*(4), 669–687. https://doi.org/10.1016/j.cell.2009.01.046

Wang, J., Mei, J., & Ren, G. (2019). Plant microRNAs: Biogenesis, homeostasis, and degradation. *Frontiers in Plant Science, 10*, 360. https://doi.org/10.3389/fpls.2019.00360

Wu, G., Park, M. Y., Conway, S. R., Wang, J.-W., Weigel, D., & Poethig, R. S. (2009). The sequential action of miR156 and miR172 regulates developmental timing in *Arabidopsis*. *Developmental Cell*, *17*(5), 712–722. https://doi.org/10.1016/j.cell.2009.06.031

Yousef, M., Saçar Demirci, M. D., Khalifa, W., & Allmer, J. (2016). Feature selection has a large impact on one-class classification accuracy for microRNAs in plants. *Advances in Bioinformatics*, 5670851. https://doi.org/10.1155/2016/5670851

Yu, L., Guo, R., Jiang, Y., Ye, X., Yang, Z., Meng, Y., & Shao, C. (2021). Degradome sequencing-based identification of phasiRNAs biogenesis pathways in Oryza sativa. *BMC Genomics*, *22*, 93. https://doi.org/10.1186/s12864-021-07406-7Meyers, B. C., Axtell, M. J., Bartel, B., Bartel, D. P., Baulcombe, D., Bowman, J. L., ... Zhu, J. K. (2008). Criteria for annotation of plant microRNAs. *The Plant Cell*, *20*(12), 3186–3190. https://doi.org/10.1105/tpc.108.064311

Yu, Y., Jia, T., & Chen, X. (2017). The 'how' and 'where' of plant microRNAs. *New Phytologist*, *216*(4), 1002–1017. https://doi.org/10.1111/nph.14834

Zhang, B. (2015). MicroRNA: A new target for improving plant tolerance to abiotic stress. *Journal of Experimental Botany*, *66*(7), 1749–1761. https://doi.org/10.1093/jxb/erv013

Zhang, W., Zhang, P., Sun, W., Xu, J., Liao, L., Cao, Y., & Han, Y. (2024a). Improving plant miRNA-target prediction with self-supervised k-mer embedding and spectral graph convolutional neural network. *PeerJ*, *12*, e17396. https://doi.org/10.7717/peerj.17396

Zhang, X., Ibrahim, Z., Khaskheli, M. B., Raza, H., Zhou, F., & Shamsi, I. H. (2024b). Integrative approaches to abiotic stress management in crops: Combining bioinformatics educational tools and artificial intelligence applications. *Sustainability*, *16*(17), 7651. https://doi.org/10.3390/su16177651

Zhang, Y., Huang, J., Xie, F., Huang, Q., Jiao, H., & Cheng, W. (2024c). Identification of plant microRNAs using convolutional neural network. *Frontiers in Plant Science*, *15*, 1330854. https://doi.org/10.3389/fpls.2024.1330854

Zhou, S., Huang, K., Zhou, Y., Hu, Y., Xiao, Y., Chen, T., ... Jiang, X. (2022). Degradome sequencing reveals an integrative miRNA-mediated gene interaction network regulating rice seed vigor. *BMC Plant Biology*, *22*(1), 269. https://doi.org/10.1186/s12870-022-03645-2

Zhu, Q.-H., & Helliwell, C. A. (2011). Regulation of flowering time and floral patterning by miR172. *Journal of Experimental Botany*, *62*(2), 487–495. https://doi.org/10.1093/jxb/erq295Zou, J., Huss, M., Abid, A., Mohammadi, P., Torkamani, A., & Telenti, A. (2019). A primer on deep learning in genomics. *Nature Genetics*, *51*(1), 12–18. https://doi.org/10.1038/s41588-018-0295-5

15

Machine Learning and Computational Biology-Based Epigenetics for Uncovering Plant Adaptive Evolution

Thiruvengadam Abarna, S. Gomathi, Shobana Devi Paulraj, Thirunethiran Karpagam, Angappan Shanmugapriya, and Ramasamy Manikandan

Department of Biochemistry, Shrimati Indira Gandhi College, Tiruchirappalli, Tamil Nadu, India

15.1 Introduction

Epigenetics is defined as the study of "molecular factors and processes around DNA that regulate genome activity independent of DNA sequence and are mitotically stable" (Skinner, 2014). Each cell type possesses a unique epigenome that dictates its specific identity and function, meaning a single genotype can give rise to a multitude of phenotypes; theoretically, infinite epigenomes can exist from one genome sequence. A primary epigenetic mechanism is DNA methylation, which influences gene expression without altering the DNA sequence itself. Other key mechanisms include histone modifications, noncoding RNA (ncRNA) activity, and higher-order chromatin remodeling.

These epigenetic changes are indispensable for normal developmental processes, such as cellular differentiation, tissue formation, and organogenesis. However, when dysregulated, they can lead to altered gene expression patterns associated with adverse clinical outcomes, including cancer, obesity, neurological disorders (e.g., Alzheimer's disease and schizophrenia), and allergies. A major focus of recent research is on how environmental exposures can induce such epigenetic dysregulation, potentially leading to transgenerational inheritance of disease states. Consequently, predicting genomic regions susceptible to these deleterious epigenetic changes is crucial for understanding the fundamental biology of health and disease (Waddington, 1953).

15.1.1 Computational Challenges in Epigenetics

A primary objective in epigenetics is to identify genomic regions susceptible to modification. To translate epigenetic theory into discovery, robust computational methods are essential. However, this pursuit is hampered by significant experimental and computational hurdles. Generating high-quality epigenetic data is both time-consuming and costly. Furthermore, the resulting datasets present a classic big data challenge; they are high-dimensional, yet the biologically significant signals, such as disease-associated differentially methylated regions (DMRs), are exceedingly rare amidst a majority of non-DMR sites, both characterized by a vast number of genomic features (Manikkam et al., 2014).

To overcome these obstacles, an integrated analytical framework is required. This has prompted the development of novel machine learning (ML) approaches that combine:

active learning (AL) to optimize resource allocation by strategically selecting the most informative data points for experimental validation;
imbalanced class learning (ICL) to improve model accuracy by effectively handling the overwhelming majority of nonsignificant sites; and
deep learning (DL) to autonomously identify complex, high-level features from raw genomic data, bypassing the need for manual feature definition.

Plants have developed exceptional regenerative abilities as a key survival mechanism. This process relies on the rapid reprogramming of somatic cells, a transition guided significantly by epigenetic regulation. Key mechanisms include histone modifications, DNA methylation, chromatin remodeling, and small RNAs, which act as dynamic signals to activate or repress gene expression (Xue et al., 2025). For instance, DNA methylation and H3K9me2 typically enforce gene silencing, while H3K4me3 and H3K36me3 are hallmarks of active transcription.

While transcriptomic and epigenetic studies have mapped broad changes during regeneration in many species (Chen et al., 2024), conventional bulk analysis methods are insufficient. They obscure the critical cellular heterogeneity within explant tissues, limiting our understanding of how totipotency and pluripotency are established at the single-cell level. Consequently, a fundamental question remains unresolved, how is epigenetic regulation precisely coordinated within specific cell types to direct fate transitions during regeneration.

15.1.2 The Challenge of Phenotypic Plasticity and Rapid Adaptation

The heritability of phenotypic variation is a fundamental driver of evolutionary change. Plants frequently exhibit phenotypic plasticity, where a single genotype can produce diverse phenotypes in response to different environmental conditions. This adaptability is crucial for survival and influences the evolution of complex traits like flowering time, drought resistance, and herbivore defense, which are typically governed by multiple genes (Lynch & Walsh, 1998).

Stable changes in epigenetic marks can create "epialleles" loci with heritable alternative epigenetic states. Research using *Arabidopsis thaliana* epigenetic recombinant inbred lines (epiRILs) has shown that such epigenetic variation can generate phenotypic diversity, enhancing the productivity and stability of plant populations (Latzel et al., 2013). However, these foundational insights come from artificially induced variation in a controlled model system. The extent to which naturally occurring epigenetic variation influences phenotypes and ecology in wild plant populations remains a significant and unresolved question.

15.1.3 Epigenetics: A Bridge Between Genome and Environment

Epigenetics serves as a critical interface, enabling a static genome to dynamically respond to a changing environment. It encompasses heritable changes in gene activity that do not involve alterations to the DNA sequence itself. This system functions as a biological bridge, converting environmental cues into stable gene expression profiles that allow an organism to adapt its physiology. This process is mediated by biochemical modifications that alter chromatin structure, effectively creating a "molecular memory" of past conditions.

A canonical example is vernalization, where a winter plant's exposure to prolonged cold triggers the accumulation of repressive histone marks, such as H3K27me3, on the flowering repressor gene FLC. This epigenetically silences the gene, "remembering" the winter and permitting flowering only when spring arrives (Sung & Amasino, 2004). This adaptive mechanism explains how plants can be "primed" by stress, enhancing their resilience. Furthermore, it provides a plausible framework for rapid population-level adaptation and the transgenerational inheritance of acquired characteristics (Weaver et al., 2004).

15.1.4 Navigating the Data Deluge

The rise of high-throughput technologies in genomics and related fields has triggered a "data deluge," generating biological datasets of such immense volume and complexity that they eclipse the capabilities of conventional analysis. For instance, modern DNA sequencers can produce terabytes of data daily, rendering manual interpretation entirely infeasible. This paradigm shift has made sophisticated computational methods encompassing bioinformatics pipelines, ML, and advanced statistical modeling absolutely essential. These tools are indispensable for processing, integrating, and extracting meaningful patterns from the data, transforming raw information into actionable biological knowledge and driving discovery in modern life sciences (Stephens et al., 2015).

15.2 Foundational Concepts in Plant Epigenetics

Due to their sessile nature, plants cannot escape environmental stressors and have consequently evolved complex systems for phenotypic plasticity. This is the capacity of an organism with a fixed genotype to alter its phenotype in response to environmental cues. Epigenetic regulation serves as the molecular foundation for this adaptability. It facilitates swift, often reversible, and occasionally heritable modifications to gene expression. This enables plants to cope with stress, regulate key developmental shifts, and preserve the integrity of their genome.

15.2.1 Key Epigenetic Marks

Plant adaptation is governed not only by genetic code but also by a crucial layer of epigenetic regulation. A primary mechanism is DNA methylation, which suppresses transposable elements (TEs) and can create stable, inherited epialleles that influence physical traits.

Histone modifications, such as H3K27me3 and H3K9ac, act as dynamic switches that alter chromatin structure to control gene activity, especially under stress. Additionally, small RNAs direct targeted gene silencing and steer epigenetic changes via the RNA-directed DNA methylation (RdDM) pathway. Collectively, these epigenetic marks enable immediate environmental responsiveness and produce a source of heritable variation that can drive evolutionary adaptation (Zhang et al., 2018).

15.2.1.1 DNA Methylation

DNA methylation, the covalent addition of a methyl group to cytosine bases, is a fundamental epigenetic mark in plants. Its functional role extends beyond maintaining genome stability by silencing TEs and is now recognized as a critical driver of rapid environmental adaptation. This mechanism provides a source of heritable phenotypic variation that complements DNA sequence evolution.

Plants dynamically reprogram their methylomes in response to biotic and abiotic stressors. This involves the targeted recruitment or inhibition of DNA methyltransferases (e.g., MET1 and DRM2) and demethylases (e.g., ROS1), leading to locus-specific methylation changes, often in promoters of stress-responsive genes or TEs. For example,

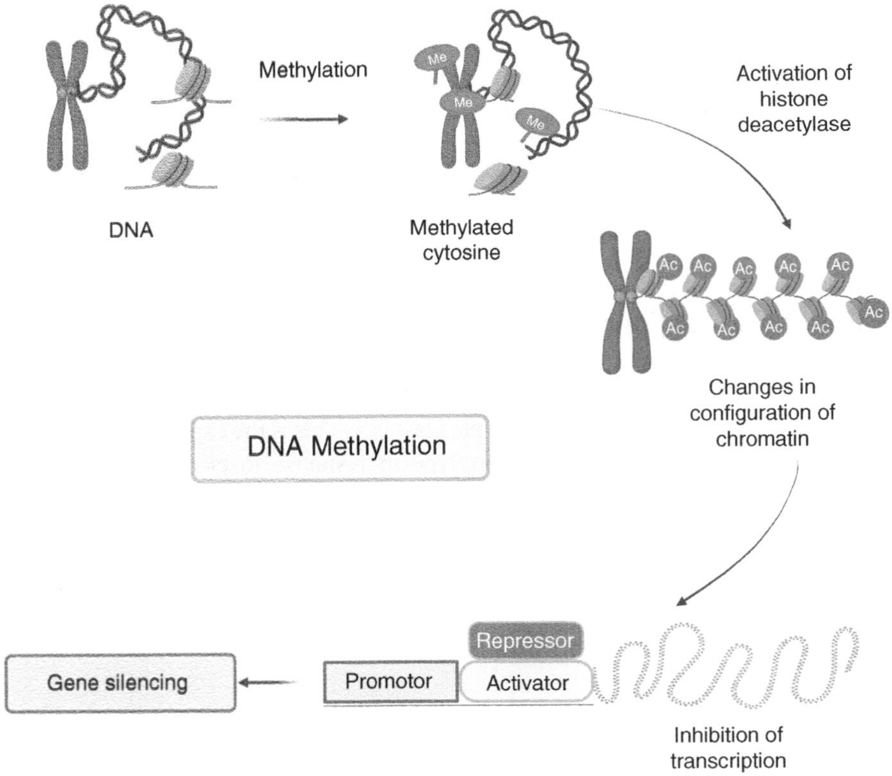

Figure 15.1 DNA methylation.

drought-induced hypomethylation can activate transcription factors, promoting survival in a clear case of rapid, nongenetic acclimation (Dowen et al., 2012).

When such methylation changes become meiotically stable, they form "epialleles," which are subject to natural selection. A classic illustration is the Lcyc epiallele in toadflax, where stable methylation causes a distinct floral symmetry phenotype (Cubas et al., 1999). These epialleles serve as a reservoir of heritable variation for adaptive traits.

Furthermore, DNA methylation plays a paradoxical role in evolution by controlling TEs. Stress-induced genome-wide demethylation can reactivate TEs, potentially causing deleterious mutations. However, this "genomic shock" can also generate a burst of novel genetic diversity, creating new variation for natural selection—a direct link between epigenetic regulation and genetic evolution (Mirouze & Paszkowski, 2011).

15.2.1.2 Histone Modifications

Histone modifications constitute a second major epigenetic mechanism, involving enzyme-mediated chemical alterations such as acetylation, methylation, phosphorylation, and ubiquitylation to the tails of histone proteins. These changes directly influence nucleosome structure and function by modulating the interaction between DNA and histones (Bannister & Kouzarides, 2011).

The functional outcomes of these modifications are diverse. Histone acetylation (e.g., H3K9ac and H3K27ac) neutralizes the positive charge of lysine residues, relaxing chromatin structure to promote an active transcriptional state (Alaskhar Alhamwe et al., 2018). In contrast, histone methylation can either activate or repress transcription, depending on the specific residue and the degree of methylation; for instance, H3K4me is associated with gene activation, while H3K27me3 is a hallmark of facultative heterochromatin and gene silencing (Esteller, 2008). Histone phosphorylation, such as on H2A(X), is primarily implicated in DNA damage signaling and repair (Rossetto et al., 2012). Histone ubiquitylation also has dual roles, with H2AK119ub linked to repression and H2BK123ub involved in transcriptional activation (Bannister & Kouzarides, 2011). Critically, the effect of a specific histone mark is not isolated but is interpreted within the broader context of other nearby epigenetic signals on the chromatin fiber, creating a complex regulatory code.

15.2.1.3 Noncoding RNA-Associated Gene Silencing

ncRNA-mediated silencing represents a more recently characterized layer of epigenetic regulation. These are functional RNA molecules that are transcribed from DNA but not translated into proteins. Once considered genomic "noise," ncRNAs are now known to be pivotal regulators of gene expression and are thought to contribute significantly to phenotypic diversity across and within species (Esteller, 2008).

Major classes include short molecules like microRNAs (miRNAs) and small interfering RNAs (siRNAs), and long ncRNAs (lncRNAs) exceeding 200 nucleotides. These ncRNAs are integral to epigenetic control, directly participating in gene silencing and guiding DNA methylation and histone modifications (Wei et al., 2017). A key mechanism involves siRNAs and certain lncRNAs recruiting enzymatic complexes to specific genomic loci, facilitating the formation of repressive heterochromatin and leading to stable gene silencing (Ching et al., 2012).

15.2.2 Mechanisms of Epigenetic Inheritance: Mitotic and Meiotic

Epigenetic inheritance serves as a critical conduit through which environmental experiences can induce rapid and heritable biological changes, independent of DNA sequence alterations. This process operates across two key scales: mitotic and meiotic inheritance.

Mitotic inheritance underpins phenotypic plasticity, enabling an organism to adapt its physiology within its lifetime. For example, a plant encountering drought or pathogens can remodel the epigenetic landscape of stress-response genes, such as through DNA demethylation, creating a molecular "memory." This altered state is propagated through cell division, priming the plant for more effective responses to future stresses (Bruce et al., 2007).

Meiotic inheritance, however, has evolutionary consequences. When epigenetic states are transmitted through gametes to subsequent generations, they create a substrate for natural selection. Notable examples include: the transgenerational increase in genetic recombination rates in Arabidopsis following heat stress, linked to stable DNA hypomethylation (Pecinka et al., 2009), and the offspring of herbivore-attacked plants exhibiting enhanced resistance, likely mediated by inherited changes in siRNA and DNA methylation (Rasmann et al., 2012).

These heritable epigenetic variants are known as epialleles. A classic example is the LINARIA flower, where a stable, meiotically inherited DNA methylation change alters a key developmental gene, resulting in a radial symmetry phenotype (Cubas et al., 1999).

15.2.3 Epigenetic Regulation of Key Agronomic Traits

Epigenetic mechanisms are central controllers of vital agronomic traits, enabling plants to synchronize development and mount resilient responses to environmental stresses.

15.2.3.1 Epigenetic Basis of Flowering Time

The transition to flowering is a precisely timed developmental shift governed by epigenetic regulation. The canonical example is the vernalization response in Arabidopsis, where prolonged winter cold triggers the PRC2 complex to deposit repressive H3K27me3 marks at the FLC flowering repressor locus. This creates a stable, mitotically heritable silenced state, allowing flowering in the favorable conditions of spring (Sung & Amasino, 2004). DNA methylation dynamics are equally crucial, as demonstrated in rice, where impaired demethylation (e.g., in ROS1 mutants) causes hypermethylation and silencing of the OsMADS51 flowering promoter, resulting in delayed flowering (Choi et al., 2021). Manipulating these pathways offers strategies for adapting crop flowering time to new environments.

15.2.3.2 Epigenetic Basis of Stress Memory

Plants can develop a "memory" of past stress, leading to enhanced responses upon re-exposure, a phenomenon known as priming. This is underpinned by persistent epigenetic marks. For instance, an initial dehydration stress can establish lasting H3K4me3 marks at stress-responsive genes like RD29B, poisoning them for rapid reactivation during a subsequent drought (Ding et al., 2012). Remarkably, this primed state can be transmitted

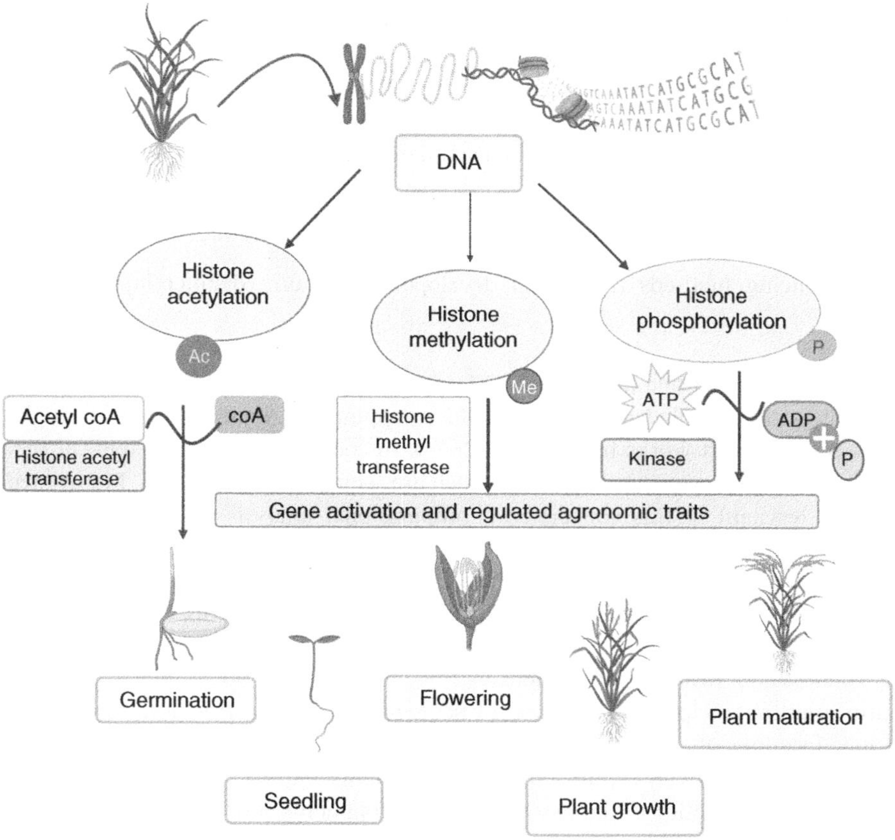

Figure 15.2 Epigenetic regulation of key agronomic traits.

transgenerationally; offspring of Arabidopsis plants exposed to pathogens or UV stress show enhanced resistance correlated with hypomethylation of defense genes (Luna et al., 2012). Harnessing this epigenetic priming holds promise for breeding more resilient crops (Slaughter et al., 2012).

15.2.3.3 Epigenetic Regulation of Disease Resistance

To balance growth and defense, the plant immune system is tightly controlled by epigenetic mechanisms. Many disease resistance (R) genes are maintained in a repressed state under normal conditions via DNA methylation and repressive histone marks like H3K9me2. Mutants in epigenetic regulators like DDM1 often exhibit autoimmunity due to the unintended activation of these R-genes (Yi & Richards, 2007). During infection, plants dynamically remodel their epigenome, such as increasing H3K9ac at defense genes to promote their expression. Conversely, the RdDM pathway silences immune components, preventing excessive activation (Dowen et al., 2012). Epigenetic engineering thus presents an avenue to stably enhance disease resistance without compromising yield.

15.3 Acquiring and Processing Epigenomic Data

The rapid advancement in plant epigenetics is driven by high-throughput sequencing technologies, which produce massive datasets. Converting this raw data into meaningful biological knowledge depends on a rigorous and standardized computational workflow.

15.3.1 High-Throughput Sequencing Technologies for Epigenomics

Specific sequencing methods have been developed to capture distinct layers of the epigenome.

15.3.1.1 Bisulfite Sequencing (BS-seq, WGBS) for DNA Methylation

This method is the established benchmark for mapping DNA methylation at single-nucleotide resolution. It works by treating DNA with sodium bisulfite, a chemical that converts unmethylated cytosines to uracils, which are sequenced as thymines. Methylated cytosines are resistant to this conversion. Computational analysis then compares the bisulfite-treated sequence to a reference genome to pinpoint methylated cytosines versus unmethylated ones. Whole-genome bisulfite sequencing (WGBS) applies this process genome-wide, generating a complete map of the methylome (Lister et al., 2008).

15.3.1.2 ChIP-seq for Histone Modifications

Chromatin immunoprecipitation followed by sequencing (ChIP-seq) is used to pinpoint the genomic locations of specific histone marks or DNA-binding proteins. The technique uses an antibody to selectively capture chromatin fragments bound to a specific target (e.g., the H3K27me3 modification). The associated DNA fragments are then sequenced and aligned to the reference genome. Genomic regions showing a statistically significant accumulation of sequencing reads indicate where that specific epigenetic mark is present, revealing its role in gene activation or repression (Park, 2009).

15.3.1.3 ATAC-seq for Chromatin Accessibility

Assay for transposase-accessible chromatin using sequencing (ATAC-seq) has emerged as the foremost technique for profiling chromatin accessibility genome-wide. The method utilizes a hyperactive Tn5 transposase enzyme that simultaneously fragments DNA and inserts sequencing adapters exclusively into "open" genomic regions devoid of nucleosomes. Sequencing these adapter-tagged fragments generates a comprehensive map of the active regulatory landscape, enabling precise identification of promoters, enhancers, and other regulatory elements with exceptional sensitivity and minimal cellular input (Buenrostro et al., 2013).

15.3.2 Preprocessing and Quality Control of NGS Data

Raw sequencing output (FASTQ format) frequently contains technical artifacts and biases that necessitate rigorous preprocessing and quality control to ensure analytical reliability. Initial quality assessment using tools like FastQC provides critical metrics on read quality, nucleotide composition, and adapter contamination. Subsequent data trimming with programs such as Trimmomatic or Cutadapt removes low-quality bases and adapter sequences, a vital step for preventing alignment errors and false discoveries. For BS-seq data, an additional computational preparation step involves generating in silico bisulfite-converted reference genomes, which are essential for proper alignment using specialized software like Bismark or BS-Seeker2 (Guo et al., 2013).

15.3.3 Core Bioinformatics Pipelines: Alignment, Peak Calling, and Differential Analysis

The fundamental computational workflow for epigenomic data involves three key stages that convert processed reads into biologically meaningful information. Quality-filtered reads are mapped to reference genomes using specialized aligners, with BWA/Bowtie2 employed for standard sequencing data and Bismark reserved for bisulfite-converted reads. For ChIP-seq and ATAC-seq data, algorithms like MACS2 identify statistically significant regions of read enrichment ("peaks") compared to control samples, demarcating protein-binding sites or accessible chromatin regions (Krueger & Andrews, 2011).

Comparative analysis between experimental conditions utilizes specialized tools, with programs like DSS and methylKit detecting DMRs in methylation studies, while software such as diffBind identifies significant changes in peak enrichment (differentially accessible regions) in chromatin accessibility.

15.4 Machine Learning for Decoding the Epigenomic Language

The epigenome constitutes a complex and dynamic regulatory layer that governs the interpretation of the genetic code. Comprising a diverse repertoire of reversible covalent modifications to histones and DNA, such as methylation and acetylation, it dynamically regulates chromatin structure and gene expression. These modifications are added and removed by specialized "writer" and "eraser" enzymes, creating a sophisticated "epigenetic language" (Dawson, 2017). Deciphering the patterns and rules embedded within these epigenetic marks is a quintessential big data challenge. ML, which enables computers to learn from data without explicit programming, has created unprecedented momentum in epigenomics. ML applications are now crucial for addressing fundamental biological questions about the role of epigenetic elements in gene regulation. (Egger et al., 2004; Holder et al., 2017).

15.4.1 Dimensionality Reduction and Pattern Discovery

Epigenomic datasets from technologies like ChIP-seq and ATAC-seq are inherently high-dimensional, featuring measurements across millions of genomic bins. When the number of features (p) vastly exceeds the number of samples (N), a scenario known as the "curse of dimensionality" arises, leading to computational inefficiency, multicollinearity, and reduced predictive power (Bellman, 1957). Dimensionality reduction is therefore a critical first step to simplify this complexity, uncover underlying structures, and avoid overfitting (Skinner, 2014).

Feature extraction transforms original data into a lower-dimensional space using linear or nonlinear operations. The goal is to preserve maximal information with fewer variables. Common methods in epigenomics include principal component analysis (PCA), t-distributed stochastic neighbor embedding (t-SNE), and Uniform Manifold Approximation and Projection (UMAP) (Altman & Krzywinski, 2018). Feature selection directly selects a subset of the most relevant features from the original set, discarding the rest to improve model performance and interpretability.

15.4.1.1 Unsupervised Learning for Epigenome Exploration

Unsupervised learning techniques are used to explore high-dimensional epigenomic data without predefined labels, with two primary goals: clustering to identify natural subgroups and dimensionality reduction for visualization (James et al., 2013).

k-Means Clustering: A centroid-based method that partitions data points into a predefined number (k) of clusters. It is widely used to segment the genome into functionally distinct regions, such as promoters, enhancers, and repressed domains, based on histone modification patterns (Murtagh & Legendre, 2014).

Hierarchical Clustering: Builds a tree of nested clusters, revealing relationships between samples or cell types and often recapitulating developmental lineages based on epigenetic similarity.

Principal Component Analysis (PCA): A linear technique that identifies orthogonal axes (principal components) that capture the maximum variance in the data. It is excellent for visualizing broad trends, such as separating cancer subtypes based on DNA methylation profiles.

t-SNE and UMAP: These nonlinear techniques excel at preserving local data structure and revealing discrete subpopulations. They are particularly powerful in single-cell epigenomics (e.g., scATAC-seq) for visualizing clear separation of cell types that may be blurred by PCA (Cusanovich et al., 2015; McInnes et al., 2018).

15.4.2 Supervised Learning for Predictive Epigenomics

While unsupervised learning explores data structure, supervised learning builds predictive models using labeled data, where each sample has a known outcome. This paradigm learns a mapping function from epigenomic features to a target variable, falling into two subcategories (Aref-Eshghi et al., 2018):

Classification Models: These predict categorical labels (e.g., disease versus healthy, cancer stage). Algorithms like random forest and support vector machines (SVMs) can classify phenotypic states from epigenetic data. A key advantage of models like random forest is their ability to provide feature importance scores, identifying the most influential genomic loci for the prediction (Capper et al., 2018).

Regression Models: These predict continuous values (e.g., gene expression levels and biological age). Methods like elastic net and support vector regression (SVR) quantify the relationship between epigenetic variation and quantitative traits. For instance, "epigenetic clocks" use regression on specific CpG sites to accurately estimate biological age, providing insights into aging and disease risk (Horvath, 2013).

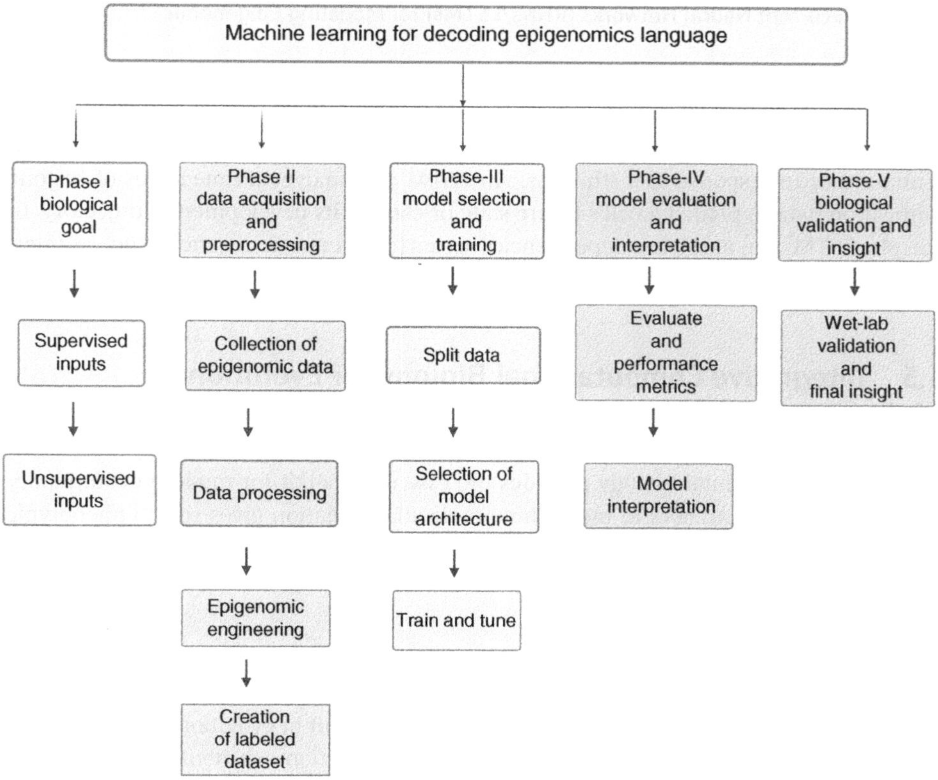

Figure 15.3 Machine learning for decoding the epigenomic language.

These models have transformative applications, from predicting gene expression levels based on histone modifications to linking epigenetic changes with physiological traits like drug response across a population (Kaur et al., 2019).

15.4.3 Deep Learning Architectures for Sequence and Function

DL, a subfield of ML using complex artificial neural networks (ANNs), excels at modeling intricate, nonlinear relationships in high-dimensional data. While powerful, these models are typically data-hungry, computationally intensive, and can be less interpretable than traditional ML (LeCun et al., 2015).

15.4.3.1 Convolutional Neural Networks (CNNs) for cis-Regulatory Element Detection

CNNs use filters to scan input data for local patterns, making them ideal for analyzing genomic sequences and epigenetic signals. They can predict chromatin accessibility or histone marks directly from DNA sequence alone (e.g., models like Basset or DeepSEA), learning the cis-regulatory code, including novel transcription factor binding motifs. CNNs can also be applied to epigenetic data tracks to precisely pinpoint enhancer and promoter boundaries with high accuracy (Kouzarides, 2007).

15.4.3.2 Recurrent Neural Networks (RNNs/LSTMs) for Modeling Epigenomic Dynamics

Recurrent neural networks (RNNs) and their advanced variant, long short-term memory (LSTM) networks, are designed for sequential data where context and order are critical. They are used to model how the epigenome changes over time, such as during cellular differentiation or in response to a stimulus. An LSTM can analyze a time series of histone modification data to predict a cell's future state or classify its developmental trajectory. In principle, LSTMs can also learn dependencies across long genomic distances, such as interactions between promoters and distal enhancers.

15.5 Integrative Computational Biology for Evolutionary Insights

Integrative computational biology provides the essential toolkit for modern evolutionary inquiry, enabling researchers to model how molecular variation gives rise to phenotypic diversity. It bridges scales to identify the genetic and epigenetic mechanisms that have been shaped by natural selection over time (Schmid et al., 2018).

15.5.1 Multiomics Data Integration

The integration of multiomics data represents a paradigm shift in evolutionary biology, enabling a systems-level understanding of how epigenetic variation contributes to phenotypic diversity and adaptation. This approach moves beyond single-data-type analyses to correlate patterns across the epigenome (e.g., DNA methylation and histone modifications), transcriptome (gene expression), and phenome (morphological, physiological, or behavioral traits).

Key computational integration strategies include:

Concatenation-Based Methods: These methods simply merge diverse datasets into a single feature matrix for input into ML models designed to predict complex phenotypes.

Matrix Factorization: Techniques like joint non-negative matrix factorization (jNMF) simultaneously decompose multiple omics matrices to identify shared latent factors representing coordinated molecular programs.

Network-Based Integration: This involves constructing multilayered networks where nodes represent genes or genomic regions, and edges are inferred from correlations across different data types (e.g., an eQTL-like analysis for epigenetic marks, or "meQTL").

The primary goal is to move from correlation to causation, distinguishing between epigenetic changes that are drivers of transcriptional and phenotypic innovation and those that are merely passengers or consequences. For instance, analyzing matched methylome and transcriptome data across divergent populations can identify conserved associations where specific methylation changes are consistently linked to expression changes in key developmental genes. This suggests a functionally important and potentially adaptive regulatory mechanism (Gomez-Chavez et al., 2021).

15.5.2 Phylogenetic Comparative Methods for Epigenetics

Phylogenetic comparative methods (PCMs), long a cornerstone of evolutionary biology, are now being adapted to dissect the evolution of epigenetic traits. These methods account for the shared evolutionary history among species, which induces non-independence in comparative data and helps prevent spurious correlations. Key applications in epigenetics include:

Phylogenetic Signal Quantification: Metrics like Pagel's λ or Blomberg's K test whether the distribution of an epigenetic mark (e.g., global methylation levels) across a phylogeny is consistent with a Brownian motion model of evolution or if it is more labile.
Models of Trait Evolution: Fitting different evolutionary models (e.g., Brownian motion and Ornstein–Uhlenbeck) to epigenetic data helps infer the mode and tempo of epigenetic change. For example, an Ornstein–Uhlenbeck process can suggest stabilizing selection on an optimal epigenetic state.
Correlated Evolution: Methods like phylogenetic generalized least squares (PGLS) test for evolutionary correlations between epigenetic marks and other traits (e.g., body size, mating system, and environmental variables) while controlling for phylogenetic history.

These methods allow researchers to determine whether epigenetic states are evolutionarily conserved, how quickly they diverge after speciation, and whether epigenomic shifts are associated with major adaptive radiations or ecological transitions (Revell & Harmon, 2022).

15.5.3 Identifying Epigenetic Footprints of Selection and Domestication

A central goal in evolutionary epigenetics is to identify genomic regions where epigenetic variation has been a target of natural or artificial selection. These regions bear distinctive "footprints" that can be detected computationally:

Population Epigenetics Statistics: Analogous to population genetics, metrics are calculated from epigenomic data across populations. For DNA methylation, these include:

Epigenetic Differentiation: This identifies loci with significantly different methylation frequencies between populations, suggesting local adaptation.
Within-Population Epigenetic Diversity (π): Scanning for regions of reduced epigenetic diversity in a population can indicate a selective sweep that has fixed a favorable epigenetic allele.
Selection Scans in Domesticated Species: Comparing the epigenomes of domesticated crops or animals with their wild progenitors is a powerful approach. Regions showing convergent methylation changes across independent domestication events provide strong evidence for their role in domestication syndromes (e.g., changes in seed size, plant architecture, or tameness).
Environmental Association Analysis: Using mixed models or redundancy analysis (RDA), researchers can identify specific epigenetic loci that are strongly associated with environmental gradients (e.g., temperature, salinity, and altitude), implicating them in local adaptation.

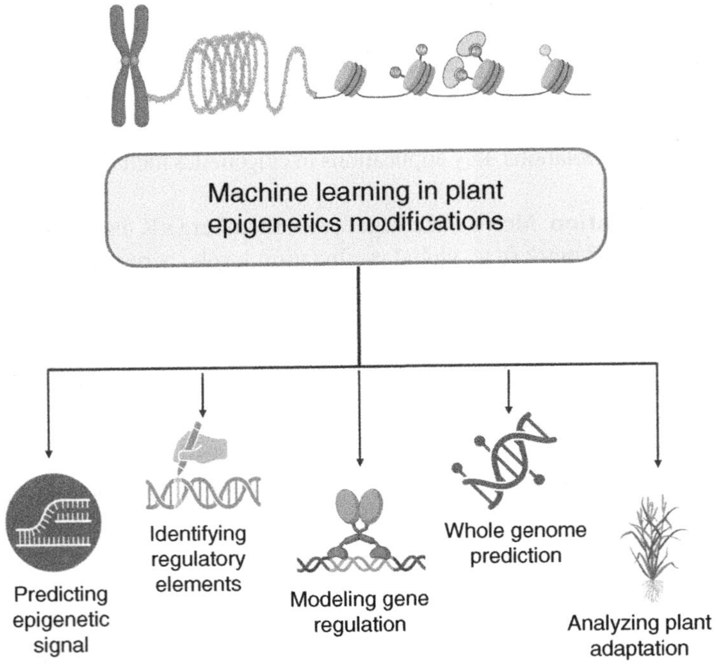

Figure 15.4 Machine learning in plant epigenomics.

These footprints pinpoint specific regulatory elements that have been pivotal in evolutionary processes, revealing how epigenetic mechanisms contribute to rapid adaptation and phenotypic diversification without changes to the underlying DNA sequence (Schmid et al., 2018).

15.6 Challenges and Future Directions

Beyond chromatin modifications, RNA modifications like m^6A have emerged as significant epigenetic regulators. Recent research demonstrates that RNA m^6A is instrumental in various plant processes, including development, stress response, and hormone signaling (Li & He, 2025). Investigating its specific role in plant regeneration, therefore, represents a highly promising frontier.

A critical future direction involves shifting from bulk-tissue epigenetic profiling to single-cell resolution. This will be essential for deciphering the unique epigenetic landscapes that distinguish different regenerative processes, such as tissue repair, de novo organogenesis, and somatic embryogenesis. Enhancing temporal resolution is equally important to establish causality between epigenetic reprogramming and subsequent changes in gene expression.

For example, studies in Arabidopsis indicate that the acquisition of pluripotency and the commitment to a shoot fate are epigenetically primed before the activation of shoot-specific genes (Wu et al., 2022). Similarly, research on protoplasts shows that founder cells responsible for regeneration undergo a “cellular evolution” driven by a genome-wide increase in chromatin accessibility, which leads to a phase of transcriptomic chaos (Xu et al., 2021).

Ultimately, the field will be revolutionized by defining the epigenetic prerequisites for designating these founder cells at a single-cell level, while precisely mapping the timeline of these epigenetic events against locus-specific transcriptional changes. This approach will illuminate the pioneering epigenetic mechanisms that initiate cell fate specification during regeneration.

15.7 Conclusion

Plant epigenetics has evolved from a gene-centric discipline to a systems-level science, now defined by a deluge of genome-wide, multiomics data. Within this complex landscape, the integration of ML and computational biology has become indispensable, enabling a critical transition from merely observing correlations to predicting outcomes and uncovering mechanistic insights.

The power of this synthesis is its capacity to manage the immense scale and interdependence of epigenetic data. Computational biology provides the foundational framework, the pipelines for processing raw sequencing data, quantifying DNA methylation from bisulfite-seq, and mapping histone modifications from ChIP-seq. However, it is ML that unlocks the deeper, nonlinear relationships within this data. For instance, supervised models like random forests or deep neural networks can be trained on integrated datasets to predict complex agronomic traits such as stress resilience or flowering time directly from a plant’s epigenomic profile, effectively serving as a powerful digital phenotyping tool.

Beyond prediction, unsupervised ML techniques like clustering and dimensionality reduction can reveal novel, functionally significant epigenetic states without prior bias. These methods can identify distinct chromatin signatures that prime genes for activation or uncover subtle patterns of transgenerational inheritance that elude conventional analysis. Most profoundly, ML models facilitate knowledge discovery. By interpreting a model’s decision-making process, for example, identifying which specific histone modification most strongly predicts a trait, researchers can generate novel, testable hypotheses about the fundamental regulatory logic encoded within the epigenome.

References

Alaskhar Alhamwe, B., Khalaila, R., Wolf, J., von Bülow, V., Harb, H., Alhamdan, F., Hii, C. S., Prescott, S. L., Ferrante, A., Renz, H., Garn, H., & Potaczek, D. P. (2018). Histone modifications and their role in epigenetics of atopy and allergic diseases. *Allergy, Asthma & Clinical Immunology*, *14*, 39. https://doi.org/10.1186/s13223-018-0259-4

Altman, N., & Krzywinski, M. (2018). Points of Significance: Statistics versus machine learning. *Nature Methods*, *15*(4), 233–234. https://doi.org/10.1038/nmeth.4642

Aref-Eshghi, E., Schenkel, L. C., Ainsworth, P., Lin, H., Rodenhiser, D. I., Cutz, J. C., & Sadikovic, B. (2018). Genomic DNA methylation-derived algorithm enables accurate detection of malignant prostate tissues. *Frontiers in Oncology*, *8*, 100. https://doi.org/10.3389/fonc.2018.00100

Bannister, A. J., & Kouzarides, T. (2011). Regulation of chromatin by histone modifications. *Cell Research*, *21*(3), 381–395. https://doi.org/10.1038/cr.2011.22

Bruce, T. J. A., Matthes, M. C., Napier, J. A., & Pickett, J. A. (2007). Stressful "memories" of plants: Evidence and possible mechanisms. *Plant Science*, *173*(6), 603–608. https://doi.org/10.1016/j.plantsci.2007.09.002

Buenrostro, J. D., Giresi, P. G., Zaba, L. C., Chang, H. Y., & Greenleaf, W. J. (2013). Transposition of native chromatin for fast and sensitive epigenomic profiling of open chromatin, DNA-binding proteins and nucleosome position. *Nature Methods*, *10*(12), 1213–1218. https://doi.org/10.1038/nmeth.2688

Capper, D., Jones, D. T. W., Sill, M., Hovestadt, V., Schrimpf, D., Sturm, D., Koelsche, C., Sahm, F., Chavez, L., Reuss, D. E., Kratz, A., Wefers, A. K., Huang, K., Pajtler, K. W., Schweizer, L., Stichel, D., Olar, A., Engel, N. W., Lindenberg, K., & Pfister, S. M. (2018). DNA methylation-based classification of central nervous system tumours. *Nature*, *555*(7697), 469–474. https://doi.org/10.1038/nature26000

Chen, C., Hu, Y., Ikeuchi, M., Jiao, Y., Prasad, K., Su, Y. H., Xiao, J., Xu, L., Yang, W., Zhao, Z., Zhou, W., Zhou, Y., Gao, J., & Wang, J. W. (2024). Plant regeneration in the new era: From molecular mechanisms to biotechnology applications. *Science China Life Sciences*, *67*(7), 1338–1367. https://doi.org/10.1007/s11427-023-2441-3

Ching, T., Pei, J., O'Kelly, J., Sasaki, T., Lahti, J., Kuo, C. J. C. W., Wan, Y., Li, R., Montgomery, K. D., Chang, H. Y., & West, R. B. (2012). Detection of long non-coding RNA in archival tissue: Correlation with polycomb protein expression in primary and metastatic breast carcinoma. *PLoS One*, *7*(10), e47998. https://doi.org/10.1371/journal.pone.0047998

Choi, Y., Gehring, M., Johnson, L., Hannon, M., Harada, J. J., Goldberg, R. B., Jacobsen, S. E., & Fischer, R. L. (2021). Demethylation of a gene body is essential for the activation of *OsMADS51*, a key flowering promoter in rice. *The Plant Cell*, *33*(4), 1180–1195. https://doi.org/10.1093/plcell/koab024

Cubas, P., Vincent, C., & Coen, E. (1999). An epigenetic mutation responsible for natural variation in floral symmetry. *Nature*, *401*(6749), 157–161. https://doi.org/10.1038/43652

Cusanovich, D. A., Daza, R., Adey, A., Pliner, H. A., Christiansen, L., Gunderson, K. L., Steemers, F. J., Trapnell, C., & Shendure, J. (2015). Multiplex single-cell profiling of chromatin accessibility by combinatorial cellular indexing. *Science*, *348*(6237), 910–914. https://doi.org/10.1126/science.aab1601

Dawson, M. A. (2017). The cancer epigenome: Concepts, challenges, and therapeutic opportunities. *Science*, *355*(6330), 1147–1152. https://doi.org/10.1126/science.aam5412

Ding, Y., Fromm, M., & Avramova, Z. (2012). Four distinct types of dehydration stress memory genes in arabidopsis thaliana. *BMC Plant Biology*, *12*(1), 229. https://doi.org/10.1186/1471-2229-12-229

Dowen, R. H., Pelizzola, M., Schmitz, R. J., Lister, R., Dowen, J. M., Nery, J. R., Dixon, J. E., & Ecker, J. R. (2012). Widespread dynamic DNA methylation in response to biotic stress. *Proceedings of the National Academy of Sciences of the United States of America*, *109*(32), E2183–E2191. https://doi.org/10.1073/pnas.1209329109

Egger, G., Liang, G., Aparicio, A., & Jones, P. A. (2004). Epigenetics in human disease and prospects for epigenetic therapy. *Nature*, *429*(6990), 457–463. https://doi.org/10.1038/nature02625

Esteller, M. (2008). Epigenetics in cancer. *New England Journal of Medicine*, *358*(11), 1148–1159. https://doi.org/10.1056/NEJMra072067

Gomez-Chavez, F., Cortez, D., & Smith, J. M. (2021). Integrative multi-omics analysis in evolutionary biology: A review. *Frontiers in Ecology and Evolution*, *9*, 706647. https://doi.org/10.3389/fevo.2021.706647

Guo, W., Fiziev, P., Yan, W., Cokus, S., Sun, X., Zhang, M. Q., Chen, P. Y., & Pellegrini, M. (2013). BS-seeker2: A versatile aligning pipeline for bisulfite sequencing data. *BMC Genomics*, *14*, 774. https://doi.org/10.1186/1471-2164-14-774

Holder, L. B., Haque, M. M., & Skinner, M. K. (2017). Machine learning for epigenetics and future medical applications. *Epigenetics*, *12*(7), 505–514. https://doi.org/10.1080/15592294.2016.1265710

Horvath, S. (2013). DNA methylation age of human tissues and cell types. *Genome Biology*, *14*(10), R115. https://doi.org/10.1186/gb-2013-14-10-r115

James, G., Witten, D., Hastie, T., & Tibshirani, R. (2013). *An introduction to statistical learning with applications in R*. Springer. https://doi.org/10.1007/978-1-4614-7138-7

Kaur, H., Bhalla, S., & Raghava, G. P. (2019). Classification of early and late stage liver hepatocellular carcinoma patients from their genomics and epigenomics profiles. *PLoS One*, *14*(9), e0221476. https://doi.org/10.1371/journal.pone.0221476

Kouzarides, T. (2007). Chromatin modifications and their function. *Cell*, *128*(4), 693–705. https://doi.org/10.1016/j.cell.2007.02.005

Krueger, F., & Andrews, S. R. (2011). Bismark: A flexible aligner and methylation caller for Bisulfite-Seq applications. *Bioinformatics*, *27*(11), 1571–1572. https://doi.org/10.1093/bioinformatics/btr167

Latzel, V., Zhang, Y., Karlsson Moritz, K., Fischer, M., & Bossdorf, O. (2013). Epigenetic variation in plant responses to defence hormones. *Annals of Botany*, *112*(4), 613–621. https://doi.org/10.1093/aob/mct149

LeCun, Y., Bengio, Y., & Hinton, G. (2015). Deep learning. *Nature*, *521*(7553), 436–444. https://doi.org/10.1038/nature14539

Li, S., & He, Y. (2025). RNA m6A modification meets plant hormones. *Nature Plants*, *11*, 686–695. https://doi.org/10.1038/s41477-025-01926-w

Lister, R., O'Malley, R. C., Tonti-Filippini, J., Gregory, B. D., Berry, C. C., Millar, A. H., & Ecker, J. R. (2008). Highly integrated single-base resolution maps of the epigenome in arabidopsis. *Cell*, *133*(3), 523–536. https://doi.org/10.1016/j.cell.2008.03.029

Luna, E., Bruce, T. J. A., Roberts, M. R., Flors, V., & Ton, J. (2012). Next-generation systemic acquired resistance. *Plant Physiology*, *158*(2), 844–853. https://doi.org/10.1104/pp.111.187468

Lynch, M., & Walsh, B. (1998). *Genetics and analysis of quantitative traits*. Sinauer Associates. https://doi.org/10.1093/genetics/148.4.2033

Manikkam, M., Haque, M. M., Guerrero-Bosagna, C., Nilsson, E., & Skinner, M. K. (2014). Pesticide methoxychlor promotes the epigenetic transgenerational inheritance of adult onset disease through the female germline. *PLoS One*, *9*(7), e102091. https://doi.org/10.1371/journal.pone.0102091

McInnes, L., Healy, J., & Melville, J. (2018). UMAP: Uniform manifold approximation and projection for dimension reduction. *arXiv:1802.03426*. https://doi.org/10.48550/arXiv.1802.03426

Mirouze, M., & Paszkowski, J. (2011). Epigenetic contribution to stress adaptation in plants. *Current Opinion in Plant Biology*, *14*(3), 267–274. https://doi.org/10.1016/j.pbi.2011.03.004

Murtagh, F., & Legendre, P. (2014). Ward's hierarchical agglomerative clustering method: Which algorithms implement Ward's criterion? *Journal of Classification*, *31*(3), 274–295. https://doi.org/10.1007/s00357-014-9161-z

Park, P. J. (2009). ChIP–seq: Advantages and challenges of a maturing technology. *Nature Reviews Genetics*, *10*(10), 669–680. https://doi.org/10.1038/nrg2641

Pecinka, A., Rosa, M., Schikora, A., Berlinger, M., Hirt, H., Luschnig, C., & Mittelsten Scheid, O. (2009). Transgenerational stress memory is not a general response in arabidopsis. *PLoS One*, *4*(4), e5202. https://doi.org/10.1371/journal.pone.0005202

Rasmann, S., De Vos, M., Casteel, C. L., Tian, D., Halitschke, R., Sun, J. Y., Agrawal, A. A., Felton, G. W., & Jander, G. (2012). Herbivory in the previous generation primes plants for enhanced insect resistance. *Plant Physiology*, *158*(2), 854–863. https://doi.org/10.1104/pp.111.185884

Revell, L. J., & Harmon, L. J. (2022). *Phylogenetic comparative methods in R*. Princeton University Press. https://doi.org/10.1515/9780691227755

Rossetto, D., Avvakumov, N., & Côté, J. (2012). Histone phosphorylation: A chromatin modification involved in diverse nuclear events. *Epigenetics*, *7*(10), 1098–1108. https://doi.org/10.4161/epi.21950

Schmid, M. W., Heichinger, C., & Grossniklaus, U. (2018). The epigenome in evolutionary ecology. *Molecular Ecology*, *27*(11), 2411–2424. https://doi.org/10.1111/mec.14588

Skinner, M. K. (2014). Endocrine disruptor induction of epigenetic transgenerational inheritance of disease. *Molecular and Cellular Endocrinology*, *398*(1–2), 4–12. https://doi.org/10.1016/j.mce.2014.09.007

Slaughter, A., Daniel, X., Flors, V., Luna, E., Hohn, B., & Mauch-Mani, B. (2012). Descendants of primed arabidopsis plants exhibit resistance to biotic stress. *Plant Physiology*, *158*(2), 835–843. https://doi.org/10.1104/pp.111.191593

Stephens, Z. D., Lee, S. Y., Faghri, F., Campbell, R. H., Zhai, C., Efron, M. J., Iyer, R., Schatz, M. C., Sinha, S., & Robinson, G. E. (2015). Big data: Astronomical or genomical? *PLoS Biology*, *13*(7), e1002195. https://doi.org/10.1371/journal.pbio.1002195

Sung, S., & Amasino, R. M. (2004). Vernalization in arabidopsis thaliana is mediated by the PHD finger protein VIN3. *Nature*, *427*(6970), 159–164. https://doi.org/10.1038/nature02195

Waddington, C. H. (1953). The epigenetics of evolution. In J. B. S. Haldane, C. N. Hinshelwood, P. B. Medawar, & P. M. Sheppard (Eds.), *Evolution: The second international symposium on evolution* (pp. 186–199). Society for Experimental Biology.

Weaver, I. C., Cervoni, N., Champagne, F. A., D'Alessio, A. C., Sharma, S., Seckl, J. R., Dymov, S., Szyf, M., & Meaney, M. J. (2004). Epigenetic programming by maternal behavior. *Nature Neuroscience*, *7*(8), 847–854. https://doi.org/10.1038/nn1276

Wei, J. W., Huang, K., Yang, C., & Kang, C. S. (2017). Non-coding RNAs as regulators in epigenetics (review). *Oncology Reports*, *37*(1), 3–9. https://doi.org/10.3892/or.2016.5236

Wu, L.-Y., Shang, G.-D., Wang, F.-X., Gao, J., Wan, M.-C., Xu, Z.-G., & Wang, J.-W. (2022). Dynamic chromatin state profiling reveals regulatory roles of auxin and cytokinin in shoot regeneration. *Developmental Cell*, *57*(5), 526–542. https://doi.org/10.1016/j.devcel.2022.02.017

Xu, M., Du, Q., Tian, C., Wang, Y., & Jiao, Y. (2021). Stochastic gene expression drives mesophyll protoplast regeneration. *Science Advances*, *7*(28), eabg8466. https://doi.org/10.1126/sciadv.abg8466

Xue, Y., Cao, X., Chen, X., Deng, X., Deng, X. W., Ding, Y., Dong, A., Duan, C.-G., Fang, X., Gong, L., Gong, Z., Gu, X., He, C., He, H., He, S., He, X. J., He, Y., He, Y., Jia, G., ... Liu, Q. (2025). Epigenetics in the modern era of crop improvements. *Science China. Life Sciences*, *68*, 1570–1609. https://doi.org/10.1007/s11427-025-2818-2

Yi, H., & Richards, E. J. (2007). A cluster of disease resistance genes in arabidopsis is coordinated by transcriptional activation and RNA silencing. *The Plant Cell*, *19*(9), 2929–2939. https://doi.org/10.1105/tpc.107.051821

Zhang, H., Lang, Z., & Zhu, J. K. (2018). Dynamics and function of DNA methylation in plants. *Nature Reviews Molecular Cell Biology*, *19*(8), 489–506. https://doi.org/10.1038/s41580-018-0008-9

16

Ethical and Regulatory Considerations of Artificial Intelligence in Agriculture

Harshit Mishra[1], *Fredrick Kayusi*[2], *Rashmi Mishra*[3], *Ioannis Adamopoulos*[4], *and Arkan A. Ghaib*[5]

[1] *Department of Agricultural Economics, College of Agriculture, Acharya Narendra Deva University of Agriculture and Technology, Ayodhya, Uttar Pradesh, India*
[2] *Department of Environmental Studies, Geography and Planning, Maasai Mara University, Narok, Kenya*
[3] *College of Economics and Business Administration, University of Technology and Applied Sciences–Al Musannah, Muscat, Oman*
[4] *Department of Public Health Policy, Sector of Occupational & Environmental Health, School of Public Health, University of West Attica, Athens, Greece*
[5] *Information Technology Department, Management Technical College, Southern Technical University, Basrah, Iraq*

16.1 Introduction

Agriculture, the foundational pillar of human civilization and sustenance, is currently undergoing one of its most profound transformations since the Green Revolution, catalyzed by the rapid and pervasive integration of artificial intelligence (AI) technologies. This paradigm shift, often referred to as Agriculture 4.0, involves deploying complex algorithms, machine learning models, and autonomous robotics to enhance productivity, optimize resource management, and bolster resilience against climate volatility and geopolitical supply chain disruptions. The imperative for this transformation is clear: global food demand is projected to rise by 70% by 2050, necessitating a dramatic increase in efficiency without commensurately expanding arable land use (Uddin et al., 2024). AI offers promising avenues to bridge this gap, delivering precision farming capabilities that allow for hyper-localized decision-making, from nutrient application and pest detection to automated harvesting. Indeed, the global AgTech market, heavily reliant on AI and machine learning applications, is anticipated to exceed $35 billion by 2030, reflecting a compound annual growth rate (CAGR) of over 11% (Ryan, 2023). This aggressive adoption is visible worldwide, where, as noted, over 75% of farmers in large agricultural economies like India are now engaging with digital platforms for critical crop advisory services, and a significant 30% of United States farms had incorporated some form of AI-driven automation, such as autonomous tractors and robotic weeders, by 2023 (Dara et al., 2022). However, this impressive technological progress is fundamentally intertwined with novel and often unsettling ethical and regulatory challenges that must be rigorously addressed to ensure the sustainable and equitable future of the global agroecosystem.

The primary locus of ethical concern resides in the colossal volume and intimate nature of the data generated by AI-enabled agricultural systems. Unlike consumer data, farm-level data encompasses highly sensitive, proprietary information about yield history, soil composition, irrigation patterns, seed performance, and economic projections. An average large-scale farm can generate several terabytes of data annually, a valuable commodity that is often aggregated, analyzed, and monetized by technology providers (Sparrow et al., 2021). This necessitates a critical examination of data ownership and consent protocols, particularly regarding who controls the outputs and intellectual property derived from farmer-contributed data. The conventional model of "click-wrap" agreements is proving insufficient to guarantee farmer autonomy and informed consent when dealing with intricate data contracts. Furthermore, the imperative for data privacy is increasingly challenging, as granular farm data, even when seemingly anonymized, possesses unique spatiotemporal characteristics that can potentially be re-identified and exploited, raising significant security risks. The convergence of farm data with wider biometric and market data also escalates the cybersecurity threat landscape, exposing complex agricultural supply chains to systemic vulnerabilities, thus demanding robust governance frameworks that secure not only the technology but also the sovereignty of the farmer's operational information (Kumar, 2024; Mishra & Srivastava, 2025).

Moving beyond data concerns, the moral and sociological ramifications of integrating AI into the agricultural value chain demand profound philosophical introspection. Traditional ethical frameworks—including utilitarianism, which seeks to maximize the greatest good by optimizing yield and minimizing resource use; deontology, which mandates algorithmic accountability and adherence to moral duties regardless of outcome; and virtue ethics, which focuses on fostering responsible human character traits in the development of sustainable technology—each offer distinct and occasionally conflicting lenses through which to evaluate the moral acceptability of AI deployment (Srivastav & Das, 2025). One of the most contentious debates centers on the displacement of human labor. While AI promises to alleviate physically strenuous and dangerous tasks, its pervasive adoption in areas like automated harvesting and predictive management poses significant socioeconomic risks to rural employment and livelihood stability, particularly in low- and middle-income countries where agricultural work remains a critical source of income. This raises profound ethical dilemmas concerning the trade-off between economic efficiency, which may benefit corporate interests, and social equity, which protects the dignity and stability of the farming community. The challenge is not merely to measure the short-term economic gains (e.g., the potential 15–20% yield gain attributed to precision agriculture) but to evaluate the long-term societal cost of marginalizing vast swathes of the agrarian workforce (Ryan et al., 2023).

The technical development of AI is inextricably linked to issues of bias, fairness, and equity. Algorithmic bias, often originating from nonrepresentative or imbalanced training datasets—for instance, those predominantly featuring large-scale commercial farms from the Global North—can systematically disadvantage and marginalize smallholder and subsistence farming communities. If yield prediction or crop recommendation models are trained primarily on data reflecting highly specific technological inputs or soil conditions unavailable to resource-poor farmers, the resulting recommendations will inherently be unfair, exacerbating existing rural inequities and creating a technological poverty gap.

For example, estimates suggest that fewer than 10% of publicly documented AI models for agriculture are tailored for diverse and complex agroecological systems, illustrating a profound training gap. Ensuring algorithmic fairness requires a deliberate shift toward inclusive data collection methodologies and model designs that explicitly account for varied socioeconomic realities and ecological heterogeneity. This necessity underscores the parallel demand for model transparency and explainability AI (XAI), as complex "black-box" algorithms must be rendered auditable and interpretable to farmers and regulators alike, fostering trust and enabling them to understand *why* an AI system arrived at a particular decision, thereby allowing for contestability and redress (Ali et al., 2024; Tzachor et al., 2022).

Complementing these ethical debates is the pressing need for robust and adaptive regulatory mechanisms. The development of agricultural AI has substantially outpaced the capacity of existing legal frameworks, leaving large gray areas regarding liability, accountability, and compliance. Current international guidelines and national digital agriculture policies often provide broad ethical principles but lack the binding, sector-specific mandates necessary for enforcement, particularly in areas like autonomous machinery operation and the deployment of AI-driven biotechnologies. Furthermore, there is a pronounced regulatory disparity: while regions like the European Union have introduced comprehensive AI Acts that cover various sectors, only a nascent fraction, perhaps less than 15% globally, of national regulatory efforts specifically target the unique risks inherent in agricultural AI, such as systemic risk in crop failure predictions or the environmental impact on biodiversity from highly precise chemical applications (Harikrishnan et al., 2025). The vacuum created by the absence of clear statutes complicates issues of redress and liability in cases of AI-induced harm, making it difficult to determine legal responsibility when an autonomous system fails or provides a flawed recommendation that leads to crop loss or environmental damage. Effective governance, therefore, requires a dynamic, multistakeholder approach that can evolve in parallel with the technology, ensuring that AI deployment remains aligned with socio-environmental sustainability goals and ethical benchmarks.

The holistic challenge of governing AI in agriculture requires a concerted, global effort to institutionalize ethical oversight and responsible design practices. Moving forward, the integration of ethical considerations must transition from a retrospective audit to a proactive, Ethics-by-design approach, embedding principles of accountability, transparency, and fairness into the very genesis of AI products. This involves establishing multidisciplinary ethical review boards (ERBs) that bring together agronomists, ethicists, policymakers, and farmers to govern project approval and implementation. Ultimately, realizing the enormous potential of AI in achieving global food security necessitates navigating these complex ethical and regulatory waters with diligence, foresight, and a commitment to justice. This involves fostering global cooperation to harmonize ethical standards and ensure that technological progress serves all stakeholders, particularly the most vulnerable farming populations, thereby ensuring that the future of agriculture is not only smart but also just and sustainable (Devasia et al., 2024; Okengwu et al., 2023).

The primary objective of this chapter is to establish a comprehensive ethical foundation for evaluating AI in agriculture by examining core philosophical frameworks and their application to specific dilemmas like labor displacement and data ownership; and critically analyze the existing global regulatory landscape concerning data privacy, algorithmic bias, and accountability mechanisms; and finally, to propose integrated governance and

Ethics-by-Design strategies necessary for fostering a future where AI systems are transparent, equitable, and demonstrably aligned with the goals of sustainable agriculture and rural equity.

16.2 Ethical Foundations and Philosophical Underpinnings of AI in Agriculture

The ethical landscape surrounding AI in agriculture is both intricate and evolving. As agriculture transitions from a primarily labor-intensive practice to a technologically augmented domain, ethical considerations have assumed paramount importance. These considerations are not merely peripheral academic debates but rather form the moral foundation upon which sustainable, equitable, and responsible agricultural modernization must stand. The moral legitimacy of AI's application in this sector depends on its capacity to harmonize technological progress with social justice, ecological balance, and human dignity.

This section delves deeply into the philosophical and ethical frameworks that inform the moral evaluation of AI in agriculture. It explores foundational ethical theories, contextual dilemmas specific to agricultural AI systems, and the broader ecological impacts on agroecosystems. Together, these discussions illuminate how ethics provides a guiding compass for the responsible deployment of AI technologies in a domain as fundamental as food production (Sangwa & Mutabazi, 2025).

16.2.1 Ethical Theories and Frameworks Relevant to AI

The ethical appraisal of AI in agriculture requires a systematic understanding of moral theories that have historically guided human decision-making. Classical ethical theories such as utilitarianism, deontological ethics, and virtue ethics serve as the conceptual scaffolding upon which the moral responsibilities of AI developers, policymakers, and agricultural practitioners may be evaluated. Each of these theories provides a distinctive moral perspective—focusing variously on outcomes, duties, or moral character—all of which are essential for understanding the ethical foundations of AI-driven agricultural practices.

16.2.1.1 Utilitarianism and Its Role in Agricultural Decision-Making

Utilitarianism, associated with philosophers like Jeremy Bentham and John Stuart Mill, evaluates moral actions based on their consequences, particularly their capacity to maximize collective happiness or welfare. In the agricultural context, utilitarian reasoning supports the deployment of AI technologies that increase productivity, improve food distribution, and minimize waste—as these outcomes ostensibly promote the "greatest good for the greatest number."

AI-driven solutions such as predictive weather modeling, pest detection algorithms, and precision irrigation systems exemplify utilitarian ethics in action. They enhance yields, conserve scarce resources, and contribute to food security—objectives that align closely with maximizing overall societal welfare. However, a critical limitation of pure utilitarianism emerges when technological benefits are distributed unevenly. Large-scale agribusiness enterprises often have the financial capacity to adopt AI technologies, while smallholder farmers may be left behind, exacerbating inequality. Therefore, an ethically refined utilitarian approach must also incorporate principles of distributive justice, ensuring that the welfare gains are equitably shared.

From a practical perspective, the utilitarian principle in agricultural decision-making must evolve beyond mere yield maximization to include considerations of social well-being, economic equity, and environmental sustainability. The moral worth of AI technologies, therefore, depends not only on their efficiency but also on their inclusivity and fairness in serving the diverse constituencies that comprise the agricultural sector (Benefo et al., 2022).

16.2.1.2 Deontological Ethics in Algorithmic Accountability

Deontological ethics, rooted in the philosophy of Immanuel Kant, emphasizes moral duties, obligations, and adherence to ethical principles irrespective of outcomes. In the context of AI in agriculture, deontological reasoning demands that AI systems adhere to standards of fairness, transparency, and respect for autonomy—not because they produce good results, but because such conduct is inherently right.

Algorithmic accountability becomes a central concern in this ethical paradigm. AI systems designed to advise farmers on crop management or to automate decision-making must be transparent and explainable. Black-box algorithms, which generate outputs without intelligible reasoning, pose significant ethical risks, as they obscure accountability and erode trust. A deontological framework requires that all data collection and algorithmic decisions respect farmers' autonomy, particularly through informed consent and the right to data ownership (Mark, 2019).

In agricultural AI, adherence to moral duty also extends to the developers and institutions responsible for algorithmic governance. Their ethical responsibility lies not merely in maximizing efficiency but in ensuring that every technological process upholds the dignity, rights, and freedoms of agricultural stakeholders.

16.2.1.3 Virtue Ethics and Sustainable AI Development

Virtue ethics, grounded in Aristotelian philosophy, shifts ethical evaluation from actions and duties to the moral character and virtues of the agents involved. In agricultural AI, virtue ethics encourages the cultivation of traits such as prudence, responsibility, humility, and compassion—virtues that foster a deeper sense of stewardship over both the natural environment and the human communities that depend on it.

A virtuous agricultural innovator is not one who merely pursues profit or efficiency but one who exercises *phronesis*—practical wisdom—in balancing technological ambition with ecological and social responsibility. For instance, AI-driven innovation that prioritizes short-term yield gains at the expense of soil health or biodiversity would be considered ethically deficient. Conversely, systems that integrate long-term sustainability and community well-being into their design reflect virtuous intent.

Virtue ethics thus bridges the moral gap between technology and humanity by emphasizing ethical mindfulness. It reminds developers, farmers, and policymakers that true innovation in agriculture should not merely transform how food is produced but should also elevate the moral consciousness of those who produce it (Gikunda, 2024).

16.2.2 Ethical Dilemmas Specific to Agricultural AI Systems

Beyond theoretical discussion, agricultural AI presents concrete ethical dilemmas that affect individuals, communities, and institutions. These dilemmas revolve primarily around issues of data ownership, labor displacement, and equitable access to technological resources. Each dilemma reflects a tension between technological potential and ethical responsibility—a tension that must be carefully navigated to achieve just and sustainable outcomes.

16.2.2.1 Data Ownership and Consent in Agricultural Datasets

AI systems depend heavily on large volumes of data derived from farm operations, soil properties, climatic conditions, and crop yields. However, ambiguity surrounding data ownership raises profound ethical questions: Who owns this data—the farmer who generates it, the technology provider who processes it, or the institution that stores it?

In many developing economies, farmers often lack awareness of how their data is being collected or monetized. This absence of informed consent violates both autonomy and justice. Ethical governance of agricultural data should therefore ensure that farmers retain ownership and control over their data, with the right to consent, withdraw, or share in the financial benefits derived from it (Brini, 2024).

For example, if a company uses aggregated farm data to train AI systems for pest prediction and later sells that service to agribusinesses, the originating farmers deserve recognition and fair compensation. Transparent data-sharing agreements and community-based data cooperatives can serve as mechanisms for equitable benefit-sharing, ensuring that AI does not perpetuate digital exploitation.

16.2.2.2 Moral Implications of Replacing Human Labor

Automation and mechanization—enabled by AI technologies such as autonomous tractors, robotic harvesters, and drones—raise ethical concerns about human displacement in rural economies. While these technologies enhance efficiency, they also threaten employment opportunities and alter the social fabric of agricultural communities (Sparrow & Howard, 2021).

From an ethical standpoint, replacing human labor must be approached with sensitivity and foresight. Job losses without compensatory measures—such as reskilling programs or employment transition strategies—can lead to economic marginalization and social distress. Moreover, the ethical value of human labor extends beyond economic productivity; it encompasses dignity, cultural identity, and emotional connection to the land.

Hence, ethical AI adoption in agriculture requires policies that preserve human welfare while embracing automation. Balancing technological progress with social justice is not merely an economic challenge but a moral imperative (van Hilten et al., 2025).

16.2.2.3 Issues of Equity and Fair Access to AI Tools

Equity lies at the heart of ethical agricultural transformation. Advanced AI systems, while promising efficiency, are often accessible only to wealthy farmers and agribusiness corporations. Small and marginal farmers, especially in developing regions, face prohibitive costs and infrastructural barriers.

This digital divide transforms technology into a tool of exclusion rather than empowerment. Ethical fairness requires that AI tools be made accessible through open-source platforms, government subsidies, or cooperative ownership models. Equal access to AI ensures that innovation benefits all stakeholders, rather than reinforcing existing hierarchies of wealth and power (Manning et al., 2022).

In essence, the ethical legitimacy of AI in agriculture depends on its capacity to democratize opportunity and reduce inequality rather than intensify it.

16.2.3 AI and the Ethical Impacts on Agroecosystems

Ethics in agricultural AI must also transcend anthropocentric considerations to encompass ecological and environmental responsibilities. Agroecosystems—comprising soils, plants, animals, and microorganisms—represent living networks whose integrity must be preserved for the long-term viability of human civilization. AI technologies, therefore, bear an ethical obligation not only to enhance productivity but also to sustain ecological harmony.

16.2.3.1 Balancing Technological Advancement with Biodiversity

While AI enables precision agriculture and yield optimization, it may also inadvertently promote monocultures and intensive farming practices that threaten biodiversity. The ethical challenge lies in ensuring that AI technologies are designed to enhance—rather than diminish—ecological diversity.

AI can, and should, be harnessed to monitor species diversity, predict ecosystem degradation, and guide regenerative agricultural practices. Ethical stewardship requires that the advancement of AI be aligned with the preservation of biodiversity, emphasizing coexistence between technological and natural systems.

For instance, an AI system that recommends crop rotation strategies based on soil microbiome data can help maintain ecological resilience while sustaining productivity. In this sense, ethics and ecology converge to form a unified framework of sustainable progress (Gupta et al., 2024).

16.2.3.2 Ethical Trade-offs in Resource Allocation and Sustainability

AI systems inherently involve trade-offs in resource allocation—for example, between maximizing output and conserving limited inputs such as water, energy, and soil fertility. These trade-offs are not purely technical; they are deeply ethical. Decisions about which crops to prioritize or which regions to irrigate raise questions of fairness, justice, and intergenerational responsibility.

An ethically sound AI design must therefore integrate principles of sustainability, ensuring that current agricultural benefits do not compromise the needs of future generations. Moreover, resource allocation decisions guided by AI must incorporate local knowledge and participatory ethics, allowing communities to influence how technology interacts with their ecosystems.

As illustrated in Table 16.1, each ethical framework contributes a distinctive lens for assessing the moral dimensions of AI in agriculture. Utilitarianism emphasizes outcome-oriented efficiency, deontology underscores adherence to duty and rights, and virtue ethics focuses on moral integrity and sustainability. Together, these frameworks provide a

Table 16.1 Ethical frameworks and their applications in agricultural AI.

Ethical framework	Core principle	Application in agriculture	Ethical risk addressed
Utilitarianism	Maximizing welfare and collective benefit	AI-driven optimization of crop yields, resource use, and food distribution	Risk of unequal benefit distribution and neglect of marginalized farmers
Deontological ethics	Duty, rights, and moral obligation	Algorithmic transparency, data privacy, and informed consent	Violation of autonomy, privacy, and fairness
Virtue ethics	Moral character, responsibility, and sustainability	Responsible innovation, environmental stewardship, and community empowerment	Over-industrialization, loss of moral accountability, and ecological neglect

comprehensive ethical matrix that can guide policymakers, researchers, and practitioners in aligning technological innovation with moral responsibility.

16.3 Data Privacy, Security, and Intellectual Property Rights

The integration of AI within agriculture has brought transformative potential to farming systems, supply chain management, crop monitoring, and decision-making processes. However, as data emerges as the lifeblood of AI-driven innovation, ethical, legal, and regulatory challenges concerning data privacy, security, and intellectual property rights (IPR) have gained significant importance. The proliferation of Internet of Things (IoT) devices, remote sensing technologies, and machine learning algorithms in farms necessitates a careful examination of how agricultural data is collected, stored, shared, and monetized. The issues of farmer autonomy, informed consent, cybersecurity, and proprietary ownership form the foundation of this complex ethical landscape.

The subsequent sections provide an in-depth exploration of these dimensions, focusing on how privacy frameworks, data protection laws, and intellectual property regimes can be effectively balanced with the advancement of AI applications in agriculture, particularly in developing economies such as India, where data governance mechanisms are still evolving (Kosior, 2020).

16.3.1 Data Collection and Privacy in Agricultural Settings

The agricultural ecosystem generates enormous volumes of data—ranging from soil moisture levels, weather conditions, and crop yield estimates to financial and transactional information of farmers. This data is primarily collected through sensors, drones, smart irrigation systems, and mobile-based platforms that utilize AI algorithms to predict and optimize agricultural outcomes. However, the unregulated or unethical collection of such information can compromise farmer privacy and autonomy. As a result, data privacy in agriculture is not merely a technological challenge but a socio-ethical imperative that demands transparent, participatory, and legally compliant frameworks.

16.3.1.1 Informed Consent and Farmer Autonomy

The cornerstone of ethical data collection lies in the principle of informed consent, which ensures that data subjects—in this case, farmers—understand and willingly agree to the ways in which their data will be used. In agricultural settings, however, obtaining informed consent poses unique challenges. Many smallholder farmers may have limited digital literacy, making it difficult for them to comprehend lengthy privacy policies or data-sharing agreements. In such contexts, informed consent must be simplified, culturally contextualized, and communicated in local languages to ensure genuine understanding.

Farmer autonomy further extends beyond consent to include the right to opt out, data portability, and access to derived insights. Ethical AI systems should empower farmers by granting them control over their data, enabling them to decide whether and how their data can be shared with agri-tech companies, research institutions, or government bodies. For instance, when AI platforms utilize farm-level data to generate predictive models on pest infestations or yield forecasts, farmers should have the right to review, contest, or withdraw their data contributions.

Moreover, as large corporations increasingly collect agricultural data, there is a growing concern about data asymmetry—where technology providers possess more knowledge and bargaining power than the farmers whose data they exploit (Zangana et al., 2025a). This imbalance risks reducing farmers to passive data providers rather than active beneficiaries. Thus, ethical frameworks must incorporate data stewardship principles that safeguard farmer interests and ensure equitable benefit sharing.

16.3.1.2 Anonymization and Aggregation of Agricultural Data

To protect farmer privacy, anonymization and data aggregation are crucial mechanisms. Anonymization involves removing personally identifiable information (PII) such as land registration numbers, geolocation coordinates, or personal identifiers that could trace data back to individual farmers. Aggregation, on the other hand, combines datasets to generate regional or sectoral insights without exposing individual-level details.

However, despite these safeguards, re-identification risks persist, especially when anonymized datasets are cross-linked with other databases. Hence, robust technical safeguards and policy frameworks must govern data sharing. For instance, multilevel anonymization protocols, differential privacy techniques, and secure multiparty computation models can enhance protection.

Furthermore, regulatory authorities may establish Agricultural Data Trusts—neutral entities responsible for managing data on behalf of farmers and ensuring that data usage aligns with privacy norms and national data protection laws such as the Digital Personal Data Protection Act (DPDP), 2023 in India. These mechanisms can build public confidence and foster responsible innovation (Mishra & Dwivedi, 2025; Smith, 2019).

16.3.2 Data Governance and Security Protocols

While privacy frameworks address the ethical dimensions of data collection and use, data governance focuses on the institutional and technological systems that ensure data accuracy, integrity, and accountability. AI-based agricultural ecosystems rely heavily on data-driven decision models, which are vulnerable to cybersecurity breaches, unauthorized access, and data manipulation. Effective governance must therefore integrate technological safeguards with organizational and legal structures to protect the digital agricultural infrastructure.

16.3.2.1 Cybersecurity Risks in AI-Enabled Agricultural Systems

Agriculture's digital transformation has expanded its vulnerability to cyber threats. Smart irrigation systems, automated tractors, and remote sensing networks connected via cloud platforms are all potential targets for cyberattacks. A breach could result in theft of proprietary algorithms, tampering with crop data, or disruption of food supply chains, leading to both economic and environmental consequences.

Cybersecurity risks can emerge from multiple sources—ranging from malware and ransomware attacks to insider threats and inadequate encryption protocols. For instance, a compromised AI model predicting pest outbreaks could yield false forecasts, leading farmers to misuse fertilizers or pesticides. Moreover, with increasing data exchange between private corporations, governments, and farmers, the surface area for cyber exploitation expands considerably (Abdul-Al et al., 2025).

To mitigate these risks, multilayered cybersecurity frameworks should be instituted. These may include:

- End-to-end encryption for data transmission.
- Multifactor authentication for platform access.
- Regular security audits of AI and IoT infrastructures.
- Adoption of ISO/IEC 27001 standards for information security management.

Additionally, building cyber resilience capacity among farmers and agricultural workers is vital. Training programs supported by agricultural ministries and rural development agencies can enhance digital literacy, ensuring that farmers recognize and respond appropriately to cyber threats.

16.3.2.2 Blockchain and Distributed Ledger Technologies for Data Integrity

One of the most promising innovations for ensuring data integrity and transparency in AI-enabled agriculture is the use of blockchain and distributed ledger technology (DLT). These technologies record data across decentralized nodes rather than relying on a single central database, thereby minimizing the risk of tampering and unauthorized data manipulation.

As illustrated in Figure 16.1, there are several types of distributed ledger technologies, including Blockchain, Hashgraph, Directed Acyclic Graph (DAG), Holochain, and Radix. Each has a unique method for storing and validating transactions. Blockchain uses a sequential chain of blocks to record every transaction, ensuring complete traceability. Hashgraph focuses on consensus-building among nodes, eliminating the need for each transaction to be validated by all participants. DAG, on the other hand, allows multiple transactions to occur simultaneously, verified by referencing previous ones, which enhances scalability. Holochain takes a more decentralized approach by shifting from data-centric to agent-centric structures, while Radix organizes transactions by events rather than chronological order, improving speed and efficiency.

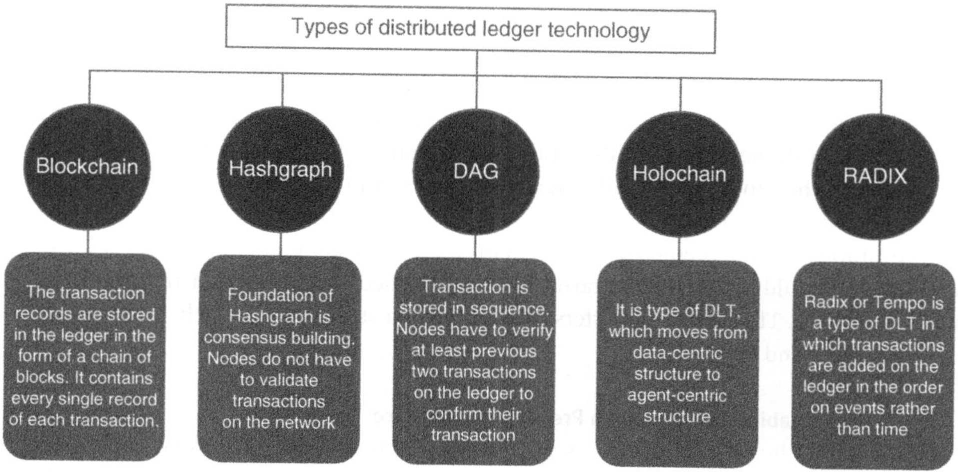

Figure 16.1 Overview of major types of distributed ledger technologies (DLTs) including Blockchain, Hashgraph, DAG, Holochain, and Radix. *Source:* GeeksforGeeks, 2025.

In agricultural data management, blockchain-based systems can ensure secure and transparent data provenance, where every change or transaction within a dataset is permanently recorded and easily auditable. This capability supports the use of smart contracts—self-executing agreements that automatically enforce terms such as data-sharing conditions or payments. For instance, a blockchain-enabled platform could automatically distribute royalty payments to farmers whenever their field data contribute to training AI models for yield prediction or pest control.

Moreover, the immutability of blockchain records builds trust among stakeholders such as government agencies, agri-tech firms, and cooperatives, allowing data sharing without fear of manipulation. Despite these benefits, challenges such as high energy consumption, limited interoperability, and insufficient rural digital infrastructure continue to hinder widespread adoption. Addressing these barriers will require collaborative efforts between the public and private sectors to create scalable, affordable, and farmer-friendly blockchain ecosystems that promote data-driven transparency in agriculture (Imran & Li, 2025).

16.3.3 Intellectual Property and Proprietary Algorithms

As AI continues to redefine agricultural innovation, questions of ownership, control, and fair distribution of benefits have become increasingly complex. IPR play a central role in determining who holds legal claims over AI-generated outputs, data-driven insights, and underlying algorithms. Balancing innovation incentives with equitable access to technology is essential to avoid monopolistic practices that could marginalize smallholder farmers and researchers.

16.3.3.1 Ownership Rights over AI-Generated Outputs

The ownership of AI-generated content or decisions—such as yield predictions, crop disease classifications, or optimized planting strategies—remains a gray area in intellectual property law. While human authorship is a traditional prerequisite for copyright protection, AI-generated works challenge this notion.

In agricultural contexts, disputes often arise regarding who owns the value created by AI—whether it is the technology provider, the data contributor (i.e., the farmer), or the institution that trained the algorithm. Ethically, a shared ownership model could be more appropriate, wherein benefits and revenues derived from AI insights are proportionally distributed among contributors.

For example, when a predictive AI tool helps increase farm productivity, part of the economic gain should return to the farmers who provided the input data that made such insights possible. This approach fosters data equity and strengthens trust between technology developers and end-users.

16.3.3.2 Patentability of AI Tools in Precision Agriculture

AI-driven agricultural tools—such as machine learning-based irrigation systems, robotic harvesters, or predictive analytics software—often involve inventive technical solutions that may qualify for patent protection. However, determining the patentability of such inventions is challenging due to the non-human involvement in their creation and the algorithmic nature of AI.

Patent offices globally, including the Indian Patent Office (IPO), typically require that an invention involve a human inventor. Thus, AI itself cannot be recognized as an inventor, though AI-assisted innovations may be eligible for patenting if they exhibit novelty, inventive step, and industrial applicability.

A key concern lies in ensuring that patent protection does not lead to technology monopolies, which could hinder agricultural inclusivity. Governments and international agencies should therefore promote balanced patent policies that encourage innovation while maintaining public access to essential agricultural technologies, especially for small-scale farmers.

16.3.3.3 Licensing and Open-Source Frameworks

An alternative to strict proprietary control is the adoption of open-source and collaborative licensing frameworks, which allow wider dissemination of AI technologies for agricultural development. Open-source AI tools facilitate innovation, reduce costs, and foster global collaboration among researchers, universities, and agri-tech startups.

For instance, open-access crop classification algorithms can be adapted by local developers to suit regional soil conditions or climate variations without violating intellectual property laws. Similarly, Creative Commons and General Public License (GPL) models can ensure that software remains freely available while protecting the moral rights of creators (Gwagwa et al., 2021).

Nevertheless, open-source frameworks also require sustainable funding mechanisms to support long-term maintenance, security updates, and ethical compliance. Hybrid models—combining open access with selective licensing for commercial use—may offer a balanced solution, ensuring both innovation and economic viability.

As shown in Table 16.2, the ethical and regulatory architecture of AI in agriculture is multidimensional, encompassing issues of data privacy, cybersecurity, blockchain integration, and intellectual property. Each dimension demands coordinated responses combining technological safeguards, legal compliance, and stakeholder participation.

Table 16.2 Ethical and regulatory considerations for data and IP management in AI-driven agriculture.

Ethical/legal dimension	Concerns	Potential solutions	Regulatory reference (India)
Data privacy	Unauthorized data collection, farmer consent	Informed consent frameworks, anonymization	Digital Personal Data Protection Act, 2023
Cybersecurity	System vulnerabilities, data theft	Encryption, audits, awareness training	National Cyber Security Policy, 2024
Blockchain	Scalability, interoperability	Public-private partnerships, rural connectivity	NITI Aayog Blockchain Strategy, 2021
Intellectual Property	Ownership ambiguity, monopolies	Shared ownership, open licensing	Indian Patent Act, 1970 (as amended)

For instance, the DPDP Act (2023) provides a statutory basis for ensuring informed consent and data anonymity in agricultural AI systems. Similarly, the National Cyber Security Policy (2024) establishes a framework for digital resilience, mandating regular audits and incident reporting in critical information sectors, including agriculture.

Meanwhile, the NITI Aayog Blockchain Strategy (2021) emphasizes decentralized transparency and promotes blockchain applications for agricultural value chains and supply traceability. Lastly, the Indian Patent Act (1970, as amended) underpins the regulation of AI-assisted innovations, ensuring that ownership frameworks remain fair and non-exploitative.

Taken together, these legal references demonstrate India's progressive but evolving approach to managing ethical complexities at the intersection of AI, agriculture, and data governance. Their effectiveness, however, will depend on continuous institutional vigilance, stakeholder education, and alignment with global ethical standards (Ijaz et al., 2024; Mishra, 2024).

16.4 Bias, Fairness, and Transparency in Agricultural AI Models

AI has emerged as a transformative technology within the agricultural domain, reshaping decision-making processes, crop management strategies, and policy formulation mechanisms. However, the increasing reliance on AI systems also brings forth profound ethical questions surrounding bias, fairness, and transparency. In the agricultural context—where socioeconomic disparities, environmental unpredictability, and diverse cultural practices coexist—the implications of biased or opaque AI models can be far-reaching. Fairness in AI-driven agricultural systems is not merely a computational concern but a moral imperative that determines whether technological innovation serves as an equalizer or exacerbates existing inequalities among farming communities.

16.4.1 Algorithmic Bias in Agronomic Predictions

Algorithmic bias refers to systematic and repeatable errors in AI outputs that create unfair outcomes for certain groups or regions. In agriculture, such bias manifests when AI systems misinterpret or undervalue the agricultural realities of smallholder farmers, women farmers, or those operating in non-industrialized rural regions. Unlike other sectors, agronomic predictions often depend on intricate interactions between climatic variables, soil chemistry, irrigation infrastructure, and socioeconomic factors—each of which is context-sensitive. Hence, when training data are disproportionately derived from large-scale or technologically advanced farms, AI models tend to privilege those systems, generating inaccurate or inequitable predictions for underrepresented regions.

16.4.1.1 Sources of Bias in Agricultural Datasets

Bias in agricultural datasets stems from multiple, often overlapping sources. Firstly, data representational bias occurs when datasets overrepresent certain geographies, crop types, or climatic conditions while neglecting others. For instance, machine learning models trained on data from technologically advanced regions of Punjab or Maharashtra may not generalize effectively to the rain-fed, resource-constrained farms in Jharkhand or Odisha. Secondly, measurement bias emerges when sensors, drones, or IoT devices collect data with variable accuracy depending on environmental conditions. Inconsistent calibration of soil sensors or incomplete satellite imagery can create skewed data inputs, leading to erroneous predictions.

Thirdly, historical bias is embedded in legacy datasets that reflect existing inequalities. For example, historical yield data may underreport the productivity of women-led farms or informal agricultural cooperatives due to systemic under documentation. Additionally, algorithmic amplification—a phenomenon where AI systems reinforce existing patterns found in data—can perpetuate these inequities. A biased crop recommendation algorithm, for example, may continue to suggest high-value crops to already prosperous farmers while offering less profitable options to marginalized groups (Kumar et al., 2024).

In summary, the quality, diversity, and contextual relevance of training datasets play a defining role in determining whether AI models serve as tools of inclusion or instruments of exclusion in the agricultural ecosystem.

16.4.1.2 Impacts on Marginalized Farming Communities

The repercussions of biased AI models are most acutely felt by marginalized farming communities. These include smallholder farmers, landless agricultural laborers, indigenous cultivators, and women farmers who often lack access to digital infrastructure and financial capital. When predictive systems fail to capture their local realities—such as unique soil compositions, traditional crop varieties, or unconventional irrigation practices—the resulting AI outputs can mislead farmers into adopting unsuitable agricultural strategies.

For instance, an AI-driven pest management system trained predominantly on data from irrigated farmlands might underestimate pest risk in arid zones, leading to severe yield losses for smallholders. Similarly, biased credit scoring algorithms used by rural banks to assess agricultural loan eligibility might assign lower scores to farmers from regions with incomplete data representation, thereby restricting access to financial services (Mishra &

Mishra, 2024). The cumulative effect of these biases deepens the digital divide, marginalizing those already disadvantaged and further concentrating agricultural benefits among affluent and well-documented regions.

16.4.2 Fairness in Decision-Making and Resource Allocation

Ensuring fairness in agricultural AI systems requires that decision-making processes and resource distribution mechanisms remain equitable across social, geographic, and economic boundaries. AI systems are increasingly being deployed to inform critical decisions—ranging from fertilizer subsidies and crop insurance premiums to water distribution schedules and yield predictions. If these systems are trained or deployed without adequate fairness constraints, they may inadvertently reproduce human biases, institutionalize regional disparities, and marginalize already vulnerable populations.

16.4.2.1 Equity in Yield Prediction and Crop Recommendation Systems

Yield prediction and crop recommendation models are among the most widely adopted AI applications in agriculture. They guide farmers in selecting optimal crop varieties, planting schedules, and nutrient management plans. However, fairness in these systems necessitates equitable access to predictive insights, unbiased data processing, and regionally contextualized recommendations.

For example, a yield prediction algorithm that relies primarily on satellite imagery may fail to account for microclimatic variations affecting small, fragmented plots. This oversight may result in underestimation of yields in certain areas, thereby influencing government procurement prices and subsidy allocations. Similarly, when AI systems recommend crops based solely on profitability metrics, they may overlook traditional, climate-resilient crops vital for food security in vulnerable regions (Calvacante et al., 2025).

Fairness-oriented AI frameworks propose algorithmic balancing techniques such as reweighting and adversarial debiasing to mitigate such disparities. However, these techniques must be contextualized within the socio-agricultural diversity of India and other developing economies, where equitable outcomes cannot be achieved merely through mathematical adjustments but also through participatory data collection and community-based validation.

16.4.2.2 Inclusion of Smallholder Farmers in AI Model Training

A fair agricultural AI ecosystem cannot be realized without the active inclusion of smallholder farmers in data collection, model design, and evaluation processes. Smallholders constitute over 85% of the global farming population, yet their voices remain underrepresented in digital agriculture initiatives. Inclusion ensures that models not only reflect their environmental and agronomic realities but also align with their socioeconomic aspirations (Mishra et al., 2025).

Participatory AI frameworks encourage the co-creation of datasets, where farmers contribute localized knowledge about soil fertility, pest patterns, and seasonal variations. Such inclusion enriches data diversity and fosters trust between AI developers and rural communities. Moreover, involving smallholders in feedback loops allows continuous refinement of models, ensuring they evolve with real-world agricultural dynamics.

Governmental and non-governmental organizations can promote inclusion through AI literacy programmes, subsidized sensor networks, and open data platforms accessible in regional languages. In India, for example, integrating farmer cooperatives into national agricultural data grids could enhance representativeness while ensuring that small-scale cultivators derive tangible benefits from AI adoption (Zangana et al., 2025b).

16.4.3 Model Explainability and Transparency

Transparency and explainability form the ethical backbone of trustworthy AI in agriculture. As AI models grow increasingly complex—particularly with the use of deep learning architectures in agronomic forecasting—understanding how these systems arrive at specific decisions becomes challenging. Farmers, policymakers, and regulatory authorities require clarity regarding the factors influencing predictions, especially when such outputs determine access to subsidies, loans, or inputs. Lack of transparency may erode trust, hinder accountability, and obscure the presence of bias within models.

16.4.3.1 Interpretable AI Approaches in Plant Epigenetics

Plant epigenetics—the study of heritable changes in gene expression that do not involve alterations in DNA sequence—has recently witnessed the integration of AI tools for analyzing vast genomic datasets. However, deep learning models used in this domain often operate as "black boxes," making it difficult to interpret how specific genetic or environmental factors influence observed phenotypic traits.

Interpretable AI approaches such as decision tree ensembles, attention-based neural networks, and SHAP (SHapley Additive exPlanations) offer pathways to understand model decisions in this context. These methods enable researchers to trace back predictions to identifiable genomic or epigenetic markers, thus enhancing scientific validity and ethical transparency. Moreover, interpretable AI supports bioethical compliance, ensuring that data from indigenous crop varieties—often preserved by local communities—are not exploited without acknowledgment or equitable benefit-sharing.

16.4.3.2 Auditable AI Systems and Ethical Benchmarks

For agricultural AI systems to be trusted, they must be auditable and benchmarked against ethical standards. Auditing refers to the systematic examination of AI algorithms, data flows, and decision processes to ensure compliance with ethical norms, regulatory guidelines, and societal expectations. Auditable systems enhance accountability, allowing independent evaluators to assess whether the model's operations align with principles of fairness, non-discrimination, and sustainability.

Ethical benchmarks can include indicators such as data diversity ratios, bias detection scores, and explainability indices, which collectively evaluate the ethical robustness of AI tools. Table 16.3 below outlines an illustrative ethical benchmarking framework for agricultural AI systems.

Establishing auditable systems anchored in such benchmarks will not only promote ethical compliance but also strengthen public trust in AI-driven agricultural policies. Transparent documentation of model development processes, data provenance, and algorithmic updates can ensure that AI remains a tool of empowerment rather than exploitation.

Table 16.3 Ethical benchmarking indicators for agricultural AI models.

Ethical dimension	Indicator	Description	Evaluation method
Data fairness	Dataset diversity ratio	Measures representativeness of regions, crops, and farmer demographics	Statistical sampling and variance analysis
Transparency	Model explainability index	Evaluates interpretability using SHAP or LIME methods	Algorithmic traceability assessment
Accountability	Bias detection score	Detects systemic disparities in predictions	Error parity and fairness metrics
Inclusivity	Farmer participation rate	Proportion of smallholder farmers involved in data collection	Field-based participatory monitoring
Ethical Compliance	Regulatory Conformity level	Degree of adherence to AI ethics frameworks	Independent audit and reporting

16.5 Legal and Regulatory Frameworks Governing AI in Agriculture

The intersection of AI and agriculture represents a transformative convergence of technology and traditional livelihoods. While the adoption of AI in agriculture has the potential to revolutionize production efficiency, environmental sustainability, and food security, it also introduces a multitude of ethical, legal, and regulatory challenges. These concerns span issues of data governance, privacy, liability, transparency, and equitable access to digital resources. As such, the development of a comprehensive and adaptive legal and regulatory framework is crucial to ensuring that AI systems deployed in agriculture operate responsibly, align with societal values, and foster inclusive growth.

The agricultural sector, particularly in developing economies such as India, faces unique regulatory imperatives because of its dependence on diverse stakeholders—farmers, agritech companies, government agencies, and research institutions. The legal frameworks must thus not only safeguard the interests of small and marginal farmers but also promote innovation and attract investments in agricultural technology. This balance necessitates a multitiered approach that includes international guidelines, national policy instruments, sector-specific regulations, and adaptive governance mechanisms.

16.5.1 Overview of Existing Legal Structures

Legal frameworks governing AI in agriculture are still evolving across the globe. While AI technologies have gained prominence in recent years, most legal systems were originally designed to regulate conventional agricultural practices, biotechnology, and digital data management separately. Consequently, the current legal landscape is characterized by a patchwork of overlapping rules and guidelines rather than an integrated system.

Nevertheless, several international and national initiatives provide a foundation upon which agricultural AI governance can be built.

16.5.1.1 International AI Ethics Guidelines and Declarations

The global community has recognized the necessity of establishing common ethical principles for the governance of AI technologies. Various international organizations have proposed guidelines to ensure that AI systems are designed and deployed in ways that uphold human rights, promote sustainability, and minimize harm.

For instance, the UNESCO Recommendation on the Ethics of Artificial Intelligence (2021) serves as one of the most comprehensive global frameworks outlining ethical standards for AI development and application. It emphasizes inclusiveness, accountability, environmental stewardship, and the avoidance of bias. In the agricultural context, these principles translate into ensuring that AI-based decision-making tools—such as those used for yield prediction, pest detection, or weather forecasting—do not marginalize smallholder farmers or reinforce existing inequalities (Wei et al., 2024).

Similarly, the OECD Principles on AI (2019) promote transparency, fairness, and human-centric AI. They encourage governments to create enabling regulatory environments that stimulate innovation while safeguarding public interests. For agriculture, these principles stress the importance of transparent algorithms that can be audited and understood by both developers and end-users, especially where decisions affect resource allocation or market prices (Marda, 2018; Mishra & Mishra, 2023).

The European Union's Ethics Guidelines for Trustworthy AI (2019) also identify seven key requirements—human agency, technical robustness, privacy, transparency, diversity, societal well-being, and accountability. Although these guidelines were primarily designed for general AI applications, their relevance to agriculture is substantial. AI systems used for soil monitoring, irrigation management, or supply chain logistics must adhere to these principles to prevent unethical manipulation of data or monopolistic control by large agri-tech corporations.

Collectively, these international declarations provide a moral and policy compass for nations formulating their domestic AI regulations. However, the challenge remains in contextualizing these principles to local agricultural realities, particularly in developing regions where digital literacy, infrastructural support, and economic stability vary widely (dos Santos, 2024).

16.5.1.2 National Policies on Digital Agriculture and AI

At the national level, several countries have taken proactive steps to integrate AI into their agricultural strategies through targeted policy frameworks. In India, for instance, the National Strategy for Artificial Intelligence (AI for All) released by NITI Aayog identifies agriculture as one of the five priority sectors for AI intervention. The strategy outlines measures for developing AI-driven precision farming solutions, improving crop monitoring, and optimizing supply chains. It also highlights the importance of establishing ethical and regulatory standards to manage issues such as data ownership, IPR, and algorithmic bias.

The Digital Agriculture Mission (2021–2025) launched by the Government of India further strengthens this commitment by promoting the use of digital tools, data integration, and machine learning for evidence-based decision-making. However, while these initiatives foster technological adoption, the legal and regulatory instruments governing AI-specific risks—such as liability for erroneous AI predictions or misuse of farm data—remain underdeveloped.

Countries such as Japan, Australia, and members of the European Union have similarly implemented digital agriculture policies emphasizing interoperability, cybersecurity, and responsible innovation. Nonetheless, harmonizing these national frameworks with international standards remains a work in progress. A critical aspect of such harmonization involves ensuring that domestic regulations encourage collaboration between public and private entities, allowing innovations to scale while maintaining accountability and fairness (Mutsuddi, 2024).

16.5.2 Sector-Specific Regulations and Compliance

AI applications in agriculture operate at the intersection of multiple regulatory domains, including environmental law, biotechnology regulation, intellectual property, and data protection. Therefore, sector-specific compliance mechanisms are essential to address the distinctive risks posed by AI tools used in precision agriculture, genetic crop analysis, and automated farm management.

16.5.2.1 Regulatory Frameworks for Precision Agriculture Tools

Precision agriculture relies heavily on AI-based systems that collect, process, and analyze large volumes of farm data through sensors, drones, and satellite imaging. These technologies necessitate a robust regulatory architecture addressing data ownership, privacy, and liability.

In India, the Personal Data Protection Bill (PDPB)—though still evolving—provides a foundational structure for regulating data use. When applied to agriculture, it implies that farmers should retain ownership and control over their data, such as soil composition, yield rates, and crop health metrics. Agri-tech firms collecting such data must obtain informed consent and ensure that data is anonymized to prevent misuse.

Moreover, the lack of specific liability regulations for AI-driven decisions—such as those made by autonomous irrigation systems or pest detection algorithms—poses legal ambiguities. Questions arise regarding who bears responsibility if an AI system makes an incorrect prediction, leading to crop failure or financial loss. Addressing such liability concerns requires clear statutory guidance or model contractual provisions to allocate accountability among manufacturers, software developers, and users.

Internationally, similar concerns are addressed under the EU Machinery Directive and proposed AI Act, which categorize high-risk AI systems and prescribe stringent compliance requirements, including transparency obligations and conformity assessments. Adopting comparable standards in developing economies could enhance consumer protection and promote safer AI deployment in agriculture (Ndiema, 2024).

16.5.2.2 Compliance with Environmental and Biotechnological Laws

AI technologies increasingly intersect with environmental and biotechnological regulations, especially when integrated into crop genetics, soil conservation, and resource management initiatives. AI-based models predicting fertilizer application, for instance, must comply with environmental laws such as the Environment (Protection) Act, 1986 in India. Overuse or mismanagement of AI-generated recommendations could lead to environmental degradation, necessitating clear regulatory oversight (Yang et al., 2025).

Similarly, the integration of AI with biotechnology—for instance, in genomic editing or seed optimization—requires compliance with biosafety regulations such as the Rules for the Manufacture, Use, Import, Export, and Storage of Hazardous Microorganisms (1989) under the same Act. Ensuring that AI systems used in genetic prediction models are transparent, explainable, and scientifically validated is essential to maintaining ecological safety and public trust.

16.5.3 Need for Dynamic and Adaptive Regulatory Mechanisms

AI technologies are inherently dynamic, evolving at a pace that often outstrips legislative and institutional capacity. Traditional regulatory models—static and prescriptive by design—are ill-suited to manage technologies that learn, adapt, and modify their own functioning over time. Therefore, governance of AI in agriculture demands adaptive, iterative, and participatory regulatory frameworks that evolve alongside technological progress.

16.5.3.1 Challenges in Regulating Evolving AI Technologies

One of the foremost challenges in AI regulation lies in the unpredictability of machine learning outcomes. Unlike conventional agricultural machinery, AI systems generate outputs based on probabilistic models that can evolve through continuous learning. This makes establishing fixed accountability and compliance criteria extremely complex.

Moreover, regulatory agencies often face a shortage of technical expertise to assess AI algorithms or audit their ethical performance. The opaque nature of certain AI models—commonly referred to as the "black box problem"—further complicates oversight. For instance, if an AI-driven crop forecasting system recommends an unsuitable crop variety, regulators may find it difficult to determine whether the error resulted from flawed data, algorithmic bias, or systemic malfunction.

Another challenge is maintaining regulatory flexibility without stifling innovation. Excessive legal constraints may deter startups and researchers from developing new agricultural AI applications, while insufficient oversight could lead to misuse or exploitation of farmers' data. Thus, regulators must strike a delicate balance between innovation facilitation and ethical assurance (Aijaz et al., 2025).

16.5.3.2 Stakeholder Participation in Policy Formation

Effective AI governance in agriculture cannot rely solely on top-down regulatory mandates. It must instead embrace a participatory approach involving all relevant stakeholders—farmers, policymakers, technologists, civil society organizations, and academic institutions. Multi-stakeholder engagement ensures that regulations are contextually appropriate, socially equitable, and technologically sound. Farmers, as primary users and beneficiaries of AI systems, must be given an active role in shaping policies that affect their livelihoods. Their experiential insights into field-level challenges can guide policymakers toward realistic and inclusive governance models.

Furthermore, collaborations between public research institutions and private agri-tech firms can enhance regulatory compliance through shared standards, data interoperability frameworks, and ethical certifications. Governments could also establish agricultural AI regulatory sandboxes, allowing controlled experimentation of new technologies under regulatory supervision before full-scale deployment. Such adaptive and participatory regulatory mechanisms will ensure that AI in agriculture evolves in harmony with ethical values, environmental sustainability, and socioeconomic justice (Santana et al., 2025).

16.6 Ethical Considerations in Automated Decision-Making Systems

The rapid integration of AI within agricultural systems has significantly transformed the processes of cultivation, monitoring, harvesting, and distribution. Yet, alongside the undeniable advantages, there emerge profound ethical questions that must be addressed to ensure responsible and equitable deployment of AI technologies. Automated decision-making systems, when introduced in agriculture, influence not only production efficiency but also the livelihood, autonomy, and dignity of farmers. Ethical reflection thus becomes an indispensable counterpart to technological advancement.

16.6.1 Autonomy and Control in AI-Driven Agricultural Decisions

The introduction of AI into agricultural systems inherently modifies the locus of control in decision-making. Traditionally, farmers have relied on experiential knowledge, local wisdom, and environmental cues to make cultivation choices. However, with the adoption of automated platforms—ranging from smart irrigation systems to AI-based crop health monitors—the capacity for independent decision-making becomes intertwined with algorithmic processes. The ethical concern here revolves around autonomy: who truly makes the decisions when AI systems are embedded within farming practices—the farmer, the system designer, or the algorithm itself?

16.6.1.1 Human-in-the-Loop Versus Fully Automated Systems

A critical ethical distinction arises between human-in-the-loop (HITL) systems and fully automated systems. HITL frameworks allow human supervision and intervention in AI operations, thereby retaining a level of human judgment and moral reasoning. This configuration ensures that final decisions—such as pesticide application or irrigation timing—can be validated by human actors who understand contextual factors that algorithms may overlook (Riczu, 2023; Soundariya et al., 2025).

Conversely, fully automated systems execute decisions without human intervention, relying entirely on algorithmic interpretation of data. While such systems promise precision and efficiency, they also pose the ethical risk of disempowering farmers, transforming them into passive observers of their own agricultural processes. Overreliance on automation can diminish human agency, leading to what ethicists describe as "technological paternalism," wherein systems dictate optimal behaviors based solely on pre-set computational logic rather than human values or situational realities.

An ethically balanced framework demands maintaining human oversight where critical moral or environmental consequences are at stake. Even the most sophisticated algorithms lack contextual empathy—the human ability to weigh economic hardship, cultural priorities, or long-term sustainability over immediate efficiency.

16.6.1.2 Accountability in Autonomous Agricultural Machinery

Accountability represents another fundamental ethical dimension in AI-driven agriculture. Autonomous agricultural machinery—such as driverless tractors, robotic harvesters, and drone sprayers—can act independently based on real-time sensor data. However, when errors occur, identifying responsibility becomes complex. Is the liability to be placed on the farmer who deployed the system, the manufacturer who built it, or the software developer who coded the decision logic?

In ethical governance, transparency and traceability are central to accountability. Systems must be designed with clear audit trails so that actions taken by AI components can be reconstructed and evaluated. This is particularly important in the agricultural domain, where harm may not be immediately visible. For instance, an autonomous pesticide sprayer misidentifying healthy crops as diseased may lead to economic loss and ecological damage, yet the chain of responsibility might remain ambiguous. Therefore, ethical agricultural AI must embed mechanisms for responsibility attribution, ensuring that the benefits of automation do not overshadow the moral duty to answer for harm caused by technological actions (Ashfi, 2024).

16.6.2 Risk Assessment and Unintended Consequences

Ethical evaluation of AI systems cannot be confined to their intended functionality alone. Instead, a comprehensive risk assessment must include unintended consequences that emerge from algorithmic opacity, data bias, and systemic dependency. Agriculture, being a dynamic and context-dependent field, is particularly susceptible to cascading errors, where a single incorrect recommendation may disrupt an entire production cycle.

16.6.2.1 Systemic Risks in Crop and Soil Management Algorithms

Algorithms governing crop and soil management often depend on extensive datasets drawn from remote sensing, weather stations, and IoT-based soil sensors. However, these datasets may not always capture local variations in climate, soil texture, or traditional practices. The ethical dilemma arises when such generalized data lead to systemic risks—for example, over-irrigation in semi-arid regions or nutrient mismanagement due to incorrect soil classification.

Moreover, the predictive nature of these algorithms can create feedback loops, wherein erroneous data become reinforced through repeated model training. Over time, such algorithmic path dependency may propagate inaccurate agricultural recommendations across regions, leading to widespread inefficiencies or even crop failures (Ali et al., 2026).

An ethically grounded approach thus requires periodic algorithmic auditing and contextual data calibration, ensuring that the models are continuously refined to reflect regional realities. It also underscores the moral duty of developers to test systems under diverse agro-climatic conditions before mass deployment.

16.6.2.2 Ethical Issues in Predictive Failure and False Recommendations

Predictive systems in agriculture are designed to forecast crop yield, pest infestation, and market trends. While beneficial, they carry an ethical burden: incorrect predictions can cause severe financial harm. A false pest alert, for example, may prompt unnecessary pesticide use, harming both the environment and the farmer's budget. Conversely, a missed alert can result in large-scale crop damage.

When AI systems fail in such predictive contexts, the moral question extends beyond technical malfunction—it becomes a matter of justice. Small-scale farmers, who often invest limited capital (e.g., ₹15,000–₹50,000 per season), may not have the resilience to absorb algorithmic misjudgments. Hence, ethical design must ensure that predictive recommendations are accompanied by confidence levels and risk disclosure statements. Table 16.4 below summarizes the key ethical dimensions of algorithmic risk in AI-based agricultural decision-making systems (Mayuravaani et al., 2024; Obeng-Ofori et al., 2025).

16.6.3 Redress and Liability Mechanisms

The ethical sustainability of AI in agriculture also depends on the robustness of its redress and liability mechanisms. As AI systems assume decision-making authority, they must also be accompanied by frameworks that enable farmers to seek justice and compensation in cases of harm. Ethical governance thus intersects with legal accountability, necessitating clear pathways for dispute resolution and farmer empowerment.

16.6.3.1 Legal Responsibility for AI-Induced Harm

Legal systems globally are grappling with the question of who bears responsibility when AI fails. In agriculture, this challenge intensifies due to the distributed nature of technology supply chains. A malfunction in a machine learning model could originate from sensor inaccuracy, coding oversight, or incorrect data interpretation. Assigning liability, therefore, demands a multi-stakeholder responsibility framework that recognizes the shared roles of developers, manufacturers, distributors, and users (Abbas Khan et al., 2024).

Table 16.4 Ethical dimensions in AI-based agricultural risk management.

Ethical concern	Description	Potential consequence	Ethical mitigation
Data bias	Incomplete or non-representative training data	Misleading recommendations for local conditions	Inclusion of region-specific datasets
Predictive failure	Inaccurate forecasts of yield or pest events	Financial loss and resource wastage	Incorporate uncertainty estimates
Systemic dependency	Overreliance on automated systems	Loss of traditional knowledge	Farmer education and hybrid systems
Algorithmic opacity	Lack of transparency in model logic	Reduced accountability	Mandatory audit and explainability standards

From an ethical standpoint, it is unjust to burden farmers—often the least technologically equipped stakeholder—with liability for algorithmic errors beyond their control. Instead, developer liability models and insurance-backed compensation schemes could serve as practical instruments to distribute risk more equitably. For instance, government-supported agricultural insurance could incorporate clauses addressing AI-induced crop losses, ensuring that farmers are not financially devastated by system errors (Biswas et al., 2023; Tariq, 2024).

16.6.3.2 Dispute Resolution and Farmer Rights

Ethical AI deployment must also safeguard farmer rights, including the right to transparency, informed consent, and accessible grievance redressal. Dispute resolution mechanisms must be farmer-centric, offering timely remedies without imposing excessive procedural burdens.

Digital agriculture platforms should integrate built-in reporting channels where farmers can contest decisions or flag malfunctions. Moreover, local agricultural cooperatives and regulatory bodies should act as intermediaries to facilitate fair resolution between farmers and AI vendors. Ethical fairness also requires that these mechanisms be accessible in regional languages and structured in ways that respect cultural sensitivities and literacy levels (Oyeniran et al., 2025).

Ultimately, a human-centered redressal system reinforces trust and ensures that technological innovation remains aligned with the ethical principle of justice. Farmers must not only benefit from AI's predictive power but also retain the right to challenge, question, and seek remedy when technology errs.

16.7 Governance, Ethics Integration, and Future Directions

The rapid integration of AI into agricultural systems presents both transformative opportunities and profound ethical challenges. As AI-driven technologies such as precision farming, predictive analytics, and autonomous machinery become central to modern agrarian practices, the governance of these systems requires careful ethical reflection and institutional structuring. Effective governance mechanisms must ensure that technological advancement aligns with societal values, equity, environmental sustainability, and the rights of all stakeholders—including farmers, agricultural laborers, and consumers. This section explores the institutional frameworks necessary for ethical oversight, the concept of ethics-by-design in AI development, and the imperative of international cooperation for establishing globally coherent ethical standards in agricultural AI.

16.7.1 Institutional Frameworks for Ethical Oversight

The governance of AI in agriculture cannot be left to technological progress alone. Instead, it requires deliberate and structured ethical oversight through institutional mechanisms capable of ensuring transparency, accountability, and inclusivity. Institutional frameworks for ethical oversight serve as the backbone for policy implementation and safeguard mechanisms to prevent misuse, bias, and social exclusion in AI-driven agricultural systems.

Such frameworks must be both adaptive and anticipatory, recognizing that the ethical implications of AI often evolve faster than conventional regulatory responses. The institutionalization of ethics thus involves both formal structures—such as ERBs and cross-disciplinary committees—and informal mechanisms like community consultations and farmer associations (Victoire et al., 2023).

16.7.1.1 Ethical Review Boards for Agricultural AI Projects

ERBs represent a fundamental mechanism for ensuring the responsible deployment of AI in agriculture. These boards are specialized committees established within universities, research institutions, private companies, and government agencies to evaluate AI projects before their implementation. Their primary objective is to ensure that the project design, data collection methods, algorithmic decision-making processes, and outcome measures comply with accepted ethical principles such as beneficence, nonmaleficence, autonomy, and justice (Kumar & Kumar, 2024).

In the context of agriculture, ERBs should not only focus on data ethics—such as privacy and consent—but also on broader implications like environmental sustainability, labor displacement, and the socioeconomic impact on smallholder farmers. For example, an AI-based irrigation system might improve water efficiency but could inadvertently marginalize small farmers who cannot afford to access the technology. ERBs would be responsible for identifying such potential inequities and recommending appropriate mitigations (Farhud & Zokaei, 2021).

Moreover, these boards must operate with multistakeholder representation, including ethicists, AI developers, agricultural scientists, rural sociologists, and farmer representatives. This inclusion ensures that the ethical assessment process is holistic and contextually grounded. Importantly, ERBs should also develop standardized evaluation criteria for agricultural AI, taking into account the unique challenges of rural data environments and local socioeconomic diversity.

16.7.1.2 Cross-disciplinary Ethical Committees

While ERBs focus on project-level evaluations, cross-disciplinary ethical committees function at a more strategic and policy-oriented level. These committees serve as advisory bodies to governments, research councils, and industry groups, ensuring that ethical considerations are integrated into long-term agricultural innovation strategies. Given that AI in agriculture intersects with multiple domains—such as biotechnology, data science, climate science, and rural development—these committees must incorporate expertise from varied disciplines. Their role extends beyond ethical evaluation to include ethical foresight, anticipating long-term consequences of technological trends (Al-Hadrami et al., 2025).

For instance, a cross-disciplinary committee might assess the ethical implications of predictive yield models that rely on farmer-generated data. Questions about data ownership, privacy, and equitable profit-sharing naturally arise. Should farmers receive financial compensation for their data contributions? How can algorithmic transparency be maintained when proprietary models are owned by private corporations? Such complex questions require interdisciplinary dialogue and consensus-based ethical guidance (Leal Filho & Gbaguidi, 2024). In addition, these committees should work collaboratively with regional and international organizations to ensure policy coherence. They can also propose ethical certification schemes for AI technologies, similar to environmental certification programs, to ensure that agricultural AI products meet defined ethical benchmarks (Schaefer, 2023).

16.7.2 Ethics-by-Design and Responsible AI Development

Ethics-by-design refers to the deliberate integration of ethical principles into every stage of AI system development—from conception and design to deployment and evaluation. Rather than addressing ethical issues reactively, ethics-by-design ensures that ethical reflection becomes a structural feature of innovation, guiding decisions about data sourcing, model architecture, user interaction, and impact assessment (Han et al., 2025).

This proactive approach is particularly relevant in agriculture, where AI tools often interact with living systems, natural ecosystems, and vulnerable human populations. Ethical design in this domain must therefore balance technological efficiency with social justice, inclusivity, and ecological balance.

16.7.2.1 Embedding Ethical Principles in AI System Design

Embedding ethical principles within the architecture of AI systems requires developers to move beyond technical optimization and incorporate moral reasoning into algorithmic structures. This process involves translating ethical values—such as fairness, transparency, and sustainability—into computational objectives and design constraints.

For example, algorithms designed for yield prediction should avoid reinforcing existing disparities between large-scale agribusinesses and smallholders. This can be achieved through bias auditing tools, equitable data sampling, and the inclusion of socioeconomic variables in training datasets. Similarly, decision-support systems must be designed to enhance farmer autonomy rather than replace human judgment (Leal Filho & Gbaguidi, 2024).

Developers can utilize frameworks such as value-sensitive design (VSD) and Responsible Research and Innovation (RRI) to guide this process. Both approaches emphasize the integration of ethical reflection, stakeholder engagement, and iterative evaluation throughout the design lifecycle. The ultimate goal is not only to ensure compliance but to cultivate ethical resilience—the capacity of AI systems to remain aligned with evolving social and environmental values.

16.7.2.2 Participatory Design Approaches in Agricultural Technology

Participatory design represents a critical dimension of ethical AI development, especially in agriculture, where technology adoption directly impacts livelihoods and local cultures. By involving farmers, community leaders, agricultural extension workers, and local policymakers in the design process, AI developers can ensure that technologies are contextually relevant, socially acceptable, and culturally sensitive (Alexander et al., 2024).

Participatory models empower rural stakeholders to articulate their needs, preferences, and concerns during the innovation process. For instance, before deploying an AI-based pest detection system, developers can conduct workshops and pilot programs to gather feedback from farmers about usability, language preferences, and the local ecology.

This inclusive approach not only enhances ethical legitimacy but also improves technological efficacy, as systems designed with user input tend to have higher adoption and success rates. Participatory methods also mitigate the risk of technological determinism, where technology dictates social practices without accommodating local realities (Natsir et al., 2025; Ogunyiola, 2021).

16.7.3 Global Cooperation and Ethical Standardization

The ethical governance of AI in agriculture cannot be confined within national borders. Agricultural systems are inherently global—interconnected through trade, data exchange, environmental processes, and shared technological infrastructure. Consequently, ethical frameworks must be harmonized at the international level to prevent ethical fragmentation and to promote global equity.

Global cooperation facilitates mutual learning, the exchange of best practices, and the establishment of shared ethical baselines. Moreover, it helps developing economies—where agricultural livelihoods are most vulnerable—benefit from AI innovations without being subject to exploitative or opaque technological practices (Cath, 2018).

16.7.3.1 Harmonization of Global Ethical Standards

The harmonization of ethical standards across nations ensures that AI systems adhere to consistent principles regardless of their origin or deployment location. This is particularly significant given the transnational nature of agricultural data flows and AI service providers.

Efforts toward ethical harmonization can build upon existing international instruments such as the Food and Agriculture Organization (FAO's) Code of Conduct for the Responsible Use of Technology in Agriculture, the OECD Principles on AI, and UNESCO's Recommendation on the Ethics of AI (2021). These documents collectively emphasize human rights, sustainability, and inclusivity (Rayhan, 2023).

Harmonization also involves establishing mutual recognition frameworks, where ethical certifications and compliance assessments conducted in one country are recognized in another. Such a framework would reduce duplication and ensure trust in cross-border collaborations (Dara & Azarpira, 2025). For example, an AI-based crop forecasting system developed in India could be deployed in another South Asian country under shared ethical validation protocols, provided that both adhere to a common standard.

Furthermore, standardization supports the creation of ethical metrics and indicators, allowing policymakers and international agencies to monitor the ethical performance of agricultural AI systems quantitatively—similar to environmental sustainability indices (Wang, 2021).

16.7.3.2 Role of International Bodies and Consortia

International organizations and multi-stakeholder consortia play an instrumental role in driving ethical governance at the global scale. Institutions such as the Food and Agriculture Organization (FAO), the World Bank, the International Telecommunication Union (ITU), and regional alliances like SAARC or the African Union, possess the convening power and technical expertise to develop shared ethical frameworks and capacity-building programs (Holzinger et al., 2024).

These bodies can facilitate cross-national dialogues, organize training for policymakers, and fund ethical AI research in low- and middle-income countries. Moreover, international consortia—comprising academia, private sector representatives, and civil society organizations—can act as platforms for co-regulation, where voluntary codes of ethics complement formal legislation (Aderibigbe et al., 2023).

Importantly, such cooperation must be inclusive and equitable. Ethical governance should not become another form of technological imperialism where standards are dictated by developed nations without considering local realities. Hence, the active participation of developing countries in agenda-setting processes is essential for ensuring global fairness and legitimacy.

16.8 Conclusion

The pervasive integration of AI represents a pivotal yet complex evolution for global agriculture, driven by the urgent need to meet escalating food demand through unprecedented efficiency. However, this technological leap into "Agriculture 4.0" heavily qualified by significant ethical and governance deficits. At the core of the challenge lies the sheer volume of sensitive farm data, which necessitates robust protocols for ownership, privacy, and informed consent to protect farmer autonomy from technological commodification. Furthermore, the reliance on traditional ethical frameworks must be tempered by socio-economic realities, particularly the critical threat of human labor displacement in vulnerable rural economies. The pursuit of algorithmic fairness is paramount, as demonstrated by the potential for bias in models trained on non-representative datasets to exacerbate rural inequities and marginalize smallholders. Consequently, the fragmented global regulatory environment, characterized by a lack of adaptive, sector-specific mandates, must be urgently strengthened to resolve issues of liability and redress in autonomous systems. Ultimately, realizing AI's promise requires a concerted shift toward an Ethics-by-Design approach and global cooperation, ensuring technology is developed and deployed responsibly to secure a just and sustainable future for all agricultural stakeholders.

References

Abbas Khan, M., Khan, H., Omer, M. F., Ullah, I., & Yasir, M. (2024). Impact of artificial intelligence on the global economy and technology advancements. In A. R. Alshehadeh, I. A. El-Qirem, & G. A. Elrefae (Eds.), *Artificial general intelligence (AGI) security: Smart applications and sustainable technologies* (pp. 147–180). Springer Nature Singapore. https://doi.org/10.1007/978-981-97-3222-7_7

Abdul-Al, T., Kanaan, R., & Kanaan, G. (2025, April). Farming ethical artificial intelligence: Emerging issues and regulatory challenges. In Abdul Razzak Alshehadeh, Ihab Ali El-Qirem, and Ghaleb Awad Elrefae Conference on sustainability and cutting-edge business technologies (pp. 295–301). Springer Nature Switzerland. https://doi.org/10.1007/978-3-031-96631-6_29

Aderibigbe, A. O., Ohenhen, P. E., Nwaobia, N. K., Gidiagba, J. O., & Ani, E. C. (2023). Artificial intelligence in developing countries: Bridging the gap between potential and implementation. *Computer Science & IT Research Journal, 4*(3), 185–199. https://doi.org/10.51594/csitrj.v4i3.629

Aijaz, N., Lan, H., Raza, T., Yaqub, M., Iqbal, R., & Pathan, M. S. (2025). Artificial intelligence in agriculture: Advancing crop productivity and sustainability. *Journal of Agriculture and Food Research, 20,* 101762. https://doi.org/10.1016/j.jafr.2025.101762

Alexander, C. S., Yarborough, M., & Smith, A. (2024). Who is responsible for 'responsible AI'?: Navigating challenges to build trust in AI agriculture and food system technology. *Precision Agriculture, 25*(1), 146–185. https://doi.org/10.1007/s11119-023-10063-3

Al-Hadrami, A., Al-Khatib, K. M., Bennour, A., Al Saqri, M., & AlAjmi, Q. (2025). Possible risks and ethical considerations of artificial intelligence considering sustainability. In A. Al Qamashoui & N. Al Baimani (Eds.), *AI integration for business sustainability: For a resilient future* (pp. 1–18). Springer Nature Singapore. https://doi.org/1007/978-981-96-3464-4_1

Ali, G., Mijwil, M. M., Buruga, B. A., Abotaleb, M., & Adamopoulos, I. (2024). A survey on artificial intelligence in cybersecurity for smart agriculture: State-of-the-art, cyber threats, artificial intelligence applications, and ethical concerns. *Mesopotamian Journal of Computer Science, 2024,* 53–103. https://doi.org/10.58496/MJCSC/2024/007

Ali, D., Iqbal, S., Rafique, S., Hanane, M., & Khalil, I. (2026). Ethical implications and data privacy concerns in AI and ML applications within food science. In T. Sarkar & A. Haldorai (Eds.), *Artificial intelligence in food science* (pp. 763–780). Academic Press. https://doi.org/10.1016/B978-0-443-26468-9.00043-6

Ashfi, S. A. (2024). Ethical implications of emerging technologies: A comparative analysis of AI, biotechnology, and robotics. *Journal of Research and Innovation in Technology and Management (JRITM), 1*(1), 80.

Benefo, E. O., Tingler, A., White, M., Cover, J., Torres, L., Broussard, C., ... Patra, D. (2022). Ethical, legal, social, and economic (ELSE) implications of artificial intelligence at a global level: A scientometrics approach. *AI and Ethics, 2*(4), 667–682. https://doi.org/10.1007/s43681-021-00124-6

Biswas, A., Kumari, A., Gaikwad, D. S., & Pandey, D. K. (2023). Revolutionizing biological science: The synergy of genomics in health, bioinformatics, agriculture, and artificial intelligence. *Omics : A Journal of Integrative Biology, 27*(12), 550–569. https://doi.org/10.1089/omi.2023.0197

Brini, M. (2024). Future perspectives: Emerging technologies and ethical considerations in data science for agriculture and natural resources. *Harnessing Data Science for Sustainable Agriculture and Natural Resource Management*, 335–355, Springer Nature Singapore. https://doi.org/10.1007/978-981-97-7762-4_14

Calvacante, S. T., Cristiane, G., Héliton, P., Brauer, V. R., Delfino, B. F. J. A., Ferreira, S. R. G., ... da Silva, S. P. C. (2025). Ethics, animal welfare, and artificial intelligence in livestock: A bibliometric review. *AgriEngineering, 7*(7), 202. https://doi.org/10.3390/agriengineering7070202

Cath, C. (2018). Governing artificial intelligence: Ethical, legal and technical opportunities and challenges. *Philosophical Transactions of the Royal Society A: Mathematical, Physical and Engineering Sciences, 376*(2133), 20180080. https://doi.org/10.1098/rsta.2018.0080

Dara, M., & Azarpira, N. (2025). Ethical considerations emerge from artificial intelligence (AI) in biotechnology. *Avicenna Journal Of Medical Biotechnology*, *17*(1), 80–81. https://doi.org/10.18502/ajmb.v17i1.17680

Dara, R., Hazrati Fard, S. M., & Kaur, J. (2022). Recommendations for ethical and responsible use of artificial intelligence in digital agriculture. *Frontiers in Artificial Intelligence*, *5*, 884192. https://doi.org/10.3389/frai.2022.884192

Devasia, J. V., Deepanramkumar, P., Helensharmila, A., & Yenduri, G. (2024). Ethical considerations and social implications. In S. Satpathy, B. K. Paikaray, M. Yang, & A. Balakrishnan (Eds.), *Sustainable farming through machine learning* (pp. 267–279). CRC Press. https://doi.org/10.1201/9781003484608-19

Farhud, D. D., & Zokaei, S. (2021). Ethical issues of artificial intelligence in medicine and healthcare. *Iranian Journal of Public Health*, *50*(11), i–v. https://doi.org/10.18502/ijph.v50i11.7600

GeeksforGeeks. (2025, September 29). *Blockchain and distributed ledger technology (DLT). Geeksforgeeks*. https://www.geeksforgeeks.org/software-engineering/blockchain-and-distributed-ledger-technology-dlt/

Gikunda, K. (2024). Harnessing artificial intelligence for sustainable agricultural development in Africa: Opportunities, challenges, and impact. *arXiv preprint arXiv:2401.06171*. https://doi.org/10.48550/arXiv.2401.06171

Gupta, K., Rajpoot, A. K., & Gupta, A. (2024, October). Ethical AI in smart agriculture 4.0: A consortium of IoT and AI for sustainable farming. In A. Bhattacharya, S. Dutta, S.-L. Peng, & X.-S. Yang (Eds.), *International conference on data mining and information security* (pp. 575–602). Springer Nature Singapore. https://doi.org/10.1007/978-981-96-6046-9_36

Gwagwa, A., Kazim, E., Kachidza, P., Hilliard, A., Siminyu, K., Smith, M., & Shawe-Taylor, J. (2021). Road map for research on responsible artificial intelligence for development (AI4D) in African countries: The case study of agriculture. *Patterns*, *2*(12), 100381. https://doi.org/10.1016/j.patter.2021.100381

Han, M., Benson, A., & Abon, J. E. (2025). Navigating ethics in wireless sensor networks for sustainable agriculture. *Frontiers in Sustainable Food Systems*, *9*, 1634643. https://doi.org/10.3389/fsufs.2025.1634643

Harikrishnan, S., Kaushik, D., Rasane, P., Kumar, A., Kaur, N., Reddy, C. K., ... Kumar, M. (2025). Artificial intelligence in sustainable food design: Technological, ethical consideration, and future. *Trends in Food Science & Technology*, *163*, 105152. https://doi.org/10.1016/j.tifs.2025.10.152

van Hilten, M., Ryan, M., Blok, V., & de Roo, N. (2025). Ethical, legal and social aspects (ELSA) for AI: An assessment tool for agri-food. *Smart Agricultural Technology*, *10*, 100710. https://doi.org/10.1016/j.atech.2024.100710

Holzinger, A., Fister, I., Kaul, H. P., & Asseng, S. (2024). Human-centered AI in smart farming: Toward agriculture 5.0. *IEEE Access*, *12*, 62199–62214. https://doi.org/10.1109/ACCESS.2024.3395532

Ijaz, S., Iqbal, J., Abbasi, B. A., Ullah, Z., Ijaz, N., Yaseen, T., ... Mahmood, T. (2024). Regulatory and ethical concerns of nanotechnology in agriculture. In R. Shahzad, S. Fiaz, A. Qayyum, M. U. Islam, & I.-J. Lee (Eds.), *Revolutionizing agriculture: A comprehensive exploration of agri-nanotechnology* (pp. 395–427). Springer Nature Switzerland. https://doi.org/10.1007/978-3-031-76000-6_18

Imran, & Li, J. (2025). UAV technology in agriculture: Legal, ethical, and future perspectives. In Imran & J. Li (Eds.), *UAV aerodynamics and crop interaction: Revolutionizing modern agriculture with drone* (pp. 431–462). Springer Nature Singapore. https://doi.org/10.1007/978-981-96-8402-1_13

Kosior, K. (2020). Economic, ethical and legal aspects of digitalization in the agri-food sector. Zagadnienia Ekonomiki Rolnej/Problems of Agricultural Economics. https://doi.org/10.30858/zer/120456

Kumar, D. (2024). Ethical considerations of AI and IoT in farming. In S. Gupta, W. Hasan, S. Singh, D. Kumar, M. J. Ansari, & S. Nisar (Eds.), *Agriculture 4.0* (pp. 319–333). CRC Press. https://doi.org/10.1201/9781003570219-16

Kumar, A., & Kumar, L. (2024). Navigating the future: The ethical, societal and technological implications of artificial intelligence. Journal homepage, https://gjrpublication.com/gjrecs

Kumar, R., Farooq, M., & Qureshi, M. (2024). Advancing precision agriculture through artificial intelligence: Exploring the future of cultivation. In A. Hamadani, N. A. Ganai, H. Henna, & J. Bashir (Eds.), *A biologist s guide to artificial intelligence* (pp. 151–165). Elsevier. https://doi.org/10.1016/B978-0-443-24001-0.000.0-5

Leal Filho, W., & Gbaguidi, G. J. (2024). Using artificial intelligence in support of climate change adaptation Africa: Potentials and risks. *Humanities and Social Sciences Communications*, *11*(1), 1–5. https://doi.org/10.1057/s41599-024-04223-7

Manning, L., Brewer, S., Craigon, P. J., Frey, J., Gutierrez, A., Jacobs, N., ... Pearson, S. (2022). Artificial intelligence and ethics within the food sector: Developing a common language for technology adoption across the supply chain. *Trends in Food Science & Technology*, *125*, 33–42. https://doi.org/10.1016/j.tifs.2022.04.025

Marda, V. (2018). Artificial intelligence policy in India: A framework for engaging the limits of data-driven decision-making. *Philosophical Transactions of the Royal Society A: Mathematical, Physical and Engineering Sciences*, *376*(2133), 20180087. https://doi.org/10.1098/rsta.2018.0087

Mark, R. (2019). Ethics of using AI and big data in agriculture: The case of a large agriculture multinational. *The ORBIT Journal*, *2*(2), 1–27. https://doi.org/10.29297/orbit.v2i2.109

Mayuravaani, M., Ramanan, A., Perera, M., Senanayake, D. A., & Vidanaarachchi, R. (2024). Insights into artificial intelligence bias: Implications for agriculture. *Digital Society*, *3*(3), 48. https://doi.org/10.1007/s44206-024-00142-x

Mishra, H. (2024). Nanobiostimulants and precision agriculture: A data-driven approach to farming and market dynamics. In V. Singh, R. A. Bhat, & G. H. Dar (Eds.), *Nanobiostimulants: Emerging strategies for agricultural sustainability* (pp. 365–398). Springer Nature Switzerland. https://doi.org/10.1007/978-3-031-68138-7_16

Mishra, H., & Dwivedi, S. (2025). Economic analysis of AI integration in internet of drones (IoD). In J. Hassan, S. Khalifa, & P. Misra (Eds.), *Machine learning for drone-enabled IoT networks: Opportunities, developments, and trends* (pp. 169–190). Springer Nature Switzerland. https://doi.org/10.1007/978-3-031-80961-3_9

Mishra, H., & Mishra, D. (2023). Artificial intelligence and machine learning in agriculture: Transforming farming systems. *Research Trends in Agricultural Sciences*, *1*, 1–16.

Mishra, H., & Mishra, D. (2024). AI for data-driven decision-making in smart agriculture: From field to farm management. In S. S. Chouhan, A. Saxena, U. P. Singh, & S. Jain (Eds.), *Artificial intelligence techniques in smart agriculture* (pp. 173–193). Springer Nature Singapore. https://doi.org/10.1007/978-981-97-5878-4_11

Mishra, H., & Srivastava, A. B. (2025). Sustainable agriculture: Economic perspectives on AI and ML in crop yield estimation. In S. S. Chouhan, U. P. Singh, A. Saxena, A. Tiwari, & S. Jain (Eds.), *Artificial intelligence and computer vision for ecological informatics* (pp. 348–392). CRC Press. https://doi.org/10.1201/9781003518778-20

Mishra, H., Mishra, R., Olayinka, O. T., Ghaib, A. A., Raza, M. Y., & Maroof, K. (2025). The role of digital agriculture in mitigating climate change and ensuring food security: An overview. In M. Ahmed (Ed.), *Climate resilient and sustainable agriculture: Volume 2: Social and transformative strategies* (pp. 287–312). https://doi.org/10.1007/978-3-032-04141-8_11

Mutsuddi, I. (2024). Ethical considerations in the intersection of AI and environmental science. In B. Singh, C. Kaunert, K. Vig, & S. Dutta (Eds.), *Maintaining a sustainable world in the nexus of environmental science and AI* (pp. 283–310). IGI Global. https://doi.org/10.4018/979-8-3693-6336-2.ch011

Natsir, M. H., Mahmudy, W. F., Tono, M., & Nuningtyas, Y. F. (2025). Advancements in artificial intelligence and machine learning for poultry farming: Applications, challenges, and future prospects. *Smart Agricultural Technology*, *12*, 101307. https://doi.org/10.1016/j.atech.2025.101307

Ndiema, K. W. (2024). Implications of artificial intelligence (AI) in land management. *Journal of Computer Science and Technology (JCST)*, *2*(1), 1–9. https://doi.org/10.51317/jcst.v2i1.490

Obeng-Ofori, D., Atianashie, M., & Kuffour, M. K. (2025). The role of artificial intelligence and robotics in shaping the future of sustainable agriculture. *Academia Engineering*, *2*, 1–3. https://doi.org/10.20935/AcadEng7626

Ogunyiola, A. (2021). *Interrogating the socio-ethical dilemmas of precision agriculture technologies*. South Dakota State University.

Okengwu, U. A., Onyejegbu, L. N., Oghenekaro, L. U., Musa, M. O., & Ugbari, A. O. (2023). Environmental and ethical negative implications of AI in agriculture and proposed mitigation measures. *Scientia Africana*, *22*(1), 141–150. https://doi.org/10.4314/sa.v22i1.13

Oyeniran, K. A., Oyeniran, O. H., & Oladunmoye, K. M. (2025). Artificial intelligence, innovation, and safety in biotechnology research. In M. Chadyiwa & O. Oladimeji (Eds.), *Cases on AI innovations in occupational health and safety* (pp. 73–98). IGI Global. https://doi.org/10.4018/979-8-3693-9301-7.ch004

Rayhan, A. (2023). Artificial intelligence in robotics: From automation to autonomous systems. *IEEE Transactions on Robotics*, *39*(7), 2241–2253.

Riczu, Z. (2023). Recommendations on the ethical aspects of artificial intelligence, with an outlook on the world of work. *Journal of Digital Technologies and Law*, *1*(2), 498–519. https://doi.org/10.21202/jdtl.2023.21

Ryan, M. (2023). The social and ethical impacts of artificial intelligence in agriculture: Mapping the agricultural AI literature. *AI & Society*, *38*(6), 2473–2485. https://doi.org/10.1007/s00146-021-01377-9

Ryan, M., Isakhanyan, G., & Tekinerdogan, B. (2023). An interdisciplinary approach to artificial intelligence in agriculture. *NJAS: Impact in Agricultural and Life Sciences*, *95*(1), 2168568. https://doi.org/10.1080/27685241.2023.2168568

Sangwa, S., & Mutabazi, P. (2025). Artificial intelligence in agriculture: Ethical stewardship, responsible innovation, and governance for sustainable food systems. *Responsible Innovation, and Governance for Sustainable Food Systems (September 22, 2025)*. https://doi.org/10.54963/ia.v1i2.1684

Santana, T. C., Guiselini, C., Pandorfi, H., Vigoderis, R. B., Barbosa Filho, J. A. D., Soares, R. G. F., ... Santos, P. C. D. S. (2025). Ethics, animal welfare, and artificial intelligence in livestock: A bibliometric review. *AgriEngineering*, *7*(7), 202. https://doi.org/10.3390/agriengineering7070202

dos Santos, T. F. (2024). A legal approach: Artificial intelligence, agribusiness and the 2030 agenda. *Observatorio De La Economía Latinoamericana*, *22*(11), e7892–e7892. https://doi.org/10.55905/oelv22n11-184

Schaefer, L. (2023). An emerging era of artificial intelligence research in agriculture. *Journal of Robotics Spectrum*, *1*, 36–46. https://doi.org/10.53759/9852/JRS202301004

Smith, M. J. (2019). Getting value from artificial intelligence in agriculture. *Animal Production Science*, *60*(1), 46–54. https://doi.org/10.1071/AN18522

Soundariya, A., Behera, N. K., Selvakumar, P., Chandel, P. S., Gupta, S., & Manjunath, T. C. (2025). AI and environmental ethics: The role of AI in addressing climate change ethical considerations in AI-driven environmental solutions. In S. Chakraborty (Ed.), *Human values, ethics, and dignity in the age of artificial intelligence* (pp. 321–346). IGI Global. https://doi.org/10.4018/979-8-3693-9035-1.ch013

Sparrow, R., & Howard, M. (2021). Robots in agriculture: Prospects, impacts, ethics, and policy. *Precision Agriculture*, *22*(3), 818–833. https://doi.org/10.1007/s11119-020-09757-9

Sparrow, R., Howard, M., & Degeling, C. (2021). Managing the risks of artificial intelligence in agriculture. *NJAS: Impact in Agricultural and Life Sciences*, *93*(1), 172–196. https://doi.org/10.1080/27685241.2021.2008777

Srivastav, A. K., & Das, P. (2025). Ethical considerations and regulatory challenges. In A. K. Srivastav & P. Das (Eds.), *Biotechnology and IoT in agriculture and food production: Green innovation* (pp. 145–152). Apress. https://doi.org/10.1007/979-8-8688-1469-3_16

Tariq, M. U. (2024). Leveraging artificial intelligence for a sustainable and climate-neutral economy in Asia. In P. O. de Pablos, M. N. Almunawar, & M. Anshari (Eds.), *Strengthening sustainable digitalization of Asian economy and society* (pp. 1–21). IGI Global. https://doi.org/10.4018/979-8-3693-1942-0.ch001

Tzachor, A., Devare, M., King, B., Avin, S., & Ó hÉigeartaigh, S. (2022). Responsible artificial intelligence in agriculture requires systemic understanding of risks and externalities. *Nature Machine Intelligence*, *4*(2), 104–109. https://doi.org/10.1038/s42256-022-00440-4

Uddin, M., Chowdhury, A., & Kabir, M. A. (2024). Legal and ethical aspects of deploying artificial intelligence in climate-smart agriculture. *AI & Society*, *39*(1), 221–234. https://doi.org/10.1007/s00146-022-01421-2

Victoire, T. A., Karunamurthy, A., Sandhiya, S., & Yuvaraj, S. (2023). Leveraging artificial intelligence for enhancing agricultural productivity and sustainability. *Quing: International Journal of Innovative Research in Science and Engineering*, *2*(2), 141–156. https://doi.org/10.54368/qijirse.2.2.0016

Wei, H., Xu, W., Kang, B., Eisner, R., Muleke, A., Rodriguez, D., ... Harrison, M. T. (2024). Irrigation with artificial intelligence: Problems, premises, promises. *Human-Centric Intelligent Systems*, *4*(2), 187–205. https://doi.org/10.1007/s44230-024-00072-4

Wong, A. (2021). Ethics and regulation of artificial intelligence. In E. Mercier-Laurent, M. Ö. Kayalica, & M. L. Owoc (Eds.), *IFIP international workshop on artificial intelligence for knowledge management* (pp. 1–18). Springer International Publishing. https://doi.org/10.1007/978-3-030-80847-1_1

Yang, C. X., Baker, L. M., Mattox, A., Diehl, D., & Honeycutt, S. (2025). The forgotten factor: Exploring consumer perceptions of artificial intelligence in the food and agriculture systems. *Future Foods, 11*, 100553. https://doi.org/10.1016/j.fufo.2025.100553

Zangana, H. M., Amelia, P., Luckyardi, S., Mustafa, F. M., & Li, S. (2025a). The role of language in regulating AI and agricultural technologies: Legal perspectives from a globalized world. In H. M. Zangana, J. Al-Karaki, & M. Omar (Eds.), *Navigating law and policy in STM enterprises: Ethical governance, regulation, and innovation strategy* (pp. 193–224). IGI Global. https://doi.org/10.4018/979-8-3373-4862-9.ch008

Zangana, H. M., Luckyardi, S., Mustafa, F. M., & Li, S. (2025b). Ethical and secure AI applications in agricultural research: Challenges and opportunities. In H. M. Zangana & M. Omar (Eds.), *Ensuring Secure and Ethical STM Research in the AI Era* (pp. 43–76). IGI Global Scientific Publishing. https://doi.org/10.4018/979-8-3373-4252-8.ch002

Index

Note: *Italics* refer Figure, **Bold** refer Table.

Printed and bound by CPI Group (UK) Ltd, Croydon, CR0 4YY

17/09/2026

14964300-0005